The Manual of Scientific Style

The
Manual of Scientific Style

A Guide for Authors, Editors, and Researchers

First Edition

Edited by

Harold Rabinowitz and Suzanne Vogel

Amsterdam • Boston • Heidelberg • London
New York • Oxford • Paris • San Diego
San Francisco • Singapore • Sydney • Tokyo
Academic Press is an imprint of Elsevier

Academic Press is an imprint of Elsevier
30 Corporate Drive, Suite 400, Burlington, MA 01803, USA
525 B Street, Suite 1900, San Diego, California 92101-4495, USA
84 Theobald's Road, London WC1X 8RR, UK

This book is printed on acid-free paper. ∞

Library of Congress Cataloging-in-Publication Data
Application Submitted

British Library Cataloguing-in-Publication Data
A catalogue record for this book is available from the British Library.

ISBN: 978-0-12-373980-3

For information on all Academic Press publications
visit our Web site at www.books.elsevier.com

PRINTED IN THE UNITED STATES OF AMERICA
09 10 11   9 8 7 6 5 4 3 2 1

Dedication

To Mr. Murray Glass, who launched my ship,
Rabbi Mark Cogan, who demanded that the ship be seaworthy,
Dr. Ralph E. Behrends, who taught me how to row, and
Dr. David Finkelstein, who taught me to love the ocean.

HR

To My Mom, Irene Greenstein

My teacher in science, and in life

SV

Produced by The Reference Works, Inc.
New York, New York

Harold Rabinowitz, Editorial Director

Andrew Chappell and Jonathan J. Londino
Managing Editors

Mehera Bonner, Jennifer Muscarella, Jane Kim,
Patrick Clay Neesham, Mark Tanguin, Christine Wang
Editors

Ellen Creig, Copy Editor
Robert Swanson, Index

Preface

So begins our journey. It is a journey that others have taken before us; in fact, in reaching our destination we will rely on the efforts of those who came before. Just as Columbus retraced steps taken by others before him (perhaps as long as two millennia before he sailed), so we gratefully acknowledge the work of those who went before. Yet, like the voyage of Columbus, there is a sense of beginning, a tenor to the enterprise that makes it a voyage of discovery.

Those who came before Columbus came for their own benefit, to find wealth and riches in an untapped land. But Columbus came for other purposes as well: he sailed for king (and queen) and country and to establish trade dominance over the seas, and ultimately the globe. The sudden appearance of a continent barring the way to China was not a disappointment; it was an opportunity—to extend an empire and to provide a place for colonization. Columbus realized that his voyage would make history and that he would return, or that others would follow him. However well visited the Western Hemisphere may have been before Columbus, it was now indeed a New World.

We have a similar sense of newness. The guidebooks on scientific style and writing that have appeared have grappled with many of the issues covered in this book and have provided much instruction of the ways of scientific discourse. The approach that we have adopted, though respectful, grateful, and admiring of previous efforts, differs from them somewhat in ways that bespeak a different set of values and guidelines. What is new is presented mainly in Chapter 1 of this work, and it may be summed up as follows: In addition to the importance of precision, clarity, and veracity in scientific reporting and discourse, there must also be a profound sense of **reality**—a connection to the genuine human thought processes that gave rise to theories; to the details and vicissitudes of the experiments that support one contention or another; to the real life circumstances of science and to the very human concerns that color what on the surface seem to be highly theoretical concerns, even when dealing with the hardware and measurements of the laboratory.

This view of science was first explored in theory in Peter Galison's *How Experiments End* (1987)*,* and later (1997) demonstrated in his detailed history of microphysics, *Image and Logic.* (References for all chapters may be found in **Appendix I**.) The underlying point—that journal articles report what researchers *believe* happened in some idealized sense, and not what actually took place in the laboratory or in the field; or that theory is more often driven by hunches, inspirations, even dreams, than by the hard mathematical demonstrations on journal pages would allow one to believe—is now being understood as responsible for providing a much-needed corrective to the relationship between science and society.

On the one hand, with so much at stake, personally and institutionally, in the assessments made in what constitutes a productive avenue of research and what does not, it is vital that scientists convey their beliefs and findings with a clarity that goes beyond the mere formal requirements of journal publication. If, for example, the Large Hadron Collider, (LHC) which has just begun operation beneath the French-Swiss border, corroborates the predictions of String Theory, then the decision not to build the Superconducting Super Collider (SSC) in Texas will be viewed as having been short-sighted and detrimental to American leadership in high-energy physics. And if the rings of the LHC produce the largest null result in human history, then the discussion on the advisability of the SSC will begin anew, but with the severe disadvantage of the argument for its construction not having been made effectively in the early 1990s.

On the other hand, public discourse on issues in which science has important things to say, such as the extent and severity of global warming, to take one of many possible examples, needs to be informed by the most precise and cogent scientific writing possible if necessary steps (whatever they may be) are to be taken to deal with the issue.

This is the new territory to be charted and which we explore here: how to navigate the human dimension of science—an enterprise that has often suppressed the humanity of the scientist, thus compromising or at least limiting the extent and richness of communicating science, both to the public and to other scientists.

We hope readers will find the structure of the book straightforward and useful. Each chapter begins with a table of contents for that chapter. In **Part I**, we examine the elements of science writing, first regarding creating engaging, effective prose (**Chapter 1**); then preparing work for the various publication outlets for scientific material, with special emphasis on preparing work for science journals and research-level publications (**Chapter 2**); and then (in **Chapter 3**) presenting the general elements of style for English, with a focus on science writing, and ending with a list of words and phrases that are often misused or confused in science narrative at many levels of scientific sophistication. **Part I** ends with two chapters—**Chapter 4**, on the proper forms of citations and referencing of sources (unfortunately, still inconsistently framed, even in other style guides); and **Chapter 5,** on the legalities and practices of copyright protection and permission procurement. The concluding part of **Chapter 2** contains guidelines on the design and creation of tables and other graphic material that may enhance or clarify the points being made in the writing. Though we have endeavored to present a helpful set of guidelines, the experience of working on this book has convinced us of the need for a thorough examination of this subject in a work with greater production values than the present volume—something to be addressed in sequels, we hope.

Part II contains eight chapters on the style conventions and practices relevant to eight areas of science writing: mathematics; physics; astronomy; chemistry; organic chemistry; earth and environmental sciences; life science; and medical science. Each chapter in Part II begins with a detailed Table of Contents for the chapter and ends, first with a list of the tables contained in the chapter, and then a list of the relevant tables contained in the Appendix chapter for that discipline in Part III.

Part III then presents Appendices, one for each discipline, labeled **Appendix A** through **Appendix H**, and containing tables, lists, glossaries, and diagrams that authors in these disciplines might find helpful. Some readers may argue that a list of journal abbreviations need not have been so extensive and others will wonder why the style guide to the spelling of proper names used to identify mathematical theorems is not longer. We acknowledge that both opinions may be correct. The final appendix, **Appendix I**, contains guidance on sources and further reading. The work ends with an **Index**.

By "scientific writing" we mean the physical sciences, as opposed to the technological areas (usually subsumed under the rubric of "engineering"), and the social sciences. There, too, other volumes would seem to be in order, so that we hope we will have the opportunity to continue with manuals of technological and social science style, as well as scientific illustration. The editors and publishers would be most grateful to readers who point out any corrections or failings that have managed to appear in this work in spite of our best efforts to eliminate any errors. This may be sent to the editors care of the publisher (see the contact information on the copyright page), or readers may feel free to communicate with the editors directly at **msseditor@thereferenceworks.com**. We welcome any criticism, corrections, information, suggestions, or advice that readers may offer, and we thank them in advance for taking the trouble of corresponding with us.

One of the people to whom this work is dedicated was a fifth grade teacher in a small, Orthodox Jewish day school in the Williamsburg section of Brooklyn. He noticed a young boy's interest in language and writing and he encouraged him; he even urged the principal of the school (another dedicatee) to "fund" a class newspaper the boy wanted to produce. The teacher impressed upon the boy the need to "make every paragraph a home for ideas," and to insist that every paragraph "earn its address"—which, of course, meant that every paragraph had to *have* an address. Thus began the practice (with this writer, at least) of numbering each paragraph, making certain that every sentence that "dwelled" in that paragraph was well-behaved; that every sentence and clause in it had its place there and was consonant with every other part of the dwelling; and that the paragraph made clear to everyone who visited it what the paragraph was saying and what sort of a "house" he or she was in. It was just a small leap from there to seeing how important it was to use these dwellings to create a street, a neighborhood, a town, a city.

For the next five years, that boy and three like-minded friends produced a class newspaper (the only publication produced by the students other than a yearbook), and would dutifully submit it to the principal for review on the first Monday of the month. The principal would correct any mistakes (which in those days meant retyping the entire page), but never once asked that any article's message or content be changed. The principal, the most impeccably tailored rabbi that boy was ever to encounter (in a life densely populated with rabbis of all stripes), remained a mentor and then friend to the boy for the next thirty years.

In high school, during a hospital stay of several weeks, the boy discovered Isaac Asimov. At one point, the boy had convinced himself that "Isaac Asimov" was (like "Nikolas Bourbaki") actually a group of people publishing under this collective name, for no one human being could possibly produce so much on so many different subjects. During that month of convalescence at the beginning of the school year, the boy continued reading Asimov (there seemed to be no end!) and tackled the opening chapters of an introductory college physics textbook borrowed by a friend from the Williamsburg branch of the Public Library. A month into his senior year, still in bandages, the boy returned to high school; it was the day of the physics midterm, and the teacher excused the boy from taking it. The boy asked if he could see it ("Let's see what I've been missing," he quipped) and instantly recognized the problems as those he had worked on from the physics textbook. Barely able to hold a pencil, the boy zipped through the exam and turned it in halfway through the period. The teacher smiled dismissively and told the boy to sit down as he glanced at what he was certain would be a paper filled with meaningless scribbling. As the boy, in pain and groggy from pain medication, lay on a bench in the hallway, the teacher looked over the boy's papers. As he read, the look on his face changed (the boy's classmates told him later) into one of horror, as if written on the paper was either the Kabbalistic formula for the creation of the universe, or a death threat. The teacher raced into the hallway and confronted the supine boy, demanding to know, "How did you do this?"—in full view of the principal, who was about to scold the boy for lying on the bench during class hours. The boy had only enough strength and clarity to call out one word: "Asimov!" After a frozen moment, both men turned and left. That would have been a wonderful opportunity for the boy to make great strides in physics, but the teacher found more joy in playing basketball with the boys of his class (on a court hidden from the principal's view—different school; different principal), and when some students yelled down at the teacher the word, "Regent's," reminding him of the state exam we were obliged to take at the end of the year, the teacher would yell back, "Asimov!"—or he'd just yell out the boy's name.

The boy's involvement with physics would have to wait for college, where, through an accident of either poor or brilliant planning, the physics department boasted an ivy-league-caliber roster of great physicists. Some were to become famous in scientific circles: Yakir Aharanov; A.G.W. Cameron; Leonard Susskind; Aage Petersen; and Leon Landovitz—and two in particular: Ralph Behrends and David Finkelstein.

They had been originally engaged to staff a graduate school, but when not enough students attended that school, they were asked to teach undergraduates. Much to their surprise, they enjoyed these chores; perhaps because it gave them an opportunity to teach a new generation of physicists the way (in their view) they were supposed to be trained. The most advanced textbooks were used (Feynman's Lectures and the Berkeley Physics Course were background reading), and when those were not good enough, the professors provided translations (nearly always from Russian) of material they thought the students really ought to read. Undergraduates were invited to seminars, colloquia, and special lectures by Nobel Laureates (or soon to be), and were encouraged, prepared (and even fed!), so that the invited notables would not be speaking to empty rooms. (The physics version of "papering the house," one might call it.)

The boy—now a young man—became a devoted student, first of Dr. Ralph Behrends, who drove home the point that no physics problem is solved until it yields a number that can be read on a gauge. Dr. Behrends conducted a private four-year seminar with the young man on mechanics—including a page-by-page study of Ralph Abraham's *Foundations of Mechanics*. Then with Dr. David Finkelstein, already widely known as an innovative theoretician, in topics in quantum theory.

On the day of the young man's graduation, his mother suddenly said to him, "Who is that bearded man running toward us and waving?" The young man turned just in time to see a car just miss hitting Dr. Finkelstein as he jogged casually across Amsterdam Avenue. The professor reached the young man out of breath and said, "I saw you from my office window. I just wanted to tell you that I've decided that the question you once asked [months earlier!] in class—what is 'is'?—is the key question in physics." And with that, he shook the young man's hand and left, saying not a word to the two puzzled parents standing there.

In years to come, the (rapidly aging) young man pursued several careers with varying degrees of success, but each united by the conviction that being crystal clear about what is being said and believed, be it in science, religion, Talmud, the arts, or public affairs, is the key to knowing the truth and knowing what course of action to take. It was once thought that all of the "big" questions of religion and philosophy were going to boil down to questions of science and logic. Now it seems these questions, and quite a few others, will hinge on the clarity of what is said and the precision with which we argue. The intellectual course has come full circle—it seems that in the end, the big questions in science will boil down to questions in philosophy. It will all come down to language—not hair-splitting semantics, but saying what we mean, no more and no less.

In light of the above, it would not be an overstatement to say that the underlying message of this work is that a preface such as this, personal as it is, is appropriate to a work purporting to be about writing for science.

We gratefully acknowledge the assistance rendered to us over the years it has taken to produce this volume: Robert Ubell, who first saw its usefulness; Dr. Jasna Markovac and Tari Broderick of Elsevier, who saw this work as a worthy addition to the Elsevier/Academic Press list; to Lisa Tickner, the publisher, editor April Graham, and André Cuello, production liaison, all of Elsevier, for generously and patiently tolerating our timetable (and our commitment to "getting it right"); and to Mitch Pessin of MP Computer Services, for use of his facilities and for keeping our equipment humming.

Finally, we thank our spouses, Ilana and Daniel, for their unwavering confidence in us, and for their ongoing support of our work, even when we ourselves were uncertain of the eventual completion of this project.

Harold Rabinowitz
Suzanne Vogel

New York City
October 2008

Contents

Note: A detailed Table of Contents for each chapter appears at the beginning of the chapter.

The Manual of Scientific Style

Part I.

The Elements of Scientific Style

Chapter 1. Elements of Science Writing

Contents

Contents, *continued*

4

Chapter 1. Elements of Science Writing

1.1 The Importance of Science Writing

People engaged in scientific research often believe that proper and effective writing lies outside their skill requirements for a successful career in science. This belief is usually engendered by the sense that writing skills properly belong to the humanities, or at most to the social sciences. Shouldn't science, they ask, speak with its data, or, to put it another way, shouldn't scientific data speak for itself? While it is true that a great many abilities are necessary for the successful pursuit of a career in science, the notion that careful and effective writing is merely an adjunct to these abilities is now understood to be deeply flawed for several reasons that arise from a clear understanding of what science, at its core, is and what role it plays in our society.

The image of the lone scientist observing natural phenomena or creating systems and theories in ("splendid") isolation is now understood to be an unrealistic image—a myth, now viewed as an idealization even in the science of previous eras. Newton, for instance, developed his mechanics during a period of isolation while Cambridge University was closed because of the Great Plague of 1665, but we know that he had contact with the leading figures of his day in many areas of science, both in Britain and on the European continent. (How else could so many priority disputes have arisen if there had not been a robust exchange of ideas and information at the time?)

This situation stands in stark contrast to that which prevailed in pre-Enlightenment times (say, before the sixteenth century and going back to antiquity). In pre-modern times, what scientific knowledge existed was safeguarded and kept secret, shared only with initiates and protégés who were honor-bound to maintain confidentiality and refrain from disseminating the details of the discipline they had been taught. The transition from this system to one in which scientists are encouraged to share their findings and insights as widely as possible (for both self-serving and altruistic reasons) is primarily responsible for the flourishing of science and its development over the past several centuries.

As a result of the growth of science communication and its centrality in the entire scientific enterprise, we can now make the following statements about the nature of science that make clear the importance of effective communication in its growth and well-being:

i. Science is a social enterprise, demanding the participation of many people and their interaction with one another if the accumulation of knowledge and human understanding of the natural world are to grow. Research nearly always requires the participation of many collaborators and an operational support structure, plus the professional institutions that enable individuals to acquire training (at a university, for example) and to pursue research in a laboratory or in the field. Even in antiquity, early scientists and naturalists relied on the assistance and collegiality of others who assisted them in their investigations and served as sounding-boards and advisors. The growth of modern science owes as much to the development of organizations and institutions that allowed for colla-boration and cooperation as on the genius of individual scientists.

ii. Science is a political enterprise, and in almost all instances, has poli-tical ramifications. At the very least, scientific consensus will determine the allocation of resources and many issues in public policy. Decisions will routinely be made regarding which research programs to support financially and who is to receive which grant, but the impact of science on politics is far greater (and growing with each passing year), as scientists are being called upon to address and solve a number of difficult and vexing problems that humanity faces today.

iii. Science is an educational enterprise that depends on the continuous influx of talented and conscientious new practitioners to carry forward its ongoing effort to understand and harness the forces of nature and the resources of the physical environment. At the forefront of science, researchers must convey (and in no small measure convince) their col-leagues of the value of their findings and conclusions. For this, effective writing is essential; the most successful scientists have almost univer-sally been recognized as much for the clarity and effectiveness of their prose as the constructs and consequences of their theories.

But at a more fundamental level, every practitioner of science is a member of a community that has an obligation to convey the essence of science and the important role it plays in human affairs to the public. The training of scientists begins at an early age when the interest and imagi-nation of young people are captured by compelling and inspiring popular science writing. The same sort of talent and dedication is required in creating the instructional materials used in classrooms at all levels. Well-written and well-designed science materials encourage young people to consider a career in the sciences. The same kind of engagement must be maintained with the general population if the aims and welfare of science and its practitioners are to be maintained, and if science is to be deployed for the betterment (and survival) of humankind.

iv. Science is a cultural enterprise that has enormous influence on what the great mass of humanity believes about the world and our place in it. This is not to say that the scientific worldview (itself an abstract idealization) is subscribed to by everyone, or even the majority of people. But the ongoing search and conversation regarding the great issues that confront humanity, both in practical matters and in areas of metaphysics (the so-called big questions), are informed by the findings and assumptions of science. No longer is science carried out exclusively in a hermetic "ivory tower" or in the unlit confines of the laboratory. The reliance of modern society on the technology that derives from the findings and constructs of modern science is so great that it is no exaggeration to say that the entire future of the human race depends on the wise and effective use of this body of knowledge and its technological capability. In this respect, the notion (ascribed to C.P. Snow) that there are "two cultures" —science and the humanities—that are doomed to isolation from one another, has been brushed aside by the ubiquitous and unavoidable presence of science and its technological product in our daily lives.

In all of these areas, science depends on effective communication, internally (among scientists), as well as in its relationship with society at large. Sound internal communication—which is dependent on clear and effective writing—is critical to the proper functioning of the scientific enterprise. Sound communication to the "outside" (meaning, non-scientific) world, however, is also critical for science in maintaining the support of the public and its representatives, and in inspiring confidence in science as a source of insight and policy in public matters great and small. In the largest context, the public application of science communication is carried out by authors of books, papers, and articles; producers of films, television programs, and documentaries; and materials in the many new media addressed to the general public.

The same interaction between the scientific community and the civilization in which we all live takes place at least thousands of times every day—in newspapers; magazines; television programs; classrooms; lecture halls; public lectures; museum exhibits; etc., at all age levels in virtually every setting. It behooves the community of scientists and of people who support science and the role it plays in promoting the welfare of human civilization to support, promote, and even *demand* the most exacting and rigorous manner of science communication in *all* settings and contexts. (Readers are directed to **Appendix I** for an annotated list of sources and supplementary reading for each section of this chapter. A similar guide to further reading and resources for each chapter appears in **Appendix I**, which contains a cumulative bibliography for the work as a whole.)

1.2 The Meaning and Nature of "Scientific Style"

The term "style" is ambiguous owing to an accident of publishing history. While in ordinary usage, the word "style" would be used to signify the characteristics of a mode of speech, dress, or expression, in publishing, the word specifically denotes the rules of grammar and usage to which published material must conform. This use of the term probably arose from its inclusion in the title of an informal booklet created by the proofreaders at the then fledgling University of Chicago Press—that was in 1896! Thus, in modern parlance, both in the title of books described as "manuals of style" and while speaking of "style issues" in the course of writing and editing, the word "style" is used in this restrictive sense. Yet, we believe any work that aims to guide and improve scientific writing must address both meanings of style, and must therefore provide guidance on *both* the methods of producing more effective and useful science writing, as well as on the strictures of grammar and usage.

This is especially true of the sciences for two reasons:

i. Correct language and correct science. In science, correct "style" (narrowly construed) is an important factor in creating effective prose. Plain and straightforward formulations of science have been valued since the time of Francis Bacon in the sixteenth century and in the period afterwards, during which the Royal Society was formed in England (in 1660), setting the standard for scientific discourse and investigation in Europe. Bacon urged scientists (in his day called "natural philosophers") to concern themselves with "things," and not with the host of elements that cluttered and obscured the science contained in much of the writing about the natural world of his day. This clutter included: the erudition and station of the author (which Bacon deemed irrelevant); the authority of the systems of the past to which the author appealed (which Bacon considered outmoded); rhetorical flourishes and emotional appeals to cherished human notions (which Bacon considered misleading); and imprecise concepts and terms that had no clear definition and no observational meaning (which Bacon dismissed as nonsense). The development of a straightforward standard of scientific writing made it possible to reproduce experiments, to verify or disprove results and hypotheses, and to crystallize the substance of any piece of scientific writing.

The transition from the "pre-Baconian" style of rhetoric that typified all writing on nature and science, to the fact-based and unadorned manner of writing that is characteristic of science writing today (and has been so for the past two centuries) was a gradual one, and not without its periods of backsliding, retreats into obscure writing, and appeals to arguments more rhetorical than logical or observational. Yet, articles in the science journals of a century ago (as Nobel chemist Roald Hoffman points out and demonstrates in his work, *The Same and Not the Same*) are linguistically accessible to scientists today, thanks to the insistence by the Royal Society and similar overseeing organizations in France, Germany, the United States, and other countries where the strictures of style are adhered to without compromise.

What has become clear over the past half-century is that biases of all sorts—personal, political, religious, and psychological—have a way of creeping into scientific writing in a way that contradicts the claim of the writing as being factual and unencumbered. It was once thought that clarity, simplicity, and precision, the values that are being espoused in this guide and the hallmark of the most influential science writing of the past two centuries, was enough to ensure correctness. William Blake meant something of this sort when he wrote (in his "Proverbs of Hell"), "Truth can never be told so as to be understood, and not be believed." It's a noble thought, but this notion is now regarded as naïve, if only because readers at every level have shown themselves capable of convincing themselves that they understand an illogical argument or an obscure piece of writing. (And, as in any human enterprise, science is subject to the same human ingenuity that allows the unprincipled to advance personal and ideological agendas in the guise of reporting or espousing pure science.) Blake's sentiment has been replaced by the aphorism propounded by H.L Mencken (and which newscaster Harry Reasoner was fond of quoting): "For every problem there is a solution which is simple, clean,…and wrong!"

For these reasons, the strictures of style—grammar; usage; word choice; sentence structure; paragraph and chapter design—all stand as watchtowers that safeguard (though not guarantee) the meaningfulness and clarity of what appears in scientific journals and in the popular and polemical writing about science that is ubiquitous in modern culture. The same may be said for the guidelines that appear in this chapter regarding word selection, sentence and paragraph construction, and paragraph and chapter design, though experience will allow a writer to know when the rules may be broken or bent—when deviation from this advice will improve communication rather than hinder it.

ii. "Science as writing." The distinction between *writing* science and *doing* science has become blurred, particularly at the frontiers of many disciplines. We owe this development, first, to the realization (arrived at relatively recently in spite of how clearly true it is) that the report of a scientific experiment or the elaboration of a scientific hypothesis are really the concluding phases of processes that include failed attempts; infuriating bouts with recalcitrant equipment (and obstreperous administrators—and sometimes the other way around); and many false leads and misguided thinking, all leading in unpredictable ways to insight and conclusions. In the past, such "blind alleys" were considered inappropriate for scientific discourse and were not found in the articles of leading scientific journals. Increasingly, however, such information is included in serious and cutting-edge articles (either as addenda or as supplementary electronic and online material, or in the body of the articles) as a means of allowing other researchers to faithfully reproduce and verify results, and, further, to allow others to retrace the steps taken in the thinking and expectations of the researchers. The desirability of this information leads naturally to the second reason it is so valuable.

The pathways of science lead through the thoughts and psychical meanderings of scientists investigating the structure and phenomena of nature, which means that many conclusions will be the result of thought processes that go beyond the strictly logical and mathematical. These processes include metaphysical underpinnings, social and cultural presuppositions (or biases), artistic and aesthetic values, and even spiritual and religious undercurrents—all playing often inscrutable and unfathomable roles. Einstein was fond of saying that "the whole of science is nothing more than a refinement of everyday thinking." We understand today that the term "everyday thinking" is packed with much more than the naïve notion of "common sense." It includes the specific everyday notions of not-so-everyday people, who have assumed the task of observing, investigating, explicating, and manipulating the world around us, to wit, the scientific community.

Out of this realization has come the idea of "science as writing" (the title of David Locke's landmark work), in which the presence of the author is palpable because the research and thought processes described are the work, words, and thoughts of a *person* or a group of people who bring their intellectual baggage with them in everything they do. Just as it would be misguided to believe that Newton's psychical life was irrelevant to his scientific work, no scientist working today (or arguably *ever*) produced scientific writing except as a human endeavor informed by his or her beliefs and predilections. This not only provides a new standard and tool for understanding and evaluating scientific writing, but it offers new means of communicating science at all levels, namely, through the art of writing. (Consult the references listed in **Appendix I**.)

1.3 Some Guidelines for Writing Effective Scientific Prose

One of the first things a writer of scientific material of any kind must realize is that there is virtually never any instance when judgment is not required. Rules are fine as general guidelines, but they should never be viewed as rigid and inviolable. Recalling the famous comment attributed to Winston Churchill on the rule that sentences may not end with a preposition ("That is the sort of English up with which I will not put."), creative and sound violation of rules can, when judiciously practiced, result in clearer and more effective scientific prose.

In developing general guidelines for creating effective scientific prose, two respected teachers who have trained writers in many areas, and specifically in science—George D. Gopen of Duke University and Judith A. Swan of Princeton University—looked carefully at the needs and expectations of readers. Their conclusions, formulated as a series of guidelines and included in an influential paper, "The Science of Scientific Writing" (*American Scientist*, Nov.-Dec. 1990; Volume 78; pp. 550–558—available online at: www.amstat.org/publications/jcgs/sci.pdf), provide direction that is general enough to be applicable to a wide variety of writing situations, yet specific enough to improve the effectiveness of nearly any kind of expository writing. The methods and conclusions of Gopen and Swan are also used and demonstrated in Robert Goldbort's *Writing for Science* (Yale University Press, 2006). Also consult the "Further Reading and Resources" in **Appendix I**. Consulting these sources will repay readers, researchers, scientists, and writers of all sorts of material immeasurably.

The essence of Gopen and Swan's guidelines is to ask what readers expect when approaching any body of text, and what reading habits guide them, even if unconsciously, as they make their way through any piece of prose. Their method recognizes the fact that the act of communication from author to reader is a cooperative and collegial act in which the reader is just as important as the writer. Gopen and Swan take this concept a step further by claiming that addressing the quality of writing is a means of improving the quality of *thought*; the act of writing and revising is conducive to clarifying ideas and argument in the writer's mind as assuredly as it is in conveying those ideas and arguments to the reader's. As we pointed out in the previous section, *writing* science is a form of *doing* science, and writing science well inevitably leads to improved scientific thinking and practice.

In **Chapter 2**, we will present the details of organization and preparation of material for various settings for scientific writing (including elucidating the classic IMRAD construction), but here we focus on the *units of communication* for scientific prose (or, in our view, prose of any kind), which is in the first instance the sentence, and in a larger context, the paragraph. To clarify, while the unit of thought may be a word, a word appearing on a page, or leaving a speaker's lips is not itself an act of communication. It is simply an utterance, an iteration in need of other words and punctuation—a syntactic context, if you will—that will turn it into a communicative act. Write the word "help" on a piece of paper and leave it at that, and you are not communicating; write it on the sand of a desert island, even misspelled and without an exclamation point, and the reader (in the airplane overhead) is entitled to regard it as an act of communication, or at least an attempt at one (and would be wise to suggest to the authorities that they investigate).

The Gopen and Swan approach, therefore, is to look at what readers expect when dealing with a sentence and with a group of sentences conjoined to form a paragraph. "Good writers," Gopen and Swan point out, "are intuitively aware of these expectations." Attending to these expectations in one's prose is likely to inculcate these habits and practices in one's writing—in due time, without even being conscious of it.

Here are some of Gopen and Swan's guidelines:

i. Verb placement. Place the verb of a sentence as close as possible to the grammatical subject. Readers expect the verb that informs what the subject of the sentence did to come soon after the subject is identified. Anything of length that separates subject and verb is regarded as an interruption and leaves the reader with a sense of unfulfilled expectations. The reader may forget just what the subject of the sentence is by the time the verb appears—or worse, the reader may imagine or invent another action that will either replace the verb or create in the reader's mind actions that are variations of the one the writer offers. In any case, placing material between the subject and the verb—particularly extraneous material—lessens the chance that the reader will understand the sentence or paragraph to mean just what the author intended to convey.

ii. "Point" placement. The "new information"—the *point*—the writer wishes to convey should be placed in the latter portion of the sentence (or paragraph). This is known as the "stress position" of the piece of prose and it reflects the simple observation that readers expect a later position to be the place where the "payoff" or the new idea—the writer's point—will be revealed. By way of example, there may be some mystery novels that "work" (that is, engage readers right to the last page) even when the culprit is revealed early in the story, but that requires a special mastery of the form. Gopen and Swan point out that this cyclical quality of reader attention is consistent with the way people apportion their energy on a task through time. Readers instinctively sharpen their attention and prepare for the climax or the point that the writer wishes to convey as they sense that they are nearing the end of the sentence or paragraph. They can see this by the simple graphic structure of the sentence or paragraph—the looming period or the imminent beginning of a new paragraph, indicating that a resolution of the author's communication is in the offing and a new point is about to be presented.

This expectation of resolution, triggered by the impending culmination of the sentence or paragraph, is also one of the tools that a printed book or journal uses to enhance and clarify the communication process (and which is lacking for a digitized text on a computer screen). The chapter structure and the clear way a reader has of knowing where in the book any given passage lies, allows the reader to asses the weight of the information being presented in the context of the message of the book or article as a whole. It is therefore even more important that material likely to be read in digitized form be structured properly if the author's information and thoughts regarding its import is to be accurately conveyed.

iii. Subject placement. Place the subject of the sentence or paragraph in the early portion of the sentence or paragraph. This is known as the "topic position" and it is the place where the reader expects the subject of the communication to appear. Readers expect the subject of the writing to appear early and perceive this positioning as a prompt to prepare themselves for information or observations later in the sentence or paragraph. So strong is this expectation, that tables that fail to place the subject material on the left and the findings or conclusions on the right become virtually indecipherable (as Gopen and Swan demonstrate). To use Gopen and Swan's narrative example, "Bees disperse pollen" is a sentence about bees; "Pollen is dispersed by bees" is a sentence about pollen. If what follows the first sentence is about pollen, or if what follows the second is about bees, readers are certain to be confused and will miss the point that the author wishes to convey.

This formulation of sentence and paragraph structure—placing the topic early; placing the new information late; keeping the subject and the action verb close—is a basic design that an author abandons only when absolutely necessary (and with due attention to compensating for confusion that such a move can cause). It also provides a guide to determining when a sentence of a paragraph is too long, and, in fact, suggests a guide for determining what a paragraph *is* in the first place and how paragraph lengths are to be determined. The decisive criterion for determining when a sentence or paragraph is too long is *not* the number of words in the sentence or the number of words or sentences in the paragraph. Style manuals and guide books that offer arbitrary numbers by which to determine if a sentence or paragraph is too long ignore the fact that short sentences can be indecipherable in spite of their brevity, and long sentences and paragraphs can, if properly constructed, read effortlessly and be perfectly clear to virtually every reader.

A sentence or paragraph is too long, according to Gopen and Swan, if "it has more viable candidates for stress positions than there are stress positions available," or as paraphrased by Robert Goldbort in *Writing for Science*, if the sentence or paragraph "cannot accommodate all the items requiring stress." This formulation expresses the observation that effecttive writing conveys information through the judicious use and construction of paragraphs. Knowing what to put in a paragraph and where it should be placed is a skill that often requires long practice. (Some writers are, it seems, born with this skill; these virtuosos are, indeed, the lucky ones.) How to construct an elegant and persuasive paragraph out of clean and concise sentences is the art of good writing, but such paragraphs will more often than not follow the three rules presented here: they will have the subject of the paragraph placed early in the topic position; they will have the point of the paragraph placed toward the end of the paragraph; and they will place as little material between the two as possible, ensuring that the point is not "lost" amid all the verbiage. These rules are helpful in effective communication because they conform to the expectations of the vast majority of readers whenever they approach *any* piece of writing. (Readers will recall that we noted in the Preface that some students are instructed early in their education to number paragraphs and to insist that, "every paragraph earns its number.")

iv. Context placement. Place "old information"—material that will provide a context for the new information—before or near the topic position. Of all the rules provided by Gopen and Swan, this is the one that requires the greatest use of intuition and a skill that may be expected to improve with experience.

What a reader must fully understand is that points being made in a paragraph should include both background information supplementary to the subject of the paragraph, and a contextual connection to the points. The agronomist Martha Davis, in her work, *Scientific Papers and Presentation*s, compares a piece of scientific writing to a house, and the reader to someone visiting that house. In addition to the utilitarian items that a house requires, a visitor to a house needs to feel comfortable with the surroundings and must be able to navigate the house almost as if he or she actually lived there. Upon entering (the house or the paragraph), there should be a vestibule or foyer that sets the tone and establishes the style. Parts of the structure should lead naturally into one another without the sudden or unexpected appearance of extraneous elements (in the form of an unexpected room in the case of a house, or an extraneous remark or anecdote in the prose). There should be a natural inevitability in the journey into the house/paragraph that provides a resting place where the visitor/reader can pause for a moment and take in the décor/point being made before continuing. Strange and unwieldy constructions and arrangements may make for innovative design, but a person is not going to ever feel totally "at home" if the structures and elements of the environment do not flow naturally into one another. This is often referred to as the "flow" of the narrative and it allows the reader to make the journey toward the point with clarity and ease. This practice also derives from the way readers react to material as it is presented in a paragraph. A reader likes to "get his bearings" and feel familiar with the surroundings before embarking on new territory.

One way of determining if the paragraph has the flow that will ensure the reader understands the point being made is to try it out verbally on someone. A gap in the logical connection between one element of the narrative and another—particularly between the subject of the narrative and the point being made about it—will become clear when that puzzled "lost" look appears on the face of a listener. In a sense, a writer must be able to imagine a listener or reader responding to a piece of prose; it is not enough that the sentence sounds good to its author. (The same may be said about constructing or decorating a house.) Gopen and Swan report that in their many years of teaching and evaluating scientific prose, the single most common error and flaw they encounter is misplacement of the elements of the paragraph—placing new information too early; placing clarifying connective text too late; interrupting the flow of the text with asides and irrelevant material. In their paper, they provide several examples of ineffective text, analyze where the prose fails to communicate effectively, and suggest ways of improving the paragraphs.

v. Verbs and action. Articulate the action of every clause or sentence in its verb. Readers expect that the action that is attributed to the subject of a sentence or paragraph is going to be described by a verb, and that the connection between the subject and the verb will be clear and manifest. Writers often allow the complexity of the writing (presumably reflecting the complexity of what they are writing about) to obscure the connection between the subject and the verb, which leaves the reader wondering exactly what is being described and what new information is being provided. It is sometimes useful to bracket the subordinate and qualifying clauses in a paragraph and highlight the subject and the action verb. When analyzing a paragraph, a writer might ask several questions to make certain that the text conveys just what the writer wants it to:

• Is the verb appropriate to the subject?

• Is it clear from the text that the verb applies only to the subject and not to another element of the text?

• Has intervening material diverted the reader's attention from the connection between the subject and the action verb?

• Does the text "pile on" additional actions by attaching more verbs than a reader can comfortably handle?

These are questions that an experienced writer will usually ask without thinking, but even the most experienced author can lapse into errors of this kind when dealing with complex material like that which is typical of much scientific discourse.

vi. Relative placement of context. Provide the reader with context and background *before* presenting new information. Writers will often be so eager to share their new findings or conclusions, that they will place the new information early in the text and will only later provide the background material that a reader needs to understand and to evaluate the new finding. If the reader is to both comprehend the point being made and evaluate it, he must have all the information in hand when the point is made. Failing that, the reader is apt to provide background and support that is contrary to what the author believes is germane or supportive of the point, and is thus likely to either misinterpret the author's intent, or come to the conclusion that the author's point is simply incorrect.

This principle is also a consequence of what readers expect when they are reading, but one way of demonstrating the logic of this principle is to consider a lecturer who is presenting a mathematical proof or derivation on a blackboard. The act of presenting the material piecemeal and in a logical sequence (and with the benefit of an audience that is responding in ways that signal their comprehension—that they are following the argument, or that they are perplexed by one particular step or another) allows the audience to become comfortable with each step in the proof and each element in the argument before proceeding to the next step. A proof presented out of sequence is likely to confuse a student, even if all the steps will eventually appear on the blackboard in their correct place at the end of the lecture. The benefit of seeing the proof built "brick by brick" is one of the advantages that a proof presented in steps on a blackboard has over the same proof laid out in its entirety on the page of a textbook (and is one reason a textbook cannot, we submit, ever totally replace a lecture). Like the clues that the pagination of a book provides in telling readers where in the author's argument they are at any given moment, the logical presentation of the proof allows the audience to travel the lecturer's same path. This provides a degree of comfort and familiarity, which allows the audience to both understand and evaluate the presentation.

In practice, an author does well to stop often in the composition of the text and ask if enough information has been provided to the reader with which to evaluate what has just been presented. Following this principle amounts to little more than recognizing that the text is not simply a repository of the information; it is an instrument that brings the writer and the reader into proximity and into a relationship that makes communication (and hence the entire enterprise of science) possible. From this observation, Gopen and Swan draw an important conclusion that could be viewed as the essential point of their essay. "It may seem obvious," they write, "that a scientific document is incomplete without the interpretation of the writer; it may not be so obvious [but is no less true] that the document cannot 'exist' without the interpretation of the reader."

vii. Emphasis and structure. Match the relative emphasis of the substance of the sentence, paragraph or chapter with the expectations of the reader created by the structure of the writing. When a paragraph is effectively constructed, it has a quality of being able to present an idea with a uniformity and coherence that readers and critics have described as "musical"—that is, a unity that presents the substance of the prose whole in a manner that Hawthorne described as "words disappearing into thought." In much the way a musical composition uses notes, chords, sounds, alternating passages loud and soft, slow and fast, tense and relaxing (in minor and major keys); so prose has a structure in which the reader's processing of the information and ideas conveyed in the text creates reactions of puzzlement and understanding, confusion and enlightenment, surprise and explanation, dramatic tension, and satisfying resolution, to name a few. The principle in bold at the beginning of this paragraph may be viewed as an expression of this quality of prose in terms of the structural directives provided by Gopen and Swan's earlier principles. That is, following the basic structural directives of subject placement, verb proximity, reader preparation, point placement, and overall paragraph design, the writer places himself or herself in a position to communicate the intention contained in the text to the reader interpreting it, and to do so with an ease that belies the work and care that went into creating the text in the first place.

1.4 Guidelines for Effective Word Selection in Science Writing

In the previous section, we have focused our attention on the structure of scientific prose and the means of "designing" a sentence, paragraph, and chapter so that the reader understands what the author wishes to convey, and does so with an ease that allows for evaluation of the import and the correctness of what is being communicated. Much of the rest of this volume will be dedicated to matters of usage and style (in the narrow, technical sense of the word), and the conventions and practices that apply to English prose in general and to the usages accepted and prescribed in the various scientific disciplines. This process begins with gaining a facility for selecting the right word or phrase, and just as importantly, avoiding and eliminating words and phrases that confuse the reader or obscure the author's true intent.

Many of the principles set forth here have been taught at all levels of instruction in rhetoric and composition for centuries. The changing values and standards of English composition demand that these be reviewed to reflect current thinking, but there is another reason for reviewing these in a guide to scientific writing. Writers of science often believe they do not have the obligation to convince and educate their readers that writers of, say, history or philosophy have. The idea is often expressed as, "If you have to ask, you can't afford it." No form of scientific writing (we hope we have made abundantly clear in what has preceded), no matter how complex or advanced, is free of the need to convince and enlighten prospective readers. Every piece of writing about science—whether it is for a local newspaper; a magazine; a book for the educated public; a textbook for students; a monograph for advanced researchers; a review or critique for colleagues; a report on an observation or an experiment; or a theoretical construct or hypothesis—is always directed at convincing a reader on the correctness and importance of what the piece is saying. Such is the reality, and anything that thwarts that objective should be avoided. That includes any elements of language and structure that create distance between writer and reader.

i. Be clear.

There is a direct connection between being clear and being simple. The foremost practitioners of writing, be it in science or in any other field, repeatedly counsel to avoid the "elegant phrase," and to present ideas in as simple language as possible. Physicist Michael Alley begins the chapter on clarity in *The Craft of Scientific Writing* with a memorable quote from Einstein: "When you are out to describe the truth, leave elegance to the tailor." Alley identifies two elements that subvert the clarity of prose: **complexity** and **ambiguity**.

a. Keep it simple. Complexity takes the form of words that are unnecessarily complicated and phrases that add little, or which can be replaced with much more direct and simpler terms. Words formed by adding *-ize* to a verb are a lazy substitute for a proper and perfectly standard word. These should be avoided unless the usage has become accepted ("familiarize" may be acceptable, whereas "particularize"—as a substitute for "specify"—is generally not). Using words because they are more formal or longer, when shorter, simpler words convey ideas more directly, not only obscures meaning, but has a distancing effect on the reader. Given how important a sense of cooperation and, indeed, collaboration between author and reader is in scientific communication, anything that widens the gulf between the two is to be avoided. No single word or phrase will do great damage, but the cumulative effect of an entire paper filled with such words and phrases will take their toll. Using "utilize" instead of "use"; "finalize" instead of "end"; "hitherto" instead of "until now" or "previously"—all create barriers between what should be a friendly and collegial exchange between author and reader. The same may be said of unnecessary words and phrases (described in guides as "useless," "empty," or "zero") that add no information and give writing a pretentious and overly-formal quality. They do nothing but further estrange the reader from the writer and discourage the reader from participating in the interpretive communication process. (Examples and suggested alternatives are provided below, *see* **1.4, iii. Be direct.**)

The same applies to phrases and sentences: needlessly complex wording may not compromise meaning, but it will almost certainly interfere in the communication of ideas. This is why Alley and others suggest that a good way to eliminate or avoid overly complex writing is to imagine that the material is being read to the most important and informed reader of the work.

Care should be taken to avoid the following:
- strings of modifiers (adjectives and nouns that serve the same purpose);
- packing sentences with prepositional phrases and subordinate clauses;
- packing a sentence with too many ideas to be effectively communicated.

This last problem is often difficult to identify, even in editing, because it can still exist even after one has eliminated run-on sentences during the revision stage. Just as there are spoonfuls of food that are too large to allow a diner to enjoy the taste of the food or even to digest it, there are sentences and paragraphs that are too laden with material to allow for sober and considered comprehension and evaluation of the ideas being expressed.

b. Keep it unambiguous. Ambiguity is often best addressed in the editing and revision process. It is often difficult to identify ambiguity until the entire piece of writing is finished, or until the writer has some "distance" from the piece so that he or she can evaluate it objectively.

An ambiguous word or phrase is not simply a word or phrase that has uncertain or multiple meanings, it is a word or phrase that can be understood by a reader in many ways because of other problems with the prose. Most writers know that "as" should not be used to mean "because," since the word "as" can also have other meanings (such as "like").

Ambiguity can also creep in through inexact *syntax*—the placement and construction of the phrase or sentence. To illustrate, Alley provides five sentences that differ only in where the word "only" is placed.

> *Only* I tested the bell jar for leaks yesterday.
> I *only* tested the bell jar for leaks yesterday.
> I tested *only* the bell jar for leaks yesterday.
> I tested the bell jar *only* for leaks yesterday.
> I tested the bell jar for leaks *only* yesterday.

The difference in the meaning of these sentences illustrates how important it is to be certain that sentences are constructed to convey what the author wants them to convey.

Pronouns also often introduce ambiguity into a piece of writing. Fowler's classic work on English usage advises that there should never be "even a momentary doubt" about to which element of a sentence a pronoun refers. Alley's example is as instructive as it is amusing: "In low water temperatures and high toxicity levels of oil, we tested how well the microorganisms survived." (Alley adds: "I hope that everyone conducting the tests survived as well.")

ii. Be precise.

In scientific prose, one expects words to be used carefully, so that, for example, a writer of a piece on physics would not confuse "mass" and "weight," and should not do so even if the level of writing is informal enough to make the distinction less important, or if the writing is directed at an audience that will still understand the point of the piece even without knowing the distinction. The value of precision in science may well be the highest value that a science writer (or a scientist, for that matter) can espouse, and a science writer compromising on precision, even for what is perceived to be a greater good of more emphatic and persuasive communication, does so in peril of rendering the writing neither forceful nor persuasive.

Often, scientists and science writers are placed at a disadvantage by pseudoscientists or those who advance a political or religious agenda under the guise of science. The discussion is manipulated by those who argue using terms imprecisely and ambiguously, a practice that a scientist or science writer would never countenance. In a sense, a science writer is in somewhat the same position as a prosecuting attorney. A prosecutor is not permitted to be anything but truthful: he or she is not permitted to withhold exculpatory evidence; not permitted to argue a position that he or she believes to be false; and not permitted to call on witnesses to give testimony that he or she suspects might be untruthful. Defense attorneys, on the other hand, are permitted to do all this (to one extent or another) in the course of providing their clients with the most vigorous possible defense. At least that is how the Western legal system is supposed to operate, and while attorneys on both sides will often bend the rules and standards to gain a momentary advantage, experience teaches that such transitory successes are gained at a cost that ultimately undermines the system as a whole. The same may be said of science writing. A momentary lapse in precision may win a point, but will also undermine the enterprise of science communication and its vital role in the culture. Words that frequently cause problems in this area are absolutes and unqualified assertions. Careful writers are loath to use words like "always" or "never." They will cringe when reading a bold statement that is in desperate need of qualification; they should cringe even more when writing that way.

a. Repetition is not a sin. The belief that repetition of words is the mark of poor writing is often the root cause of imprecision because writers sacrifice precision and use synonyms in order to refrain from repeating words. While repetition may interfere the flow of certain types of writing (in which the very reappearance of the word interrupts the narrative flow), it is important to realize that synonyms are often only approximately equivalent in meaning, and using them carelessly runs the risk of subverting the entire meaning the writer wishes to convey. If a word or phrase perfectly describes a situation, writers should not hesitate to use it repeatedly in the text—with caution, however, and not with wild abandon. Mark Twain, who was a judicious stylist and mindful of the way a repeated word can annoy and distract a reader, nevertheless advised using the right word (once found) whenever necessary. The difference between the right word and "almost the right word," he would say, is the difference between "lightning" and "lighting bug."

b. Connotation. Be aware of what "baggage" a word brings with it. Words or phrases have acquired connotations in the course of common parlance. This stands in opposition to the word's "denotation," which refers to the clear and "unvarnished" meaning of the word. Writers must be sensitive to the connotation of words for the simple reason that a connotation may lead a reader to understand the exact opposite of what the writer intends. An oft-cited example is the word "adequate." Although the denotation—the straightforward definition—of "adequate" ("1. Able to satisfy a requirement" according to *The American Heritage Dictionary*) would make it a perfectly suitable word to use in describing the condition of an airplane or surgeon, the word has the connotation that what is described is *barely* satisfactory for the task at hand. Most travelers or patients would hope to find an aircraft or physician much better than merely adequate.

The connotation of words change frequently and often unexpectedly (sometimes as a result of the use of a word in the public arena or media) and thus a competent writer must stay abreast of what is happening to words in the minds of readers—or be aware of such nuances of meaning that may be newly reported in dictionaries and style guides. (That same *American Heritage Dictionary* offers as the second definition of "adequate": "2. Barely satisfactory or sufficient.")

Here are three examples of language that carry unwanted "baggage" that often prevent the reader from understanding the author's intent:

Value-laden words or phrases—terms that on the one hand seem to describe or modify in a straightforward way, but which, on the other hand, are used so frequently with decidedly positive or negative associations that they take on similarly positive or negative connotations—may also color a piece of prose in ways that are contrary to what the writer intends. For example, saying that a drug has been observed to "exacerbate" or "aggravate" a bodily function will lend a negative connotation to the report, just as applying the words "alleviate" or "enhance" to the same observation will give the report a positive connotation. These descriptive terms should certainly not be used if they convey the opposite of what the writer wishes to convey, but such value-laden terms should be avoided in all cases because they introduce an imprecision in the text. If the finding reported is, in fact, salutary in the opinion of the writer, then he or she should say so explicitly and not through the connotation of the words used or through innuendo. Such use of language can leave readers with the sense that they are being manipulated and not dealt with honestly, and this creates a barrier between writer and reader. (It is Hawthorne, again, who warns that, "imprecision is a 'blood relative' of dishonesty.")

Jargon and slang present the same sort of problems when used in scientific writing: connotations and associations with words and phrases introduce meanings that are irrelevant or contrary to the intent of the writer. Many readers will not be able to avoid understanding such words and phrases in light of the connotations and associations that these words and phrases bear with them. While linguists will point out that many terms, particularly in the sciences and technology, go through a stage in which they develop from jargon and slang into members of the standard vocabulary, such words should not be used until their "flowering" is complete and they are accepted in public discourse and in the parlance of practitioners of a discipline. In **Section 1.7**, below, we offer lists and guidance on specific examples and usages of jargon and inappropriate language for scientific writing.

Finally, "**anthropomorphism**" refers to the use of words that imply human agency or intention to inanimate objects, abstract concepts, or physical phenomena. In addition to being inaccurate and imprecise—animal subjects do not "prefer" or "detest" one food over another in any way that approximates human choice and decision signified by these words; studies do not "ignore" or "praise" other studies (authors of studies do); and molecules do not "dance" or "crowd" into potential wells—such usage introduces an element in the writing that invites the reader to explore associations and to decipher what the author meant by introducing this material. This further separates the reader's understanding from the writer's intent, and interferes (needlessly, in our view) in the communication process. While we have argued (in **Sections 1.2, and 1.4, iii**) that scientific writing should still be human and that the humanity of the author should be present if relevant to the subject and the substance, anthropomorphic language is not what we had in mind.

c. Level of detail. The level of detail chosen for any particular piece of scientific writing is among the most crucial decisions an author can make. The inclination to pack a piece of writing with all available information at the most advanced level of technical language and mathematical representation must be overcome and controlled; such writing can drive away even the most erudite and interested reader. It will certainly not engage general readers or even readers working in nearby fields. Following are seven suggestions, made by several guidebooks, on how to address the question of determining the right level of detail that a piece of scientific writing should have:

1. Strike a balance between the specific and the general. Clear and persuasive science writing uses both specific statements and generalities that support and explain one another. The level of detail of a piece of scientific writing is the net effect of these two kinds of statements—an "average" between the specific and the general.

2. Use examples to illustrate abstract concepts. Judiciously chosen examples are often the best way to communicate a subtle or complex concept. An experienced teacher will regard the examples he or she presents to students to be critical to successful education. The same goes for communication in any sort of scientific writing.

3. Use analogies to paint a picture in the reader's mind. The right analogy or descriptive image can not only elucidate an idea, but can also give the reader an image that will linger and with which to remember and ponder the concept.

4. Use comparisons to place information in perspective. Measurements and dial readings reported in scientific literature do not generally come with scales or ranges, so that many readers will not be in a position to know if a reading is remarkably high or low. Additional information that explains what typical or "normal" readings or measured quantities are, can help readers relate to and understand highly technical and specific statements.

5. Review the piece and be on the lookout for needless complexity or obscure terminology. We will indicate later (**"1.5 Getting Started"**) how important editing a draft of a work is, and we will emphasize that the best editing is done by someone *other than* the author of the work. But the first review and editing of a work is best done by the person most familiar with it, namely, the author. Time spent reviewing (which means at a minimum genuinely rereading the work, preferably with fresh eyes, and asking how a new reader will understand each part of the work) will pay dividends in better reader comprehension.

6. Simplify sentences and phrases whenever possible—particularly clusters of nouns and adjectives. Evaluate the piece "microscopically"—phrase by phrase, sentence by sentence, and paragraph by paragraph. Address needlessly complicated phrases or sentences and be certain that terms are used properly and with ample explanation and preparation. Construct sentences as simply and as plainly as the level of technical sophistication and complexity will allow.

7. Step back and ask yourself, "Have I made my point?" This step entails looking at the piece "globally"—meaning, does the piece convey the most important point that you, as the author, wanted to convey to your ideal or typical reader? Clearly, other readers may be better able to answer that question than you, the author, but making a first attempt at it may well be beneficial. Sometimes, a single phrase or sentence, near the beginning (announcing your intent) or near the end (summarizing your conclusions) is what is necessary to drive home the point you wish to convey.

iii. Be direct.

Characteristic of scientific writing (perhaps its most distinguishing characteristic) is that it is *direct*, or, as some guidebook authors put it, "forthright" (although the word "forthright" has the connotation of "being honest," and one would hope the author's honesty is not in question). As was noted in **Section 1.2**, one of the milestones in the development of modern science was the emergence of a manner of communication that could be shared across otherwise unbridgeable chasms of time, geography, language, social environment, culture, political system, and even religious conviction. Scientific writing is simply about what it purports to be about, and should not be advancing any hidden agenda or subliminal message. It should make no appeal to special, hidden or esoteric knowledge; should make no appeal to authority, nor to any source of fact and truth other than logic and observation.

The road to the prevalence of scientific style was long and difficult, and it did not suddenly appear full-blown. It was arrived at one step at a time. Today, attempts are being made to recover some of the beneficial elements of scientific discourse that the development of science writing has sacrificed. But these efforts are not attempting to replace or compromise what has been gained in the past two to four centuries.

There are two main areas in which the direct quality of scientific discourse and science writing are manifest, and it is these two areas that should inform one's writing:

a. Avoid pretentious, arrogant, and clichéd language. Pretentious words and phrases are those that are needlessly long, but which express simple ideas. There are certainly times when an unfamiliar or complex word will be necessary to convey a similarly unfamiliar or complex idea. But to use such words to convey simple information or express simple ideas is certain to alienate the reader and stand in the way of a reader's comprehension of the work. The tone of the writing fairly screams at the reader that the author believes the reader is fortunate to even be allowed to read what the author has written. Readers are understandably insulted by this insinuation, and they are less likely to read what the author has to say with any degree of sympathy.

The list of pretentious words is a long one and below is a small sampling. A rule of thumb that may be helpful is: if the word or phrase is one you would not use in speaking to a close relative or loved one, it is likely to be inappropriate for scientific writing.

Instead of	Use
component	part
facilitate	cause
implement	use
in close proximity to	near
on two occasions	twice
conduct an investigation	investigate

Arrogant language takes the insult a step further by raising an insinuation to a bald assertion. Phrases like "as is well known," "it is obvious to any reader," and "clearly demonstrate" (and others of that kind) convey to the reader the sense that only a fool would dispute the statement made or the conclusion drawn. Readers are generally perfectly capable of seeing through such language. The result is likely to be a begrudging and resistant reading of the piece, and a jaundiced attitude toward its author and the idea the piece contains.

Clichés and instances of extremely colloquial language are likely to also result in a "disconnect" between author and reader. Phrases like, "sticks out like a sore thumb" and "back to the drawing board" convey an attitude on the part of the author that he or she thinks either the subject is not deserving of serious attention, or the reader does not deserve a serious discussion. The impression may be subtle and the reader may be only mildly or unconsciously aware of these messages, but if the reader is to have an important role in successful communication of the author's point, then these message will be picked up by many readers and will, at the very least, result in far fewer of them understanding or accepting the author's thesis.

b. Strong nouns and verbs. Use direct and "strong" nouns and simple, active verbs to convey information. Finally, being direct in one's writing means couching ideas in terms that convey the subject clearly and that convey the action taken by the subject emphatically. This suggestion is often conveyed in terms of using the "active" voice in favor of the "passive" voice, and this has in turn given rise to the notion, which may be found in older style guides, that writers should always use the active voice and never use the passive voice. As with most absolutes, this is an overstatement and there may well be instances where the passive form is the best way of describing a situation—readers will recall the example given above, in **Section 1.3, iii**, in which "Pollen is dispersed by bees" is a preferable form to "Bees disperse pollen"—*if* the subject of the writing is pollen and not bees. Writers often find that the most effective writing (in science and elsewhere) strikes a balance between the active and the passive voice, a balance that is acquired through experience.

c. Concrete vs. abstract. The same is true regarding the old stricture that favored the concrete over the abstract: each case has to be looked at on its own merits. As a general guideline, however, direct, active, and concrete writing, as reflected in the choice of words and the description of action, arrests the reader's attention more than indirect, passive, and abstract writing, because it forces the reader to stop and reflect. This could also, however, cause the reader to lose the train of the argument.

d. Pronouns and tense. Under this rubric, we caution that pronouns and tense be used carefully when describing the author and the work or research performed. Authors will often use the passive voice in presenting their work because they believe that this lends the work an air of objectivity and seriousness. Goldbort points out that arguably the most important paper in the history of modern life science, Watson and Crick's report on the structure of DNA, begins with "We wish to suggest a structure for the salt of deoxyribose nucleic acid (D.N.A.)." Earlier (in **Section 1.2, ii)** we suggested that the elimination of the author as a participant from scientific writing was an unnatural conceit that gave readers the impression that the effects and experiments happened by themselves without any human effort, which further insinuates that the experiments *had* to happen just as they did.

The same misleading implication can result from the misuse of tenses in reporting results or hypotheses. It is correct to say that an experiment conducted *found* something to be the case; reporting that the experiment or the experimenter *finds* something to be the case implies that the experiment or phenomenon is ongoing, and that the observation is a universally expected phenomenon of nature. Use of past tense when reporting conclusions likewise represents the fact that a hypothesis is being made and a conclusion is being drawn, as opposed to a phenomenon being observed.

iv. Use shared language.

A challenge that every writer of scientific material faces, regardless of level, is making unfamiliar terminology accessible to the reader. Scientists often disparage popular writing for bowing to this problem, but even technical writing can suffer from improper use of technical language because it can cause the reader to stop and evaluate the term, and that will interrupt even the most accomplished and competent reader in his or her participation in the communication process.

a. Define technical terms. Avoid using technical terminology without giving due thought to defining or explaining the terms used. There are many ways to use technical terms that vary across time, settings, cultures, languages, and disciplines. We therefore couch this advice in the negative, urging the writer to *always* ask if the term used has been adequately defined (either in a footnote or a glossary if too many definitions will intrude on the flow of the text).

Another reason to be extremely careful with the use of technical terms is that often what is assumed in one setting to be a properly defined and understood technical term is actually **jargon**, specially used by a limited number of people and practitioners in a field. This is particularly prevalent in technical fields, where ordinary daily discourse is laced with newly-minted terms that reflect the fast pace of technical development of the field. Thus, terms in information and computer technologies are created with the same dizzying rapidity as the technology itself. Writers must be careful to use terminology that readers will understand if they are to communicate effectively—this truism is never more true than in the areas of contemporary technology.

In other areas, the burst of creativity at the very cutting-edge of fields encourages the coining of new terms—**neologisms, abbreviations,** and **acronyms** that are often designed to convey the unfettered use of imagination and even a note of whimsy. These have to be tightly controlled by writers if they are to be understood. It is difficult to imagine contemporary particle physics, for example, advancing without the creative and inventive use of language to describe structures and processes at the very edge of human comprehension. It is important, however, that these terms be well grounded and concretely defined so that the terminology and the syntax do not run roughshod over meaning and understanding. In these areas, it is well to remember physicist Niels Bohr's advice, that "one's writing should never be clearer than one's thinking." Behind the evocative labels and abbreviations, there must be a firm sense of concrete definition if anything written about these terms is to be meaningful.

b. Use examples, analogies, and comparisons. Explain unfamiliar words and concepts in terms of words and phenomena familiar to the reader. Researchers and educators for whom imparting knowledge to the next generation and to their fellow human beings is of paramount importance value highly the apt and telling example, the evocative and illuminating analogy, and the memorable and striking comparison. Selecting these tools effectively in the course of educating anyone in any setting is the hallmark of a good teacher, and (it may be argued, as it was in **Section 1.2.ii**) a quality of a good scientist.

In selecting and fashioning **examples,** care must be taken to maintain the focus of the example on what is being explained and illustrated. General statements that are not illustrated with an appropriate example will not only be forgotten, but will leave most readers uncertain of whether they have correctly understood the statement in the first place. Examples work best when they are concrete and specific; when they are whimsical or contain tangential material, they divert the reader's attention and focus. A well-crafted example is plain and to the point.

An **analogy,** on the other hand, is designed to evoke a picture in the reader's mind—an image that will stay with the reader as he or she continues on through the text. The author hopes the image will remain vivid as the subject is developed; the author also hopes that the image of the analogy will not import new associations and complexities into the text with which the reader will have to grapple further on in the piece. Analogies, therefore have to be vivid and simple, and they require imagination and creativity if they are to be artfully crafted. Einstein was a master of the well-wrought analogy and pointed example, and much of the success of relativity by the scientific community owed to the vivid quality of his writing. (When, in the early days of relativity, it was joked that only eight people understood the theory, it would have been more accurate to say that only a few people had something original and illuminating to *say about* the theory. The new physics took some time to develop its language and imagery tools, and for this Einstein and his supporters called upon their literary talents as much as their scientific expertise, just as modern physics has had to undergo the same "linguistic evolution"—using many of these same literary tools—over the last decades of the twentieth century.)

A third device that may be used to explain an idea so that it is clear, vivid, and memorable is **comparison**. A famous example (cited by Alley) is given by the physicist Richard Feynman in the opening of his lectures on electromagnetism. Feynman describes the magnitude of the electrical force by comparing it to the force of gravity:

> If you were standing at arm's length from someone and you had one percent more electrons than protons, the repelling force [on you] would be incredible. How great? Enough to lift the Empire State Building? No. To lift Mount Everest? No. The repulsion would be enough to lift a weight equal to that of the entire earth.

In his subsequent discussion of the difference between the forces of gravity and electricity, this comparison is in a class by itself and is remembered by virtually every physics student who has ever heard or read it.

v. Be concise.

In today's world, the value of conciseness is arguably greater than ever. The flow of information is accelerated and coming from many different directions; it is no wonder we have had to develop the ability to grasp the essence of things quickly—to integrate information in a "low-attention-span" environment. Concise writing is best achieved in the revision and editing process; an author is wise to review anything he or she writes at least once with no other purpose in mind but the elimination of unnecessary language. Here are some of the language forms that one should watch for:

a. Redundancy. Simply eliminate it. Phrases that are filled with words that add nothing to the basic idea being conveyed are, in today's communication environment, dated and annoying. Perhaps there will come a day when ornate shirt sleeves will once again be fashionable, but ornate, flowery writing is passé and subverts the communication process. To say that something is happening "at the present time" instead of saying it is happening "at present," dates whatever is being said. Both Alley and Goldbort provide examples of such language, but these redundant styles are holdovers of earlier ages, so it is likely that any list of this kind will be different (hopefully shorter) in the future.

An old English instructor once advised students to imagine that the text is being inscribed in stone, as written material was in antiquity, and to then ask themselves if every word was necessary. Now that more and more material appears on computer screens, the burden that was once placed on the writer is placed on the eye of the reader. Appropriate advice today would be: imagine you have to read the text on a computer screen after a long day of eye-straining work, and then ask if every word is necessary.

b. "Deadwood." Consign it to the flames. This refers to empty verbiage and unnecessarily ornate language; words and phrases that offer no information to the reader. This language is usually designed not to convey information, but to leave the reader with an impression about the writer—that he or she is: erudite ("as a matter of fact"); cultured ("it should be pointed out that"); authoritative ("it is significant to note that"); sophisticated ("it is generally conceded that"); or likable ("I might be forgiven if I say"). These phrases add no information, and using them runs the risk of alienating the reader.

c. "Fat." Cut it out. These are phrases that use many words to say what can be said in a word or two. Earlier, we suggested that "at present" was a more efficient way of communicating than "at the present time," but "now" would be an even better way of saying the same thing.

The impressions these examples of "inconcise" language leave with the reader are either that the point being made is so important that it must be adorned with elaborate language, that the author is to be admired simply for making the point, or, worst of all, that this may well be the last valid or interesting point the author will ever have. None of these impressions is conducive to a sympathetic reading. Concise language, by contrast, conveys to the reader that the author values the reader's time and attention.

vi. Be fluid.

In the previous section (**1.3**), we looked at how to structure sentences and paragraphs so that the points they contain are communicated effecttively. Important as that is, it will not be enough to ensure that readers will absorb from a piece of writing just what an author wishes to convey. This is because no matter how well constructed a piece of writing is, and no matter how well chosen its words and phrases are, if the writing is dull, readers will have difficulty following the argument. In addition to being well crafted and well designed, writing has to be interesting to read. The most scintillating piece of prose will fail to connect with a reader if its read monotonously in a drone voice. When writing has a structure that keeps the reader's attention as it moves from sentence to sentence and point to point, it is said to be "fluid"—simply because it flows easily and is read by the reader ("goes down") with ease.

The great literary stylists develop a fluid style with much practice and by applying their innate talent, but there are a number of mechanical practices and habits that any writer can apply to make the writing (of *anything*, from a technical article for a research journal to an opinion piece for a local newspaper) more fluid. We call these practices "mechanical" because they amount to little more than introducing variety in the writing. Variety alone will not guarantee that the writing will be interesting, any more than a uniform style is sure to result in dull, lifeless writing. The great stylists applied rules such as these instinctively in bringing their writing to life. In time, applying these devices may inculcate a sensitivity to what is dull prose, and what is vibrant writing.

a. Vary sentence rhythm and length. Sentences of nearly the same length through a piece of writing have a numbing effect on the reader and make it impossible for most readers to stay focused and engrossed in the piece. Variety can be introduced by the simple device of varying the length of the sentences, or by varying their rhythm. This can be done by: introducing prepositional phrases of varying length; using (judiciously) an introductory clause that sets the stage for what follows; and using variety in subject placement.

Too much variety comes across as chaotic, just as too little runs the risk of being monotonous. As with most elements of fluid style, another pair of eyes (i.e., another reader or editor) may be better able to assess what needs to be done.

b. Vary sentence style. Not every sentence should be a simple declarative sentence. Insert an occasional aside or a rhetorical question. Even a parenthetical remark (that does not divert the reader's focus on the argument) can be useful in making the writing more fluid.

c. Vary opening sentences of paragraphs. The opening sentence of a paragraph sets the tone of the entire paragraph, so it is important that it sets the proper tone. Since not all paragraphs will be the same with respect to the overall argument, it is important that the opening sentence "announce," through its structure and tone, the fact that the crux of the argument and the "new information" of the piece is imminent.

d. Clear up the "logjams." Be aware of sentences and paragraphs that lead the reader down a tangential path away from the main thrust of the argument. Such diversions may play a useful role in setting the context or preparing the reader for what is coming, but irrelevant material should be avoided in the immediate area of the point of the piece. Tangential diversions destroy reader attention, and ultimately reader comprehension.

e. Use surprise and the unexpected. "Artistry," Beethoven often said, "is knowing when the 'mistake' is better; genius is being able to *always* make those 'mistakes'." With all the advice and rules provided above, the ultimate arbiter of whether a piece of writing "works"—whether it conveys the author's point effectively and convincingly—is the personal reaction of readers. Do not hesitate to experiment. Be prepared for the moment of serendipitous inspiration—the "typo" that seems to convey what is meant better than the conventional, rule-obeying writing you are trying to compose. Don't hesitate to compose a passage just to see how it reads—you can always erase it (or delete it) and try again. Hemingway labored long and hard to produce his 600 words a day, paring it down from text two or three times as long. Writing clear and effective prose is something that (like almost anything else in life) comes through effort and with practice—and no small measure of good fortune.

vii. Follow correct usage.

Communicating difficult, complex, and unfamiliar ideas to a wide variety of readers with great variation in their backgrounds demands that the writing be carefully wrought and precisely worded. For researchers and practicing scientists, this is no small task, especially when preparing material for a popular and general readership. What readers know and what readers don't know are key questions that authors must continually ask themselves and address if they are to produce meaningful and effective writing. But even when addressing colleagues and initiates within a discipline, adherence to the standards and rules of style are what allows scientists to converse across time and nationalities; what allows laboratories in different countries and epochs to produce and reproduce experiments and scientific processes; what permits scientists and researchers of different languages and different points in the history of science to converse with each other across chasms that are all but unbridgeable in other intellectual disciplines.

1.5 Getting Started (and Dealing with "Writer's Block")

In spite of their experience and the instruction of inspiring teachers, there will be some scientists who will look upon writing and communicating as a bothersome necessity undeserving of their time or attention. This attitude will be a primary cause of scientists devoting less than their best efforts to science writing and communication. ("I'm not good at it" and "It's a waste of my time" become two notions that reinforce one another in the minds of many scientists.) Then there will also be many scientists who recognize full well the importance of good science writing and its contribution both to their effectiveness as science researchers, their careers in science, and to the community at large. These people will find the time and resources to craft their writing so that it is clear, precise, and effective, no matter whether they are writing a review for a science journal or a book for a trade publishing house. (More often than not, these are individuals who simply cannot bear to see their names associated with anything shoddy or second rate, a quality that often translates into a meticulous attention to detail in their scientific work.)

There will be still others who have written successfully in the past or who have appreciated the value of good writing in advancing science (and perhaps, as has been argued above, even in *doing* science), but who are sufferers of "writer's block"—that is, they are simply unable to get started—unable to "put pen to paper."

To the best of our knowledge, only one work that may be categorized as a manual or guidebook devoted to science writing contains a brief discussion of "Avoiding Writer's Block" (*The MIT Guide to Science and Engineering Communication*, by James G. Paradis and Muriel L. Zimmerman; MIT Press: 1998; pp. 11-13). Whether or not writer's block qualifies as a legitimate psychological syndrome is arguable; one day it may be regarded as a disorder in need of, and responsive to, therapy and treatment. For now, however, we draw upon Paradis and Zimmerman's recommendations, which, if not methods for addressing a legitimate syndrome, are at the very least sound advice on how to prepare for a writing assignment and how to overcome certain obstacles to the writing process that come from within.

i. Gather sufficient data.

In many cases, an author will not have conducted sufficient research, or will *believe* he or she has not conducted sufficient research in order to start writing. Often, such research requirements are unrealistically large, making the fear and anxiety of failure to complete the research self-fulfilling. The recommendation is to set a definable and proscribed program of research and to begin writing as soon as it is completed. The writing process will inform not only what thought the author has, it will also engender further thinking and may direct the course of additional research.

ii. Define the task of the writing specifically.

Authors will sometimes be confused about the exact nature of their writing aims. They wonder: for whom are they writing; at what level should they be writing; how comprehensively should their writing be; and at what level of technical sophistication should their writing aim? In this case, a preliminary memo (to themselves) may help remove this stumbling block.

iii. Organize the material.

The complexity and the scope of the subject may be unrealistically large and may prove too daunting a task. Creating a detailed outline that organizes the main points of the work—a process that should be given ample time, but is best created under a strict time limit—can help advance the writer to the next stages of the work.

iv. Discuss the work.

While some writers find discussing a work in progress impedes the work rather than helps it, if one is having difficulty going forward, what is there to lose by talking about the piece? Discussing the work—its main argument; its tone and audience; the plan for its execution and completion—with a colleague, supervisor, or a confidant, either face to face in conversation, or in the form of a preliminary draft or notes, can lead to productive work on the project.

v. Sketch the graphic components of the work.

Some people organize material best orally; some in written form; and some in graphic form. To exploit the graphic proclivity that a writer might have, preparing sketches and "story boards" of the graphics program of the work may stimulate and crystallize thought to the extent that the writing will become easier and more forthcoming.

vi. Create a conducive environment.

Writing is an activity that frequently requires sufficient time, an environment free of distractions, and the materials necessary for the simple act of writing. Without being overly attentive to such details, set a time of day for writing when you will not be disturbed; arrange for a comfortable place with no distractions; and have sufficient writing material on hand.

vii. Don't insist on writing a perfect first draft.

Justice Brandeis often said, "There is no great writing; only great rewriting." The revision process is critical in the creation of effective prose of any kind; this is especially true of scientific prose. While rewriting, a writer should be prepared for the following:

• The ideas—sometimes central to the thesis and argument of the work—may change in the course of, possibly even as a *result* of, writing.

• Research conducted during the course of writing (or previous research first understood during the course of writing) may alter elements of the work, perhaps radically.

• The process of articulating ideas and arguments will, by itself and by its very nature, crystallize and clarify ideas and connections in the writer's mind that would otherwise have gone unnoticed.

viii. "Get thee an editor."

Leading scientists have discovered what great writers of fiction and non-fiction have known for a long time: editing is a critical part of the process that results in the production of an excellent piece of writing of any kind. We hesitate to say that editing is an integral part of the writing process, because ingrained in our mind is the notion that writing is a solitary process, performed by a solitary individual pounding on a keyboard in an office, an attic, or a basement. Editing is an essential part of writing or artistic creation; only Mozart was able to produce perfectly constructed scores without so much as crossing out an errant note.

The only question is: can one edit one's own work, or must it be done by someone else? A writer who edits his or her own work is in much the same position as a lawyer who defends himself or herself. Such a lawyer (lawyers say) has a fool for a client. The art of editing entails being able to read a piece of writing with sufficient detachment to spot confusion and miscommunication and to suggest means of improving the writing—but with sufficient involvement to comprehend and resonate with the author's style and intent, without imposing the editor's own style on the writing. This would seem to preclude the possibility of an author being able to edit his or her own writing.

1.6 Words Often Misused or Confused

a, an – *A* is used before any word starting with a consonant sound (*a* house, *a* union, *a* European, *a* B.S. Degree). *An* is used before any word beginning with a vowel sound (*an* hour, *an* igloo, *an* M.S. Degree).

abduct, adduct – *Abduct* is used in physiology to mean "moving a limb or body part away from the midline of the body (the median axis) or from another part"; *adduct* means the opposite—drawing near to the median or to another body part.

ability, capability, capacity – *Ability* is used to describe the physical or mental skills of an individual ("His *ability* to climb trees is impressive"). *Capability* refers to the specific skill or power, not the individual ("She is *capable* of producing children"). *Capacity* is used to describe the amount that a vessel can hold or contain ("The ship was filled to max *capacity*"), or it can be used to describe an individual's learning abilities ("He has the *capacity* to learn many new languages"). See also **capacitance**.

about, approximately – These terms are synonyms, however in the sciences *approximately* is preferred, while *about* is more common in daily or informal discourse.

absorbance, absorptance, absorptivity – *Absorbance* is a measure or logarithm of light, which enters or passes though a liquid or a solid. *Absorptance* and *absorptivity* can be defined as the ratio of energy absorbed by a material.

absorption, adsorption – *Absorption* is a process in which atoms, molecules, or ions are taken up by volume in a capillary, osmotic, chemical, or solvent action. *Adsorption* occurs when atoms, molecules or ions are held at the surface of a solid or liquid.

abstruse, obtuse – If something is *abstruse* it is extremely difficult to understand. *Obtuse* refers to someone who has difficulty understanding something.

accident, injury – *Accident* should not be used in scientific studies when referring to an *injury*, this is because in most cases an *accident* can be prevented or predicted, and implies that there is nothing or no one at fault. For example, instead of referring to a "car *accident*," words like "crash" or "collision" are preferred.

accord, accordance – *Accord* is used to define an agreement ("They were in *accord* on the theory"). *Accordance* means to conform to a standard or agreement ("Their experiments were in *accordance* with practical standards").

accuracy, precision – *Accuracy* refers to a degree to which a measured amount is correct. *Precision* refers to the act of measurement—how carefully it was done.

ACE, Ace – "ACE" is an acronym for "angiotensin-converting enzyme"; "Ace" is a brand-name for a a kind of flexible bandage.

acute, chronic – These terms should only be used to describe the duration and severity of a patient's symptoms or conditions; they should not be used to describe medications, treatments or patients themselves.

adapt, adopt – To *adapt* means to modify, to *adopt* means to take unchangingly as one's own.

addicted, dependent – To be *addicted* is a physical condition, while to be *dependent* is a psychological condition.

adduce, deduce, induce – To *adduce* is to bring forward evidence, arguments, or proof; or to cite an example or passage ("The evidence she *adduced* showed that the claims were false"). To *deduce* is to come to a specific conclusion based on a general idea ("He *deduced* a date of entry by observing movements around the site"). To *induce* is to come to a generalization based on detailed facts ("After studying migratory patterns, they *induced* that the tribe were hunter/gatherers"). To *induce* can also mean to cause or to force (to *induce* labor).

adequate, enough, sufficient – *Adequate* is used when referring to the quality or sufficiency of an explanation or idea. *Sufficient* refers to the appropriate amount of material. *Enough* refers to count nouns (*enough* patients) and mass nouns (*enough* air).

adherence, compliance – These can be used synonymously, however *adherence* is used more frequently when a patient is voluntarily acting in accordance with medical advice he or she has received: taking medication; seeking treatment; watching diet; etc. *Compliance* usually implies that a patient has been forced to *adhere* to the medical treatment given.

administer, administrate, administration – To *administer* is to apply a therapy, treatment, or drug; to *administrate* is to manage. The noun form for both is *administration*, which means caution must be exercised not to confuse the two meanings.

adrenalin, Adrenaline – *Adrenaline* is the term used for the chemical epinephrine outside the United States, but is not used clinically in the United States because the term "Adrenalin" is a trademarked pharmaceutical. (In the U.S., *epinephrine*—also spelled "epinephrin"—is used.)

adsorption – see **absorption**

adverse, averse – *Adverse* is used to describe an unfavorable or unfortunate condition; it is usually used to describe an object or thing, not a person ("The medication had *adverse* effects"). *Averse* is used to describe a person who is opposed to, or has negative feelings about, a subject ("He was *averse* to having a discussion about that subject").

affect, effect, impact – *Affect* is a verb used to describe the influence or cause of an outcome ("They wondered how the election would *affect* their society"); it is usually applied to something that already exists. In psychology, it is used as a noun defining an emotion, or a mood. *Effect* is a noun that refers to an outcome or a result ("The *effects* of the experiment had positive results"). *Impact* is often used as a synonym for *affect* but it should preferrably be used to describe the physical hitting or striking of an object or body.

afflict, inflict – *Afflict* is used when referring to sufferers upon whom are visited disease or other troubles ("The patient was *afflicted* with a number of ailments"). *Inflict* is used when referring to something causing pain or suffering to someone or something ("The murderer *inflicted* fatal wounds on the victim").

after, following – These are synonyms, but *after* should be used as a term meaning "later" ("*After* the surgery, the patient had many side effects"), while *following* should be used to indicate a sequence or position ("The *following* artifacts should be examined").

after having – A redundancy. Instead of "*After having* completed the experiment" use "After completing the experiment."

afterward, afterword – *Afterward* is an adverb meaning later, while an *afterword* is an epilogue to a published work or a piece of writing.

age, aged – When a precise *age* cannot be given, *aged* should be used instead (teen*aged* not teen*age*).

aggravate, irritate – To *aggravate* means to make worse or intensify an existing condition (*aggravate* the assault), while *irritate* means to bother, annoy, or create a new condition.

agonist, antagonist – *Agonist* refers to a drug that stimulates a reaction by a cell that is ordinarily caused by a naturally-occurring substance; *antagonist* is used in biochemistry to refer to a chemical substance that inhibits or counteracts the action of another substance.

albumen, albumin – *Albumen* is defined as egg whites, while *albumin* is a protein found in blood or plasma that is manufactured by the liver.

aliquant, aliquot, sample – An *aliquot* is a portion of a gas, liquid, or solid that divides evenly into a whole, an *aliquant* does not divide evenly into a whole, while a *sample* is a portion taken to represent the whole.

all right, alright – *All right* is the correct term ("the answers were *all right*" and "I'm feeling *all right*"); *alright* is a slang term meaning satisfactory, but is not considered standard English.

all together, altogether – see **altogether**

allude, elude – To *allude* is to suggest or to refer to something indirectly ("She *alluded* to the issues"). To *elude* is to avoid, escape, or evade ("He tried to *elude* their advances").

allusion, reference – An *allusion* is an indirect reference often used to refer to a well-known work ("Her selling of apples was an *allusion* to the Garden of Eden"). A *reference* is a specific example or mention of a previous work ("He cited many *references* to confirm his theories"). After something has been alluded to, it is incorrect to allude to it again later.

altar, alter – An *altar* is a table or stand used in religious service. To *alter* means to change.

alternate, alternative – *Alternate* means a substitute ("Since she was sick, she called in an *alternate*"), or to take turns back and forth ("We decided to *alternate* writing chapters"). *Alternative* implies a choice or option between things ("He decided to take an *alternative* route").

alternation, alteration – *Alternation* means the occurrence of two things by turns; *alteration* means simply a change. Both are applied to gene science—*alteration* is sometimes used as a synonym for mutation, while *alternation* is used to refer to a particular kind of mutation in DNA sequencing.

although, though, while – Both *although* and *though* mean in spite of the fact, and they can be used interchangeably; however *although* should be used more often since *though* is an abbreviation of *although*. *Though* should be used only as a synonym for however ("*Although* my stomach hurt, I still ate ice cream" or "My stomach was hurting, though I still ate the ice cream"). *While* means during the same time as ("*While* I was eating my dinner, she was eating cupcakes").

altogether, all together – *Altogether* refers to the entirety or whole of a subject or object ("*Altogether* her thesis represented new ideas in the field"). *All together* refers to a group or unity ("The family tried to remain *all together* for the holiday").

always – This word and other absolutes should be avoided in scientific papers whenever possible.

ambiguous, ambivalent – *Ambiguous* refers to having more than one definition or interpretation ("His results were *ambiguous*"). *Ambivalent* refers to having conflicting opinions or mixed emotions ("She was *ambivalent* about which direction to follow").

amend, emend – To *amend* means to change or to add to something ("They decided to *amend* the law"). To *emend* means to correct ("He had to *emend* his student's paper").

among, between – *Between* should be used when referring to a group of two objects or people; *among* should be used with a group of three or more ("They found several similarities *among* the patients").

amount, number – *Amount* is used with mass nouns (*amount* of knowledge) and *number* is used with count nouns (*number* of studies).

an – see **a**

analog, analogous, homolog, homologous, homoeolog, homoeologous
– *Analog* (n.) should be used when referring to electronics or computer equipment; *analogous* (adj.) is used to define different compounds or organs that are similar to each other but differ in their structure and original compounds. *Homolog* and *homologous* refer to organs that are similar in structure and origin but differ in function. *Homoeolog* and *homoeologous* are partially *homologous* chromosomes.

anatomy, morphology, structure – *Anatomy* is defined as the study of the structural make up of living things. *Morphology* is the study of the form and *structure* of living things. *Structure* is defined as the parts of a living or non-living thing that form a pattern to make up the whole.

and/or – Considered non-standard usage. Replace with "*and*" or "*or*" when possible. Instead of "the surgery can cause swelling *and/or* bruising" use "the surgery can cause swelling, bruising, or both."

anuresis, enuresis – *Anuresis* refers to the condition of not being able to urinate or lacking urine altogether; *enuresis* refers to bedwetting.

ante, anti – Both are prefixes; neither can stand alone. *Ante* means "before" and *anti* means "against." Most often, *ante* does not require a hyphen; *anti* often does (on a case-by-case basis).

anymore, any more – *Anymore* refers to time ("that will not happen *anymore*"); *any more* to quantity ("We cannot have *any more* of that").

anytime, any time – *Anytime* as an adverb that means "at a variety of points in time" ("we can do the experiment *anytime*"); *any time* is an adjective and a noun, and means "at a particular point in time" ("We do not have *any time* to perform the experiment").

anyway, any way – *Any way* refers to a path or method of accomplishing something; *anyway* means "in any case."

anywhere, any place – *Anywhere* usually refers to an indefinite location ("The samples can be found *anywhere*"). *Any place* is used more specifically ("She couldn't find *any place* to conduct her experiments").

appertain, pertain – *Appertain* is to belong, relate, or be relevant by right ("I know my rights *appertaining* to the contract"). *Pertain* is to belong, relate, or be relevant to something ("This book on Art Deco buildings *pertains* to me because I have a degree in Architecture").

appose, oppose – *Appose* means "to place next to" ("He *apposed* the specimen to the X-ray film"); *oppose* means "to be against, of a contrary opinion" ("He was *opposed* to that course of treatment").

appraise, apprise – To *appraise* is used to evaluate or give a value to something ("He had his grandmother's ring *appraised* before giving it to his fiancée"). To *apprise* is to give information, instruction, or notification ("They were *apprised* of where they could park their car").

approve, endorse – *Approve* is used when there is a positive agreement or thought on a subject ("The senator *approved* the advertisement"). *Endorse* is used when there is a positive action along with an agreement on a subject ("The Senator *endorsed* his candidacy").

approximately – see **about**

apt, likely, liable – *Apt* is used in reference to habitual tendencies or inclinations ("Dogs are *apt* to be loyal to their owners"), while *likely* is used to indicate a higher degree of probability or expectation ("When dogs are not properly socialized, they are *likely* to attack"). *Liable* connotes something is likely to happen because it fits patterns of experience.

are found to be, are known to be – These phrases contain too many unnecessary words; instead just use "*are.*"

are in an agreement – This phrase contains too many unnecessary words; instead just use "*agree.*"

article, manuscript, paper, typescript – A *manuscript* is the physical incarnation of a *paper* or *article* (the completed pages). A *paper* is the intellectual document itself, and the *article* is the published incarnation of the paper. *Manuscripts, papers,* and *typescripts* are studies that have not yet been published.

as, because, since – All are conjunctions, but *as* should be used only to show a sense of time. *As* should never be used in place of *because.* Use *because* in a causal sense, and use *since* to show a relation in time.

as per – Instead of "*As per* her decision" use "as she decided."

assemblage, assembly – An *assemblage* refers to a collection or group of people or things. An *assembly* is a group of people that come together for a specific reason.

assent, consent – To *assent* is to agree enthusiastically. To *consent* means to give permission or allow.

assess, determine, evaluate, examine, measure – *Assess* should be used in a monetary sense or in estimating the value of an item. To *determine* is to establish or set a limit for something. To *evaluate* is to find an item's value after carrying out a study. To *measure* is to *examine* an object in numerical values.

association, relationship – Use *relationship* when describing two objects or variables that show a cause and effect; *association* implies that one object or variable does not cause or effect the other.

assumption, presumption – An *assumption* is a hypothesis not usually drawn from evidence. A *presumption* is based on evidence or fact.

assure, ensure, insure – To *assure* is to give a promise, to affirm, or to guarantee that something is sure. To *ensure* is to remove any sense of doubt or to make certain. To *insure* is used primarily to indicate monetary protection against loss or failure (fire *insurance*; life *insurance*).

at present, at the present time, at this point in time – Instead of these wordy phases, simply use "now" or "currently."

aural, oral – *Aural* means pertaining to the ear or the sense of hearing; *oral* means pertaining to the mouth or "communicated by speaking."

attenuate, attenuation – *Attenuate* means to reduce, and is used in CT scans when referring to the absorption of x-rays by a patient's body. Levels or areas of black on the scan are defined as low or hypo*attenuation*, while levels of whiteness are defined as high or hyper*attenuation*.

average, characteristic, typical – *Average* should be used only for statistical findings. *Characteristic* and *typical* can be used as adjectives for showing a representation in any place aside from statistical figures ("The patient showed *typical* symptoms, *characteristic* of her age range").

averse – see **adverse**

avocation, vocation – An *avocation* is a leisure activity or a hobby, while a *vocation* is a person's career, profession, or calling.

axenic, gnotobiotic – *Axenic* is used to describe cultures that are free from other organisms, or organisms kept sterile or in isolation. *Gnotobiotic* refers to animals raised in laboratories that have been kept unexposed to any agents or infections other than those induced purposefully.

basis – *Basis* means foundation. It is frequently preferable to omit using *basis* in such terms as "on a daily *basis*"; "Daily" is adequate and less pretentious.

because – see **as**

because of, caused by, due to, owing to – *Because of* and *owing to* should only be used in place of "as a result of" ("The picnic was postponed *because of* the weather"). While *due to* and *caused by* should only be used in place of "attributed to" ("The picnic's postponement was *due to* the weather"). These terms cannot be used interchangeably. Also, *due to* should be used to modify a preceding noun or pronoun, or following the verb form "to be."

before, prior to – *Before* should refer to an event or situation that precedes another, but in which the event does not hold importance over the one following it ("I went to school *before* 10 o'clock"). *Prior to* should refer to an event that occurs *before* another event due to its increased importance over the other event, or influence over the other event's effectiveness ("*Prior to* baking the cake, preheat the oven").

believe, feel, think – To *believe* is to have a firm and definite opinion on a view regardless of the strength of the evidence supporting that view. To *feel* is to have an instinctive or not fully reasoned conviction. To *think* is to have a view based on evidence and knowledge.

between – see **among**

bi-, semi-, quasi- – *Bi-* means two (*Bi*monthly or every two months) while *semi-* means half (*semi*monthly or twice a month). *Quasi-* is a prefix that means "to some extent" ("This food is only *quasi*-good").

biannual, biennial, semiannual – Both *Biannual* and *semiannual* mean twice a year, but *biennial* means once every two years. To avoid confusion, instead of using *biennial* use "every other year" or "every two years."

billion, trillion – In the U.S. *billion* refers to 1,000,000,000 and a *trillion* is 1,000,000,000,000. In Great Britain and some other countries however, a *billion* is 1,000,000,000,000 (or the American trillion) and a *trillion* is 1,000,000,000,000,000,000 (or the American "quintillion").

biopsy – *Biopsy* is defined as the removal of tissue or cells for examination, it can also be used as a verb ("The mole was *biopsied* for cancer").

blinding, masking – *Blinding* is a term used in studies when the person conducting an assessment is unaware of the assignment of treatments. It is also known as *masking* in some journals and disciplines such as ophthalmology.

born, borne – *Born* is used as an adjective, as in a *born* mathematician, or as a past participle verb, as in "he was *born* to royalty." In science, *borne* is often used as a suffix (air*borne*) or it can be used as a past participle of to bear, as in " having *borne* a child" or "the diagnosis was *borne* out by the test performed."

breach, breech – A *breach* can either refer to a failure or violation ("This affair is a *breach* of our agreement"), or to a gap or opening. A *breech* is the low end or bottom of something.

breastfeed, nurse – *Breastfeed*ing should only be used when describing human lactation; *nursing* is used for any other mammalian lactation.

breech – see **breach**

bring, take – If an action is towards you, use *bring* ("*Bring* in the paper"). If the action is away from you, use *take* ("I want to *take* you out to lunch").

by reason of – This is established phraseology in legal discourse (*By reason of* insanity), though "because" or "because of" normally suffice in other contexts.

cadaver, donor – *Cadaver* should be used only when describing a body that is used for anatomical dissections. A *donor* or deceased *donor* should be used in reference to organs and tissues that are used for transplants.

calyx, calix – In botany, *calyx* is always used to refer to the outer sheath of a flower—the sepals. In zoology, either spelling is used to refer to various cuplike structures: a portion of the pelvis of the mammalian kidney; the cavity in a calcareous coral skeleton that surrounds the polyp; or the plated body of a crinoid excluding the stalk and arms.

can, could, may – *Can* means "to be able to" and usually expresses certainty ("I *can* finish it tomorrow"). *Could* usually expresses uncertainty ("I *could* finish it tomorrow"). *May* is used in forms of permission ("*May* I finish it tomorrow?") or possibility ("I *may* finish it tomorrow").

capability, capacity – see **ability**

capacitance, capacity – *Capacitance* refers to the amount of electrical charge a capacitor can hold in a given electrical circuit. *Capacity* is a more general term more often used for other physical systems, such as the "heat capacity" of a system or substance.

carat, caret, karat – A *carat* is the measurement of a gemstone's weight. A *caret* is an editorial mark indicating the location of an insertion or addition. A *karat* is the measurement of the purity of gold.

carotene, creatine, creatinine, keratin – *Carotene* is the yellow-red pigment found in egg yolk, carrots, etc. *Creatine* is a compound formed in protein metabolism in living tissue. *Creatinine* is an anhydride of creatine produced as cells metabolize creatine, the product of which is excreted in urine. *Keratin* is a fibrous protein that supplies the main structural component of hair, nails, hoofs, feathers, horns, claws, etc.

carry out, conduct, do, execute, perform – The verbs *carry out, execute,* and *perform* can often be omitted and replaced by another more specific verb. Instead of "he *executed* the operation on the patient" use "he operated on the patient." *Conduct* and *do* are used as synonyms, but *do* is preferred (because it is generally more direct). *Perform* should be used only in reference to ceremony or entertainment.

case, client, participant, patient, subject – A *case* is an example or a particular instance, not a person. A *client* is not a patient or participant; however *client* is sometimes used in the fields of psychiatry or substance abuse treatments. A *participant* is a person who participates in either research or control studies. A *patient* is a person under medical care. A *subject* is defined as a discipline or a study; it is not a person.

catatonic, manic, psychotic, schizophrenic – These terms refer to severe levels of psychiatric disease; they may often be replaced with adjectives for non-clinical descriptions. Instead of *catatonic,* use motionless. Instead of *manic,* use overactive. Instead of *psychotic,* use senseless. Instead of *schizophrenic,* use contradictory or disorganized. Even in clinical contexts, a person should not be referred to as a *schizophrenic,* but as "a person with schizophrenia." (See **Section 1.8**, below.)

caudate, chordate – *Caudate* means "having a tail" in the manner of sperm cells, many protozoa, and bacteria. In anatomy, the word is used to refer to several features that have a tail-like appearance, such as the *caudate nucleus* in the cerebrum, the *cauda equina* structure at the end of the spinal column, and the *caudate lobe* of the liver. *Chordate* refers to a wide variety of animals (vertebrate and invertebrate) that have "notochords"—cartilaginous skeletal rods in embryonic stages.

cause, etiology – A *cause* is an explanation or reason for an occurrence. *Etiology* is defined as the study of a *cause.*

caused by – see **because of**

Celsius, centigrade – Both terms refer to the same temperature scale, but *Celsius* is the preferred nomenclature.

censer, censor, censure, sensor – A *censer* is a container that holds burning incense. A *censor* is a person who suppresses obscene or objectionable material; it also can be used as a verb in the act of suppressing the material. To *censure* is to criticize, reprimand, or to disapprove. A *sensor* is an electronic detector.

center around – The *center* is at the middle of something, therefore it cannot be *around* anything. Instead use "center on" or "revolve around."

centigrade – see **Celsius**

cesarean delivery, cesarean section – *Cesarean,* or abdominal delivery, should be used in place of the incorrect *cesarean section.*

certainty, certitude – *Certainty* can be used when describing facts and people, while *certitude* is used only for describing people.

chief complaint, chief concern – *Chief concern* is the preferred phrase because *chief complaint* can be seen as confrontational.

childish, childlike – *Childish* is used to describe negative attributes (a *childish* tantrum) while c*hildlike* is used to describe positive attributes (a *childlike* curiosity).

chitin, chiton – *Chitin* is the polysaccharide-filled fibrous part of the arthropod exoskeleton and fungal cell wall. *Chiton* is a kind of marine mollusk.

chronic – see **acute**

circadian, diurnal – *Circadian* describes a 24-hour period or interval; *Diurnal* describes a process or cycle that occurs every 24 hours. In botany, *diurnal* usually refers to a flower that opens in the morning and closes at night.

circumduction, sursumduction – *Circumduction* refers to the rotational movement of the eye or of an extremity; *sursumduction* refers to the upward movement of only one eye in a test for vertical divergence.

cite, sight, site – *Cite* refers to a citation or source of information. ("She *cited* her references in the bibliography"). *Sight* is the ability to see things using the eye, and also refers to something worth viewing ("I saw all the *sights* in Paris"). A *site* is a place or location ("Wear a helmet in the building *site*").

classic, classical – In science, *classic* refers to something of importance or continuing value. *Classical* refers to the humanities, languages, art, work, or characteristics being traditional in a cultural or historical perspective. In science, *classical* can also refer to the best, or earliest characterized form.

claustrum, colostrum – The *claustrum* is the thin layer of gray matter in the cerebral hemisphere between the gray matter of the lentiform nucleus and the insula. *Colostrum* refers to the initial secretion of milky fluid from the mammary glands at parturition.

clench, clinch – *Clench* is the physical act of tightening the hand in a fist, usually an indication of anger; while *clinch* is used to describe a confirming or winning argument or event. (*Clinch* also describes fighting at close quarters, which gives rise to the confusion between the terms.)

client – see **case**

climactic, climatic – *Climactic* means leading to, or culminating in, a climax; *climatic* means related to weather and climate.

clinician, practitioner – These terms describe a person working in a clinical healthcare practice, psychology, dentistry, nursing, etc. They are not used for people in the fields of research, theory, writing, etc.

collegial, collegiate – A *collegial* answers to a colleague, a *collegiate* answers to a college.

commendable, commendatory – *Commendable* describes something that is admirable, worthy of praise, or done for a worthy cause. *Commendatory* is the object, speech, or gesture that serves to praise the thing or person being commended ("a *commendatory* plaque").

common, frequent, regular – *Common* describes something that appears often or frequently. *Frequent* is defined as occurring often or in short intervals. *Regular* can mean "normal," occurring in fixed time, routine, or consistent.

common, mutual – *Common* describes a characteristic or trait that is shared by two or more people or things. *Mutual* describes something that is reciprocal or exchanged (*mutual* respect).

compare to, compare with, contrast, versus – To *compare to* is to note only similarities between multiple things, while to *compare with* is to note similarities and differences between multiple things ("What is Aspirin like compared with Tylenol as a treatment for headaches?"). To *contrast* is to note only the differences between multiple things. *Versus* means "against," and is commonly used legally (Roe *versus* Wade).

compelled, impelled – To be *compelled* implies being forced to take a certain action. To be *impelled* means that one is driven to do something even if one does not agree with it.

compendious, voluminous – *Compendious* can be described as brief, compact, concise, abridged, or summarized. *Voluminous* can be defined as vast, bulky, lengthy, or literally large in volume.

complacent, complaisant, compliant – To be *complacent* means to be contempt or at peace. To be *complaisant* is to be cheerful or easygoing. To be *compliant* means to follow or obey the rules.

complementary, complimentary – The word *complementary* is used to describe something that completes, matches, or corresponds with something else (*complementary* colors). In science, it can also refer to a group of proteins that are active in the immune system. To be *complimentary* is to offer praise or to give something for free.

compliance – see **adherence**

compliant – see **complacent**

compose, comprise, constitute – To *compose* is to create, construct, or form something. To *comprise* is to be made up of, or to contain something. *Constitute* is often used as a synonym for *compose* but can also mean, "to amount to."

concept, conception – Both terms refer to an abstract idea or thought, however *conception* also describes the act of thinking of an idea.

condole, console – To *condole* is to sympathize. (Hence, mourners are offered *condolences*.) To *console* is to comfort.

conduct – see **carry out**

confidant, confident – A *confidant* is a person in whom one can confide. *Confident* means being certain in one's beliefs or actions or being generally self-assured.

congenital, genetic – *Congenital* is used to describe a condition or disease with which someone is born. *Genetic* is used when describing the determination of characteristics by the genes.

congruent, congruous – Both terms are synonyms for being in agreement or being equal, *congruent* is also used in geometry to describe two figures that are equal in size, shape, and measure.

conjecture, hypothesis, law, theory – A *conjecture* and a *hypothesis* are both speculations that are meant to be the potential explanation for a specific phenomena or occurrence. They are tested through experimentation, observation, and study. A *theory* is a concept based on observations and experiments, but which has has not been proven true or false. A theory that has been proven (to a high degree of certainty) by experimentation becomes a *law*.

connive, conspire – To *connive* is to pretend to ignore (or to deliberately ignore) a malicious act in order to escape blame. To *conspire* is to make plans with another to perform a malicious act.

connote, denote – To *connote* is to imply or express an additional meaning in one's speech or writing beyond the specific meaning of the words used. To *denote* is to indicate the specific meaning of something.

consent – see **assent**

consensus of opinion – This phrase is redundant, since *consensus* means an agreement on an opinion. One should simply use *consensus*.

consequent, subsequent – Something that is *consequent* implies that there is cause and effect. To say that something is *subsequent* means that it happened after an event, but that is not necessarily an effect.

conservative, conserved – *Conservative* refers to a method, treatment, or principle that has been widely accepted; *conserved* refers to quantities that remain unchanged in closed systems (in physics), or to the replication of genetic material unchanged from generation to generation.

constant, continual, continuous – *Constant* refers to something that holds true invariably and unceasingly. *Continual* refers to a sequence that is frequently and regularly repeated. *Continuous* refers to something that is completely steady, uninterrupted, and unbroken. *Continuous* can also refer to both time and space, while *continual* refers only to time.

constitute – see **compose**

contagious, infectious – Something that is *contagious* is spread by contact with the infected. Something that is *infectious* is caused by and harbors an infection; it is not necessarily always *contagious*.

contemporaneous, contemporary – Both terms refer to time, however *contemporary* is used when referring to people ("He is my *contemporary*"); *contemporaneous* is used when referring to actions or things that occur during the same general period.

contemptible, contemptuous – To be *contemptuous* is to have feelings of contempt towards other people or things. To be *contemptible* is to have others feel contempt towards you.

content, contents – *Content* refers to a specific topic or message within a written or expressed piece. *Contents* refers to the actual items or ingredients that make up a whole.

continual, continuous – see **constant**

contrast, contrast agent, contrast material, contrast medium – *Contrast* can be defined as the whiteness or blackness found on an image. *Contrast agent, contrast material,* and *contrast medium* can be defined as certain substances that are applied to an image to enhance certain structures. (See also **compare to**).

contravene, controvert – To *contravene* is to conflict, deny, or to go against. To *controvert* is to contradict or oppose.

conventional, customary, norm, normal, traditional – *Conventional* refers to a practice that has been established and agreed upon. *Customary* refers to a practice that has become a habit or that has been in use for a long period of time. *Normal* describes something that conforms to the majority. *Norm* can describe something that is *normal,* but can also describe a desire for that which is considered normal or what is expected. *Traditional* refers to a practice that has been agreed upon and has been in use for a long period of time.

convince, persuade – *Convince* refers to making someone believe something, while *persuade* refers to making someone do something.

corollary, correlation – A *corollary* is a statement or a theorem that follows an already proven theorem, and therefore requires no proof. A *correlation* is a complementary association or relation between things.

corporal, corporeal – Something that is *corporal* relates to or affects the body. Something that is *corporeal* has a body.

correlation – see **corollary**

could – see **can**

councilor, counselor – A *councilor* is someone who serves on a council or some governmental or official body. A *counselor* is someone who gives advice.

credible, creditable, credulous – *Credible* describes a person or story that is trustworthy or believed to be true. *Creditable* describes a person who is respected because of his or her many merits. *Credulous* describes a person who is gullible or will believe anything.

crevasse, crevice – A *crevice* is a small crack, while a *crevasse* is a large fracture—usually used to describe cracks in glaciers.

criteria, criterion – *Criteria* is the plural of *criterion*.

customary – see **conventional**.

cyst(o)-, cyt(o)- – *Cysto* is a prefix meaning "relating to the urinary bladder" (thus, *cystotomy* is an incision in the bladder); *cyto-* is a prefix meaning relating to the cell (thus, *cytology* is the study of cell function and structure).

damp, dampen – To *damp* means to moisten or to lessen with moisture ("*damp* a fire"). To *dampen* similarly means to moisten, but can also mean to lessen in a figurative sense ("*dampen* a spirit").

data, data set, database – *Data* is the plural of datum and in scientific writing it should be used as such; however, in relation to computers, *data* is now defined as a mass of information, so it is acceptable when used as a singular. Many publications in science accept *data* as both singular and plural. A *database* is a structure that stores, organizes, and retrieves *data*. A *data set* is a body of *data* that is maintained in a *database*.

deadly, deathly – If something is said to be deadly, it means that it can cause death (*deadly* chemicals). If something is *deathly*, it means that it is like death in its tone or in how it is regarded (a *deathly* fear).

decision, discission – *Decision* is an ordinary English term for coming to a conclusion, whereas *discission* refers to surgically cutting into tissue such as a cataract or the cervix uteri (as an older treatment of stenosis of the cervix).

deduce – see **adduce**

definite, definitive – *Definite* means to be exact, precise, clear, and firmly established (*definite* answer). *Definitive* means to be final, unquestionable, and authoritative (a *definitive* guide).

delegate, relegate – A *delegate* is a person who acts on another person's behalf ("The *delegate* spoke for the state"); to *delegate* is to allow that person to act on another's behalf. To *relegate* is to move a person or thing to a position of lesser importance, or to pass something on to somebody else to be dealt with ("This report needs to get done, so I'm going to *relegate* it to you").

demonstrate, exhibit, reveal – These terms are all often used synonymously for "show." However, to *demonstrate* is to illustrate how to complete an action or procedure. To *exhibit* is defined as the action of making something visible. Instead of saying "the patient *demonstrated* or *exhibited* the following symptoms" it is better to say, "the patient had the following symptoms." To *reveal* is to uncover or make visible something that was hidden.

denote – see **connote**

denounce, renounce – To *denounce* something is to criticize or to speak out against a person or an action. To *renounce* is to reject or give up something (usually, to which one is entitled).

describe, report – Use the term *describe* when explaining both patients and cases. Use *report* when describing or explaining only cases.

desirable, great, important, influential, major, significant, useful, valuable – In the sciences, *significant* is used when indicating a sign of an important outcome, though not necessarily beneficial or desired ("The increase in white blood cell count was *significant*"). All the other terms should be used when indicating a desired outcome and one that is beneficial. In measurement, *significant* figures refer to a degree of exactness that is ceratinly within an instrument's capability.

determine – see **assess**

diabetes mellitus – Type 1 *Diabetes mellitus* is now used to refer to juvenile diabetes, juvenile-onset diabetes, and insulin-dependent *diabetes mellitus*. Type 2 *Diabetes mellitus* is now used to refer to maturity-onset diabetes, adult-onset diabetes, and non-insulin dependent diabetes mellitus. Impaired glucose tolerance is now used to refer to chemical diabetes, borderline diabetes, and latent diabetes.

diagnose, evaluate, examine, identify – When determining conditions, symptoms, and diseases are identified, only use the terms *diagnose*, *evaluate*, and *identify*. The term *examine* is used only with patients. A doctor does not *diagnose* a patient; a doctor *diagnoses* a disease.

die from, die of – Using *die from* is incorrect; a person can only *die of* a disease or other medical complication.

different, differing, disparate, diverse, varying – To be *different* means to be unlike, or to have dissimilar characteristics or traits. ("Sarah and Tim's ideas about politics are *different*."). To be *diverse* is defined as two or more people or things having a large range of differences. To be *disparate* means to be incongruous or distinctly *different* from something else. *Differing* should be used when describing things with *different* characteristics. ("Sarah and Tim have *differing* ideas about politics"). *Varying* should be used to mean changing (*varying* weather patterns). In the United States, the preposition with *different* usually used is "from," though it is acceptable to use "than" in such sentences as "The desert is *different* in California than it is in Arizona."

digit, number, numeral – *Number* is the quantity or count of a specific group or class. A *numeral* is a symbol that represents an arabic character for a *number*. A *digit* has the same definition as a *numeral,* but is also used to refer to the amount of *numerals* in a number, for example, 310 is a 3 *digit number. Digit* can also be used to refer to a finger or a toe.

dilate, dilation, dilatation – To *dilate* means to open, become wider or expand. *Dilation* is the act of dilating. *Dilatation* is the state or condition of being stretched ("dilatation of the pupil"; "a venous dilatation").

disburse, disperse – *Disburse* is only used in the distribution of money. *Disperse* is used when describing the distribution of all other things; it can also mean to break up or to scatter.

disc, disk – In computer terms, *disk* is most often used (floppy *disk*, *disk* drive, etc); however *disc* is also used to a lesser extent (compact *disc*). In anatomy, *disk* is used (invertebral *disk*), while in ophthalmology, *disc* is used (optic *disc*).

discreet, discrete – *Discreet* describes being trustworthy, judicious, and prudent (a *discreet* editor). *Discrete* describes something that is separate, individual, and distinct (having three *discrete* parts).

discriminating, discriminatory – *Discriminating* can have both positive and negative connotations. For example *discriminating* means being tasteful, and showing careful judgment (having a *discriminating* ear for music). However, *discriminatory* means being biased or prejudiced.

disinterested, uninterested – To be *disinterested* is to be unbiased or impartial to a particular outcome ("They chose a *disinterested* person to listen to their positions"). To be *uninterested* is to be inattentive, unconcerned, or have no interest in a particular subject or outcome ("The children were *uninterested* in finishing their homework").

disk – see **disc**

disorganized, unorganized – *Unorganized* means simply not being organized. *Disorganized* refers to a group that has been assembled in a confused manner and will never be organized.

disperse – see **disburse**

distinctive, distinguish, distinguished, distinguishable – To *distinguish* means to point out a certain thing, characteristic, or trait. *Distinctive* describes a trait that is easy to *distinguish*. *Distinguished* is a positive term used to describe a person who is honored, exalted, or elegant. *Distinguishable* is used more negatively to describe a person or thing that is different or abnormal.

diurnal – see **circadian**

diverse – see **different**

do – see **carry out**

doctor, physician – A *physician* is a specific person who holds a degree in medicine or osteopathy. A *doctor* can have a degree in a number of fields (PhD, DDS, EdD, DVM, PharmD).

doctrinaire, doctrinal – To be *doctrinaire* is to be stubborn, arrogant or devoted to dogmatic theories. *Doctrinal* means relating or concerning a doctrine.

donor – see **cadaver**

dosage, dose – A *dose* is a specified amount or measurement of a medicine to be taken at one time. A *dosage* refers to the quantity of medicine, but also to the frequency or regimen in which the medicine is to be taken ("Take a 20mg *dose* twice a day").

drunk, drunken – *Drunk* is defined as an intoxicated state of mind; it can also be used to describe the person who is intoxicated, as in a drunkard. *Drunken* is used to describe the traits or actions of someone who is *drunk* (*drunken* slurs of speech).

due to – see **because of**

dumb – *Dumb* is often used to describe someone who is deemed unintelligent, but it can also describe the inability to speak (though in that case, it is preferable to use "mute").

dyeing, dying – *Dyeing* is the present participle of *dye* (*dyeing* one's clothes). *Dying* is the present participle of *die* (*dying* from cancer).

dyskaryosis, dyskeratosis – *Dyskaryosis* is a misarrangement of nuclei and cell structure often found in malignant cells; *dyskeratosis* refers to aberrant structure of keratin in hair, feather, and bone tissue.

dysphagia, dysphasia – *Dyspahgia* refers to difficulty in swallowing (by the organism, but possibly also at the cellular level of phagocytes); *dysphasia* refers to impairment of the power to speak or to understand speech, as a result of brain injury, stroke, or disease.

each other, one another – *Each other* should be used only when there are two persons or things being discussed; *one another* should be used when there are more than two persons or things under discussion.

eatable, edible – Something is *eatable* if it can be eaten, regardless of its effect on the eater. Something is *edible* if it is fit for human consumption.

economic, economical – *Economic* refers to the economy or finances on a large scale. To be *economical* means to be thrifty or to manage personal finances wisely.

effect – see **affect**

effective, effectiveness, efficacious, efficacy – In pharmaceutical terms, *efficacy* and *efficacious* refer to the capacity of medication or treatment to produce the desired results. *Effective* and *effectiveness* refer to the extent that medication or treatment produce intended results.

e.g., i.e. – *E.g.* stands for *exempli gratia,* and means "for example." *I.e.* stands for *id est,* and means "that is." In most cases, it is better to use simply "for example" and "that is."

elemental, elementary – When something is *elemental* it means that it is essential; it can also refer to the chemical elements. When something is *elementary* it is considered basic or introductory—a fact or principle that is so obvious as to require little formal training or knowledge.

elicit, illicit – To *elicit* is to bring out an answer or a reaction. If something is *illicit* it is dishonest or unlawful.

elude – see **allude**

emend – see **amend**

empathy, sympathy – *Empathy* is the feeling or thought of putting one's self in another's position, or identifying with the other's feelings or life. *Sympathy* is feeling compassionate and sorry for another person.

employ, use, utilize – To *employ* means to put a person to work or to put an object to use. To *use* and to *utilize* is to apply or to put into service. In terms of consumption (*use* drugs, etc.), it is better to use *consume.*

endemic, epidemic, epiphytic, epizootic, hyperendemic, pandemic – An *epidemic* is an outbreak of disease in humans. Epidemics only target or affect a certain group of people—for example, people living in a certain country—but the disease eventually lessens or ends over time. An *endemic* breaks out in a certain group of people but is continuously present. A *hyperendemic* is an endemic that affects a high number of those at risk. A *pandemic* can affect a large number of people, including the whole world. An *epiphytic* is an outbreak of disease among plants. An *epizootic* is an outbreak of disease among animals.

endorse – see **approve**

enervate, innervate – To *enervate* is to drain something of energy, while to *innervate* is to provide something with more energy.

enormity, enormousness – *Enormousness* should only be used when referring to something that is great in size. *Enormity* should be used when describing something that is extremely evil or immoral.

enough – see **adequate**

ensure – see **assure**

enumerable, innumerable – *Enumerable* means that something is able to be counted. *Innumerable* means that there are too many to count.

erectile dysfunction, impotence – *Erectile dysfunction* is now the preferred term over *impotence,* though technically, *impotence* includes conditions in which there is a failure to copulate other than the inability to achieve erection.

et al., etc. – *Et al.* is short for *et alii,* meaning, "and others." It should only be used when referring to people. *Etc.* is short for *et cetera,* meaning, "and other things" or "and so on," and implies that an extensive list of (obviously) like items is indicated. It should only be used when referring to objects, not people. It should only be used at the end of a list or sentence. Putting "and" in front of *etc.* is redundant, hence unnecessary.

etiology – see **cause**

evaluate – see **assess**

every one, everyone – *Every one* means each member of a group of items; *everyone* means everybody or every person.

examine – see **assess**

exhibit – see **demonstrate**

examine – see **diagnose**

exceptionable, exceptional – If something is *exceptional* it is extraordinary, outstanding, or stands out from the rest. If something is *exceptionable* it causes offense, and is a cause for objection.

execute – see **carry out**

facial, facie, fascial – *Facial* means relating to the face or to the facie, which in geology are the characteristics of a rock as expressed by its formation, composition, or fossil content. In ecology, *facie* are the characteristic set of dominant species in an environment. *Fascial* refers to the thin fibrous sheath covering a muscle or organ (the fascia).

fast, fasting – In medicine, *fast* means to abstain from food; it can be used in a variety of forms. *Fasting* may be used as a verbal adjective (the *fasting* patient); it can also be used as a verbal noun (the effects of *fasting*). *Fasted* can be used as the past tense of *fast* ("The patient *fasted* overnight"), or it can be used as a past participle (three *fasted* patients).

faze, phase – To *faze* is to disturb, disconcert, or to put off ("She was *fazed* by the idea of looking for a new job"). To *phase* (v.) is to perform or plan a task in stages. A *phase* (n.) is the stage or period in a process.

Feel – see **believe**

fever, temperature – A *temperature* is the actual degree of heat ("Everyone has a bodily *temperature*"). A *fever* is a condition in which the body *temperature* is abnormally high. If someone has a *temperature* of 101°F, it is incorrect to say that he or she has a *fever* of 101°F.

fewer, less – *Fewer* is used when referring to a number of people or things (*fewer* patients). *Less* should be used when referring to mass, volume, or things—quantities that cannot be counted (*less* water). *Less* can also be used in reference to time and money (*less* than a week ago).

fictional, fictitious – If something is *fictional,* it is imaginary. *Fictitious* refers to something that is counterfeit such as a *fictitious* name.

film, radiograph – In radiography, *film* is generally considered to be an outdated term and should only be used when referring to actual *film* that is being exposed (not digitally) to produce an image. The images being produced should be called by their specific name (*mammogram, radiograph*, etc.) rather than *film.*

flammable, inflammable – These words are synonyms, however *flammable* is the preferred term because *inflammable* (which means able to be set aflame) is often incorrectly thought to mean non-flammable.

flexor, flexure – A *flexor* is a muscle that flexes a joint; a *flexure* is the bent portion of an organ or structure (such as the sigmoid flexure).

follow, follow up, observe – *Follow* should only be used in reference to cases, not patients. Patients are *observed*. *Follow up* can be used in reference to patients in cases in which the patients either cannot be found or contacted (lost *follow-up* or unavailable *follow-up*).

following – see **after**

forbear, forebear – To *forbear* is to restrain from doing or straining something. A *forebear* is an ancestor.

forego, forgo – To *forego* is to go before; to *forgo* is to give up or to do without something. (Thus, a *foregone* conclusion is one that requires no additional argument or support.)

foreword, preface – Both are introductions in a book, however, a *preface* is written by the author of the book and a *foreword* is written by someone other than the author.

frequent – see **common**

fungus, fungal, fungous, fungoid – A *fungus* (n.) is an organism without chlorophyll that has rigid walls and reproduces through its spores. *Fungal* and *Fungous* are adjectives that describe something that is caused by a *fungus*. *Fungoid* describes something that resembles a *fungus*.

galactorrhea, glacturia – *Galactorrhea* refers to the abnormal flow of breast milk; *galacturia* refers to urine that has a milky appearance.

gauge, gouge – In physics, *gauge* refers to symmetry groups used in quantum theory and particle physics; in electronics it refers to measures of equipment such as wire. In medicine, a *gouge* is a hollow chisel used to hollow out bone or cartilage.

gender, sex – *Sex* is the classification of living things as male or female based on their reproductive organs. *Gender* is the sociological representation of how a person is classified, or sees himself or herself as a man or a woman.

general, generally, generic, generically, usual, usually – *General* and *generally* are used when describing a broad or shared trait or action among a group. *Generic* and *generically* refer to items that are in the same category as one another. *Usual* and *usually* refer to expected or normal situations.

genetic – see **congenital**

gibe, jibe – A *gibe* can be defined as an insult. To *jibe* is to fit in or to conform.

global, international, worldwide – *Global* and *worldwide* are used when referring to the world as a whole. *International* is used when referring to two or more nations.

gnotobiotic – see **axenic**

-gram, -graph – *-gram* refers to the recording made; *-graph* refers to the apparatus making the recording. (Thus, an *electrocardiogram* is the recording made by an *electrocardiograph*.) Exceptions include: *photograph* and *radiograph*.

grateful, gratified – To be *grateful* is to be thankful, while to be *gratified* is to be satisfied.

great – see **desirable**

hanged, hung – Use *hanged* only when referring to the killing of a person by suspension from the neck. *Hung* is used when referring to the suspension of another body part, besides the neck.

healthy, healthful – *Healthy* describes a living thing that has good health. *Healthful* refers to something that promotes or supports good health (a *healthful* diet).

historic, historical – *Historic* refers to an important or momentous event that had an affect in history. *Historical* refers anything that happened in the past. For example, a *historical* map shows where *historic* wars took place.

homolog, homologous, homoeolog, homoeologous – see **analog**

homogenous, homogeneous – Both terms refer to two or more things that are similar in elements or structure. *Homogenous* also describes two or more structures that are similar and have common origins. *Homogeneous* refers to having the same traits and qualities throughout.

humeral, humoral – *Humeral* refers to the humerus bone; *humoral* is an adjective that refers to any bodily fluid generally (such as hormones).

hung – see **hanged**

hyper-, hypo- – *Hyper-* is a prefix that means excessive ("*hyper*tensive"), above ("*hyper*sonic"), or beyond normal ("*hyper*thyroidism"). *Hypo-* is a prefix that means below normal ("*hypo*glycemic"), slightly ("*hypo*manic"), or unusually low (as in chemistry, where "*hypo*chlorus" means having an unusually low valence).

hyperendemic – see **endemic**

hyperintense, hypointense – Areas of whiteness that appear on a magnetic resonance (MR) image are called *hyperintense*. Areas of blackness on an MR image are *hypointense*.

hypothecate, hypothesize – To *hypothesize* is to form a hypothesis. To *hypothecate* is to promise property or goods as a security without giving up rights or ownership.

hypothesis – see **conjecture**

i.e. – see **e.g.**

identify – see **diagnose**

ileum, ilium – The *ileum* is the third part of the small intestine (between the jejunum and cecum); the *ilium* is the large broad bone that forms the upper part of each half of the pelvis (hipbone).

illegible, unreadable – *Illegible* refers to the quality of the print or handwriting and means it is of such poor quality that it cannot be read or deciphered. *Unreadable* refers to a piece of writing's poor content or composition ("The book was so boring that it was *unreadable*"), though it may still have been legible.

illicit – see **elicit**

immunize, inoculate, vaccinate – *Vaccinate* means to purposefully inject an animal with a vaccine comprised of specific antigens in the hopes of producing antibodies to protect the animal from sickness. To *immunize* means to make an animal immune to a particular disease through exposure to antigens. *Immunization* is the result of these *inoculation* procedures.

impelled – see **compelled**

imply, infer – To *imply* is to suggest something indirectly that is not necessarily based on fact. To *infer* is to deduce or conclude something on the basis of evidence and fact.

important – see **desirable**

impotence – see **erectile dysfunction**

incidence, period prevalence, point prevalence, prevalence – In epidemiology, *incidence* is the number of new cases of a disease diagnosed in a certain amount of time or a certain place. *Prevalence* is the total number of cases in a certain amount of time or a certain place. *Point prevalence* is the number of cases recorded on a specific date. *Period prevalence* is the number of cases recorded during a specific period.

incredible, incredulous – *Incredible* refers to something that is unbelievable, usually used as a positive adjective. *Incredulous* refers to someone who is skeptical or unwilling to believe something.

individual, person – *Individual* can be used as a noun or adjective, and represents an independent unit or organism that is separated from a group. A *person* is an *individual* human being, although it is best to use a more specific term such as man, woman, adult, child, etc.

induce – see **adduce**

infected, infested – In medicine, if something is *infested* it is harboring parasites or contains a large number of insects, worms, etc. But the insects do not cause an immunological consequence. To be *infected* is to harbor a virus or bacteria that does have immunological consequences.

infectious – see **contagious**

infer – see **imply**

inflammable – see **flammable**

inflict – see **afflict**

influential – see **desirable**

infra-, intra- – *Infra-* is a prefix that means under, beneath, or below; *intra-* is a prefix that means within (as in "*intra*ocular," which means within the eye).

inherent, intrinsic – These are synonyms describing something characteristic or innate. In anatomy, *intrinsic* can also be defined as belonging entirely to an organism or to a part or system of the body.

injury – see **accident**

innervate – see **enervate**

innumerable – see **enumerable**

insulin, inulin – *Insulin* is a pancreatic hormone; *inulin* is a fructose-based polysaccharide derived from plants and used to test kidney function (as in "inulin clearance test").

insure – see **assure**

international – see **global**

intrinsic – see **inherent**

irritate – see **aggravate**

its, it's – *Its* is the possessive form of it (*its* habitat), *it's* is the contraction of it is (*it's* improving).

jibe – see **gibe**

karat – see **carat**

kind, type – *Kind* is not a synonym for *type*. *Type* should be used in scientific writing to describe an object, plant, or animal that is representative of a larger group (a *type* species). *Kind* is a group of individuals or things that share characteristics.

knot – A measure of speed, used especially at sea. A knot is defined as one nautical mile per hour, so the phrase "*knots* per hour" is incorrect.

law – see **conjecture**

lay, lie – *Lie* is an intransitive verb: "we *lie* in the snow," "we are *lying* on the bed." *Lay* is a transitive verb: "we *lay* the pillow on the bed now," "we *laid* our bodies on the bed last week." The past participle of *lie* is *lay*; the past perfect is *has lain*. The past participle of *lay* is *laid*; the past perfect is *has laid*.

leach, leech – To *leach* is to separate solids from liquids in a solution through percolation. A *leech* is a bloodsucking animal; the term is used both figuratively and literally.

less – see **fewer**

liable – see **apt**

likely – see **apt**

loath, loathe – A *loath* person is someone who is reluctant in some capacity ("Kenneth is *loath* to eat bacon"). To *loathe* is to hate, detest, and be disgusted by something or someone ("Kenneth *loathes* bacon").

localize, locate – *Localize* means to restrict to a particular location or place ("The infection *localized* in the root of the tooth"). *Locate* means to specify a precise place. *Localize* should not be used in the place of *locate*. Incorrect: "we *localized* the infection in the root of the tooth." Correct: "we *located* the infection in the root of the tooth."

lucency, opacity – *Lucency* refers to the black areas on an image in radiology. *Opacity* refers to white areas on an image in radiology.

majority, most – *Majority* is a synonym for *most,* but *most* is favored when not speaking in quantitative terms.

malarial, malarious – *Malarial* means "pertaining to malaria" (a *malarial* mosquito"; *malarial* fever). *Malarious* means "being infected or infested with malaria" (a *malarious* region; a *malarious* population).

malignancy, malignant neoplasm, malignant tumor – *Malignant neoplasm* and *malignant tumor* should be used when referring to a specific tumor. *Malignancy* is the state of being *malignant*.

maltreatment, mistreatment – *Maltreatment* is the more severe form of *mistreatment,* and implies abuse, cruelty, and malice.

man, mankind, humankind, staff – Gender bias should be avoided in using such terms. A laboratory is *staffed,* not *manned* with technicians; and *humankind* is considered preferred over *mankind* when referring to the whole of humanity. (*See* **Section 1.8**, below, for more on "**Bias-Free Language and Descriptions.**")

management, treatment – *Management* should generally be used when referring to a particular case or a disease, and not to refer to a patient. A patient is not managed; their case is. Thus, one refers to the "*treatment* of the patient." The use of *management* is acceptable when referring to a class of patients ("the *management* of patients with cervical cancer"), since it is presumed to be referring to care strategies for a disease.

manic – see **catatonic**

manuscript – see **article**

masking – see **blinding**

may, might – *May* determines what is potential or possible ("She *may* have left the lights on"). *Might* expresses what is uncertain or possible ("I *might* have left my lunch at home"). See also **can**.

mean, median – The *mean* is the average of a set of measurements or quantities. The *median* is the midpoint in a sequence of values.

measure – see **assess**

media, mediums – In science, the plural of *medium* is *media.*

median – see **mean**

meiosis, miosis, mitosis – *Meiosis* is cellular division in which cells with a diploid number of chromosomes are divided into cells with a haploid number of chromosomes. *Mitosis* is cellular division in which cells are divided to create new cells with a diploid number of chromosomes. *Miosis* is excessive constriction and smallness of the pupil.

melanotic, melenic – *Melanotic* refers to the excessive presence of melanin as indicated by darkening of the skin; *melenic* refers to the dark sticky feces as a result of partially digested blood in the feces. (Because the root of both terms is *melas,* Greek for "black," the terms are often confused.)

method, methodology, methodical, technique – *Method* and *technique* are both procedures of doing something according to a definite quantitative plan. *Technique* refers more specifically to the skills involved in incrementing a plan. *Methodology* is a set of *methods*, rules, and *techniques* in any given procedure. *Methodical* means "done with precision and a planned order."

might – see **may**

militate, mitigate – *Militate* means to have an effect, and is usually used in conjunction with the word "against" ("The freezing temperature *militates* against us having a nice day outside"). *Mitigate* means to alleviate, moderate, or make less severe.

miosis – see **meiosis**

mistreatment – see **maltreatment**

mitigate – see **miligate**

mitosis – see **meiosis**

mutual – see **common**

mucus, mucous, mucoid – *Mucus* is a thick discharge produced by glands and membranes in the body that is designed to lubricate and protect. *Mucous* is a term used when something produces *mucus,* such as a *mucous membrane*. *Mucoid* means "*mucus*-like," and is used when something resembles *mucus*.

mutant, mutation – A *mutant* is an organism, the DNA of which is carrying a genetic *mutation*. A *mutation* is a change in DNA or RNA sequence caused by a change in chromosome, which may or may not cause an observable mutation of the organism.

need, require – To *require* is to be in *need* of something or someone for a particular reason. *Require* has a stronger meaning than *need* and carries with it a greater sense of urgency ("The doctor *requires* two nurses to aid her in the procedure"). *Need* is used when something is desired in order for something to achieve success or fulfillment. It should not be used for a passive agent ("People *need* food to survive"), though one may speak generally of an organism's *needs* and *requirements* (in which case the distinction between the two concepts is less).

norm, normal – see **customary**

notable, noticeable, noteworthy – *Notable* is used when a person, place, thing, or attribute is deserving of immediate notice. *Noticeable* is an attribute of a person, place, or thing that is physically apparent. *Noteworthy* should be used when something is deserving of notice and attention (with a generally positive connotation).

number, numeral – see **digit**

nurse – see **breastfeed**

nutrition, nutritional, nutritious – *Nutrition* refers to the science of food, health, and nourishment. *Nutritional* means of, or pertaining to, *nutrition*. *Nutritious* refers to food that contains substances that promote healthy nourishment.

observe – see **follow**

obtuse – see **abstruse**

one another – see **each other**

opacity – see **lucency**

operation, surgeries, surgery, surgical procedure – An *operation* is a *surgical procedure* or the term describing the time that a patient is induced, incised, dissected, excised, closed, and emerges from anesthesia. *Surgery* is the classification of procedures done by a surgeon and can refer to surgical care, treatment, or therapy. *Surgeries* is used in Great Britain when referring to a physician's or dentist's office.

oppress, repress – To *oppress* is to subjugate or persecute. To r*epress* is to control or restrain.

oration, peroration – *Peroration* is the conclusion of a discourse or *oration*. *Oration* is a speech, lecture or other instance or example (referring also to text) of formal speaking.

osteal, ostial – *Osteal* means "relating to bone" and is used often in combination, as in "periosteal." *Ostial* refers to any opening (or *ostium*) leading into a vessel or body cavity.

ought, should – *Ought* indicates a sense of obligation to do something. *Should* indicates a sense of duty that is not as strong as *ought*. *Ought* is said in conjunction with an expressed infinitive ("You *ought* not go to that house").

outbreak – A term often used (though considered imprecise) to describe the sudden appearance of a disease or affliction of some kind. Suggestions for more precise alternatives are: sudden occurrence, sudden appearance, or sudden development.

owing to – see **because of**

palpation, palpitation – *Palpation* refers to examination by touch or tapping; *palpitation* is a rapid flutter or throbbing (as in "heart palpitations").

pandemic – see **endemic**

Pap smear/test, PAP, pap – *Pap smear* and *Pap test* are named after George N. Papanicolaou, and are thus always capitalized. *PAP* is used as an acronym for several chemical subjects: "peroxidase-antiperoxidase"; "protatic acid phosphatase"; "positive airway pressure"; and other laboratory tests, procedures, and symptoms. It is therefore important that the context makes clear the exact meaning of the acronym whenever it is used. When the term *pap* appears in all lowercase, it usually means soft food (such as baby food).

paper – see **article**

parameter – A variable to which a value can be given to establish the value of other variables. (*Parameter* word should not be used as a synonym for "variable"). *Parameter* can also be used to mean a condition or quality that limits how something is performed or done.

part, portion – A *portion* is a specific *part* that is separated from a whole ("I will have a small *portion* of mashed potatoes"). A *part* is a subdivision of the whole and should be used in less specific instances. For example, rather than "China forms a huge *portion* of Asia," write, "China forms a huge *part* of Asia."

participant – see **case**

partly, partially – Both *partly* and *partially* mean "to some extent," but *partially* also conveys a sense of incompleteness ("Finish your *partially* eaten sandwich") as well as a preference, favoritism, or bias for one thing over another ("He is *partial* to that restaurant").

pathology – *Pathology* is the scientific study of the origin, cause, and development of diseases, disorders and abnormalities in humans, plants, and animals, and the changes produced by them. It should not be used as a synonym for "disease," "abnormality," or "disorder."

patient – see **case**

peak, peek, pique – A *peak* is an apex ("I climbed the highest *peak* in the Rocky Mountains"), a *peek* is a quick and secretive look at something ("I couldn't wait to open my present, so I *peeked* under the wrapping paper"), and *pique* is to stimulate a feeling of interest or curiosity ("What she said about Picasso's paintings *piqued* my interest").

penultimate – *Penultimate* means the second to last in a sequence.

people, persons – *People* refers to a collective group of individuals, or *persons*, with something in common, be it community, ethnicity, or location ("The *people* by the court house are waiting"). *Persons* is also a collective term, but is less general and more specific ("*Persons* with impaired vision sometimes use a guide dog"). *Persons* can also be used when referring to a group consisting of a specific number of *people* ("Five *persons* were taken into custody by the police").

percent, percentage – *Percent* is the quantity of units in an amount of units expressed in hundredths, which is represented by the symbol % (25% is 25 units per 100 units). *Percentage* is a rate expressed as a percent (25% is a *percentage*). The difference between 2 *percents* should be expressed in *percentage* points (the difference between 25% and 30% is 5 *percentage* points).

perform – see **carry out**

period prevalence – see **incidence**

peroration – see **oration**

person – see **individual**

persons – see **people**

persuade – see **convince**

pertain – see **appertain**

phase – see **faze**

phenomenon, phenomena – *Phenomenon* is an occurrence that is out of the ordinary and stimulates excitement among people who observe it. *Phenomena* is the plural form of *phenomenon.*

physician – see **doctor**

place on, put on – Both these terms are jargon. A patient is neither *placed on* or *put on* a drug, but is prescribed or given medication.

point prevalence – see **incidence**

portion – see **part**

possible, practicable, practical – If something is *possible,* it is theoretically capable of being carried out or done. If something is *practicable,* it is capable of being carried out or done. If something is *practical* is expected to be effective and useful.

practitioner – see **clinician**

precision – see **accuracy**

predominant, predominate – *Predominant* is an adjective that means the most common or frequent, as well as the most important. *Predominate* is a verb that means to be the most common or frequent, as well as the most important.

preface – see **foreword**

prescribe, proscribe – *Proscribe* is to prohibit or denounce something. *Prescribe* is to direct a course of action or to advise a medical remedy.

presumption – see **assumption**

prevalence – see **incidence**

preventative, preventive – *Preventative* and *preventive* both mean to stop something from taking place, but *preventive* is preferred.

principal, principle – As a noun, *principal* refers to the director of a school, a character in a dramatic production, or someone acting on his or her own behalf in a business transaction. A *principle* is a law, a precept, or proposed assumption for a line of thought, assumed to be true for the sake of argument.

prior to – see **before**

proved, proven – *Proven* should be used when something has been established and made factual ("He was *proven* guilty"). *Proved* is the past-participle form of *proven*.

provider – *Provider* can mean an organization that provides a service. In science, it is commonly a health care professional or a medical institution. When using the term *provider,* include the specific type of *provider* meant (for example, "pediatric *provider*").

psychotic – see **catatonic**

put on – see **place on**

quasi- – see **bi-**

radical, radicle – In chemistry, a *radical* is a group of atoms behaving as a unit in a number of compounds. In medicine, *radical* surgery or *radical* treatment means procedures directed at the root cause of a problem or dysfunction (and not necessarily implying any danger or severity). In mathematics, a *radical* is the root of a number or quantity. In anatomy, a *radicle* is the smallest branch of a vessel or nerve.

radiography, radiology – *Radiography* is the process of making a *radiograph*—an image created by the exposure of tissue to radiation such as X-rays or gamma rays. *Radiology* is the branch of medicine that uses imaging and radioactive substances to diagnose and sometimes to treat disease. (See also **film**.)

rare, unique, unusual – *Unique* means one of a kind, *unusual* means uncommon or out of the ordinary, and *rare* means seldom occurring.

reference – see **allusion**

refrain, restrain – To *refrain* is to hold someone back from doing something. To *restrain* is to keep something under control, or limit it.

regime, regimen – A *regime* is an established system of doing things in a regular pattern; the term can also refer to a government ("The government's political *regime* was aggressive at best and oppressive at worst"). A *regimen* is a program or schedule for the management or treatment of something ("I am on a strict dietary *regimen*").

regular – see **common**

relation, relationship – *Relationship* means the connection between two or more persons. *Relation* means the connection between two or more objects or things. See also **association**.

relegate – see **delegate**

reluctant, reticent – *Reticent* is not a synonym for *reluctant*. *Reticent* means to be uncommunicative. *Reluctant* means to be unwilling, uncooperative, or disinclined.

remarkable, marked – *Remarkable* is used to refer to an observation that is significant; *marked* usually refers to observations or changes that are measurable or noticed, but not necessarily significant.

renin, rennin – *Renin* is a renal enzyme that promotes the production of the protein angiotensin. *Rennin* is an enzyme derived from calf rennet that is used to curdle milk.

renounce – see **denounce**

repetitive, repetitious – Both terms mean "to occur over and over again," but *repetitious* has an association with tediousness, as though something that is *repetitious* is tiresome.

repress – see **oppress**

reproducible, reproductive – *Reprodicible* means capable of being repeated, as in *reproducible* experiments or results. *Reproductive* refers to biological processes or systems where organism produce offspring.

report – see **describe**

require – see **need**

reticent – see **reluctant**

reveal – see **demonstrate**

sample – see **aliquant**

scatoma, scotoma – *Scatoma* refers to a tumor-like mass in the rectum formed by the accumulation of fecal matter; *scotoma* is a "blind spot" in the visual field or an area of diminished vision.

schizophrenic – see **catatonic**

section, slice – *Section* is used to refer to a portion of a radiological image. *Slice* is used to refer to a cross-section or portion of tissue.

semi – see **bi**

semiannual – see **biannual**

sex – see **gender**

should – see **ought**

sight, site – see **cite**

significant – see **desirable**

since – see **as**

slew, slough, slue – *Slew* is an informal equivalent to many or several and should be avoided. A s*lough* is a marshy or swampy area of land. *Slue* is a verb meaning to pivot around.

stanch, staunch – *Staunch* means to be loyal or dependable. *Stanch* is a verb meaning to stop the flow of liquid, and is often used in regard to bleeding.

-stomy, -tomy – *-stomy* is used to indicate a surgical opening (stoma) into a part of the body (as in *colostomy; appendicostomy,* etc.). *-tomy* refers to the operation or cutting itself (*colotomy; appendectomy,* etc.).

strata, stratum – *Strata* is the plural form of *stratum,* meaning any of several layers or levels of something.

structure – see **anatomy**

subject – see **case**

subsequent – see **consequent**

sufficient – see **adequate**

super-, supra- – *Super-* is a prefix meaning "in excess" (*superinfection*); *supra-* is a prefix meaning "above or over," usually physically or geometrically (hence its preferred use in anatomy).

surgeries, surgery, surgical procedure – see **operation**

sympathy – see **empathy**

table – In the United States, to *table* something means to postpone it or remove it from an agenda; in Britain, it means to bring the matter up for immediate consideration and discussion.

technique – see **method**

temperature – see **fever**

that, which – *That* is a relative pronoun used to identify or indicate someone or something being talked about ("*That* building was built in the gothic style"). *Which* is a relative pronoun that is used when adding a clause that provides information about a subject or item already mentioned ("I went to the gym today, *which* was a big effort for me"). *Which* is preceded by a comma, parentheses or dash, unless being used restrictively ("She found herself in a situation in *which* she was sad").

theory – see **conjecture**

therefor, therefore – *Therefor* means "in return for" or "for it" ("I returned my dress and received a refund *therefor*")—a somewhat antiquated way of speaking. *Therefore* is a term meaning "because of" or "as a consequence of" ("The bread is moldy; *therefore* I won't eat it.").

think – see **believe**

titrate, titration – To *titrate* is the process of *titration*. *Titration* is the measurement of the concentration of a substance through adding small quantities of that concentration until a reaction occurs.

torpid, turbid, turgid – *Torpid* means idle, dormant and sluggish ("My dog is *torpid*"). *Turbid* means unclear, confused or opaque ("My teacher gave a *turbid* response to my question.") *Turgid* means swollen ("Her fingers were *turgid*") and can also mean pompous ("He used so many unnecessarily big words that I found his speech incredibly *turgid*").

toward, towards – Both terms mean the movement of a person or thing in the direction of another person or thing, but the preferred form is *toward.*

toxic, toxicity – *Toxic* means relating to or caused by a toxin or poison. *Toxicity* is a measure or degree of being poisonous.

traditional – see **conventional**

transcript, transcription – A *transcript* is a written record of oration or speech. A *transcription* is the act of making a *transcript*.

transplant, transplantation – In science, *transplant* means to transfer an organ or tissue from one body to another. Include the organ being transplanted when using this term to describe surgery ("The patient is having a liver *transplant*"). *Transplantation* is the term used to describe the overall procedure in a non-specific way ("There were 20 *transplantations* at the hospital today").

treatment – see **management**

trillion – see **billion**

tubercular, tuberculous – *Tubercular* means relating to or covered with tubercles; *tuberculous* refers to the disease tuberculosis. (Often used interchangeably.)

turgid – see **torpid**

type – see **kind**

typescript – see **article**

ultrasonography, ultrasound – An *ultrasound* refers to the high frequency sound waves that penetrate the body during an *ultrasonography*; also, the procedure itself.

uninterested – see **disinterested**

unique – see **rare**

unorganized – see **disorganized**

unreadable – see **illegible**

unusual – see **rare**

-urea, -uria – *-urea* is used to specify a particular kind of urea, as in "nitroso*urea*" (urea with a –NH group); *-uria* is used in relation to a urinary condition, as is "nitit*uria*" (nitrites in the urine), or "noct*uria*" (the need to urinate during the night, interrupting sleeping).

use – see **employ**

useful – see **desirable**

usual, usually – see **general**

utilize – see **employ**

vaccinate – see **immunize**

valuable – see **desirable**

varying – see **different**

venal, venial – To be *venal* is to be easily persuaded, bought, and open to bribery ("Judas was a venal disciple"). *Venial* means pardonable or easily forgiven ("Laura argues a lot, but her other faults are *venial*").

versus – See **compare to**

vertex, vortex – A *vertex* is the top of something, such as an organ; a *vortex* is a whirled pattern as may be found in a fingerprint or a weather pattern or a hair growth.

vesical, vesicle – *Vesical* is an adjective that means "referring to or affecting the urinary bladder; a *vesicle* is a small fluid-filled sac, blister, or cyst in the body ("vesicular" when used as an adjective).

viscous, viscus – *Viscous* refers to the thickness of a fluid; *viscus,* the singular of *viscera,* refers to any main organ in the abdominal cavity.

vocation – see **avocation**

voluminous – see **compendious**

wheal, wheel – A *wheal* is a raised, discolored patch on the body, often a result if a blow or an allergic reaction to an injection or sting (a "wheal-and-flair reaction"). A *wheel* is a round instrument.

which – see **that**

while – see **although**

who, whom – *Who* is used to ask a question about the identity of a particular person or group ("*Who* is going to vote in this year's election?"). It is also used to give information about a particular person or group ("This house was built by my father, *who* is an architect"). *Whom* is an objective pronoun that can appear as the object of a verb ("I didn't get the name of that girl *whom* I met") or the object of a preposition ("This is my person with *whom* I'm going to spend the rest of my life").

who's, whose – *Who's* is a contraction of "who is." *Whose* is a possessive ("*Whose* coat is this?") and refers to things as well as individuals ("The publishing house, *whose* book we are printing").

worldwide – see **global**.

X-ray – This term appears in this form (which is preferred, in keeping with the German provenance of the term, *X-strahl*), as well as in the forms: *x-ray, x ray,* and *X ray*. In physics literature, *X ray* is most frequently used.

your, you're – *Your* is the possessive form of you. *You're* is a contraction of "you" and "are" ("*You're* a bully").

1.7 Jargon and Inappropriate Language

Words or phrases that may be easily understood in the course of everyday conversation are often inappropriate and unsuited for scientific writing. Scientific writing has its own preferred set of common scientific words and concise general phrases that take the place of jargon that may otherwise be confusing and unspecific.

The use of jargon in scientific writing depends on the author's readership. If the author is addressing people in his or her own field, it is best to exclude jargon. The primary objective is to be as clear as possible in order to insure that readers understand the language and phrasing of a text.

A scientific text should not be filled with unneeded phrases. They do not add information and they make a sentence harder to understand. For example, "despite the fact that" should be omitted from a sentence and replaced with "although."

The following chart lists the preferred forms of scientific jargon as well as the more concise form of commonly used phrases that are unnecessarily long and wordy.

Jargon/Circumlocution	Preferred Form
A majority of	most
A number of	few, many, several, some
Accounted for the fact that	because
Along the lines of	like
An increased number of	more
An order of magnitude	ten times
Are in agreement with	agree
Are of the same opinion	agree
As a consequence of	because
Ascertain the location of	find
At the present moment	now
Blood sugar	blood glucose
By means of	by, with
Cardiac diet	diet for a patient with cardiac disease
Carry out	perform, conduct
Caused injuries to	injured
Completely filled	filled

Jargon/Circumlocution	Preferred Form
Conducted inoculation	inoculated experiments on
Chart	medical record
Chief complaint	chief concern
Circular in shape	circular
Commented to the effect that	said, stated
Conduct an investigation into	investigate
Congenital heart disease	congenital cardiac anomaly
Definitely proved	proved
Despite the fact that	although
Draws to a close	ends
Due to the fact that	because
During the course of	during
During the time that	while, when
Emergency room	emergency department
Exam	examination
Expired	died
Fall off	decline, decrease
Fewer in number	fewer
File a lawsuit against	sue
For the purpose of examining	to examine
For the reason that	because
Future plans	plans
Gastrointestinal infection	gastrointestinal tract infection
Genitourinary infection	genitourinary tract infection
Give rise to	cause
Goes under the name of	is called
Has the capability of	can, is able
Has the potential to	can
Have an effect on	affect
Heart attack	myocardial infarction
Hyperglycemia of 250 mg/dL	hyperglycemia (blood glucose level of 250 mg/dL)
If conditions are such that	if, when
In a satisfactory manner	satisfactorily, adequately
In all cases	always, invariably
In case	if
In close proximity to	near
In connection with	about, concerning
In my/our opinion	I/we think

Jargon/Circumlocution	Preferred Form
In order to	to
In regard to	about, regarding
In terms of	in, of, for
In the course of	during, while
In the event that	if
In the near future	soon
In the vicinity of	near
In those areas where	where
In view of the fact that	because
Is in a position to	can, may
It has been reported by Smith	Smith reported
It is apparent, therefore that	apparently
It is believed that	[OMIT]
It is often the case that	often
It is possible that the cause is	the cause may be
It is this that	this
It is worth pointing out that	note that
It would thus appear that	apparently
Jugular ligation	jugular vein ligation
Lab	laboratory
Labs	laboratory test results
The labs have not	the laboratories have not
Lacked the ability to	could not
Large amounts of	much
Large in size	large
Large numbers of	many
Left heart failure	left ventricular failure
Lenticular in character	lenticular
Located in, located near	in, near
Look after, take care of	watch, care for
The majority of	most
Make an adjustment to	adjust
Masses are of large size	Masses are large
Necessitates the inclusion of	needs, requires
Normal range	reference range
Of a reversible nature	reversible
On account of	because
On behalf of	for
On the basis of	from, by, because
On the grounds that	because

Jargon/Circumlocution	Preferred Form
On two separate occasions	twice
Original source	source
Orthopod	Orthopedic surgeon
Owing to the fact that	because, due to
Pap smear	Papanicolaou test (or Pap test)
Passed away	died
Past history	history
The patient failed treatment	treatment failed
The patient was diagnosed	the patient's illness was diagnosed
The person in question	this person
Plants exhibited good growth	plants grew well
Preemie	premature infant
Prepped	prepared
Prior to	before
Produce an inhibitory effect on	inhibit
Psychiatric floor	psychiatric department, service, unit, ward
The question as to whether	whether
Referred to as	called
Respiratory infection	respiratory tract infection
Results so far achieved	results so far, results to date
Right brain	right side of the brain
Serves the function of being	is
Smaller in size	smaller
Status post	after, following
Subsequent to	after
Surgeries	operations, surgical procedures
Symptomatology	symptoms
Take into consideration	consider
The fish in question	the fish/these fish
The question as to whether	whether
The tests have not as yet	The tests have not
Therapy of [a condition]	therapy for
Through the use of	by, with
Throughout the entire area	throughout the area
The treatment having been performed	after treatment
Two equal halves	two halves

Jargon/Circumlocution	Preferred Form
Urinary infection	urinary tract infection
Was of the opinion that	believed
With a view to getting	to get
With reference to	was
With regard to	about, concerning
With the result that	so that

1.8 Bias-Free Language and Descriptions

Terminology and phrasing for issues regarding gender, sex, disabilities, race, and ethnicity develop constantly in response to changing attitudes about what is considered appropriate and acceptable. It is important to keep up-to-date on opinions regarding these issues and to realize that even relatively recent style manuals may be inconsistent in this area. For example, a manual from the late 1980s uses the term "mentally retarded" in its section on how one should avoid bias against individuals with disabilities. This term is now out-of-date and is considered inappropriate and offensive in both everyday use and in scientific writing. Using socially acceptable bias-free words and phrasing is vital to any scientific writing, and because it is such a delicate area, it is necessary to be as knowledgeable as possible about changes in appropriate terminology.

i. Gender and sex. Avoid using pronouns that are social stereotypes or habitually biased. Some sentences are clearly gender biased and should be avoided. The sentence, "We will have a new president in eight years and he will change the country," is biased because of the assumption that the new president will be male. Instead, write, "We will have a new president in eight years and he or she will change the country." To avoid the pronoun altogether write, "In eight years, our new president will change the country."

Some terms with gender reference are considered widely acceptable because of their non-gender specific definition. For example, "This research provides new information on the development of humankind" may be regarded as being habitually gender-biased because of the inclusion of "man" in "humankind." If this term is unacceptable to an author or his or her audience, an alternative to consider is, "This research provides new information on the development of men and women throughout history." Avoid terms like "poetess" or "lady doctor" because they imply that these occupations are normally male. "Poet" or "doctor" adequately covers both sexes.

Only use *man* or *men* when referring to a single man or a group made up of only men, and only use *woman* and *women* when referring to a single woman or group made up of only women. If referring to a group that includes both men and women, find a gender-neutral term. For example, do not use the words *spokesmen* or *spokeswomen* unless the entire group is made entirely up of men or women, respectively. Instead use *spokesperson*, or, if the group is made up of both men and women, use *spokespeople.*

Indicate the sexual orientation of a man or woman in the text only if and when it is relevant. When referring to a specific group of men or women, the terms *gay men, gay women,* or *lesbians* are preferred to the umbrella term *homosexuals.* Do not use the term *sexual preference* because it assumes that one chooses ones own sexual orientation, which is scientifically questionable. Terms to describe the relationship between heterosexual couples are the same as those used to describe the relationship between homosexual couples, e.g., *companion, wife, husband, partner, life-partner, girlfriend, boyfriend.* Use of the terms *same-sex marriage* and *same-sex couple* are considered an appropriate means of referring to the status of a homosexual relationship.

ii. Race and ethnicity. The term *race* is a cultural construct without a specific biological meaning. It is a term generally used when describing a person's physical traits, assuming that his or her physical traits fit in with a larger group of people sharing those traits. It has been argued that racial categories are not an acceptable way to define people because of scientific evidence that argues that human races do not actually exist. Because defining people by race can lead to generalization and stereotyping, it is important to accurately use the term and to qualify it when necessary. Rather than use race to describe a social group or population, use less ambiguous criteria such as country of birth, or self-description.

The race of an individual or group should not factor into scientific writing on health related research. An individual's genetic heritage can help in understanding certain biological tendencies, but avoid using the broader term of race in defining an individual's medical history.

Categorizing an individual's race is often only useful in providing very general information, and therefore is not scientifically accurate. For example *White* and *Caucasian* are both too broad a term to be used in a scientific text because they do not convey any substantial information about an individual's genetic history. If it is necessary to refer to racial or ethnic groups (such as "White American"), terms should be capitalized.

Regarding ethnic or racial designations that are used in contexts where that information is relevant to the scientific research or content of the material, the *AMA Manual of Style* (Tenth Edition) offers the following guidelines:

• The term *African American* is preferred to *black*—though it should be used specifically to refer to US citizens of African descent (and should not be hyphenated).

• The term *American Indian* is preferred to the term *Native American,* though the latter is acceptable. A difficulty with *Native American* is that the term is used by the US government to refer to Samoans, Alaskans, and Hawaiians, which are not what is generally meant when the term is used. Authors should therefore feel a particular responsibility to carefully identify their subjects and referents as clearly and as precisely as possible.

• The terms *Hispanic* and *Latinos* are broad terms that may be used to refer to people of Spanish descent or decent from Spanish-speaking people of Mexico, South and Central America, and the Caribbean. However, the term *Latino* is generally understood *not* to refer to people of Mexican or Caribbean ancestry. In cases where an ethnic designation is necessary for conveying scientific information, authors should avail themselves of precise terminology such as Mexican, Mexican American, Cuban, Cuban American, Puerto Rican, etc.

• Ethnic identification through the use of exclusionary negatives—for example, characterizing a group as "nonwhite"—has disquieting connotations and should be avoided. The phrase *people of color* has some acceptability in some circles, but should be avoided primarily because of its vagueness. A term such as *multiracial* is also vague and mildly dismissive, and should be used with the utmost of care.

iii. Age. Discrimination on the basis of age (known as *ageism*) is illegal in many countries; reference to people that stereotypes them as less than productive or legitimate members of society should be avoided. Thus, referring to people as *elderly* or *aged* is inappropriate in serious writing and should be avoided.

iv. Disabilities. A disability (according to the Americans with Disabilities Act (http://www.usdoj.gov/crt/ada/) is a physical condition that "substantially limits a major life activity, such as walking, learning, breathing, working, or participating in community activities." Americans as well as people of other countries have been slow to recognize that,

(a) people with disabilities so defined have often been some of society's most productive and beneficial members of society (a list that certainly bears the name of a celebrated four-term president of the United States); and

(b) the notion of a disability can be applied to a very large portion of the population who are able to function well only because society has already made accommodations for them.

It is therefore important that individuals with disabilities (of any kind) not be demeaned or depersonalized in carefully constructed scientific writing (or, for that matter, in *any* kind of written material.

Following are some guidelines on how to avoid such objectionable formulations:

• Avoid referring to a disabled person only in terms of his or her disability. Avoid such phrasing as "Smith is a diabetic," because it identifies Smith only in terms of his disability. Instead write, "Smith is a diabetic patient" or "Smith has a diabetic condition." This phrasing makes is clear that Smith is not defined only in terms of being a patient with diabetes.

• Do not describe individuals with disabilities as victims, or in any other way that might imply that their disability renders them helpless (*suffering from, afflicted with*). Currently acceptable terms for disabilities include *blind, deaf, cannot hear or speak* (instead of *deaf-mute* or *deaf-dumb*), *hearing loss, hearing impairment* (instead of *partially deaf*), and *congenital disability* (instead of *birth defect*).

• The term *handicapped* should not be used since it implies that a person with a certain disability has disadvantages when compared with a person without that disability.

• Avoid use of metaphors or figures of speech that are insensitive to disabled people: this includes such formulations as "blind to the truth," "turn a deaf ear," "lame excuse," etc. (Some publications go so far as to also discourage use of the term *double-blind experiment*. Authors are advised to ascertain the policy of the particular publication for which they are writing.)

The previously cited *AMA Manual of Style* offers the following helpful table with alternatives to problematic phrases and terms:

To be Avoided:	*Instead Use:*
the disabled; the handicapped	persons with disablities
disabled child; mentally ill person; retarded person	child with a disability; person with mental illness; person with an intellectual disability; person with an intellectual disability (mental retardation)
diabetics	persons with diabetes; study participants in the diabetes group; diabetic patients
asthmatics	people/children with asthma; asthma group; asthmatic child
epileptic	person affected by epilepsy; person with epilepsy; epileptic patient
AIDS victim; stroke victim	person with AIDS; person who has had a stroke
crippled; lame; deformed; disfigured;	physically disabled
deaf; blind	deaf person; deaf community; hearing impaired; vision impaired

The table above may well be found wanting by advocates for the disabled community, which only serves to show how much more is in need to be done to sensitize society to the values and standing of the disabled in society—and in literature of all kinds.

Chapter 2. Preparing the Manuscript

Contents

Contents, *continued*

Contents, *continued*

Chapter 2. Preparing the Manuscript

Virtually everyone involved in science, no matter at what level and no matter in what discipline, is called upon from time to time to compose and submit for publication a piece about science. Researchers wish to communicate their findings and observations to colleagues; theoreticians wish to share their hypotheses and insights with others in the field; administrators and clinicians wish to propose new experimental or clinical programs; and the scientific community at large feels a responsibility to educate the public and maintain a high level of awareness and literacy among all segments of the population. All this demands a sophisticated communication structure with which all practitioners become familiar, and which the most successful members of the scientific community master.

Writing directed at the public, either to educate or to influence policy, must be accessible to that readership. It is widely held today that the same is true of all scientific writing: the need to be clear, engaging, and persuasive is no less vital at the highest levels of scientific discourse than it is in popularizations of science. In both venues, the competition for the attention of the readers one wishes to reach is formidable, making every effort to make the reader *want* to read what has been written a necessity.

The previous chapter focused on principles for writing clearly and for effectively conveying information to a reader. Later chapters will focus on the details of English usage (**Chapter 3**), the conventions regarding citations and references in scientific writing (**Chapter 4**), and the rules regarding copyright and permissions (**Chapter 5**). Part Two will examine the practices and conventions of specific disciplines. This chapter examines the various avenues open for publication and the rules and practices for submitting material for publication in each case.

2.1 Types of Science Writing

2.1.1 Technical versus Nontechnical Science Writing

We first distinguish between two broad classes of writing about science: "technical" and "nontechnical." The former is embedded in the infrastructure of a discipline and is in reality the means by which researchers and investigators in a field become aware of the work of their colleagues. This makes it difficult for an uncredentialed author to have something published in a technical science journal—not simply because it is presumed that only a professional could produce work suitable for such a publication and the professionals it serves, but also because readers of such a publication are primarily interested in what progress is being made in their field by their colleagues

For this reason, at the highest level of technical scientific public-cation, the science journal, articles are reviewed by respected profes-sionals in the field—called "referees"—to determine if the author has (a) made a contribution to the field that others may wish to read and take note of; and (b) has done so competently and in accordance with the standards of the discipline at the moment of submission.

There are various kinds of technical science publications, as we shall soon see, but they all have one thing in common: they require the appro-val of a practitioner, the "review" of a "peer"—hence the term "peer review"—to validate the work as worthy of the attention of others work-ing in the field. The pathway of an article that appears in a peer-reviewed journal is therefore a complex one, involving preliminary judgments by the editorial staff as to whether the work is within the scope of the journal; whether it comes from a credible source; whether it is clearly written; and whether it makes its point and presents its findings cogently. It is then passed on to professionals in the field for review—respected journals will insist that every article must be approved by at least two referees—and the article may then pass several times between reviewers, editors, and the author until it is deemed acceptable for publication.

Publications that are not peer-reviewed are not considered technical scientific publications. While the authors of material that appears in those publications may be eminent scholars and researchers in the field, and the pieces that appear there may be valuable as explications of areas of science, the articles are not part of the ongoing conversation that rep-resents the advance of research and knowledge in the field.

How does one evaluate whether a given journal is indeed peer-reviewed to the standards set forth by the governing professional bodies of a discipline? Generally, this can be determined by seeing whether the journal's articles are indexed or the abstracts of its articles are published in the services that are maintained specifically for this purpose. Journals listed and referenced in such services as Index Medicus, Chemical Ab-stracts, Current Physics Index, etc., have been examined by a respected scientific society or organization and are considered to be properly peer reviewed. The Index service thus serve the dual purpose of monitoring the advance of the discipline as portrayed in its papers, and of apprising researchers in the field of what new developments may be germane to their own work. The cooperative nature of academic publishing is the core of modern scholarship and research, and informs the practices, values, and ethical standards of scientific publishing.

i. Formal and stylistic distinctions. The differences between technical and non-technical science publications translates into marked differences in the form of material submitted (and eventually published) in each sort of publication. Articles submitted to technical science journals must exhibit proficiency in the support disciplines of the subject—mathematical, taxonomic, physiological, chemical, etc.—to the standards of the discipline as then practiced. This has given rise to the criticism that journals of science do not allow for the introduction and evaluation of radical and creative ideas that may not conform to the conventional wisdom of a discipline.

The scientific community respond to this concern in two ways: First, within the confines of the technical journal, efforts are made to allow for innovative approaches to problems in the form of opinion articles, subject review articles, editorials, and the correspondence sections of journals. Some disciplines have entire publications devoted to such material to encourage the free expression of what would normally be regarded as speculative and imaginative "musings" at the cutting edge of a field. Until humankind can be certain it knows all there is to know, such work will be essential to the growth and advancement of science.

Second, and more important, however, is the fact that journal articles in many disciplines are in reality the surface layer of a wide and complex process in which scientists discuss with colleagues the details of current issues in the field. Many such discussions become part of the proceedings of meetings and colloquia at which such material is first aired. The frequent acknowledgement of the response from the audience who first hears a paper delivered and of the suggestions of colleagues, readers, reviewers, and students are not simply a matter of courtesy; it is a vital part of the scientific process. There may be instances in history where a lone thinker created a brilliant theory in isolation with hardly any contact with or influence from insightful respondents, but the advance of science over the past two centuries is the result of the sharing of ideas and the increased communication among members of the scientific community.

ii. Determining the audience. Before beginning the writing process, it is only reasonable that the author asks at what audience the material is being directed. Philip Rubens, in his classic, *Science and Technical Writing*, presents an extensive program for analyzing the audience for a piece of writing. To some, his program may be more than can or should be performed for more than a few pieces in the course of a researcher's career, but the essentials of the program offer sensible advice that a writer (not only of science, of *any* kind of material) ought to consider at the very outset. Being clear about whom one aims to reach increases the likelihood that one's ideas will be communicated effectively.

Some factors that an author does well to consider in this regard are:

- **The educational level of the audience and its professional background.** Material directed at professionals working in related or nearby areas may require more elaboration and support than material meant for those working in the same area as the author.

- **The proficiency of the audience in English.** While science writing at any professional or technical level should avoid language that cannot be understood by a generic, competent speaker of English, special care needs to be taken if the intended audience will not be native English speakers.

- **The reading context—the conditions under which the material will be read.** In this sense, it is not only important to consider what the physical conditions under which the work will be read will be, but what the reader's expectation will be; what he or she hope to derive from the piece. An instructive piece on a technique, a piece of equipment or a matter of policy will each have different tones than a hypothesis regarding a perplexing phenomenon, or a report on the results of an experiment.

2.1.2 Types of Technical Science Writing

Although there is no formal classification for scientific writing, in most instances it is fairly clear what form and in what sort of publication for which a piece of science writing is most appropriate. For most scientists, keeping abreast of the key journals in one's field occupies a significant portion of a scientist's work day.

The more productive researchers are looking for two things as they pour over these publications (or peer at the computer screens, if they are reading the papers in electronic form): they are looking for advances in the specific problems that riddle any area of research as colleagues grapple with these problems in institutions all over the world. They are also paying careful attention to the "thrust" of the journal as a whole—the directions the consensus of practitioners are taking and the avenues they are exploring, as well as the sort of material (its tone and its content) that is gaining the attention of the editors and reviewers of the journal as a reflection of the direction the field as a whole is moving.

Some (even in the scientific community) have been critical of this practice as antithetical to the spirit of free inquiry that is so cherished a part of the modern intellectual tradition. Yet, without contact and communication with others working in the field, science become insulated and insular—the private ruminations of an isolated mind, instead of the contributory ideas of a member of the community. As we shall see, science literature as presented, promoted, and preserved in journals, offers many different possible avenues for researchers and theoreticians of many stripes, so that no good or novel idea ever need go wanting for a fair airing and an attentive audience.

2.1.2.1 Scientific Journals

Science journals cover a wide spectrum of areas of scientific research and do so in a variety of styles and methods. It is important for anyone planning to submit a work to a journal to be familiar with the scope and practices of the journal as well as with the requirements the journal's editors place on any author submitting anything for consideration. It is important to realize that most people involved in the journal—from the editors to the reviewers to the referees to the publishers—are involved as an expression of their interest and support of a discipline. Members of a journal's editorial board usually serve without payment, and referees spend their time reviewing and assessing articles simply as part of their membership in the community of scientists seeking to advance a body of knowledge.

i. Journal style matters. The requirements that a journal lays down for material submitted is designed to expedite what is at its core a complex process. A typical journal may have several hundred articles under evaluation at any given time, and each published piece may undergo a dozen reviews and revisions until the reviewers and editors feel the piece meets the standards of the publication and warrants publishing.

It is therefore no more than common courtesy that the author adhere *in detail* to the manuscript and submission requirements of the journal to which one is submitting a paper. Any deviation from the protocols about which the journal has apprised prospective authors impedes the process not only for that author, but for the journal as a whole. Even so simple a matter as typing two spaces after a period—a practice common in typescript, but an unnecessary practice for published material—can add time and work by an already strapped journal staff.

ii. Common Sections of a Science Journal

Though there is no established typology for scientific papers, they can be generally characterized as falling into one of several categories, which gives rise to science journals typically containing several sections in which papers of similar type are gathered. Practiced readers of these sections in virtually all journals recognize that there is a wide latitude given to determining what is appropriate for any given section of a journal—perhaps more leeway than one might find in any popular media outlet (such as a newspaper).

Following are the main sections of a science journal and the kinds of articles and papers that one is apt to find in them:

a. Original articles and research. The mainstay of science journals are the reports on the experiments and observations that are ongoing continually in laboratories, universities, and in the natural environment all over the world. Since these papers report on experiments, observations, and finding that have actually taken place, it is vital that these papers be precise and that they conform to the form and structure that readers of such papers have come to expect. This is known as the "IMRAD" structure (short for "Introduction; Method; Results; And Discussion"), which will be analyzed in detail in **Section 2.2.1.6,** below.

Readers of such papers are evaluating elements of the presentation that go far beyond the plain meaning of the text: they are assessing the propriety of the method; the bias of the researcher; legitimacy of the conclusions; the possibility of alternative explanations or of peripheral factors that may have caused the result or skewed the observations reported. Moreover, the possibility of replicating an experiment, verifying an observation, and corroborating (or disproving) a finding, elements at the core of the scientific method (as well as its ethos), are subverted all too easily when the report of what happened does not conform to what colleagues expect of such a report, and to how others in the field have reported their findings.

In light of this, most serious scientific journals will simply return a paper purporting to report on an experiment that does not clearly follow the "template" by which others in the field evaluate new information. For this reason, educators have determined that students wishing to pursue a career in any scientific discipline must learn this format and become adept in its use and fluent in its language, even as undergraduates.

b. Review articles. Journals will frequently publish articles in which a respected practitioner offers a summary of recent work in the field and an assessment of how certain problems are being addressed. These papers are intended to be surveys of the work of many researchers and theoreticians in the field, and will often present conflicting ideas and gaps in the ongoing research program of the discipline, without promoting the author's individual opinion or personal bias. Like the referees of research papers, authors of review articles are expected to suspend their own agendas (personal as well as professional) in performing this important service. Review articles may be seen as the discipline's periodic "step back," to take stock of the accumulating data and to make some sense of the field's research strategy in broad perspective. A review article will often point to neglected areas of investigation or point out some obvious (but unnoticed) flaws in the program the community of researchers is pursuing, or opportunities that are being overlooked.

In medical science, such articles take the form of "consensus statements" that summarize the cumulative clinical experience of physicians and researchers and provide clinical guidelines for what may be considered prudent and sensible courses of therapy or treatment. No assumption is made that any such presentation is final or definitive; the history of medical science is replete with the overturning of conventional wisdom. But neither is the review article an opportunity to advance a pet theory or an approach favored by the author in opposition to the preponderance of opinion in the field. Without compromising his or her own position, the review article endeavors to present the current state of affairs fairly, fully realizing that future research may well lead in a different direction—perhaps even closer to the opinions held by the author.

c. Theoretical papers. Within any discipline, there are bound to be a set of observations and findings (sometimes a large set) that elude explanation and defy logical consistency. A theory or explanation may have already been put forth to reconcile these problems and explain difficulties within the principles by which the majority of practitioners and researches abide—or perhaps it is a theory or explanation that calls upon members of the discipline to reevaluate a cherished principle and consider an alternative because of how effectively it resolves certain difficulties. Whatever the case may be, theoretical articles are most effective (and are most likely to find an audience) if they are couched in the context of the principles of the discipline they address, not lying outside what researchers in the field have come to regard as foundational.

Contrary to a popular notion, scientists are interested in and receptive to novel formulations and new ideas—more so, one may argue, than the general public—provided the foundation for these ideas is presented in the context of what is known and in a language that is understood. Theoretical articles in science must therefore be mindful of precisely where the argument deviates from the conventional, and where it rests on what is believed to be the case. Such articles are an important part of any journal's offerings, and they are often written by some of a discipline's most creative—and disciplined—thinkers.

d. Notes on method; case studies. Like the theoretical article, journals will often publish papers that present new methods in the laboratory—either in the performance of an experiment or procedure, or in the analysis of the data collected. These reports (in the medical literature they may take the form of "case studies," in which an exhaustive report is presented on a course of treatment or the progress of a disease) allow the wider community of researchers to benefit from the laboratory, clinical, and experimental experience of other scientists, even when the focus of their research lies in a direction other than their own.

e. Editorials; opinion pieces. Again contrary to a widely held stereotype, scientists are not without opinions and beliefs that are "non-scientific." These beliefs may involve moral issues that relate to the conduct of research, or about the application of a therapy or the results of a projected course of research. Being human, however, makes scientists as subject to their intuitive and judgmental faculties as anyone else, but with the following important proviso: they may never relinquish their commitment to sober consideration of the facts or abandon the empirical underpinnings of the scientific enterprise. This means that editorials and opinion pieces that are published in science and medical journals are vetted and scrutinized as thoroughly as a research article or an article advocating a course of medical treatment—because often, such pieces do just that.

In fact, opinion in science journals often flows both ways, in that readers are encouraged to respond by correspondence to articles, including opinion pieces and editorials, as part of the review process of the journal. Such interaction becomes one of a journal's most provocative features and often stirs new ideas and lines of research. (*See* **Section 2.1.2.2, Communications**, below, immediately following this section.)

f. Other types of material. Journals have increasingly seen fit to include material that once lay outside their scope (nearly always submitted in response to an invitation from the editor), including:

• Memoirs of important figures or about important episodes in the history of a discipline. These may include appreciations of a neglected researcher, teacher, or thinker whose work is belatedly appreciated.

• News of interest to professionals in the field, particularly as it affects working conditions and public appreciation of the profession.

• Reviews of books and other media relating to the discipline, both as as professional contributions and as a measure of public perception.

• Pieces on the cultural impact of the discipline and vice versa, including art, poetry, literature, as well as reports on the state of the discipline elsewhere and in light of national and international affairs.

2.1.2.2 Communications (Letters and Responses)

An important part of the peer review process is the response professionals in the field have to the papers that appear in the journal, particular those that express opinions and allow for differences of opinion. Whereas such contributions were once considered secondary to the main work of a journal, they are now viewed as vital elements in the intra-professional communication that the journal sees as its responsibility to encourage and facilitate. No doubt the ease with which people can communicate today via the internet or by telephone has resulted in scientists maintaining close communication with colleagues in distant lands more than ever before. Yet, for the most part, such communication remains private and, as a consequence, the "marketplace of ideas" is deprived of the contributions of many who might have corresponded with the journal, which may have published their correspondence.

Journals are attempting to address this by incorporating e-mail correspondence in their pages (in print and online), without compromising the review process that has routinely been applied as assiduously to correspondence as it has been to all other contributions.

Correspondences also play an important role in the development of an idea or an approach. Many landmark theories, explanations, and concepts saw their first light in a correspondence—a letter to the editor—of a journal. The form gives an author an opportunity to propose an idea before it is supported by the full complement of data and mathematical formalism. With journal correspondence reviewed, researched and footnoted nearly as thoroughly as an article, it is not surprising to see correspondences listed on professorial publication lists.

2.1.2.3 Manuals and Handbooks

Another way in which the proliferation of the internet has altered people's reading habits, and with it publishing, is the renewed interest (and utility) of two related types of publications: one is the sharply focused instructional guide, which we place under the rubric of "manual" and which is a blood descendant of the manual that accompanied computers and electronic equipment; and the other is the "handbook"—a multi-authored collection of, again, highly focused chapters on very specific subjects, usually pitched at a level of advanced researcher or graduate-level student.

The internet has given readers the possibility of getting small portions of information virtually instantly, but has not (as yet) solved the problem of eye-fatigue when reading long tracts on screen. Until then, highly focused specialized volumes will be the preferred way of publishing material directed at a professional readership. While omnibus reference works are virtually extinct, such handbook volumes (in all academic areas, but especially in the sciences) have found an appreciative readership.

2.1.2.4 Monographs, Books, and Proceedings

The presentation of a sustained argument or a thorough analysis of a subject still warrants a book-length treatment, so that scholars remain in need of academic book publishers (intellectually as well as professionally).

One may distinguish three types of book of interest to the academic author or scientist:

• **Monograph.** A scholarly work by a single author (as opposed to a volume with contributions by several authors) on a specific subject (as opposed to a survey history or a volume that covers a field completely). Monographs are frequently organized as a series on a general subject with titles on specific subject within that general area. There is a presumption that monographs are peer-reviewed (by the general editor or by the editorial board), but more often than not, the invitation to the author (or the acceptance of his or her work in the series) is determined more by the author's credentials and reputation than by the quality of the work.

• **Books.** A more general notion, a book covers a larger subject area than a monograph (though where the line of demarcation lies is by no means clear), and is selected for publication (by whatever procedure practiced by the publisher) independent of any series or publishing program. In the minds of many scholars and scientists, monographs are small and books are large, though counter example to both impressions abound. The fact that a single or a few titles in a monograph series ultimately stand out as of particular merit and are then republished and remain in print while the rest of the series lapses out of print should indicate that the distinction between a monograph and a book is fairly arbitrary.

• **Proceedings.** A book in which fairly precise articles covering narrow areas of a general subject are collected, presumably because they were delivered at a meeting or conference (though the paper may not have actually been delivered, nor is it certain that the conference ever took place); or in honor of a colleague (in celebration of an anniversary or birthday—a "Festschrift"); or a collection of papers—either invited especially for this publication, or assembled from previously published material, such as journals— on a subject of current interest to scholars or scientists.

Setting aside the commemorative aspect of such collections, the practice serves to make the work of important researchers in the field (past, present, and future) available to a wider audience. It has often been the case that individual essays in a collection has been republished often because it was deemed a seminal paper in the field. But whether such collections are published as free-standing volumes or as supplementary issues of a journal is frequently a decision made for reasons having little to do with the importance of the collection.

See **Appendix I** for references on the mechanics and practices of publishing that are relevant for academic and scientific authors.

2.1.2.5 Textbooks. The importance of textbooks in modern education— lower, middle, and higher—has not been dimmed by the internet as much as it has by the economics of publishing. Several million dollars are invested in the creation of a new textbook, and since textbooks that become mainstays of the educational system have guaranteed sales and undergo periodic revision in new editions, that investment may well prove to have been a wise one *if* the textbook becomes popular. Unfortunately, chance and good fortune are as operative in this field as they are in many areas of publishing, so that the best textbook in any field does not always become the leader, or even in the running.

It is unfortunate that a set of criteria have not emerged for the production of textbooks, though each new season brings enhanced design, graphics, and features—a host of "bells and whistles"—that are aimed at, not the edification of the student, but impressing the instructor, teachers, librarians, and purchasing agents for school districts, in hopes of procuring lucrative orders. As a result, areas of science that are capable of presenting an "attractive package" have encouraged administrators to schedule courses in those subjects beyond their importance in the curriculum or the discipline, while other areas of greater importance have not been the focus of educational attention simply because the textbooks available are not attractive to students and instructors.

There will be those who dismiss this analysis as attributing shallow values to the educational establishment. Consider, however, the large expenditure of energy and resources of the promoters of pseudoscientific theses in creating and disseminating attractive text-books with the veneer and accoutrements of legitimate science instruction as a cornerstone in their efforts to introduce their ideas into the educational marketplace. Those activities should underscore the importance of well-designed, accurate, precise, and engaging instructional material, beginning with textbooks, as an important enterprise within any scientific discipline.

2.1.3 Types of Nontechnical Science Writing

It is only recently that programs devoted to teaching the craft and art of science writing have been introduced to the curricula of major universities in the United States and around the world. This has been done in recognition of the need for better training of those entrusted with informing the public of recent scientific discoveries. This salutary development stands in contrast to the general difficulties being experienced in publishing as a business and as a cornerstone of modern culture.

One difficulty that science writers experience is finding publications to write for, a situation not helped by the difficult straits in which newspaper and magazine publishing find themselves. Yet, every challenge is also an opportunity, and science writers are going to have to open up some venues to science writing that may have been closed or limited in the past. Creating prose that captures public imagination and interest while clarifying scientific principle and research now becomes a vital element of the public discourse and education, and science writers must, we would argue, play a vital role in that enterprise.

2.1.3.1 The Popular ("Mass") Media

When speaking of "mass" media, we mean publications available to the general public with no formal training and without any institutional membership requirement (such as would be required for a technical journal). As such, any outlet that reaches a public without restriction should be looked upon as a suitable platform to reach a readership. Older models of racks filled with magazines must give sway to new models that now have an electronic complexion.

Not every scientist or science writer can capture the Op-ed pages of major newspapers, but a well-crafted and interesting piece on a scientific development—particularly one that relates to a subject in the general news—will find an outlet and an appreciative reader *somewhere*.

Following are some *practical* guidelines for science writers and researchers on approaching and "breaking into" nontechnical media.

i. Newspapers and newsletters. Establish an ongoing relationship with a local newspaper so that you may be counted on as providing an article on a scientific development of local interest. These may have to be done without payment, but it will establish your credentials and reliability as a source. In the current marketplace, and bearing in mind the objective of educating the public, no publication should be considered too small or unworthy of attention.

ii. Magazines and nontechnical periodicals. Though there are fewer magazines devoted to being published, there is greater interest among the editors of general interest periodicals in subjects relating to science. Whether the writer is a practicing scientist or a science journalist, editors seek material about institutions, laboratories, task forces, and groups devoted to a accomplishing a specific task (such as the creation of more efficient fuels from vegetation), or the examination of a specific phenomenon (such as the monitoring of the severity of Gulf Coast hurricanes over the past decade), or the work of a task force in addressing a pressing public need (such as the monitoring of inner-solar system asteroids for possible bodies on a collision course with earth). All of the examples given here were the subjects of cover stories of national trade magazines (*see* **Appendix I** for references), written by science journalists without advanced degrees in science, but about serious efforts and projects being undertaken by agencies, research institutions, or academic departments.

iii. Web sites and electronic media. It is a truism that the internet is the new forum where individuals interested in specific subjects (*any* subject at all) find outlet to converse and communicate with interested parties. No science communicator can afford to ignore this important venue. Science bloggers have proliferated, but the nature of the internet is such that even with the number of such blogs numbering in the tens—even hundreds—of thousands, the technology allows interested readers to find their work and engage them in back-and-forth communication.

iv. Pamphlets and booklets. The internet has also made self-publishing a viable possibility, both economically and procedurally. It is not only possible to produce books using templates provided on the websites of publishers, but it is also possible to market publications that are produced on-demand, minimizing the outlay of capital required for manufacturing, warehousing, and fulfillment. (Again, *see* Appendix I.)

Is this a major development in the world of publishing? Well, yes and no: no, insofar as it is severely limited in terms of penetration of the marketplace; but yes, in that it provides a means for a writer (particularly of non-fiction) to reach a precisely targeted audience. One result is that scientists and science writers have established a loyal and growing readership by publishing small, focused publications—pamphlets and booklets—on specific subjects in science and technology.

v. Trade publications: books and chapters. Finally, the route of traditional publishing—through agents, editors, and packagers, etc., described in sources reference in Appendix I—is still available, though becoming increasingly more difficult with each passing day.

2.2 Manuscript Preparation and Submission Requirements

2.2.1 Journals

Though we have taken pains to emphasize in the foregoing pages that science writing for a general non-professional audience is an important activity for the members of any discipline to undertake and to regard as part of their responsibility (we would go as far as to say a *sacred* part) as members of their discipline, we focus here on the science journal, where the cutting edge of work and the results of scientific research are shared with colleagues in the community.

We note, however, that much as a researcher requires years of training, practice, and experience in the laboratory or at the blackboard developing those skills that make one capable of engendering original, productive, and effective science, so must the writing skills required for effective communication at this advanced level be honed and developed through years of practice in writing material at all levels. For those in the scientific community who do not share this mission (and who would rather "stick to their knitting"), it should nonetheless be clear that the clarity and effectiveness of the most technical scientific paper will benefit from an author's experience and skills in the native and elemental elements of writing that any form of writing experience may provide.

2.2.1.1 Submitting to Science Journals

The papers that appear in every issue of science journal (meaning, a peer-review science journal) have undergone a complex process of review, consultation, revision, and reevaluation that is itself a remarkable achievement of modern science. (A chart outlining the process, adapted from the very useful *Authors' Guide of the American Meteorological Society*, appears on the next page.). The very first piece of advice that one can give an author seeking to submit a paper for publication to a science journal is for that author to become familiar with the process and with the practical and mechanical requirements of the specific journal being considered. The process is simply too complex to allow for latitude in this regard (and still permit the journal to function).

i. About peer review. The process of peer review actually begins well before any paper is submitted; it begins with the selection of the editorial board of the journal, chosen for being leading, respected, and active practitioners of the discipline and researchers in the field. The first level

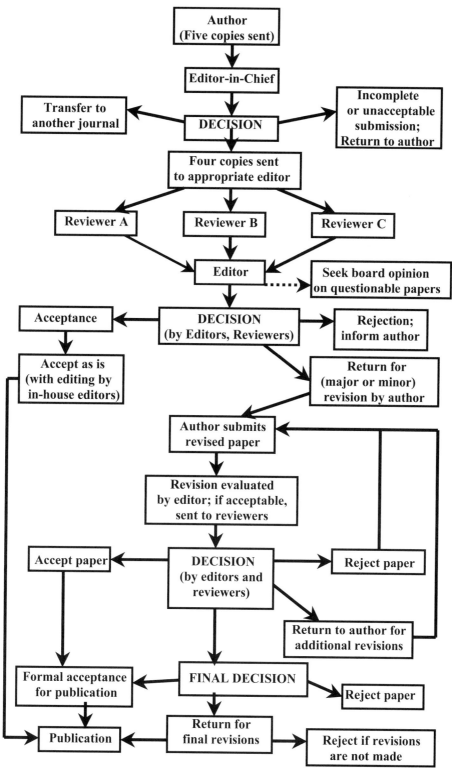

Figure 2.1 Schematic of the Journal Publishing Process

of responsibility for reviewing a paper falls upon the members of this board, who will either accept that responsibility on themselves and personally review the paper, or advise the editor of colleagues who may be called upon (in the name of the journal's sponsoring organization or simply in the name of the board and its members) to review the involved review of the paper.

When a reviewer has been assigned (and a serious review process calls upon the review of at least two, preferably three competent professionals in the field), the reviewer (also known as "referees" or "assessors") will assess the paper on the basis of several criteria, each a fulfillment of responsibilities a reviewer has to various parties in the process, among which are the following:

a. Is the paper relevant to the journals' scope and area of concentration? (This is in keeping with the reviewer's responsibility toward the journal's sponsoring institution and publisher.)

b. Does the paper make a valid contribution to the field? (How valuable or useful a contribution it might be has yet to be determined by what use others may make of it; reviewers have enough experience with findings suddenly and unexpectedly becoming useful to make such a determination.)

c. Is the presentation clear, concise, and well organized? (The reviewer is asked to be considerate of his or her colleagues' time.)

d. Are the paper's point and conclusions, as well as its support material, original when so claimed, or properly attributed and referenced when derived from other work? (This is the responsibility the reviewer has to other researchers and workers in the discipline.)

e. Does the paper use proper mathematical techniques and notation, as well as consistent and approved units (preferably, or at times by journal policy, exclusively SI units) units and nomenclature? (This will make the work accessible to researchers in the field, especially those who may wish to replicate and corroborate findings and observations. These considerations are particularly important when statistical inferences are made or when other areas are applied to finding from disciplines in which the author may not have expertise or credentials.)

f. Are the mechanical elements of the paper—title; abstract; indexing terms or key words (if required; references; notes; tables; graphics; etc.—in keeping with the guidelines and policies of the journal? (These elements will expedite the review and editorial process.)

ii. Responsibilities and ethics. In addition to professional responsibilities, there are ethical matters that need to be addressed, primarily by the journal's editorial management, which acts as the representatives of the sponsoring institution or the publisher. An editor of a journal in a discipline must thus be aware of what is going on in the community of scientists that make up that discipline and be on the lookout for possible ethical violations among reviewers. The Tenth Edition of the *AMA Manual of Style* devotes nearly 20-percent of its over 1000 pages to "Ethical and Legal Considerations" and to issues of editorial and institutional responsibility as an indication of how important these issues are in scientific publishing.

The very useful *Publication Handbook and Style Manual of the ASA-CSSA-SSSA* offers guidelines for ethical conduct by reviewers. A reviewer must recuse himself from reviewing a paper if the answer to any of the following questions (or questions like them) is "yes":

a. *Have you had significant and acrimonious disagreements with the author (or any of the authors of a multi-authored paper) in the past?*

b. *Are the authors and your co-investigators on any other research project (or serve together on any professional committee)?*

c. *Have you and the authors jointly published a paper or work of any kind on any subject in the past five years?*

d. *Are you a close personal friend or relative of any of the authors?*

e. *Were you consulted by the author regarding this paper or the work behind it, or did you previously review the work for another journal?*

e. *Are you currently working in the same area of research as the subject of the paper, so that you might be considered a competitor and may thus gain some advantage by having access to the paper and its findings?*

f. *Are you not competent to review the paper, generally or in detail— or are you unable to devote the time necessary for personal reasons?*

These last two questions pose some problems to journal editors because there may well be certain areas of science in which only a few researchers are well-enough apprised of all work in the field to review a paper, and thus are very likely to know, and possibly even be friendly with, those few colleagues in the field to whom they may converse.

It is assumed that reviewers will maintain the confidentiality of any paper submitted for review, and will refrain from contacting the author directly with regard to any matter concerning the paper, no matter its import. This at the very least violates the presumption that the author is unaware of the identity of the reviewer during the consideration process.

Periodically, various scientific disciplines have attempted to institute a "double blind" system where the reviewer is unaware of the identity of the author as well. In practice, however, with so thorough a system of communication connecting scientific communities the world over (itself a development considered salutary to the furtherance of science), it is nearly impossible to ensure that a reviewer would be unaware of the identity of at least one of the authors of any paper submitted for review.

2.2.1.2 Technical Requirements

There may yet be a few journals that accept submissions by hardcopy typescript, but increasingly, science journals insist on electronic submission of papers. (That many of the guidelines and author instructions of science journals, once available in print, are now available only online is indicative of how thorough a trend this is.)

In addition to the standard word-processing programs used in contemporary computer-based communication, each scientific discipline has developed "authoring tools" that may be used to format papers in a manner that is more likely to yield a paper in conformity with rules and strictures of specific journals in that field.

In the chapters below, such programs are identified and guidance is offered in accessing and using them to create papers. It is worth noting, however, that any paper that is submitted to a journal should be reviewed by someone conversant with these programs and with the varied requirements of individual journals. (Many departments employ such professsional assistance in the person of a "documentation manager," who is responsible for this area as well as making certain that grant proposals, compliance reports, and other official documents adhere to their stylistic requirements.)

i. Commonly accepted formatting programs. In the event an author does not use an institutional formatting program such as those offered in mathematics, physics, geoscience, and biomedical sciences, it is still advisable (if not often required) that papers be submitted as both a word-processing document and as PDF, to make certain that what is being submitted is being read correctly by the computers at the receiving office. (Those experienced with computers know that even when the machine, operating system, processing program—even the fonts are identical in the recipient office, transferring material from machine to machine can create variations in the final product due to variations in font editions and interaction with other programs on the machines. Since a PDF file provides a faithful picture of what the page actually looks like as it goes to press, such a file is useful in resolving unwanted variations that arise as a result of the transfer.)

ii. Range of submission requirements. For journals that accept submissions in manuscript form, the general practice is to require five copies of the manuscript for the sake of proper review and vetting. Virtually all journals will insist that submissions are in a state of completion before the submission will be acknowledged as received or sent on to the next steps in the publication process. Delay in meeting these requirements may have a profound effect if questions of priority should arise at a later date, so that editors are hesitant to even read a submission to see if it is appropriate to the journal until the minimal submission requirements are met. (The situation is not unlike that which obtains in most trade publishing houses, where an unsolicited manuscript that is unrepresented by a recognized agent will generally not even be opened, let alone read, because it makes the house liable to later claims of appropriation of intellectual property by authors whose work was rejected simply for being substandard.)

2.2.1.3 Submission Methods

i. Electronic submission. It is today virtually impossible for an author to submit a paper to any serious (and seriously indexed) journal without first contacting the editor and apprising him or her that the author has a paper that he or she wishes to submit for consideration. This is generally done through a query letter, and if the editor is satisfied that the author is a serious researcher in the field, an FTP site address, user name, and password (usually valid for a specific author and paper, and possibly for a limited time, to ensure the security and integrity of the site) to which the author may download the paper and all relevant materials.

On the following two pages is a checklist that details the requirements that authors are expected to meet in submitting papers to a science journal for publication. Note all these requirements are in addition to submission of a complete paper, meaning, with no gaps in the text and no missing tables, graphics, or images necessary for completion of the body proper of the paper. In some instances, an editor will initiate the review procedure with the complete text in hand, but with only some of the additional requirements set forth in the checklist fulfilled, but *never* without some statement by the author (in a written communication or an e-mail) that these requirements are forthcoming. The acknowledgment of the editor of receipt of the paper will, in such cases, include a statement that acceptance of the paper for consideration is provisional until all author requirements are met.

ii. Camera-ready copy. In the past, submission of papers in camera-ready form was appreciated by editors insofar as it expedited the composition process and allowed for quick turn-around and publication. Today, however, with the composition aided by advanced and widely-available publishing programs, camera-ready submissions are discouraged because they add time and work to the review and publishing process.

2.2.1.4 Manuscript Checklist for Authors

(The checklist below is adapted from *JAMA* Instructions to Authors, with notations on the requirements that may be in force for other journals.)

1. Authors are asked to read the instructions, which is on the journal or society website. For *JAMA,* the website is: http://manuscripts/jama.com. In some cases, authors are required to sign a waiver (or acknowledge via e-mail) that they have read the instructions and will comply with them, holding the publication blameless in the event the paper is not published due to their failure to abide by some provision.

2. A cover letter must be included as a separate attachment.

3. A corresponding author of the paper must be designated and corresponding addresses and all available contact information (postal address; telephone and fax numbers, e-mail address) must be provided.

4. The full names, degrees, institutional affiliations, and e-mail (or a postal address) for all authors of the paper.

5. A word count of the paper, exclusive of the title, the abstract, references, and legends of any figures and tables—included on the title page. (Usually, papers longer than 7500 words, or 26 double-spaced manuscript pages, require the permission of the editor for submission.)

6. An abstract of the paper that conforms to the journal's standards.

7. The pages of the manuscript must have ample margins; the text must be double-spaced, with only one space after a period; and the right justification must be ragged. The word processing program must use no unusual or unnecessary fonts; must not have hyperlinks embedded in the text unless expressly permitted by the editor; must not have graphics (diagrams, tables, drawings, maps, schematics, photographs, etc.) in the body of the text. These should be presented on separate sheets (or files), with legend material on a separate sheet or file keyed to the graphic, and the location of the graphic in the text clearly indicated in the paper itself.

8. References must be checked for accurate format, conforming to the practices of the journal; should be cited in numerical order in sequence in the text. Most journals prefer the automatic footnote feature of many word processing programs not be used; notes should be indicated with bracketed numbers and notes should appear on a separate sheet(s) or file.

Author checklist, *continued*

9. Provide each table with a title—a brief phrase of 10-15 words—and a legend that briefly explains what the table purports to show.

10. Each author must sign and submit (by mail) a copyright transfer form and any other forms that the journal requires regarding authorship contribution and responsibility.

11. Each author must sign and submit (by mail) any form the journal requires regarding conflict of interest, financial disclosure, or disclaimers of any kind as may be required by institutions where research was performed. No journal that requires such documentation will allow a paper to go to press without these forms duly executed and on file.

12. An affidavit signed by the corresponding author that all necessary permissions have been obtained in writing and that all individuals or institutions named in the Acknowledgments section has also granted written permission to be so named.

13. A statement from at least one author that he or she has had full access to and takes personal responsibility for the accuracy of all data included in the paper.

14. A statement regarding all sources of funding for the study, or any material support (fellowship; grant, financial support, research facilities, etc.) that contributed to the work or provided an opportunity to conduct research.

15. A statement of the role funding institutions or sponsors of the research had (if any) in the design or conduct of the research and in the interpretation of the data collected.

16. Written permission from any source of unpublished data.

17. Any institutional review, ethics compliance statement, or waiver.

18. Original tables, drawings, or graphics, unless expressly permitted by the journal editors. All material submitted should (ideally) be original.

19. Informed consent forms from any individuals who participated in the study, plus written consent forms for use of images or descriptions.

20. Clinical trial identification numbers and the registration site URL.

2.2.1.5 The Parts of a Scientific Paper

Researchers and professional scientists (if they are to be successful) learn to regard the architecture of a scientific paper the same way architects look at the physical restrictions of a building: as the structure within which the creative expression of the individual's art and temperament is given free reign. Just as walls create and define space, the parts of a scientific paper and the structure of its text provide the common language through which the scientist may communicate an astounding array of information and ideas. Though the strictures and guidelines may be dictated and derived from the practical needs of editors, reviewers, and publishers (just as architects must deal with pesky requirements of building codes, client needs, and gravity), they provide a context within which the most creative and earth-moving (and people-moving) ideas may be expressed.

In this section, we examine the features of a scientific paper; in the next we will examine more closely the structure of the expository section of the paper—the text.

i. Query and cover letters. Since the first written communication of an idea to the world is in the private communication of that idea to an editor of a journal, it is advisable that that relationship begin with a direct statement to the editor—in the form of a "query letter"—of the intention and desire on the part of the author to submit a paper on a given subject. In the query letter, an author should be asking an editor if there is any interest on his or her part in seeing the paper—is the subject of the paper appropriate to the scope of the journal? Is the author aware of papers published previously by the journal that would indicate the paper contemplated by the author would be of interest to the journal's readership? The editor certainly would be, and may either suggest a better, more effective form in which the idea the author wishes to express may be presented—a correspondence; a note; a letter to the editor; a review article—or else another journal or venue where the idea would be more appropriate and thus more widely appreciated.

At the very least, the author should express the desire to have that paper published in a cover letter accompanying the submission (which readers will note is a requirement on the author checklist above).

ii. Proposal. It is sometimes advisable that the author provide the editor with a proposal in which the salient points and much of the support material for the paper is presented. A proposal gives an editor an initial opportunity to assess the professionalism of the author and the general contours of the work behind the paper and the main points the author wishes to make. In a way, a proposal to submit a paper is a way of bringing the editor into the process of creating the paper in the first place, involving an individual in that process who is going to be an important if not decisive factor in determining if and how the paper is published.

iii. Title Page.

The opening page of a paper—the "title page"—contains several elements that prove to be critical to both the likelihood that a paper submitted will be published by a given journal, and that the paper will be read by readers who will be able to use and appreciate the work, or who may be important to the professional life and career of the author. Each element of the title page must be fashioned with care. Before examining these elements in detail, we list them:

- the title, which will often be accompanied by an explanatory "subtitle" that explains and expands on the title;
- the "author statement" (or "byline");
- the author affiliation;
- an abstract (summary) of the paper;
- bibliographic references (either numerical from an index prepared by the discipline, or in the form of "keywords" under which the paper will be indexed and cited in citation services);
- acknowledgments (of formal, institutional support—personal acknowledgements of the assistance and support of individuals are best placed in the footnotes of the paper proper);
- footnotes (regarding the authors contact information, the availability of reprints, the authors to whom comments should be sent and the means to do so—see below);
- "footline" elements regarding the journal title, its volume and issue number, copyright notice, and any DOI (see below) and internet source identification coding.

a. Title. The title of a paper should be as straightforward and as unadorned as possible; it is generally the first opportunity to apprise a prospective reader of the content of the paper, and many professionals will go on to examine the abstract only if intrigued and interested in the subject of the paper as presented in the title.

There are two general styles of title: the "headline style" announces the subject as a title or headline of an article; the "mini-abstract" or sentence style states the fundamental point or conclusion of the paper in a concise sentence. Such titles tend to over-emphasize a conclusion or a specific point in the paper and can obscure other, equally important elements. The mini-abstract style of title should thus be used with care and caution—in other words, rarely.

Following are some guidelines regarding the wording of the title (adapted from a list of 13 rules that appears in the *CSE Style Manual,* pages 457-58, and a similarly extensive list of recommendations, with examples of phrasings to avoid and to employ, in the *AMA Manual of Style,* pages 8-11):

- Being too general can make a paper sound like an entire book and discourage investigation by a researcher seeking specific information and data. Find the middle ground between the general and the specific in fashioning a title. A paper on the effects of a specific drug or therapy, or on a specific experimental analysis of a precise natural phenomenon should make specific reference to those details in the title.
- The title is not the appropriate place for literary flourishes or obscure cultural references. Aside from discouraging readers who are seeking hard information, such titles will make finding the paper through the various search engines and indexing servers that index articles difficult.
- In general, authors should be aware of the practices and limitations of search engines when they title a paper. Abbreviations are often not recognized by such programs; the same is true of acronyms, special symbols, chemical symbols, neologisms (especially when not widely shared or known).
- Couching a title for a scientific paper as a question may be an effective rhetorical device, but it does not serve many of the major interests of readers and indexers, and thus not of the author. Titles with questions indicate an opinion piece or an editorial and will be so categorized by the referencing services.
- A **subtitle** is a title that comes after a colon in the title. The function of the subtitle is to clarify, qualify, and otherwise provide the reader of information that will be important in evaluating the information presented. For example, if the paper reports on a randomized control trial, that information may be included in the subtitle, as might the length or extent of the study (if relevant), its location (again, if relevant), or the name of the group conducting the study, if it is widely known.

   (Many journals have come to prefer leaving such information out of a paper's title or subtitle, believing it to be unduly prejudicial. A study should stand on its own work and not on the reputation of work done by others connected to the study or group, regardless of its standing in the scientific community.)

b. Author statement (byline). For major papers and articles, the author's name appears right under the title; for letters and other ancillary pieces, the author's name generally appears at the end of the piece (much as a signature). The author's full name should appear on the title page (including "Sr.," "Jr." "II," etc.) in the manner in which the author would like to be listed in the indexing of the article. (To avoid confusion, it is best to decide how one is to be listed in indexes, as that is going to unify all one's professional work for many years.)

Journals have varying policies regarding the order of names for authors from countries or cultures where the family name is given first. Some will list the name as provided and consider the right-most name as the author's family name and index it accordingly; others will accept direction from an author that the family name comes first in his or her culture and should be indexed under that name. It is best to inquire on the policy of the paper and to act accordingly. (*See* the *Chicago Manual of Style,* 15th edition, pages 778–782, for more on the presentation of names from foreign cultures.)

All persons listed as authors should qualify as authors of the work, which means each person listed contributed the following to the work:

- Each person listed contributed significantly to the design, conception, and research or observations presented in the work.
- Each person contributed significantly to the drafting of the text and reviewed it for intellectual as well as stylistic content.
- Each person had final approval of the work and gave the work that approval before submission.

Individuals who provided substantial assistance to the work or to the drafting of the paper may be acknowledged (even profusely so) in the appropriate section (*see* below under "Acknowledgments"), but because authorship is not simply a matter of vanity, but entails responsibility for the work—its precision, its legitimacy, its interpretation—rigor is called for in assigning whom to credit, blame, or query regarding any piece of original scientific research. (The old practice of simply adding names of department heads and uninvolved individuals to the byline of a scientific paper is today considered a gross violation of the tenets of ethics that should guide the conduct and dissemination of scientific research.)

c. Author affiliation, degrees, and site of research. There is wide variation of practice here as well: some journals will list the author's affiliation immediately following the name below the title; others will provide this information in a footnote on the title page indicated by a footnote symbol instead of a number; yet others will include this footnote as the first of a paper's numbered footnotes and place them either on the title page or at the beginning of endnotes if that is the journal's practice.

In some journals, degrees are listed—in medical journals, the highest degrees and certification (such as Board certification) is listed with the author's name; in other fields, all an author's degrees are listed (in some with the name; in others in the affiliation footnote); and in yet others, none are listed (but there may be mention of the academic position an author maintains in the institution with which he or she is affiliated. It is exceedingly rare that a journal will permit the inclusion of honorifc titles—"Professor," "Doctor," "Admiral," etc.—with an author's name.) As with all other elements of the submission, authors should consult the guidelines set forth by the publication to which they are submitting scientific papers, and adhere to those guidelines.

    d. Abstract. The abstract of a paper—a summary of the paper designed to give researchers an indication of its content and a means of determining if it is pertinent to them and their work—also appears on the title page and is often called the second most read portion of a scientific paper (after the title). Abstracts play an increasingly important role in modern scientific research and information technology, especially with the growing practice of publishing material strictly electronically (making casually browsing through a journal a thing of the past).

Fashioning an effective abstract may be considered an art akin to composing a haiku—there is much to accomplish in very little space. Journals will each have very specific instructions on the length of abstracts that will be accepted—generally from 50 to 250 words—and other guidelines regarding the content and structure of the abstract. In most journals, an abstract may not contain paragraphs, subheadings, bold type, abbreviations, non-standard or unapproved acronyms, references, footnotes, undefined symbols or chemical symbols (even for elements or isotopes), or proprietary or trademarked names. All these embellishments are to be avoided not simply for the purpose of making the abstract read as clearly as possible, but because these elements will confound the indexing programs that will be applied to the paper. In many cases, the abstract will be archived separated from the paper with a link that will allow readers to retrieve the paper if they are interested.

From this fact we may derive the single most important suggestion that can be given an author regarding the abstract: **An abstract should be capable of standing alone—and on its own.** A consequence of this newfound utility for the abstract is to make a style of abstract once widely used now exceedingly rare. This is the "descriptive" abstract (also referred to as "topical" or "indicative"), in which the author presents the general subject of the paper without relating the precise results or conclusions. These were more summaries *about* the paper than summaries *of* the paper. In the current environment, abstracts need to accomplish more.

The more prevalent form of abstract now appearing is the "informative" abstract, which reports the details of the paper, complete with a summary of the methods, foundations, and conclusions. This can often be a great deal for an abstract to cover given the space limitations imposed by the journals and the indexing services. To assist in this process, two organizations—the National Information Standards Organization (NISO) and its counterpart in Switzerland, the International Standardization Organization (ISO)—have promulgated abstracting tools for authors that assure correct and effective production of abstracts. (*See* **Appendix I** for references to this subject.)

Using a structure developed by R.B. Haynes and his colleagues at McMaster University, the NISO/ISO have instituted the "**structured abstract**," in which each element that ought to appear in an abstract, appears in its own paragraph and under its own heading. While making the abstract space on the title page a cluttered affair in print, the electronic version becomes easier to use, more uniform, and more consistent across studies. (*See* **Appendix I** for references to this work.)

The elements of the NISO/ISO structured abstract may be considered necessary components for all abstracts, whether or not the journal allows for the structured format, and for disciplines other than the medical sciences. These elements are as follows:

- **Context**. A sentence or two offering an explanation of the importance of the paper's subject and the problem under investigation.
- **Objective**. A concise and precise statement of the primary objective of the study. If a hypothesis was being tested, state it.
- **Design**. Describe the essential elements of the experiment, observation, or clinical methodology used.
- **Setting**. Describe those elements of the study's (or experiment's) setting that might be relevant to another researcher.
- **Patients or clinical subjects.** In medical contexts, these would include a careful delineation of the disorders investigated, the eligibility requirements and sociodemographic features of patients, and any other pertinent information on the population treated. In other disciplines, this section might identify and describe the population or phenomena investigated.
- **Intervention(s)**. The methods and duration of any treatments applied—or, in the case of other disciplines, observations, experiments, and analytical tools and methods employed.
- **Outcome measure(s)**. Indicate the range of outcomes that were anticipated and the breadth of the parameters being recorded. Indicate if the hypothesis or expectations were formulated before the study, or developed during the course of the study.

- **Results**. Provide the essential and main outcomes of the study (including error estimates and confidence levels).
- **Conclusion**. What conclusions are directly supported by the data? Include only those conclusions directly supportable; avoid over-generalization or speculation. Indicate how the results may be applied, and what avenue of further research or investigation is suggested by the study and its results.

That is a great deal to pack into an abstract, which is why many journals have increased the word limit for abstracts (to 300 or 350 words), or have formalized the abstract by creating a form for authors to fill out in which space is limited by character count.

e. Acknowledgements. In addition to indications of financial support for research or material support in terms of laboratory facilities or other technical assistance, authors should acknowledge individuals and institutions (libraries, computer facilities, meetings, etc.) that lent material assistance to the substance of the paper. This may appear in a final paragraph of the paper, or may be included in the paper's opening footnote, (where some journals will included author degrees and affiliations. Authors should check the individual journals for its policies in this regard (and be aware that there is much variation on this subject).

f. Bibliographic Reference; Keywords. It is important that the terms under which a paper should be indexed be indicated by the authors; leaving this purely to the discretion of the editors or indexing service runs the risk that important words by which readers may arrive at the paper will be ignored and the paper will be improperly indexed.

Some journals may provide a space for the inclusion of the keywords, either following the abstract or at the very top of the title page. Authors are advised to consult with dictionaries and thesauruses—such as the BIOSIS Search Guide, the MeSH (Medical Subject Headings) guide of the National Library of Medicine (at: http://www.nlm.nih.gov /mesh/MBrowser.html), or similar guides at the websites of AIP (in physics); CASSI (Chemical Abstracts Service Source Index); AMS (mathematics); IAU (astronomy); or MEDLINE Index or Index Medicus (for medicine).

g. Footline. Finally, when the paper is published, the final submission of corrected proof should include at the bottom of the title page the title of the journal, the volume and issue number, the issue date and a copyright notice; some will also include the submission and acceptance dates of the paper, and some will also include the journal's URL.

2.2.1.6 Structure of the Scientific Paper (IMRAD Format)

The so-called IMRAD format for scientific papers is so prevalent in scientific literature today, that it is difficult to imagine there ever being a time when every paper was not structured according to the sequence of Introduction–Method–Result–and–Discussion. Yet, the development of the IMRAD format can be traced in the history of twentieth-century scientific literature and can be seen to have grown from being used in about a third of papers written in the early part of the century to over 70-percent today.

As with other guidelines and formats, the IMRAD structure is designed not as a rigid structure that constricts creativity, but as a means of ensuring that critical elements of a presentation are not inadvertently omitted. With practice, an author may forego strict adherence to this format and still present all the elements required of a scientific paper coherently and effectively.

Authors should also note that IMRAD format is not always appropriate or the best way to present scientific material. Clinical studies, case histories, and opinion pieces are not always given to this format. (Also note that in some guidebooks, the term is expanded and framed as "AIMRAD" to include "Abstract" in the format. While this properly apprises authors of the importance of the abstract in the paper as a whole, it could also be misconstrued as saying that the abstract should be considered part of the paper itself. This could cause problems, and the advice given here is that the abstract should be created independently both in practice and in concept. As previously noted, abstracts must stand on their own, particularly in the current electronic environment.)

We now discuss each element of the organization of the text of a scientific paper in turn.

i. Text subheadings. Under the main headings of IMRAD, authors are advised to think about the subheadings that will be used to alert readers of the subject of the upcoming narrative. Subheadings are most useful when they are clear and concise—authors frequently must exercise restraint in refraining from literary flourishes in subheadings. Such devices (puns, rhetorical questions, clever or "cute" constructions, etc.) may provide a momentary relief, but will in the end detract from the seriousness and import of the points the paper presents.

During the course of writing a paper, the organization, and thus the outline and the headings, will change. This is not an undesirable development; a paper should be continually under review and evaluation by the author, and this is likely to result in reorganization and reformulation of the points made.

Some guidebooks advise that a scientific paper should be organized around the graphic material available—the tables, the images, the graphs, and other visuals, the rational being that this material is difficult to come by and so the organization should take full advantage of all the visual resources available. As practical and as sensible as this advice might be, it introduces the danger that the images and graphics that just happen to be available, instead of the writing and the content, will dictate the presentation. While it is a good idea to be mindful of the graphics available, it seems to us a mistake to allow that to be the chief organizing factor in designing a scientific paper.

ii. Introduction. Like the preface or introduction of a monograph, the Introduction section should be composed at the outset and then revised as the writing progresses, and finalized last. Authors should look upon the Introduction as the defense of the paper—why the subject is important and should be of interest to other researchers in the field, and where the findings or substance of the paper stands in relation to other work done in the field. In formulating the Introduction, authors should consider the following items as appropriate to this section:

a. The nature, scope, and significance of the paper's subject. This should be direct and specific, as opposed to vague and general. The title of the paper should not simply be repeated; neither should any evaluation of the work as "important" or "pathbreaking" be suggested. The clearer and more concise the Introduction is, the better.

b. A review of the relevant work in the field with specific reference to how that work left certain questions unanswered or unsatisfactorily explored (and which formed the motivation of the current research). This section should not contain a comprehensive survey of the field as a whole, nor should it contain digressions into irrelevant arcane areas of past research. Readers are interested in knowing if the author is aware of the key work being done in the field and this can usually be accomplished without voluminous bibliographic references.

c. The specific purpose of the research undertaken (as an individual project, or as part of a general research program). This allows the author to place the paper and the research in the context of his or her own previous research, and provides readers with references to that work.

d. The specific point to be explored in this paper. Introductions often concentrate on the first two items and offer scant material on these last four. Authors should recall, however, that there are various reasons another scientific researcher may consult a particular paper, ranging from keeping up with the latest thinking in a field somewhat related to the reader's area of research to learning of the current work in a field of daily and vital interest to the reader. These readers are likely to read portions of the paper with varying degrees of scrutiny, so that an Introduction that allows a reader to gain a comprehensive view of the work's importance, context, methods, and conclusions will be appreciated by a wider readership, and in all likelihood, will be more widely read.

e. The justification for using a specific research method, or a particular experimental or theoretical approach, design, or set of tools. This will also allow a reader to evaluate early if the detailed portions of the paper are relevant to the particular research the reader is doing.

f. The primary results of the research and the conclusions those results suggest. Authors have to overcome the great temptation of keeping the results of a paper undeclared until later, and especially of refraining from "letting the cat out of the bag" at the very beginning of the paper. This once again ignores the reality of how papers are accessed, read, and incorporated into the research lives of practicing scientists. A reader will (gratefully) come back to a paper that does not force the reader to reread it in its entirety in order to obtain its results and its essential points, but which provides this information in the Introduction.

These results should be presented with due attention to the implications implied by the tenses used. If the paper describes a new methodology, the present tense should be used ("A new techniques *is* described..."—not "will be described"). Results of a particular experiment should indicate that a result *was* found or *has been found* to be the case—not *is* the case, which implies the finding is universally accepted as fact.

iii. Method: Experimental details or theoretical basis. The ability of researchers working in the same area as the one being explored in the paper to repeat experiments and make the same observations is critical to the scientific system of testing and validation. This requires a detailed presentation of the materials used, the way in which they were used, and the detailed analysis of how the equipment was controlled—and ultimately, how the readings were made, recorded, and analyzed.

Following are some elements to be aware of in formulating this section of any scientific paper:

a. Be aware of the journal requirements. Many journals will have precise requirements for materials used and methods, plus requirements regarding contextual matters (ethical oversight; environmental impact; informed consent; etc.) and regarding analysis (statistical analysis; recording verification; quality control; integrity management, i.e., assuring the security of data; etc.) that will be a part of any paper published. Being familiar with these requirements is essential before submitting any paper to any such journal.

b. For methodologies previously published (by the author or by others), or for widely accepted methods, a citation will suffice, but such citations must be precise and accurate. Professional researchers who consider the results important will, as a matter of course, retrace those steps in coming to a final evaluation of the work.

c. Be prepared to elaborate on or revise this area in response to peer reviews. Often, elements of an experiment that the researcher takes for granted or overlooks is pointed out by a reviewer who is coming at this material from a perspective that is not entirely "fresh" (otherwise the reviewer would not be an appropriate individual to conduct a peer review), yet from an individual perspective unburdened by the presuppositions to which people are often prone. Authors are advised to regard these criticisms as valuable contributions to the intellectual underpinnings of a line of research (even if at first the critique seems trivial).

Authors should also be prepared to bring in additional technical support in the way of engineering or statistical analysis when deemed necessary. Such analysis is a specialty area in scientific research within each discipline and may well be deemed an additional resource required by editors and reviewers in light of specifics of the area of investigation.

iv. Results. Though one may rightly consider the "Results" section of a paper its heart and its essence, it may also be the shortest section *if* the other arts of the paper are well constructed. This is true first because once the context and the methodology have been properly presented, the only question remaining is, were the expectations of the experiments met and was what was expected to be observed actually observed? It is also true because in this section, the author may use visuals—graphs, tables, photographs, etc.—to elucidate the results or, in some instances, to present them as a straightforward observation.

Journals are well aware that such material can also obscure the findings of an experiment with clutter and the appearance of thoroughness. As a result, journals will often have clear guidelines on the use and preparation of visual materials, to which authors must adhere if they intend to have the journal publish their paper. (*See* the next section, **Section 2.2.1.7** for more regarding this subject.)

Several points to consider in formulating this section of a paper are:

a. This is NOT the place for drawing conclusions. The results should be the actual measurements and findings of the research or the results of any computations or mathematical derivations that are part of the objective of the paper. Any implications and conclusions the author wishes to draw are not properly part of this section; they should be contained in the following sections.

b. It is even more important for this section to be concise and direct. The critical questions that a reviewer will ask is, does all the data collected support the result asserted by the author? This may be a more important question than whether the conclusions drawn by the author are supported by these results; researchers in science are well aware that interpretation and analysis of data is a somewhat subjective and highly personal enterprise. Different people see the same set of data differently and draw their own conclusions. The creation of a consensus of interpretation, however, depends on the reliability of the data, and this is measured in the confidence a reader may have in the Results section, which is why so much care is lavished on this section by journal editors.

c. The Internet acts as a repository of supplementary support material that is beyond the practical capability of print. As a result, there is no reason why the full picture of a research experience should not be available to all interested colleagues through electronic means. As noted earlier, there may be proprietary issues that will cause a researcher to hesitate divulging everything regarding an experiment or any area of research; adjudicating this matter may have to become the responsibility of the editor, the publisher, or the journal's sponsoring society. But absent such considerations, in the modern environment, researchers should expect to present their work in as complete a form as possible for review, evaluation, and ultimately the enlightenment of colleagues in the field.

v. Discussion. This section calls upon the author to draw conclusions from the results and to address the questions that were raised in the Introduction of the paper. Partly because this kind of material calls for reflection by the author in synthesizing the results and coming to some conclusion (a somewhat creative exercise), and partly because the real work in drawing conclusions from data will be undertaken by the discipline as a whole as different researchers and theoreticians draw their own conclusions from the results, authors are given more latitude in this section. In spite of this (or perhaps because of it), reviewers are most likely to ask for the most cutting in this section. Unwarranted conclusions in this section is a frequent reason why papers are rejected, even after revision.

For these reasons, authors should be especially careful to consider the following questions in framing the Results section of any paper:

a. Do the results and the conclusions answer the questions and address the problems posed at the outset in the Introduction? This is the first order of business in this section of the paper, and should be attended to with great care and precision. A paper with a discussion that does not succeed in answering (or at least addressing) the question posed in the Introduction is likely to be returned for revision no matter how remarkable or careful the data and its collection may be.

b. Do the conclusions follow from the result? In most cases, experimental results—even hard mathematical equations—will provide support for several conclusions and an author serves his or her own (favored) conclusions best by acknowledging this.

c. How do the results or conclusions compare with those of other researchers? Are there differences in methodology that could account for the differences in results or conclusions? These questions place the work described in the paper in the context of the current discussion in the discipline of the issue at hand. Readers familiar with the literature and with work in the field will consider such questions, so it stands to reason (according to some people's thinking) that an author does service to one's own research by offering an opinion at the outset.

d. Are there unresolved issues that should be addressed? Are there predictions that may be derived from the findings that may clarify discrepancies or uncertainties? This will not only provide grist for further discussion of the paper, but will be seized upon by other researchers as an invitation to further the research and extend the field in the direction marked by the author and the research reported in the paper. For such is the reality of the scientific enterprise: a finding must not only be enlightening in itself, it must point to further work in experimentation or in theorizing that will broaden the field of knowledge.

vi. Conclusion. A paper need not have a separate section entitled "conclusion"—this is part of the Discussion section and is often so labeled. The conclusion is the author's best opportunity to place the findings of the research in the context of the current state of the field. In this section, authors may be permitted greater use of the first person as they discuss conclusions they have drawn based on their experience and research.

vii. Summary. An abbreviated version of the paper—longer than an abstract; shorter than the paper—is also asked for by some journals and is made available online.

Goldbort ends his discussion of "The Features of a Scientific Journal Article" (a discussion which contains many illuminating examples—*see* **Appendix I**) with the following sage advice: "Simplicity and economy in expressing the meaning of one's scientific labors commands more authorial power and reader attention than verbosity and eloquence."

viii. Footnotes and endnotes. The guidelines for most journals today ask authors to (a) refrain from using the "insert footnote" feature of the word-processing programs in use; instead indicate footnotes with brackets or parentheses (depending on the journal); and (b) to place all notes in a separate file, sequentially numbered and consonant with the numbering of the text. The journal will either place these notes at the bottom of the page or at the end of the paper, depending on the house style.

ix. References. *See* **Chapter 4** for guidelines on citations and references.

126

2.2.1.7 Graphics and Other Material

Different people integrate information in different ways and that results in some people being better able to receive information as text, while others do better with images, and still others (lecturers are quick to note) do best receiving information aurally. (Some even do better with the menu-driven programmed medium of the computer.) This feature of human learning, coupled with the preponderance of images and figures as a routine feature of scientific writing, leads to the inclusion of visual elements—tables, graphs, diagrams, maps, photographs, schemata, drawings, etc.—in the panoply of tools an author may use to convey research, observations, theory, or hypotheses.

Unfortunately, a consistent and universally accepted set of values and methods for creating and presenting images in scientific literature has yet to be devised, witness the fact that some guides list as a reference for this subject (and in some cases, the *only* reference) the *Pocket Pal: A Graphic Arts Production Handbook,* the small (but amazingly useful) booklet distributed by printers for over 50 years (and now in its 18th edition). Even recent style guides cite sources 20 years old or older.

This is not to say some important work has not been done in this area: The volumes produced by Yale professor Edward R. Tufte on the visual display of scientific and quantitative information (largely self-published over the course of the last 25 years) are as beautiful and inspiring as they are instructive and innovative.

In the guidebooks, the limitations of black-and-white printing (which Tufte ignores with a vengeance) make a full presentation of criteria and methods for the presentation of visual information in science difficult. The standard guidebooks have over many editions presented some material on this subject, but the only one that we recommend are the chapters in the current (3rd) edition of the *ACS Style Guide,* authored by Betsy Kulamer—one chapter on Figures; another on Tables. The following is adapted from those chapters (with some material from other guides and author instructions, and with ideas developed during three decades of creating tables, not the least of which the 100-plus tables created for this volume), which, until our own projected "Manual of Scientific Graphics" is published, will have to do.

i. Tables. Authors are advised by virtually all journals to use tables only if the text cannot convey the information adequately, either because there is so much data, or because a relationship between the values in the table are compared in the text for the purpose of demonstrating a point. This approach gives the impression that tables are less than ideal as a form of communication, which is an outdated idea. Used effectively, a table, like any other graphic, can clarify and enhance the text and make a point instantly clear to the reader. (The denigration of figures and tables by previous guides reflects cost factors, which publishers sought to limit. Improved technology has resulted in a reduction in costs for producing and publishing graphics of all kinds, making them lest problematic. A well designed graphic, by virtue of its explanatory power, can actually reduce costs.)

Below are two tables that describe the parts of a table—they are rough in the sense that tables come in many varieties and some thought must be given to design. The word-processing software, though capable of creating tables, is not friendly to the process.

Table 2.1 The Parts of a Table

	Spanner heading			
	Subspanner heading 1		Subspanner heading 2	
Stub heading	**Column heading 1**	**Column heading 2**	**Column heading 3**	**Column heading 4**
Main stub heading 1				
Stub subheading				
Row heading	value 1	value 2	value 3	value 4
	value 5	value 6	value 7	value 8

Both tables on this page were created with the "Table" function of MS Word: Table 2.1 uses the grid lines and shading functions; Table 2.2 suppresses those functions and is designed to appear as a table created strictly with word processing and without the Table function.

Table 2.2 The Parts of a Table[a]

	Spanner heading[b]			
	Subspanner heading 1		Subspanner heading 2	
Stub heading	**Column heading 1**	**Column heading 2**	**Column heading 3[c]**	**Column heading 4**
Main stub heading 1				
Stub subheading				
Row heading	value 1	value 2	value 3	value 4
	value 5[d]	value 6	value 7	value 8

a. Footnote crediting source of data
b. Footnote explaining spanner head
c. Footnote explaining column heading or giving units
d. Footnote qualifying data in cell

The "Spanners" are header titles that cover more than a single column (in ungridded tables, a "straddle rule" beneath these titles indicates the extent of their applicability); the columns are each headed with a "column heading"; and the "stub" of a table is the left-most column that contains the headings for the rows.

Journals have a wide variety of policies regarding table design and submissions, so it is best to consult each individual journal's author instructions. Following are guidelines for creating effective tables:

a. Size. Tables of fewer than 4 rows may not be the best way of conveying the information contained in it. Consider conveying the information in text. Tables with six columns will generally fit in a single-column of a journal page; 13 columns will fit on a double-column spread. Books generally allow for 8-column tables. Wider tables may be rotated and displayed lengthwise. (If they are rotated, and if they require more than one page, place the header on the extreme left and orient the continuing table pages the same way.

b. Title. Begin each table with the word "Table" followed by the table number—numbered sequentially through an article, book, or chapter. Following the number, place the title (in either headline or sentence style, without periods). The title should be brief and explain the table without requiring reference to the text. Tables should be referred to in the text, where they are explained.

c. Units. The units used are denoted either in the column headings, or in footnotes, or in a separate line that appears above the "data field," which is the collection of values presented by the table.

d. Headings. These should be succinct (2-3 lines maximum) and appropriate to the width of the data in the cells beneath. No value should appear in the data field without clear connection to a row heading to the left and a column heading above.

e. Abbreviations. Use standard abbreviations and define non-standard abbreviations in footnotes. The abbreviation "N/A" (also as "n/a") is used to mean "not available"—sometimes an ellipsis may be used for this purpose—or "ND" for "not determined." Do not use ditto marks in a table; it is often confused with other symbols.

f. Footnotes. Use letters for footnotes—superscript numbers can be confused for exponents or other symbols.

ii. Figures, images, and visual presentations. At some point in the publishing process, an image is going to be scanned and converted into an electronic format (if it is not already in such a format when submitted), so that a fundamental question that will be asked is, of what quality or "resolution" is the image electronically? The question actually has two components: what is the dpi (dots-per-inch) of the file, and what is the quality of the scan? Electronic files can be saved into several formats, some (like TIFF) provide a high quality even at lower dpi and are preferred for print; others (like JPEG) degrade with each copy and are often unacceptable even at high dpi. Others (like GIF) are preset at a low dpi and are suitable mainly for web-based publishing and not print.

2.2.1.8 Special Sections

Articles may have a need for a special section that collects material used throughout and provides readers convenient access to this information. Special sections are placed at the end of articles; sometimes (particularly if they are comparatively large) they are placed on the on line service link of the journal. Among these sections are the following:

i. Methods and materials. Where detailed information about apparatus, design, or analysis is required (or where the information has already been presented elsewhere), a special section may be appropriate, either in print or through a link to the online service site for the journal.

ii. Abbreviations. Either non-standard abbreviations (especially if used often) or abbreviation particular to a field allied to the main area of the journal (e.g., aviation-related abbreviations in a mathematics journal).

iii. List of mathematical notation and symbols. If there is any possibility that the notation will be obscure or unrecognized by the target readership of a paper, a list of notation and symbols is advisable.

iv. Glossary. Use of specialized or proprietary designations can be clarified and properly attributed in a glossary. (Indicating in a glossary that a term is trademarked allows it to be used without further comment.)

v. Acknowledgments. Aside from funding support or an institution providing facilities or aegis for a research project, material support of the work may be acknowledged in detail in a separate section.

vi. Appendices, addenda, and supplementary material (WEOs). Special illustrative material—color images; animations and video; rotatable figures; audio; lab output; spectra; specimen slides; radiography; particle tracks; astronomical images; etc.—can be made available (in quantity) as "web-enhanced objects" archived on the journal's website and referenced in both the print and electronic editions.

Chapter 3. Elements of Style and Usage

Contents

Contents–3.4 Italics, *continued*

Contents–6. Punctuation, *continued*

Contents, *continued*

Contents, *continued*

Contents, *continued*

(More on usage conventions in mathematics may be found in **Chapter 6** *and in* **Appendix A**.*)*

Chapter 3. General Elements of Style

3.1 Spelling

3.1.1 Consistency

Many words can be correctly spelled in more than one way. In such cases, make sure the same variants or spelling conventions are applied consistently throughout a work, from the cover to the index. Failure to be consistent casts doubt on the reliability of the entire text.

3.1.2 American vs. British Spelling

One way to maintain consistency in spelling is to adopt either American or British spelling. Generally, British spelling reflects the origins of the English language, with etymological remnants from Greek, Latin, and French that Americans have simplified into more phonetically consistent forms. Some general characteristics that distinguish them are presented below, but there are exceptions for all of them, and there are dozens of miscellaneous differences that defy categorization. In case of doubt, a reliable dictionary will identify both American and non-American variants.

3.1.3 Plurals

3.1.3.1 General Principles

i. Adding *s*. Most English nouns can be made plural by adding *s*.

atom → atoms
globe → globes

ii. Adding *es*. If a word ends in a letter that cannot be pronounced easily and clearly by adding just the *s* (such as *s, x, z, sh* and *ch*), it requires *es*.

peach → peaches
waltz → waltzes
abscess → abscesses

Table 3.1 American and British Spelling Compared

Spelling Difference	American Preference	British Preference
–or vs. *–our*	color favor behavior	colour favour behaviour
e vs. *oe* *e* vs. *ae*	fetus leukemia cesium	foetus leukaemia caesium
–yze vs. *–yse* *–ize* vs. *–ise*	paralyze civilized organization realize	paralyse civilised organisation realise
–log vs. *–logue*	dialog catalog	dialogue catalogue
–ll vs. *–l*	distill enrollment skillful	distil enrolment skilful
–er vs. *–re*	liter caliber center	litre calibre centre
silent *e*	judgment sizable	judgement sizeable

iii. Ending in *f* or *fe*. Some words ending in *f* or *fe* change to end in *v* and *es* when they become plural.

wife → wives
leaf → leaves

iv. Ending in *o*. For words ending in an *o* that follows a vowel, add just an *s*.

video → videos
imbroglio → imbroglios

If the *o* follows a consonant, look up the word in a dictionary, as there is no reliable and simple rule in this case.

v. Ending in *y*. When a word ends in a *y* following a consonant (or *qu*), change the *y* to *i* and add *es*. If the *y* follows a vowel, simply add *s*.

frequency → frequencies
medley → medleys

3.1.3.2 Nouns and Compound Terms

i. Proper nouns. Most proper nouns are made plural according to the above rules for common nouns, with the exception that proper nouns ending in *y* usually take just an *s*.

a Rolex (watch) → two Rolexes
Canterbury → Canterburys
Kennedy → (the) Kennedys

ii. Irregular forms. Some nouns have irregular plural forms, and can change vowels or endings to become plural. Some words do not change at all. A few words have different plural forms with separate meanings. These words should be looked up in a dictionary if there is any doubt about their spelling or usage. In scientific writing, be especially careful with Latin- and Greek-derived words that end in *-um*, *-us*, *-on*, *-is*, and *-itis*.

goose → geese
offspring → offspring
stratum → strata
virus → viruses

iii. Plurals of compound words and terms. For hyphenated or open compound nouns, make the main noun plural, even if it is not the final word in the term.

> mother-in-law → mothers-in-law
> sergeant first class → sergeants first class
> hanger on → hangers on

If the compound term is a single word, or if it does not contain a noun, treat the whole term like a normal, non-compound noun.

> bookmaker → bookmakers
> merry-go-round → merry-go-rounds

iv. Italicized titles. When making italicized titles plural, the *s* is usually set in roman type, not in italics. If the title is in the plural form already, no *s* is needed.

> both *A Midsummer Night's Dream*s
> ten thousand *A Farewell to Arms*

v. Non-English words. Plurals of italicized foreign words should be set entirely in italics.

> *Hund* → *Hunde*
> *paloma* → *palomas*

See **Sections 3.4** and **3.5** for treatment of italicized and foreign text, respectively.

3.1.3.3 Letters, Numbers, Noun Coinages, and Abbreviations

i. Letters. Uppercase letters become plural by adding an *s* with no apostrophe; but lowercase letters require an apostrophe to distinguish them from small words.

> W → Ws
> i → i's (and not "is")

ii. Numbers. Numerals become plural with just an *s*; they do not require apostrophes.

> twelve 747s
> the 1860s

iii. Noun coinages. Noun coinages, used as nouns but made up of other parts of speech, become plural by adding *s* or *es*, with apostrophes or spelling changes if necessary.

> how-to → how-tos
> ups and downs
> yes's and no's
> haves and have-nots

iv. Abbreviations. Abbreviations may be made plural with just an *s*, but sometimes with an apostrophe, especially after abbreviations with more than one period or those with mixed uppercase and lowercase letters. An exception to the last rule is *PhDs*.

> Sorry, no c.o.d.'s.
> two KotOR's

Abbreviations for units of measurement should not change when plural.

> 1.0 μg/mL
> 25 μg/mL

3.2 Possessives

3.2.1 General Principles

The possessive case is used to express ownership, association, or agency. If following the rules in this section results in an awkward spelling or construction, the possessive case can often be bypassed with the word *of*.

To form the possessive of a singular noun, add an apostrophe and an *s*. For plural nouns ending in *s*, add just an apostrophe, but for plural nouns that do not end in *s*, both the apostrophe and the *s* are needed.

a monkey's uncle
the boss's office
the troops' morale
the mice's cheese

A separate, simpler system is to leave out the additional *s* on all words already ending in *s*, whether they are singular or plural, common or proper; but this method is counterintuitive to pronunciation.

the bus' tires
Chuck Jones' cartoons

3.2.2 Singular Nouns Ending in *s*

i. Plural in form, singular in meaning. Nouns that are plural in form but singular in meaning should be made possessive with the addition of an apostrophe only, even when used in the plural.

astrophysics' most complicated theorems
those species' genetic history

ii. Words ending in unpronounced *s*. For words that end in a silent *s*, it is optional to omit the possessive *s* and use just an apostrophe.

the Peace Corps' mission statement
Mardi Gras' historical roots

iii. Ending in "*eez*" sound. Possessives of names ending in the sound *eez* are formed with just an apostrophe.

Hercules' labors
Laertes' poisoned sword

142

iv. Of *for...sake* type expressions. When the noun ends in an *s* sound in these expressions, only the apostrophe is added.

for goodness' sake

3.2.3 Specific Terms

i. Proper names. Possessive titles take an apostrophe and an *s*, unless the title ends in a regular plural noun. Titles in quotation marks never take apostrophes; they must be reworded.

Google's page layout
Acme Test Tubes' customer service line
author of "Jabberwocky"

Proper names also take an apostrophe and *s*. If a name ends in *s*, adding just the apostrophe is also correct.

Juan Perón's presidency
Charles Dickens's beard
PZ Myers' career

ii. Italicized words. Italicized terms in the possessive case should be followed by the appropriate ending in roman typeface, not in italics.

Hitchhiker's Guide to the Galaxy's devoted fans

iii. Quoted terms. If a term is within quotation marks, it cannot be made possessive with an apostrophe. It must be reworded.

her "soul mate" ['s] advice →

advice of her "soul mate"
or
her so-called soul mate's advice

iv. Letters and numbers. Letters, abbreviations, and numbers are treated the same as normal words in the possessive case.

the letter *O*'s similarity to the numeral zero
PCP's behavioral effects
1985's hottest band

v. Diseases. Possessive forms of diseases like *Alzheimer's*, *Parkinson's*, and *Tourette's* are often favored in nontechnical contexts, but in science writing, such terms should not be written in the possessive form.

3.2.4 Grammatical Concerns

i. Followed by a gerund. When a noun is followed by a gerund (a verb ending in *-ing*), it may be written in the possessive form. To check if this is appropriate in the context of the sentence, make sure it still makes sense if you replace the noun in question with an appropriate possessive pronoun (*his, her, its, your,* or *their*).

That man's [his] talking and cigar-smoking ruined the movie.
I was not offended by your sisters' [their] falling asleep.

ii. Adjectives and possessives. When a noun phrase is modified by both an adjective and a possessive, the possessive always precedes the adjective.

Bob's discount furniture
women's denim jackets

iii. Two, as unit. When two nouns possess something together, only the second takes the possessive form; but when separate things are being possessed, both nouns require the possessive form.

Adam and Eve's fall from grace
Adam's and Eve's social security numbers

iv. Used attributively. When it is unclear whether a plural noun in front of another noun is indicating possession or is used simply as an adjective (attributively), use an apostrophe except for proper names.

the ventriloquists' convention
Upright Citizens Brigade

v. Compound expressions. Compound expressions (such as titles and honorifics) take a possessive ending on the last term.

the prime minister's pants
the actor-director's next movie
my father-in-law's pipe

vi. Expressions of duration. Genitive expressions use the same form as the possessive to quantify a length of time.

an hour's wage
two weeks' notice

3.2.5 Possessive Pronouns

i. Personal pronouns. Personal pronouns in the possessive case (*my*, *our*, *your*, *his*, *her*, *its*, and *their*) need objects to refer to.

her kazoo
my recollection

But their independent forms (*mine*, *ours*, *yours*, *his*, *hers*, *its*, and *theirs*) can stand alone and be treated as nouns. These never take an apostrophe.

Theirs is missing.
I prefer hers.

When such a pronoun is used after *of*, it is sometimes technically redundant, but it is a fixed rule for pronouns. In the same construction, a proper name can either take the possessive form or not.

a friend of mine
that cat of hers
that cat of Sue's

ii. Indefinite pronouns. Most indefinite pronouns can be made possessive by adding *'s* to the end of the term or by adding it after the adverb *else*.

everyone's shoelaces were untied
nobody else's project was complete

iii. Relative pronouns. Possessive forms of these pronouns are *of whom, whose, of which, of what*, and *of that*.

The defendant, the alibi of whom the prosecution tore down, was convicted of murder.

3.3 Word Division

3.3.1 General Principles

To prevent fully justified lines from being too sparsely or too heavily laden with text, some words will need to be divided between one line and the next. Modern word processors have made this automatic in most cases, but they are not perfect; and for many neologisms and proper nouns, guidelines are required to decide how to separate between and within words in a way that preserves the flow of the text.

i. Resources. Most dictionaries divide polysyllabic headwords into syllables with raised dots: "lu·gu·bri·ous." For a list of chemical names and their division, *see* **Appendix E, Table E2**, in **Part III**, below.

Often, the logic for where to divide a word is based on pronunciation more than on etymology. Always check a dictionary to be certain.

auton-omy
fun-gus

ii. After a vowel. When faced with a choice of where to divide a word of more than two syllables, try to divide after a vowel, unless that vowel is part of a diphthong. If the vowel stands alone as a syllable in the middle of the word, it should not be placed first on a new line.

> egre-gious
> recon-noiter
> edify-ing

iii. Two-letter word endings. A word with a two-letter ending should not be divided immediately before that ending.

> blessed (not "bless-ed")
> ga-loshes (not "galosh-es")

iv. Gerunds. Divide gerunds and present participles immediately before the *"ing"* that forms the gerund, unless it is preceded by a double consonant, in which case, separate the consonants.

> tast-ing
> encroach-ing
> bub-bling

v. Words with prefixes or suffixes. Separate prefixes and suffixes from the rest of words that begin or end with them.

> pre-disposed
> dispos-able

vi. Compound words. Compound words should be divided in between their constituent parts, whether they are hyphenated or part of a single word. Any break in a hyphenated term should occur after the hyphen.

> book-keeping
> stream-lined
> editor-in- / chief

vii. Words not appropriate for division. Some words should never be divided, including words that are only one syllable; words which, when divided, would leave a single letter separated; and words the divided parts of which could be confused for other words or are misleading in pronunciation when split. For example:

scrounged   valets
aplomb   issue

3.3.2 Abbreviations

Numerals followed by abbreviations should not be divided between one line and the next.

15 J/kg
15 s
9:15 AM

3.3.3 Foreign Languages

The same basic rules of word division that apply to English apply to most foreign languages, with a few special considerations.

Latin words should be divided between syllables. When a consonant is surrounded by two vowels, the division should occur immediately before that consonant. *Ch, ph, th, gu, qu, chl, chr, phl, phr, thl, thr, bl, br, cl, cr, dl, dr, gl, gr, pl, pr, tl*, and *tr* are never divided.

3.3.4 Proper Nouns

Dictionaries contain many word entries—some even have indexes—that suggest divisions for proper nouns. For words that do not appear in dictionaries, follow the guidelines for common nouns. If the pronunciation of the word is not evident, divide after a vowel or leave the word unbroken, if possible. For proper nouns containing multiple words, try to separate any that are hyphenated or otherwise conjoined before making divisions within words. Do not let initials or suffixes of names appear immediately after a break.

Cincin-nati
Jo- / Beth Casey
Roger / Meyers Jr. (not "Roger Meyers / Jr.")
W. E. B. / Du Bois (not "W. / E. B. DuBois" or "W. E. B. Du / Bois")

3.3.5 URLs and E-mail Addresses

When dividing a URL or an e-mail address between lines, do not use a hyphen. If there is a hyphen within the URL, do not break around it. Instead, divide after a colon, a slash, a double slash, the @ symbol, or a word, but before a period and any other punctuation marks. If it is absolutely necessary, a URL may be divided between syllables of words, according to the normal rules.

http://www.example.com/worddivision/
overlong_url/example123.html

3.3.6 Run-In Lists

Do not break the line immediately following a letter or number in a run-in list, such as that which starts with (1) or (a).

3.3.7 Typographic Considerations

To help preserve even lines, word breaks can be used even in text with unjustified right margins. No more than three lines in a row should end in divided words.

3.3.8 Numbers and Units

Break numbers only if absolutely necessary, preferably after a comma. Do not break a number after a decimal point or a single digit.

3.1415926535- / 8979323846
154,000,- / 000,000,000

3.3.9 Divisional Marks

Do not confuse the hyphen (-), the short horizontal line that separates words and parts of words, with the en dash (–), a longer mark used to indicate *through*, or with the em dash (—), a still-longer mark used like a comma, colon, or parenthesis in a complete sentence. *See* **Section 3.6.11** for other uses of hyphens and **Section 3.6.8** for the use of dashes.

3.4 Italics

Authors should bear in mind that a page of text that contains many instances of italicized type presents readers with a sense of confusion, and makes the page seem untidy.

For plurals of italicized words, *see* **Section 3.1.3.2, iv–v.**
For possessives of italicized words, *see* **Section 3.2.3, ii.**

3.4.1 For Emphasis

Italic type is used mainly for emphasis or otherwise distinguishing text from its surroundings. Rarely should more than one sentence in a row be italicized for emphasis.

3.4.2 Added to Quotations

Italics may be used to emphasize a specific part of a quotation, provided the author clarifies that the italics were not part of the original text. This may be done in brackets immediately after the italicized text, in parentheses after the entire quote and source, or in parentheses within a relevant foot- or endnote.

> I believe that monopolies … can be prevented under the power of the Congress to "regulate commerce with foreign nations and among the several States" through regulations and requirements *operating directly* upon such commerce, the instrumentalities thereof, and those engaged therein. (Theodore Roosevelt, December 1902; italics added)

3.4.3 Words and Phrases Used as Words

Words and phrases that function as instances of those words, rather than as what they represent, should either be put in quotation marks or italicized. Individual letters or groups of letters should be italicized.

> The word *marathon* has an interesting story behind it.
> I expected a better response than a shrug and a "whatever."
> Not all nouns that end in *us* have plurals ending in *i*.

3.4.4 Foreign Words

Foreign words that readers are unlikely to be familiar with should be italicized, except for proper nouns. If such a word occurs frequently throughout a text, it should only be italicized the first time. Words that appear in an English dictionary should be set in roman type, unless this creates inconsistency or is otherwise confusing. A sentence or more of foreign text should be set in roman type and treated like a normal English quote. *See also* **Section 3.5**.

> *Magischer Realismus*
> *ikkyoryōtoku*
> ipso facto
> schadenfreude
> mano a mano

3.4.5 Word's First Occurrence

One may use italics to draw attention to important terms when they are first used in a specific context, but after this use roman type.

> Two mutually exclusive philosophies of the field are *illusionism* and *supernaturalism*. Proponents of illusionism assert that...

3.4.6 Titles

Names of books, journals, synopses, plays, movies, television shows, pamphlets, works of art, poems, songs, court cases, planes, ships, and spacecraft must be set in italics.

Journal of Comparative Zoology
Roe v. Wade

3.4.7 Reverse Italics

Within italicized type, a term that would normally be italicized should be set in roman type. But for rules on handling titles within other titles, *see* **Section 3.8.3, xv.**

Rethinking thinking in Homo sapiens

3.4.8 Genus and Species Names

Italicize genus and species names of plants and animals, but only if used in the singular form. Other taxonomic classes are not italicized.

Macropus giganteus can travel quickly.
Not all bacilli are rod-shaped.
Most prokaryotes are bacteria.

3.4.9 Endnotes Keyed to Page Numbers

When endnotes arranged by page number refer to specific words or phrases taken from the text, those words or phrases should be in italics.

204 *a pair of shiny boots* For an in-depth look at the history of shoe-shining in this era, see Phineas Pendleton's *The Shoemaker's Bible* (New York: Borington Press, 1961).

3.4.10 Indexes

When cross-referencing within an index, *See* and *See also* should be in italics unless they precede a word or phrase in italics.

fault scarp, 199-200. *See also* erosion
Principia. See *Philosophiæ Naturalis Principia Mathematica*

3.5 Foreign Words

3.5.1 Italics

Use italics for foreign words only if they are appearing for the first time, are not commonly known, and are not proper nouns.

3.5.2 Quotation Marks

Foreign words should not be distinguished by quotation marks unless they make up at least a full sentence, in which case they should be set in roman type and adhere to the normal rules for English quotations.

3.5.3 Familiar

Foreign words and phrases listed in an English dictionary and familiar to English speakers need not be italicized.

3.5.4 First Use

If a foreign word or phrase occurs frequently throughout a text, it should only be italicized the first time.

3.5.5 Plurals

When an italicized foreign word is made plural, its ending should be set in italic type as well, unless it is a foreign title taking an English plural ending. (*See* **Section 3.1.3.2, v.**)

3.5.6 Proper Nouns

Foreign proper nouns are not italicized unless they would need to be italicized in English.

the Champs-Élysées
Velazquez's painting, *El triunfo de Baco*

3.5.7 Documentation

For the title of a foreign work in a bibliographic reference, capitalize only the first word and proper nouns. To insert a translation of the title, do so immediately after the original and use brackets.

Delgado, Marisol. 1992. *Gardel y la historia del tango* [Gardel and the history of tango] Buenos Aires, Argentina: BA Press.

3.5.8 Glossaries

A glossary of foreign words should define the terms on separate lines in alphabetical order. It should come before a bibliography.

3.5.9 Translations

Parentheses or quotation marks may be used to add a translation immediately after a foreign term.

Don Quixote, with his lean figure, lance, and *baciyelmo* (basin-helmet), is one of the most recognizable figures in all of literature.

Parentheses may also be used to add an original foreign term that an English term was translated from. Within a quote or within a translated work, brackets should be used instead. (For more on translations of quotes, *see* **Section 3.10.13**.)

To my chagrin, instead of a ham sandwich (*de jamón*), I inadvertently asked for a soap sandwich (*de jabón*).

3.5.10 Place-Names

Place-names that contain geographical terms in a foreign language do not need that geographical term repeated in English.

Fujiyama (*not* Mt. Fujiyama)

the Rio Grande (*not* the Rio Grande River)

3.5.11 Ligatures

Ligatures, which resemble vowels joined together, should not be used except in Old English or French words in their respective contexts.

aeon	subpoena
oeuvre	hærfæst

3.5.12 Latin Abbreviations

The following is a list of Latin abbreviations that are often found in scholarly works:

ab init.	*ab initio*, from the beginning
ad inf.	*ad infinitum*
ad init.	*ad initium*, at the beginning
ad int.	*ad interim*, in the intervening time
ad lib.	*ad libitum*, at will
ad loc.	*ad locum*, at the place
aet.	*aetatis*, aged
bibl.	*bibliotheca*, library
ca. *or* c.	*circa*, approximately, about
Cantab.	*Cantabrigiensis*, of Cambridge
cet. par.	*ceteris paribus*, other things being equal
con.	*contra*, against
dram. pers.	*dramatis personae*
D.V.	*Deo volente,* God willing
e.g.	*exempli gratia*, for example
et al.	*et alii*, and others
etc.	*et cetera*, and so forth
et seq.	*et sequentes*, and the following
fl.	*floruit*, flourished
f.v.	*folio verso*, on the back of the page
ibid.	*ibidem*, in the same place
id.	*idem*, the same
i.e.	*id est*, that is
inf.	*infra*, below
infra dig.	*infra dignitatem*, undignified
in pr.	*in principio*, in the beginning
loc. cit.	*loco citato*, in the place cited
loq.	*loquitur*, he or she speaks

Latin abbreviations, *continued*

m.m.	*mutatis mutandis*, necessary changes being made
MS (*pl.* MSS)	*manuscriptum* (pl. *manuscripta*), manuscript
n.	*natus*, born
NB, n.b.	*nota bene*, take careful note
non obs.	*non obstante*, notwithstanding
non seq.	*non sequitur*, it does not follow
ob.	*obit*, died
op. cit.	*opere citato*, in the work cited
Oxon.	*Oxoniensis*, of Oxford
PPS	*post postscriptum*, a later postscript
pro tem.	*pro tempore*, for the time being
prox.	*proximo*, next month
PS	*postscriptum*, postscript
QED	*quod erat demonstrandum*, which was to be demonstrated
q.v.	*quod vide*, which see
R.	*rex*, king; *regina*, queen
RIP	*requiescat in pace*, may he or she rest in peace
s.a.	*sine anno*, without year; *sub anno*, under the year
sc.	*scilicet*, namely; *sculpsit*, carved by
s.d.	*sine die*, without setting a day for reconvening
sec.	*secundum*, according to
s.l.	*sine loco*, without place
sup.	*supra*, above
s.v.	*sub verbo*, *sub voce*, under the word
ult.	*ultimatus*, ultimate, last; *ultimo*, last month
ut sup.	*ut supra*, as above
v.	*vide*, see
v.i.	*verbum intransitivum*, intransitive verb
viz.	*videlicet*, namely
v.t.	*verbum transitivum*, transitive verb

3.6 Punctuation

A. Intrasentence Marks

3.6.1 Period

i. Ending a sentence. Periods end sentences that are statements or commands, or, less formally, pieces of sentences.

> This is a self-referential, declarative sentence.
> Do it yourself. Or don't. I don't care.

When such a sentence ends in an abbreviation, only one period is needed, but commas may follow abbreviations with periods.

> Capitol Hill is in Washington D.C.
> Sammy Davis, Jr., of Rat Pack fame, was born in 1925.

ii. Single (not double) space after. After a period at the end of a sentence, leave only one space before beginning the next sentence. (Though commonly practiced in typescript, such double-spaces will have to be changed to single spaces when the manuscript goes to composition.)

iii. With abbreviations. Periods should be placed after abbreviations that consist of lowercase letters, but not after those that consist of uppercase letters, unless they are part of a name. Many scientific abbreviations, such as units of measurement, do not require periods (*see* **Chapter 7, Section 7.2.5**).

> a.m. PO Box
> BBC 30.4 mm

iv. In display lines. Display lines, such as titles of chapters or headings of sections, should not end in a period, except for titles that introduce content on the same line.

v. In double or multiple numeration. Double numeration separates sections and subsections, numbering each separately within its parent section. Periods may be used to divide the numbers representing these elements.

> 7 Minerals
>> 7.1 Types of Rocks
>>> 7.1.1 Igneous
>>> 7.1.2 Sedimentary
>>> 7.1.3 Metamorphic
>> 7.2 Crystal Structure

vi. In indexes. Periods should not follow index entries, except before cross-references.

> egg cell. See *ovum*

vii. In outline style. In numbered lists, a period comes after the number and before a single space and a capital letter. Periods should only be used at the ends of listed items if those items are complete sentences. Lines after the first should line up with the first letter of an item, not with their numbers.

> How to make brownies:
> 1. Combine butter, chocolate, sugar, eggs, vanilla, and flour.
> 2. Pour the mixture into a greased baking pan and bake for twenty minutes.
> 3. Let the confection cool and slice it into brownies.

If a vertical numbered list is introduced without punctuation and has elements that are separated by semicolons, a period may come at the end of the last line.

> Walters said he would agree to the deal, provided that
> 1. all changes to the constitution be submitted for his approval before being made;
> 2. derivative share requirements be indexed according to cost and percentage analysis;
> 3. "Wacky Tie Days" be limited to one instance per annum.

viii. In references. Periods separate elements of reference list or bibliography entries.

> Spunkmeyer, O. 1954. On the orbital structure of elementary
> particles. *Particle Physics Quarterly* 31:90-112.

ix. In URLs and e-mail addresses. URLs and e-mail addresses can be followed by periods or other punctuation, if necessary, but there should never be a space within the item itself.

> Here is what I learned at www.example.net: the URL
> "www.example.net" is not available for registration.

x. Periods as decimal points. Periods are used as decimal points in much of the English-speaking world. (For commas as decimal points, *see* **Section 3.6.4, xviii.**)

> 3.14159265358979323846...

xi. In publishing history on copyright page. If publishing history items are listed without line breaks, use periods to separate them, but do not end the final item with a period.

> First Edition published 1999.   Second Edition 2005

xii. Block quotations and. Complete sentences that introduce a block quotation may end in either a period or a colon, but always use a colon when the quote is introduced by *thus* or *the following*.

> FDR was a man who believed in the value of hard work, as he made
> clear during his famous inaugural address.
>
>> Happiness lies not in the mere possession of money; it lies in the joy of
>> achievement, in the thrill of creative effort. The joy and moral
>> stimulation of work no longer must be forgotten in the mad chase of
>> evanescent profits. These dark days will be worth all they cost us if
>> they teach us that our true destiny is not to be ministered unto but to
>> minister to ourselves and to our fellow men.

xiii. Permissible change of. Periods in between foreign titles and subtitles may be changed to colons when written in English.

xiv. With quotation marks. Periods should come before closing quotation marks, single and double, except in very specific instances in which the exact punctuation of quoted material is critical and the construction unavoidable.

> Philip's favorite cereal is called "Admiral Crunch."
> She clarified, "I was referring to Edgar Allan Poe's poem, 'The Raven.'"

xv. With parentheses and brackets. Material within parentheses or brackets should take a period on the inside of the closing mark only if it is a complete sentence that is separate from other sentences. When only the last part of a sentence is in parentheses, only one period is needed, belonging after the closing mark.

> Thomas frowned when he read the label. (He was allergic to peanuts.)

> The fig tree provides nourishment for hundreds of animals every time it bears fruit (several times randomly throughout the year).

xvi. Ellipses. Ellipses are made up of three or four periods with spaces in between them. They serve primarily to indicate where text has been left out of a quotation. Use three dots when omitting a word or phrase and four dots when omitting an entire sentence or more. Spaces should always come before and after a three-dot ellipsis, and after a four-dot ellipses. A word starting a sentence after an ellipsis should be capitalized, regardless of whether or not it was capitalized in the original text. (Brackets around the capitalized letter may be used to indicate such a change.)

> Four score and seven years ago our fathers brought forth on this continent, a new nation, conceived in Liberty, and dedicated to the proposition that all men are created equal....

> ... [W]e can not consecrate—we can not hallow—this ground.... It is rather for us to be here dedicated to the great task remaining before us ... that government of the people, by the people, for the people,   shall   not perish from the earth.

Ellipses can also be used to indicate hesitating or broken-up speech.

"You ... you can't be serious ... can you?"

For dashes used to indicate interrupted speech, *see* **Section 3.6.8**. For ellipses in mathematical constructions, *see* **Part II, Chapter 6, Section 6.2.2.2**.

3.6.2 Question Mark

i. Ending a sentence. Question marks are used for asking questions or for expressing uncertainty or confusion.

Why would they come to my concert just to boo me?
Who can say?

ii. Omission for courtesy question. Some polite commands are formed as questions, but these do not need question marks.

Will the owner of a blue Nissan Sentra please come to the front.

iii. Omission for indirect question. Do not place question marks after indirect questions.

How it could be turned into energy was a question for another era.

She asked if there would be enough time.

iv. In run-in quotations. Normally, periods are not needed after sentences that end in quotations; the final punctuation of the quoted material is enough. But with a sourced quote that ends in a question mark (or exclamation point), there needs to be a period placed after the source.

Dr. Jones asked no one in particular, "Snakes ... Why did it have to be snakes?" (*Raiders of the Lost Ark*, 121).

3.6.2.1 Question Mark with Other Punctuation

i. With comma and period. A question mark should never be placed next to a comma, nor should it be next to a period unless it is part of an abbreviation.

> By the time I had turned to ask "What was that?" it was too late.

> Juliet asked rhetorically, "What's in a name?"

> Did you hear about the scandal at Henderson & Co.?

ii. With exclamation points and other question marks. Question marks should not generally be doubled for emphasis or paired with exclamation points. Exclamations phrased as questions should usually take exclamation points, not question marks.

> How dare you speak to me that way!
> Who cares!

iii. With parentheses, brackets, and dashes. Like a period, a question mark should come before the second of a pair of parentheses or brackets only when it applies to what is within them.

> I counted eleven olives (only eleven left?) and four cocktail umbrellas.

> Who would do such a thing on purpose (besides a sociopath)?

They can also come before em dashes, and before en dashes when used to represent uncertainty in a date or figure.

> I didn't make the bed—why bother?—but I did manage to get some vacuuming done.

> The playwright George M. Mangostein (1702?–1777) was quite prolific.

iv. With quotation marks. Unlike a period or a comma, a question mark should come before a closing quotation mark only when it applies to what is inside the quotes. For conflicts between punctuation of quoted material and the surrounding sentence, *see* **Section 3.6.7, xxiv.**

The question I really wanted to ask was "Are you for real?"

What could he do but nod and say "Have a good evening"?

3.6.3 Exclamation Point

i. Ending a sentence. Exclamation points that end a sentence or phrase express strong emotion, loud noise, or irony. Their use in most narrative forms is discouraged; in scientific writing (outside of technical conventions, where an exclamation point is used as a symbol), it is to be used rarely and only when extremely necessary.

I don't believe it!

She promised him her future son—a hefty fee, indeed!

ii. With other punctuation. Exclamation points should appear within quotation marks, parentheses, brackets, and dashes only when they apply directly to what is inside those marks.

"Alack!" isn't the kind of exclamation you hear a lot these days.

He wasn't the most eloquent conversationalist—far from it!—but people were always interested in what he had to say.

Do not place an exclamation point next to a comma, a question mark, or another exclamation point. If an exclamation is also a question, choose which mark is more appropriate and use only that one.

iii. As parenthetical note of surprise or exceptional material. An exclamation point is sometimes used in parentheses ("(!)") to indicate that a remark just made is exceptional and deserving of special notice. This is a highly colloquial usage; stating directly in additional text or in a footnote that the point is noteworthy is much preferred.

3.6.4 Comma

i. With introductory words and phrases. Use a comma to set off an adverbial or participial phrase that introduces a sentence, but if the phrase is short and would produce no confusion without a comma, the comma may be omitted.

> Rather taken aback, she did not immediately find the words to reply.
> Glancing around, Clark was relieved that no one had seen him.
> In a pinch you can use masking tape instead.

If the phrase modifies a verb, it should not be separated with a comma.

> Into the saucepan went an entire stick of butter.

Words like *yes* and *no* can be followed by a comma to indicate a pause.

> All right, let's do it.
> No, no thank you.
> Yes yes, but do hurry up!

ii. With interjections and descriptive phrases. For minor breaks in a sentence such as interjections, commas may be used instead of parentheses or em dashes.

> The soup pot, alas, was empty.
> The "contest" was, at best, a misleading advertisement and, at worst, a dodgy scam.
> Now it was only a matter of asking for permission, but, of course, it wouldn't be that simple.

iii. Compound sentences and compound predicates. Compound sentences have two or more independent clauses, parts which could theoretically stand as their own sentences. A comma is called for when these elements are joined by a conjunction. If a sentence is short enough, however, the comma may be omitted.

> I'll prepare a 13.5 M solution of HCl for you, but you should be really careful with it.

Compound predicates occur when one subject performs two or more actions in the same sentence. The predicates will be separated by a conjunction, but they do not require a comma in between them unless they form a series of more than two actions, as in the second example of rule **iv.**, following.)

The hamster pressed the correct lever and was rewarded with a yogurt drop.

iv. Serial comma. The serial comma, or Oxford comma, is what comes before the conjunction of the last item in a series. It should be included in all cases except in company names where an ampersand is used instead of *and*.

The most important elements are hydrogen, carbon, and oxygen.

He packed his equipment, mounted his camel, and rode east into the desert.

Marley, Steinberg & Tate

v. With adjectives preceding noun. When two or more adjectives precede a noun, a comma or commas may be used in place of conjunctions.

On the table sat an old, dusty book.

Care is required for adjective–noun pairs that take on a specific meaning as a unit. Adjectives modifying these phrases should not be followed by a comma.

Any doctor would have noticed the obvious red flags. (No comma after "obvious")

vi. Conjunctions instead of. Conjunctions may be used in place of commas within a series.

Any small object can be used—a paperclip or a pushpin or a penny.

vii. Semicolon instead of a comma. Normally, commas are used to separate series of words or phrases, but when those words or phrases contain commas themselves, semicolons are needed.

> When in Europe, she visited Bath, England; Nice, France; and Bologna, Italy.

viii. In dialogue and direct address. Use a comma to separate words and names being directly addressed from the rest of a sentence.

> Harvey, come take a look at this.

> That, ladies and gentlemen, is the end of our show.

> Quiet, you!

ix. Introducing quotations. Use a comma along with quotation marks to separate quotes from the rest of a sentence, unless they are already introduced by *that* or follow a similar construction. For quotations after *thus* or *the following*, a colon is preferred.

> "Excuse me," she said, "Is this yours?"

> I believe it was Ben Franklin who quipped that "in this world nothing can be said to be certain, except death and taxes."

> Why bother, then, to "take arms against a sea of troubles"?

x. Homonyms separated by commas. When two words are used in a row that are spelled the same but have different purposes within the sentence, you may separate them with a comma if there is a minor pause when the sentence is spoken out loud.

> What is, is.

> It is good to have someone you can count on on short notice.

xi. In indexes. In indexes, commas are used to separate headings from page numbers, as well as to reverse names and add information.

clog-dancing, traditional Welsh, 296
clothespins
    one-piece, 153–154
    spring-powered, 155, 158

xii. Restrictive clauses and phrases. A restrictive phrase tells something about a noun or noun phrase and lets us know that no other type of thing is being talked about—only that to which the phrase applies. It is not usually set off by commas. A nonrestrictive phrase adds grammatically optional information about something; it may be removed without altering the fundamental meaning of the sentence, and must be set off by commas. A nonrestrictive phrase might comfortably be enclosed in parentheses or dashes instead of set off by commas, while a restrictive phrase would seem awkward or change the meaning of the sentence.

Chips made with sunflower oil cost more money than regular chips.
These chips, made with sunflower oil, are a tasty snack.

Restrictive and nonrestrictive clauses work the same way, but are introduced by relative pronouns and contain verbs.

The stapler that works is missing; all I can find are broken ones.
The stapler, which works fine on all types of paper, is on my desk.

xiii. Dangling or misplaced modifiers. Check to see that words separated by commas have a logical grammatical connection to the rest of the sentence. In science writing, where many of the sentences are written in the passive voice (omitting an explicit subject), this type of error often goes unnoticed.

Being careful not to contaminate the instruments, 12.04 mL of saline solution were collected.

In the above example, nothing within the sentence is doing the *being careful* in the introductory phrase. A suggested rephrasing might be:

A 12.04 mL sample of saline solution was collected carefully, so as not to contaminate the instruments.

xiv. Appositives. Appositives are phrases that come after a noun and refer to the same thing in different words, especially in order to clarify or add information. When such a phrase is optional, adding only extra information to the noun it modifies, it should be placed in between commas.

> "Kingda Ka," the tallest and fastest roller coaster in the world, is in New Jersey.

> Willie Mays, the baseball player, was called "The Say Hey Kid."

When the phrase in apposition is essential to the meaning of the sentence (for example, restricting something there is more than one of with something specific), commas should not be placed around it.

> *Dracula* the book is better than *Dracula* the movie.

> The baseball player Willie Mays was called "The Say Hey Kid."

xv. Parenthetical notation. A comma should never come before a closing parenthesis or bracket, but may come after, as context demands.

> You may also call my cell phone (555-5555), but only for the direst of emergencies.

xvi. Multiple adjectives. Avoid cumbersome chains of adjectives—or nouns used as adjectives—that might come up in technical writing. Instead, use abbreviations (*see* **3.6.9, vii**), leave out implied words that have already been stated, or rephrase to keep the adjectives from all appearing in a row. For punctuating adjectives before a noun, *see* **Sections 3.6.4., v.,** and **3.6.11, v.**

> *Unwieldy:*
> sensitive high resolution ion microprobe spectrometer

> *Alternatives:*
> SHRIMP (*use an abbreviation after the first use*)
> The spectrometer... (*omit the qualifying descriptions*)
> ion microprobe spectrometer of high resolution (*rephrase being careful not to lose precision*)

xvii. Elliptical constructions. Occasionally, repeated words may be left out of a sentence and replaced with a comma. If the sentence is easily understood without the comma, it too may be left out.

In September there were twelve plants left; by November, only three.

His eyes were brown, hers a brilliant blue.

xviii. Commas for decimal points. Much of Europe uses a comma for a decimal point in numbers, but in the U.S. system the comma is reserved for separating three-digit groups to the left of the decimal point. When expressing SI units, one must use spaces instead, and they extend to the right of the decimal point as well.

299,792,458.1415
452.004 142 58

xix. Addresses. Streets named after numbers should be written out in letters (unless the number is over one hundred), but address numbers themselves, which come before a street name, should remain numerals. Commas may be used between separate elements of an address, as on an envelope, along with wherever else the grammar of the sentence demands, but they should not be used within numbers.

Thirty-fourth Street

34 Forest Dr.

Stop by my office in Riverville, at 10201 Oakvale Avenue, if you want to speak with me in person.

xx. Dates. Dates in a sentence that give the month, the day, and the year (in that order) are written with two commas—one after the day and one after the year. Commas are not necessary for other systems. For more about dates, *see* **Section 3.8.9**.

December 19, 1924, was a most unusual day.

The August 2001 exhibition in Munich was especially successful.

The report was released on 8 June 2007.

xxi. Abbreviated inclusive numbers. When abbreviating the second of a pair of numbers separated by an en dash, keep together any groups of digits separated by commas. *See also* **Section 3.12.2, iv–v.**

4,276–279
101,000,664–665

xxii. With that is, namely, and similar phrases. Use a comma after phrases like *that is* or *namely*, and before *or* when it offers another word for something.

Traditional ceremonies occur whenever there is a solstice or an equinox (i.e., every three months).

His reputation for card tricks—that is, for bungling them up—preceded him.

The baselard, or Holbein dagger, was popular in the sixteenth century.

xxiii. With questions. Directly quoted questions do not always need quotation marks; a comma alone can introduce them. The question may start with a capital letter if it is long or contains commas or other punctuation itself. Indirect questions do not require commas.

Why, he wondered, would she need to do that?

The issue is simply, Can the proposal be submitted by Friday or will it require two, three, or even four weeks of additional preparation time?

xxiv. With em dash. The only time a comma should be used after an em dash is when a speaker is interrupted.

"That's just what—," he began, but Denise's urgent glance stopped him short.

If you're ever in the city again—or even if you just want to talk—give me a ring.

xxv. In run-in lists. When constructing a run-in list (a list that does not start a new line for each item), place a comma after each item. If there are already commas within the items, use semicolons to separate them.

Buy these items at the general store: 1) a bushel of onions, 2) a sharp hatchet, and 3) a bottle of laudanum.

xxvi. With *et al.*. When the Latin abbreviation *et al.* is used after a single name, it does not need to be separated by a comma, but after two or more names it may either take a comma or not.

In a fascinating study from 2005, Higgins et al. discuss the results of exactly such an experiment.

xxvii. With *etc.* and *and so forth*. The abbreviation *etc.,* along with similar phrases, should be enclosed in commas.

He dropped the tackle box, and hundreds of hooks, lines, lures, etc., scattered across the floor.

xxviii. And table alignment. Numbers that are arranged vertically within a table that do not contain decimal points but do contain commas should be lined up by the commas. *See* **Chapter 2, Section 2.2.1.6**.

xxix. In titles of works. Commas that have been removed from the end of a title's line should be replaced when that title is referred to. Double titles with *or* in between them take a mandatory comma before the *or* and an optional one after. A comma should also be inserted before *and other stories* or a similar such phrase.

15. Smith, Bonnie, *The Moon in June is a Big Balloon, and Other Silly Rhymes* (New York: Button Press, 1984), 8.

xxx. Unnecessary and undesirable uses. A comma should not appear next to a question mark or an exclamation point. In cases where it would seem to need to, the comma must be omitted, or the sentence must be rephrased.

When I watch *Jeopardy!* I know all the answers.

As with other grammar rules, the rules for using a comma should not be followed blindly. The ultimate goal of punctuation is to make text easier to read and understand, so if use of a comma that lies outside the rule accomplishes this, it may still be the better choice.

xxxi. Commas and quotation marks. *See* below, **Section 3.6.7, xxiv**.

3.6.5 Colon

i. After preposition or verb. Colons are used in sentences to signal a coming example or elaboration of what has been said. They should not usually be used to introduce a series after a preposition or a verb. After phrases like *that is* or *for example*, a comma is preferred over a colon (*see* **Section 3.6.4, xxii**).

ii. With *as follows* and similar constructions. Use a colon after phrases like *as follows*, *thus*, and *the following*.

> The school principal declared that he would not tolerate the following activities: lollygagging, buffoonery, and dawdling.

iii. Spacing after. Leave only one space after colons in running text, except in specialized technical uses, such as when separating parts of chromosomes or molecules. (*See* **Part II, Chapter 12, Section 12.2.1.4**.)

iv. Lowercase or capital letter after. When a colon is used in a sentence to introduce two or more sentences, or when it is used to introduce dialogue after a speaker, the word after it needs to be capitalized. Otherwise, the word may remain lowercased.

> The outlook was grim: If he withdrew his ships, the enemy would have time to recover or even bring in reinforcements. But if he ordered a full assault, he could be caught in the middle of two opposing armies.

> Suddenly everything was overcooking at once: the potatoes were turning black, the pasta was boiling over, and the soufflé had no chance to survive.

v. Quotation marks and. Introduce long or formal quotations with a colon instead of a comma.

> On July 7, 2002, the defendant stated: "I have no recollection of the event whatsoever."

vi. For subtitles. On title pages, where the font size and/or style of the title might differ from those of the subtitle, there does not need to be a colon in between them. Otherwise, there should be a colon, and at least the first letter of the subtitle should always be capitalized.

> *Spill the Beans: The Sordid Truth about the Bean-Packing Industry*

vii. Introducing statements and lists. Formal salutations in correspondence are followed by colons.

> Dear Sir or Madam:

viii. In URLs. Colons are used in URLs.

> http://www.merriam-webster.com

ix. After a speaker. Use a colon to introduce speech after a speaker's name.

> COSTELLO: When you pay off the first baseman every month, who gets the money?
>
> ABBOTT: Every dollar of it! Why not, the man's entitled...

x. For subsections. Sections and subsections may be separated with a colon.

> 2 Birds
>   2:1 Flightless birds
>   2:1:1 Penguins
>   2:1:2 Ostriches

xi. In cross-references. When an index entry directs a reader to a specific subentry of another entry, a colon may be used. A colon also separates a main entry from subentries that come directly after it in a run-in index.

> theramin. *See* Theremin, Léon: theremin
> cigar rolling: history of, 44; techniques, 45

xii. Before block quotations. Either periods or colons may be used to introduce block quotations after complete sentences, but if a sentence ends in *thus* or *the following* or a similar phrase and is then followed by the quotation, then a colon is preferred.

xiii. In vertical lists. Colons should be used to introduce a vertical list, unless the items of the list directly complete the sentence.

> An applicant must excel in at least three of these six character attributes:
>> strength
>> dexterity
>> constitution
>> intelligence
>> wisdom
>> charisma

xiv. With other punctuation. Colons, unlike periods and commas, always come after closing parentheses, brackets, and quotation marks.

> The results were twofold (or so they would have you believe): Model A-13 would become obsolescent, and consumer spending would increase.

> Here's what I think of your "epiphany": I think you need to lay off the sauce.

3.6.6 Semicolon

i. Separating clauses. In a sentence, a semicolon often separates two independent clauses; that is, two groups of logically connected words that could technically stand as sentences on their own. If a comma and a conjunction in place of the semicolon does not result in the sentence not making sense, then a semicolon is very likely correct. (In the first example below, replacing the semicolon with " , because" does not alter the meaning and correctness of the sentence.)

> Answer the phone already; it's been ringing for half an hour.

> There was no point in trying to fix the toaster; the machine was beyond repair.

ii. Before an adverb. When a sentence of two independent clauses is joined by an adverb like *thus, then, therefore, however*, or *indeed*, a semicolon should come before that adverb.

> The Ice Hotel is made almost entirely of ice; therefore, hairdryers are prohibited.

> We all hate mosquitoes when they choose us as hosts; however, they may teach us a lot about diseases and the immune system.

iii. With *that is, namely*, and similar phrases. Before words and phrases like *that is* or *namely*, a semicolon can be used.

> It was difficult to find someone to write on spec; that is, without any guarantee of being paid.

iv. Before a conjunction. Normally a comma is used to separate two independent clauses joined by a conjunction; but a semicolon may be preferred if one of the clauses is long or already has a comma itself, especially if the sentence cannot easily be split into two sentences.

> Usually on the weekends we would go to the park to play games, or, if we were feeling lazy, to nap on the grass; but that Saturday I had very important work to do (to the extent that a nine-year-old can be said to have very important work to do).

v. In run-in lists. Use semicolons instead of commas to separate items in a series that have commas in them already, or those that are otherwise complicated or long.

> The only chores he had left to do were milking the cows, which wouldn't take more than a few minutes; feeding the chickens, sheep, and other livestock, including his aunt's plump, prize-winning sow; and gathering firewood.

vi. In a vertical list punctuated as a sentence. Lists that complete a sentence yet are still arranged vertically, with each item starting a new line, should take semicolons at the end of each item, except for the last, which should end in a period.

> Tell the producer that I will not accept the project without
> 1. full creative control, including over casting decisions;
> 2. the monetary compensation I quoted at our first meeting;
> 3. parts for all of my friends and people to whom I owe favors.

vii. In indexes. Semicolons separate subentries in run-in indexes. If the main entry has its own locator numbers in addition to subentries, there should also be a semicolon after those locator numbers.

> carnies, 34–40; slang of, 66–72; famous carnies, 95–99; prejudice against, 111-113

Cross-references should also be separated by semicolons.

> pancakes: types of, 12; toppings, 13. *See also* crêpes; tortillas

viii. Two subtitles. When a book or article has two subtitles, a colon should come before the first one and a semicolon before the second.

ix. Separating multiple citations. When multiple sources are used to support one thing, the note should separate the references using semicolons. Also, parenthetical citations that refer to two or more works should separate them with semicolons.

> (Parker et al. 1996; Lloyd and Greene 2003)

x. With other punctuation. Semicolons come after closing parentheses, brackets, and quotation marks.

> I don't smoke (anymore); it's a revolting habit.
> Morgan said no to "Canon"; she thinks it's too clichéd for a
> wedding.

3.6.7 Quotation Marks

i. Words and phrases used as words. To distinguish words that directly represent their meanings from those used as instances of words, consistently employ either italic type, or—especially if the word occurs in the context of being spoken out loud—quotation marks.

> *Mississippi* is a word that is often misspelled.

> Mississippi is a state whose name is often misspelled.

> Some people use "Mississippi" to pace themselves when counting
> seconds.

ii. Plurals of words in. An apostrophe and an *s* can be inserted within a term in quotation marks to make the term plural, if rewording is not an option. Plural endings should not come after quotation marks.

> Plenty of carefully enunciated "How do you do's" were overheard
> at the fancy party.

iii. Possessives and. If a word or title is in quotation marks, it cannot be made possessive. The sentence must be reworded.

iv. Quoted speech, dialogue, and conversation. Quotation marks appear around direct dialogue, with a new paragraph starting if there is a change of speaker. If a single speaker goes on for more than one paragraph, opening quotation marks should appear at the beginning of each paragraph, and closing quotation marks should appear at the end of only the last paragraph of the spoken words.

v. Beginning of a chapter. When a chapter that starts with a large or decorative initial letter (a "drop cap") begins with a quote, the opening quotation mark may be omitted. Quotation marks may also be omitted in epigraphs, which introduce a chapter or text.

> Everything is theoretically impossible, until it is done.
> —Robert Heinlein

vi. Familiar expressions and maxims. Familiar expressions and maxims do not necessarily require quotation marks unless they are being referred to as sayings themselves.

> Kill two birds with one stone by taking him along with you.

> The bit of folk wisdom "An apple a day keeps the doctor away" isn't about apples; it's about taking daily, personal responsibility for one's health.

vii. Single-word speech. Unless directly quoted, words like *yes* and *no* and interrogative pronouns do not need quotation marks.

> He said yes, but I do not think he was sincere.
> They asked how, so we explained how, in detail.
> "Yes," she said, and then frowned, "But how?"

viii. Unspoken dialogue. Characters' thoughts or other dialogue which is not spoken out loud may be enclosed in quotation marks, set in italics, or left in roman type without quotation marks.

> *Great*, he thought, *another chance to prove myself incompetent.*

> Two more hours of this, she thought glumly, and she would be wanting a drink after all.

ix. Indirect dialogue. Discourse that restates speech indirectly does not require quotation marks.

> Yesterday he was told it would begin at six o'clock, but today he heard it wouldn't begin until seven.

x. With more than one paragraph. If a quote goes on for longer than a paragraph, if should probably be a block quotation; but if it is not, a new set of opening quotation marks must begin each paragraph and only one set of closing quotation marks should appear, at the very end.

xi. "Scare quotes". Quotation marks can be used to identify an unusual or ironic use of a word or term. After *so-called*, which achieves a similar effect, no quotes are needed.

> His shirt was incorrectly buttoned, and his "portfolio" was a pizza box full of old newspapers.

> The so-called miracle of modern science was not even approved by the Food and Drug Administration.

xii. Permissible changes. When quoting something that itself has quotation marks, those original quotation marks may be changed from double to single or vice versa.

xiii. Placement with citations in text. Citations enclosed in paren-theses after quotes should occur immediately after the closing quotation mark, before any other punctuation.

> Harker himself did not expect divine retribution: "I have never so firmly disbelieved in anything as much as life after death" (*Topics* 167)—despite his being raised as a Christian.

If the quote occurs at the end of a sentence, a single period should be used after the closing parenthesis of the citation. If the quote itself ends in a question mark or exclamation point, there also needs to be a period after the closing parenthesis of the citation.

> Indeed, Yorick had no answers to give for the rhetorical "Where be your gibes now? your gambols? your songs?" (5.1.253).

Block quotations do not require quotation marks, and quotes within them take double quotation marks in the same way that regular text does.

xiv. For quotations marks within quoted titles. Double quotation marks may be changed to single quotation marks within an article title. In a reference list, where article titles are not usually put in quotation marks, internal quotes should remain.

> "Classical Allusions in Campbell's 'Letter to a Patriarchy'"

xv. For translations. Double quotes can be used to add a translation after a foreign word or term.

> To my chagrin, instead of asking for a *sandwich de jamón*, "ham sandwich," I inadvertently asked for a *sandwich de jabón*, "soap sandwich."

xvi. Quotes within brackets within quotes. When quoted material occurs within brackets inside of a quote, it should be within double quotes, not single.

> "Later on when I competed against Hattie [Harriet Fieldings, the "Quahog Queen"], she told me I had real potential."

xvii. Computer writing. Quotation marks should not be used to distinguish elements related to computing, such as files, commands, or text to be typed. Italics, boldface type, or another (serif) font should be used instead. Be careful to maintain consistency.

> Under the **File** menu, click **Save As**.

> Type `format c:` and hit enter.

xviii. Reference lists. Quotation marks are not needed around article or chapter titles in reference lists.

> Kowalczyk, Peter. 1991. Defense mechanisms of *Hirundo rustica*. *Fictional Journal of Ornithology* 13:175–188.

xix. Slang or coined terms. Slang or made-up words should be enclosed in quotation marks the first time they are used. Afterwards, they may appear without quotes. (But most new words formed with prefixes do not even require hyphens, let alone quotation marks.)

xx. Double and single. When quotation marks are called for in text that is already within quotation marks, single quotes are used. Within those, double quotes are used again, and, beyond that, single and double quotes alternate as needed.

> Charlie spoke, "We had a chat about pronoun case after Sam asked me, 'Is it "this is him" or "this is he"?'"

A British convention for quotes within quotes is to alternate the same way between double and single marks, but to start with single quotes.

xxi. Smart quotes. When using a word processor, make sure quotation marks (both double and single) are angled to face the text that they enclose. (These are known to copy editors as "smart quotes.")

> "Notable Quotations" → "Notable Quotations"
> "What's that?" → "What's that?"

xxii. Single, for definitions. Single quotes are often used in linguistic and phonetic texts to clarify the meanings of words.

xxiii. Alternative to. European writers often use em dashes or guillemets (<< >>) instead of quotation marks to represent dialogue or other quoted material.

xxiv. Other punctuation with quotation marks. Closing quotation marks must always come after periods and commas (in spite of the apparent illogic).

> The article presented, called "Endosymbiosis of *Mixotricha parado-xa* and *Mastotermes darwiniensis*," was well received.

> Henry sat down at the harpsichord and played "Greensleeves."

Other punctuation marks, including question marks, exclamation points, colons, semicolons, dashes, and ellipses, go where they are logically needed in relation to the closing quotes: if they apply to what is within the quotation or title, they go before the quotation mark; if they apply to the surrounding text, they go after.

> Why do I get the feeling you've regretted this marriage from the moment you said "I do"?

> My main concern (and what he called a "minor problem") was that he did not have any legal permission whatsoever.

There should never be two terminal punctuation marks (periods, commas, question marks, or exclamation points) in a row, even if they would be interrupted by a closing quotation mark. If there is a conflict, punctuation that is part of a title should be chosen over the punctuation of a sentence, and question marks and exclamation points should be chosen over periods.

> What is the name of that rock song that begins, "Hey! Ho! Let's go!"

> Never, ever ask "How old are you"!

These rules apply even if there is more than one closing quotation mark, as in a quote within a quote. A period or comma must come before all of them, and any other punctuation mark must go exactly where it logically belongs.

> "I've never read 'The Road Not Taken,'" she admitted.

> Charlie spoke, "We had a chat about pronoun case after Sam asked me, 'Is it "this is him" or "this is he"?'"

An alternative (mainly British) system reserves the inside of the closing quotation mark expressly for punctuation that applies to what is quoted, even for periods and commas.

For commas or colons introducing quotes, see **Sections 3.6.4, ix**, and **3.6.5, vii**, above.

3.6.7.1 Quotation Marks for Titles

i. Articles. Except in some reference lists, article titles should be in quotation marks. If there are double quotation marks in the title, they may be changed to single quotation marks.

> "The Rubicon of Expression: The capitalist paradigm of narrative in Gibson's 'The Genre of Reality'" by F. Wilhelm Pickett

ii. Conference names. Conference names take quotation marks.

> The conference "Nanotechnology in Medicine and Biotechnology" was held in September 2008.

iii. Musical works. Operas are italicized, but all other titles or albums appear in quotation marks.

> *The Barber of Seville*

> "Light My Fire" by the Doors

iv. Photographs. Use quotation marks to refer to the titles of photographs.

> "Greenscape," a photograph of a Siena alleyway

v. Poems. With the exception of epic poems, the titles of most poems should be enclosed in quotation marks, whether they have official titles or are named after their first lines.

> "Do not go gentle into that good night" by Dylan Thomas
> Homer's *Odyssey*

vi. Television or radio episodes. Titles of television shows and radio shows should be set in italics, but their individual episode titles must be in quotation marks.

> His favorite episode of *Seinfeld* is "The Puffy Shirt."

vii. Unpublished materials. These must be put in quotation marks, not italics.

"Reference Library Requiem," an unpublished nonfiction book

3.6.8 Dashes

3.6.8.1 Em Dash

i. Definition. Em dashes (—) are horizontal lines that are about the width of the letter *m*.

ii. For explaining and amplifying. As markers for explanatory words or phrases, em dashes perform many of the duties of commas, colons, and parentheses. They can be used alone or in pairs.

iii. Appropriate use. For purposes of clarity, no more than one pair of em dashes should be used within a single sentence. Parentheses, commas, or colons are often appropriate substitutions.

iv. Colon replacement. Since the em dash can serve the same function as the colon, it may sometimes replace it, especially when a colon might seem too formal.

But here's the thing—it doesn't run on gas or even electricity; it's powered completely by solar energy.

v. With comma. Phrases used appositively (*see* **Section 3.6.4, xiv**) may be set off by em dashes instead of by commas. This can be useful when there are already other commas complicating the phrase in question, such as in a series.

The three different types of volcanic eruptions—Plinian, Hawaiian, and Strombolian—cause different kinds of damage to their surrounding environments.

vi. With other punctuation. Within a sentence, question marks, exclamation points, and closing parentheses or brackets may appear before an em dash, but commas, colons, and semicolons may not. Periods may do so only if they are part of abbreviations.

vii. Subject separated from pronoun by. Another use for the em dash is in the grammatical construction where a sentence is introduced by the naming of a subject or a series of subjects, followed by a pronoun and the main predicate. The em dash inserts the pause required after the subject.

> Beets—they're one of nature's sweetest foods, and right now they're more popular than ever.

> Gourmet meals, leather seats, fancy video screens—are these really worth the price of a first-class ticket?

viii. In tables. Em dashes may be used in tables to fill in a location where there is no data. They should be centered and used consistently. *See* **Chapter 2, Section 2.2.1.6**.

ix. With *that is* or *for example*. Along with commas and semicolons, em dashes or pairs of em dashes may be used with phrases like *that is* or *for example*.

> He called to reschedule because Audrey—that is, his car—was in the shop.

> Older participants chose more traditional ice cream flavors—e.g., vanilla, chocolate chip, butter pecan.

x. In titles of works. If a title has an em dash in it, what comes immediately after it does not necessarily need to be capitalized, as a subtitle (separated by a colon) would.

> *Man—the Deadliest Game*

When two chapter titles have the same subject, roman numerals are used to distinguish them, and em dashes may separate the titles from those numerals.

Chapter 3: Mathematics of Dark Energy—I
Chapter 4: Mathematics of Dark Energy—II

xi. In indexes. Subentries in an indented index that have multiple sub-subentries should be introduced with em dashes.

xii. Sudden breaks and abrupt changes. Em dashes can designate a sudden interruption, especially in speech.

"What the—ah! Get off me!"

"It's completely unbreakable, even if you twist—oops."

xiii. Double and triple em dash. Double em dashes are used to replace words or parts of words, such as confidential names, profanity, or un-readable text from a source document. Spaces should only come before or after a double em dash if what it replaces is at the beginning or end of a word.

Melissa D——e

Dr. E—— and Mr. A——'s correspondence

The triple em dash is reserved for bibliographic entries in which the same author or authors as previously named are responsible for another cited work.

Wynchman, Hubert. *Things to Do with String*. Chicago: Norseman Press, 2002.

———. *More Things to Do with String*. Chicago: Norseman Press, 2004.

xiv. In manuscript preparation. If your word processor cannot insert an em dash directly into a manuscript, type two hyphens in a row (with-out spaces) to indicate an em dash, four to indicate a double em dash, and six to indicate a triple em dash.

3.6.8.2 En Dash

i. Definition of. En dashes (–) are horizontal lines that are about the width of the letter n.

ii. To connect numbers. En dashes are most often used to connect numbers in order to indicate *up to and including*. It is incorrect to use the words *from* or *between* before numbers paired with an en dash. For more on en dashes that connect numbers, *see* **Section 3.12.3.4**.

Trials 11–20 were completed with a delay of 3–4 seconds between each application.

Anywhere from 15 to 30 percent is normal for a healthy adult.

iii. In compounds. En dashes can be used to reduce ambiguity in compound terms, especially those that consist of two or more hyphenated terms, or of a noncompound term and a compound term.

post-apocalyptic–science fiction movies

toppled Humpty Dumpty–like off the wall

In science writing, the en dash can act like the words *and* or *to* between two nouns:

reduction–oxidation reaction

RNA–DNA helicases

iv. With nothing following. En dashes that are missing a second number signify something that continues to the present, such as someone's life.

Cornelius Westfield Jr. (1933–)

v. In indexes. In indexes, en dashes indicate the range of the page or section numbers.

chess, 259, 270–285

vi. Versus minus sign. Be careful to differentiate between the minus sign and the en dash, which, though similar, are separate characters. *See* **Appendix A, Section 2** for tables of mathematical symbols.

vii. Slash to replace. Slashes can replace en dashes in dates, especially to signify the end of one year and the beginning of another.

He spent the 1996/97 term in intense study.

3.6.9 Parentheses

i. Characteristics of. Within a sentence, parentheses are used to express supplementary information, especially when that information is only indirectly connected to the surrounding text.

When I checked to see if the kitchen was swept (as Albert had assured me it had been), I found it in dusty disarray.

The world's largest producers of tungsten (China, followed distantly by Russia, Austria, and Portugal) export the element in powdered or solid bar form.

Although the precipitant collected from the first reaction was greater than that collected from the second (4.186 g versus 4.030 g), the results were consistent with the stoichiometric analysis (*see* Table 2).

ii. Font of. Parentheses should be in the same typeface as their surrounding text, not as the material they contain.

The earliest watches were large and unwieldy, but, like much technology, they soon became smaller and more efficient. (For more information, see Templeton's *Encyclopedia of Pocket Watches.*)

iii. More than one sentence in. If a pair of parentheses contains more than one sentence itself, it should not interrupt the main text in the middle of a sentence.

The idea that bees violate scientific laws in order to fly, for example, has been around since the 1930s. (This myth is based on an oversimplified mathematical analysis that treats a bee's wings like those of an airplane. Scientists have long known that subtler effects, such as non-linear motion and vortices, are at work in such systems.)

iv. Note number and placement. Note numbers should come after a closing parenthesis.

In fact, he would later briefly run a successful restaurant (but not until well after the war).[3]

v. With cross-references. Cross references after subentries should be enclosed in parentheses.

surfaces: boundaryless, 54–58; non-orientable, 60–61, 65; (*see also* Möbius strip)

vi. For glosses or translations. Use parentheses to translate or define terms in midsentence, but brackets if adding such material within a quote. *See* **Sections 3.5.9** and **3.6.10, ii,** for examples.

vii. For abbreviations. If a technical term is to be abbreviated throughout a work or a section, it should be spelled out the first time it is used, and followed by the abbreviation in parentheses.

A comparison was made of long-chain unsaturated alkenones (LCUAs) and their saturated counterparts.

viii. Subgenus. Subgenus names, when inserted between the genus and species names in a taxonomic title, should be in parentheses.

Cypraea (Cypraea) tigris

ix. In indexes. Clarifying information for index entries belongs in parentheses.

Madonna (mother of Jesus). *See* Mary
Madonna (entertainer), 124–126

x. With Italics Added. *See* **Section 3.4.2**.

xi. In lists. Numbers and letters that introduce items in a list may be enclosed in parentheses, or may each precede a single closing parenthesis.

A few things made him suspicious about the request: (1) it was unprovoked, (2) it contained an unusually casual tone for a legal passage, and (3) there was no option to deny the application or close the window.

To do:
1) laundry
2) return books
3) deposit checks

xii. In telephone numbers. Parentheses sometimes enclose the area code of a telephone number.

(555) 555-0255

xiii. For potential plurals. When speaking about something that may be either singular or plural, it is acceptable to add the plural ending in parentheses to indicate this.

The study should be double-blinded, to eliminate potential influence on the subject by the experimenter(s).

xiv. In captions. In the captions of a picture, diagram, or table, patients or cases may be referred to in parentheses.

Fig. 4.2. Bilateral nephrocalcinosis (patient 13)

xv. Two sets of. Do not put two parenthetical remarks side by side in two separate sets of parentheses. Occasionally different information may be in the same set of parentheses, separated by a semicolon.

"I stopped by the laboratory to check on the samples before *switching off the lights* and heading out for lunch." (Kenneth Brotalk, 1996; italics mine)

3.6.9.1 Parentheses with Other Punctuation.

i. Brackets. For parenthetical material already enclosed in parentheses, use square brackets (*see* **3.6.10, iii**).

ii. Comma. Parentheses are useful alternatives to commas when the parenthetical material contains commas itself.

The Three Wise Men (Caspar, Melchior, and Balthazar) are portrayed in most Nativity scenes.

Commas should never come before parentheses except in run-in lists (*see* **Secion 3.6.4, xxiv**), but they are allowed to come after closing parentheses, as is the case in this sentence.

iii. Periods, question marks, and exclamation points. Periods, question marks, and exclamation points may all come either inside or outside of a closing parenthesis, depending on whether they apply to only the text within or to the whole of the surrounding text. *See* **Section 3.6.7, xxiv**, for examples.

iv. Em dashes, colons, and semicolons. These are punctuation marks that require text to follow them, so there is no reason they should need to come immediately before a closing parenthesis. When they apply to the outside text, however, they may certainly appear after the closing parenthesis.

> This was what he had dreamed of since he had first become interested in science (in the winter of 1904): a chance to conduct his own research in a professional laboratory.

3.6.10 Brackets

3.6.10.1 Square Brackets

i. Appropriate use. Square brackets are mainly used to add information or replace words within text that are not the original author's.

> According to Pete Peterson, the team "never got the recognition [they] deserved, even after winning the [1964] championship."

ii. With Quotations. Brackets can be used in direct quotes to change the tense or pronoun case of the speaker.

> The discovery was shocking to Sanders, who described it as "beyond [her] most ludicrous imaginings."

Letters that have been changed from lowercase to uppercase in a quote may be placed within brackets to reflect that change, especially in contexts where perfect accuracy in quoting is important.

> "Yes, I was in charge, but I didn't have a title… [W]e never made it official; not until the meeting in July of 2003."

Expository text that would interrupt the quotation marks of a run-in quote may be inserted within brackets inside a block quotation.

The tides were turning in the world of salty snacks:

> People don't want their fathers' snacks anymore; [Chippos president Chester Cooke wrote] they're responding more and more to bold and gourmet flavors like "sun-dried tomato and basil" and "roasted garlic and goat cheese."

Brackets can be used to insert comments or guesses in a text with missing or illegible words.

> The aspect of alchemy that inter[ests me] most of all is [that which] involves the transmutation of lead into gold.

If clarifying information is inserted into a quote, it should be in brackets, not parentheses.

> "I admit that the subtleties of natural philosophy [i.e., physics] are largely beyond my reckoning."

iii. Brackets within parentheses. Another main use of brackets is to act as parentheses within parentheses.

> Perkins remained president until 1841, when accusations of sedition forced him to resign, despite vigorous denial on his part. (In fact, this was a main component of the organization's policy, even though they didn't [officially] recognize it until 1893.)

iv. Font of. Like parentheses, brackets should be in the same typeface as their surrounding text, not as the material they contain.

> After writing two more books about art museums (*MoMA Moment* [2004] and *Surfing the Met* [2006]), she put her writing career on hold to pursue other interests.

v. In mathematical expressions. Brackets and parentheses have special technical rules of usage in mathematical notation. For the rules regarding proper use of these marks in mathematics, *see* **Chapter 7, Section 7.2.5**. *See also* **Section 3.12**, below, for more on usage rules regarding numbers in text.

vi. For real name or pseudonym. In a reference list or index, an author's real name may appear in brackets after his or her more commonly known pseudonym.

> Twain, Mark [Samuel Clemens] 56–59

vii. For anonymous and known author's name. When the author of a work is not known for certain but is guessed, the name should appear within brackets. If the name is known to be a pseudonym but the real name is unknown, *pseud.* may follow it, inside brackets.

> Sue Dunham [pseud.]. *Recollections of a Heretic.* McGruff-Knoll Press, 1956.

viii. In phonetics. Phonetic transcriptions can be enclosed in brackets

> *Quandary* may be pronounced [kwän´-də-rē] as well as [kwän´-drē].

ix. In foreign languages. The French, Spanish, and Italian languages put brackets around their ellipses to indicate omitted material. For the English format, *see* **Section 3.6.1, xvi**.

x. For translations. In text that has been translated to English, the original foreign-language word or phrase can be inserted after it in brackets. A longer translation at the end of a block quotation, whether to or from English, is usually encased in brackets.

> It wasn't a small bird [*pájaro*] but a large bird [*ave*].

Brackets also enclose English translations of foreign titles in reference lists.

> Delgado, Marisol. 1992. *Gardel y la historia del tango* [Gardel and the history of tango] Buenos Aires, Argentina: BA Press.

xi. Other punctuation with. Brackets interact with other punctuation exactly as parentheses do. *See* **Section 3.6.9.1**.

3.6.10.2 Other Brackets

i. Angle brackets. Angle brackets are used in computing and in publishing to signify specific typesets or sectional divisions.

> Refer to the section <sc> Caring for Your Turtle </sc> for more information about turtle nutrition.

Angle brackets should not be used around web addresses, for these may contain angle brackets themselves, but they are often used to indicate a key on a keyboard.

> Press <F1> for Help.

ii. Braces. Braces are used in special contexts, such as computer programming.

```
main() {
printf("hello, world");
      }
```

B. Terms and Word Marks

3.6.11 Hyphen

i. Main uses. The main uses of the hyphen are to join and separate words and letters, especially in compound terms. For dividing a word or phrase between lines, *see* **Section 3.3,** above.

ii. Types of compounds and. There are three types of compounds: open compounds, which, though connected syntactically, are spelled as separate words without hyphens.

> tank top   giant squid   ammonium nitrate

There are hyphenated compounds, which are connected with hyphens.

> thirty-four   sister-in-law   have-nots

There are also closed compounds, which have been joined to form a single word.

breadbasket   crossbow   whippersnapper

A further distinction is made between "temporary" compounds, which are constructed for a momentary purpose (*Swahili-speaking*, *proto–double-helix*), and "permanent" compounds, which can generally be found in dictionaries (*fancy-pants*) and are considered standard usage.

iii. Before or after noun. Often a compound adjective that comes before the noun it modifies is hyphenated to avoid ambiguity, while the same compound placed after the noun (and a linking verb—usually a form of *to be*) is left open.

Only washed-up actors do infomercials.
You're all washed up.

If the compound adjective starts with an adverb ending in *ly*, it does not need a hyphen, whether it comes before or after a noun.

Both correct:
A tastefully decorated room awaited us.
This room is tastefully decorated.

iv. Phrasal adjectives. Phrasal adjectives are simply compound terms that function as adjectives.

gamma-ray burst

If a phrasal adjective contains within itself a hyphenated compound, the whole phrase may or may not be hyphenated.

day-old-bread recipes

When more than one phrasal adjective is used to modify the same noun, hyphenate carefully.

long-standing monkey-bars champion
twenty-four hour time system

v. Omission of second part of a hyphenated term. If the second part of a hyphenated or closed compound is left out, to be subsequently completed with another compound, the first word must be hyphenated.

gold- and copper-plated circuitry

vi. Hyphens and dashes. Hyphens are similar to dashes in appearance, but are shorter and have separate uses. *See* **Section 3.6.8, xviii**, for dashes used in compound terms.

vii. English names with. Hyphenated first or last names must be treated as one word and not truncated.

viii. With ethnic and national group names. Hyphens in ethnic and group names, such as *African American*, are optional, even when used as an adjective preceding a noun.

Italian-American food
popular among Irish Americans

ix. In numeration. Hyphens may separate section numbers in a work that is organized according to a multiple numeration system (*see* **Section 3.6.1, v**) For separating numbers in order to indicate a range, use an en-dash (*see* **Sections 3.6.8, xvi** and **3.12.3.4**).

x. Hyphens and readability. Hyphens should be used to make text clearer, joining or separating words or parts of words to reduce ambiguity. For instance, most prefixes do not require hyphens, but one should be added in the four-syllable word *un-ionized* so that it is not confused with the three-syllable word *unionized*. More commonly, hyphens resolve ambiguities between words, especially when there may be confusion over what modifies what.

the ugly carpet salesman → the ugly-carpet salesman

sweet potato dumplings → sweet potato-dumplings

seven foot soldiers → seven-foot soldiers

Whether to insert a hyphen or not is often a judgment call, especially when there are potentially many hyphens in a row. One must walk a line between leaving a phrase unclear or illogical on one end, and creating an unwieldy string of hyphenated words on the other. Sometimes rephrasing is the best solution. Above all, be consistent.

xi. For separating characters. Hyphens can separate numerals and letters, for example, in phone numbers or spelled-out words.

> 1-800-555-8466
> "Is *insistent* a-n-t or e-n-t?"

xii. Soft vs. hard. A hyphen that appears solely to divide a single, continuous word between lines is called a *soft hyphen*. All other hyphens are *hard hyphens. For rules about dividing words between lines, see* **Section 3.3**.

xiii. Stacks of. No more than four lines in a row should end in hyphens.

xiv. Trend away from. Many compounds, especially newly formed ones, undergo transformations from open to hyphenated to closed, some faster than dictionaries can process them. If a compound is closed in common use and presents no ambiguity, it may be acceptable to forgo the hyphen or space.

> E-mail → email
> Web site → website

xv. Unnecessary uses. Hyphens are not necessary when a letter or group of letters modifies a noun (unless the entire phrase is itself being used as an adjective).

> DNA molecule
> DNA-drug interactions

Many scientific terms have become established within their fields and have their own special usage rules. For this reason, it is advisable to look up uncertain, unfamiliar terms in a reliable scientific dictionary.

3.6.12 Slash

i. Slash with abbreviations. A slash is sometimes used for abbreviations, either serving as a period or replacing the word *per*.

one hamburger w/ pickles
earning $8.15/hour

ii. Slash with alternatives. A slash between two words may be used to indicate alternatives.

and/or
Venus/Aphrodite
beer/wine/liquor

iii. *And* replaced by a slash. Occasionally a slash replaces the word *and*.

the Hemingway/Faulkner generation

iv. Computer and internet uses of. In the context of computing, forward slashes are used in URLs, while backslashes are used in certain operating systems to separate file paths.

http://www.theonion.com/content/index
C:\Program Files\ScumSoft Inc\AC Deluxe

v. With dates. Using slashes to represent a date (as in 12/3/03) is discouraged in formal writing, since different geographical regions arrange the numbers differently. A more accepted use of slashes is to replace en dashes in ranges of dates, especially to signify the end of one year or decade and the beginning of another.

He spent the 1996/97 term in intense study.
late-60s/early-70s muscle cars

vi. With fractions. *See* **Section 3.12.2.2**, below, and **Part II, Chapter 6, Section 6.2.7.1** for the use of slashes to represent fractions.

vii. Terms for "slash". A slash (/) is also known as a *diagonal, solidus, slant,* or *virgule.* It is sometimes called a *forward slash* to differentiate it from a backslash (\).

viii. Slash with two publishers' names. Slashes separate names of two publishing companies in a single entry of a reference list.

> Wilson, Gregory. *Why is the Sky Red? A Guide to Martian Meteoro-logy.* Philadelphia: Jet Black Press / Lyrer Adams, 2008.

3.6.13 Apostrophes; the Prime Sign

i. Apostrophes for possession. For apostrophes used to indicate the possessive case, *see* **Section 3.2.**

ii. In abbreviations. Apostrophes can replace numbers in year dates, though this is not common in scientific literature. (Note that the apostrophes before the dates below are *not* single quotation marks, which would face the number to the right, but apostrophes, which are equivalent to end-single quotation marks.) When referring to a decade, no apostrophe is needed between the year and the *s*.

> summer of '89
> the '50s and '60s   (*not* '50s and '60s)

iii. Other punctuation after. Apostrophes should always be placed immediately next to whatever they are intended for, unlike double or single quotation marks.

> " 'No foolin','" he snorted, "Who says "No foolin' '?"

iv. Prime sign. The apostrophe should not be confused with the prime sign, a similar-looking mark that is used mainly for formatting latitudes and longitudes, and in mathematical formulas.
    For mathematical uses of the prime symbol, *see* **Part II, Chapter 6, Section 6.2.2.1, ii.**

v. Pluralizing abbreviations. The use of apostrophes for pluralizing abbreviations is limited to those with more than one period and those with mixed uppercase and lowercase letters.

Sorry, no c.o.d.'s.
two KotOR's

vi. Lowercase letters. Lowercase letters that are alone require apostrophes to be made plural so that they are not confused with small words.

We'll cross all your t's and dot all your i's.

vii. Noun coinages. For use of apostrophes in pluralizing noun coinages, such as *comings and goings, see* **Section 3.1.3.3, iii.**

viii. Proper nouns. *See* **Sections 3.1.3.2, i,** *and* **3.2.3.1** for information on making proper nouns plural and possessive, respectively.

ix. Words in quotation marks. *See* **Section 3.6.7, ii,** for rules on pluralizing words within quotation marks.

3.6.14 Diacritical Marks

Of the many words and phrases in English that have been borrowed from other languages, some have retained their non-English marks and some have not. Fortunately, current English dictionaries keep track of standardized forms of such terms. Any foreign expressions that are not found in a reliable dictionary should be italicized and should keep their marks. For more on foreign terms, *see* **Section 3.5,** above.

château
angstrom
piña colada
Santeria

3.6.15 Asterisk

i. For footnotes. Asterisks are used to mark footnotes, especially when superscript numbers are already in use to designate endnotes.

ii. For significance levels. Asterisks can be used to represent significance levels in a table, with notes below giving the specific probabilities. *See* **Chapter 2**, **Section 2.2.1.6**.

5.53*　(*table entry*)
- }
- }　(*other table entries*)
- }

---

*$p<.01$　(*footnote beneath the table*)

iii. For text breaks. A set of asterisks may set off a line break between sections that are distinct but do not merit separate subheadings.

3.6.16 Ampersand

i. In abbreviations. Ampersands do not require spaces when they appear in abbreviations.

AT&T　　　S&M

ii. Versus *and*. To reference a work by more than one author, use the word *and* to separate them, not an ampersand.

Pixelman, Ned, Sebastian Bjorkman and Juliette Fontaine. "Underwater rocket propulsion and its effects on marine life." *New England Journal of Marine Studies* 8 (2006): 111–140.

Ampersands in quoted and cited titles may be written out as *and*.

iii. Serial comma omitted with ampersand. Series connected with ampersands instead of the word *and*, such as those in company names, do not require serial commas.

Peabody, Sherman & Tetley
Acme Ball Bearings, Banana Peels & Roller Skates Co.

3.6.17 *At* Symbol (@)

The *at* symbol (@) is most commonly used in email addresses. Its modern use is primarily in accounting (where it means "at the rate of") and should not be used in scientific writing as a substitute for "at."

Two legitimate uses of the *at* symbol are

• In chemical formulae, the @ is used to denote trapped atoms or molecules. For instance, $La@C_{60}$ means lanthanum inside a fullerene cage.

• In genetics, an *at* symbol after a gene symbol indicates that it is part of a gene cluster.

3.6.18 Marks for Line Relations

i. Brace. Braces group items together next to a heading. Their use outside of mathematical contexts is discouraged. Instead, try to represent such a relationship in plain text or in a table. For their use in mathematics, *see* **Chapter 6, Section 6.2.5**.

ii. Ditto mark. Ditto marks (") are often used to replace text that would be the same as that on a previous line or table cell. Because of aesthetic concerns and potential ambiguity, use of ditto marks is discouraged in scientific writing.

iii. Paragraph mark. The paragraph mark (¶) identifies a new paragraph, and is seen mostly in editing, proofreading, and legal contexts.

iv. Section mark. This abbreviation for *section* (§) is rarely used outside of legal writing.

(*See* **Section 3.11.6** for proofreading symbols.)

3.7 Syntactic Capitalization

A. In Text

3.7.1 "Down" Style

With regard to capitalization, many style guides, including the *Chicago Manual of Style*, advise writers to err on the side of keeping words lowercased, especially words derived from proper nouns. However, no pattern of capitalization will be appropriate in all contexts, and much science writing is based on the European tradition, which tends to favor capitalization. Refer to **Chapters 6–13** for capitalization conventions within specific subjects, keep both a standard and science dictionary at hand for specific terms, and be aware of the practices of the publication for which you are writing.

3.7.2. Words Derived from Proper Names

Capitalize proper names used as adjectives and adjectives derived from names of famous people, but do not necessarily capitalize nouns derived from any proper nouns. When in doubt, check a dictionary and err on the side of leaving the term lowercased, unless it actually involves an original, literal definition of the proper noun.

Hawking radiation	petri dish
Newtonian	saturnine
stoic disposition	Stoic philosophy

3.7.3 Prefixes

If a prefix is joined to a proper noun, it should be lowercased and hyphenated. Permanent compounds (ones that might be found in a dictionary) have adopted varying patterns of capitalization.

Below are instances of three possibilities: (a) hyphen with the main word capitalized (two instances); (b) unhyphenated with no capitalization; (c) unhyphenated and a capitalization of the prefix.

pre-Newtonian era	anti-American sentiment
unearthly	Mesoamerican

3.7.4 Quotations. For capitalization within quotations, *see* **Section 3.10. iv,** below.

3.7.5 First Word of a Sentence

Under no circumstances should a sentence begin with a lowercase letter. If something that is usually only lowercased occurs at the beginning of a sentence—such as a web address (http://...), some company or organization names (*eBay*, *µTorrent*), or a particle attached to a name or foreign word (*duBois*)—the sentence must be reworded, if possible, or the term must nonetheless be capitalized.

3.7.6 First Word after a Colon

Capitalize the first word after a colon only if the colon introduces two or more sentences, or introduces a quotation that requires capitalization. *See* **Section 3.6.5, iv.**

3.7.7 Vertical Lists

Capitalize the first words of items in a vertical list only when the items are numbered and the terms are complete sentences, but not when the whole list is punctuated as a single sentence (*see* **Section 3.6.1, vii,** for examples).

3.7.8 Ellipses. For capitalization after ellipses within a quotation, *see* **Section 3.6.1, xvi.**

3.7.9 Question within a Sentence. For capitalizing questions within sentences without quotation marks, *see* **Section 3.6.4, xxiv.**

3.7.10 Emphasizing Words

Initial capitals should not be used to emphasize words—consider italics instead, or roman type if the surrounding text is already italicized (*see* **Sections 3.4.1 and 3.4.7**). For text requiring all capital letters, SMALL CAPS may be preferred.

3.7.11 Footnotes

A footnote should be capitalized if it is a complete sentence; otherwise, its first word may be lowercased and a final period omitted.

3.7.12 Translations

In translated text, words that were capitalized in the original language but would *not* be capitalized in English should not be capitalized. (Exceptions are instances where the author citing the translated text wishes to capture the original flavor of the work, and indicates that as a desideratum in prefatory material.)

3.7.13 Editor's Note

When capitalization has been modernized in a historical text, it should be noted in an editor's preface.

3.7.14 Physical Characteristics

Physical characteristics, such as skin color or disability, as well as people characterized by them, are not generally capitalized. For appropriate language in such cases beyond mechanical conventions, *see* **Chapter 1, Section 1.8**.

white people
the hearing-impaired
little person

3.7.15 Marking Manuscripts

To indicate that a letter or group of letters should be capitalized in a manuscript, underline three times each letter to be capitalized. Letters that should be in small caps are underlined twice. A diagonal line through a capital letter signifies that it should be set in lowercase.

See **Section 3.11.6,** below, for a list of proofreading marks.

B. In Titles

Note: This section deals with general conventions of capitalization within titles. For treatment of specific titles, *see* **Sections 3.8 and 3.9.**

3.7.16 Headline Style

Headline-style capitalization is used for many titles, including book and journal titles. It is often used to cite titles of articles in notes and bibliographies. It follows no exact method, but relies on several guidelines.

The first and last words of a title should always be capitalized, as should all other stressed or important words. Remaining prepositions, articles, and small conjunctions, along with the words *to* and *as*, should be lowercased.

> Attack of the Crazy Ants—Invasional 'Meltdown' on an Oceanic Island

However, prepositions used as adverbs or parts of verb phrases should be capitalized.

> How to Wake Up a Sleepwalker
> Putting Out: Why Golfers May Be More Promiscuous

Specific names of species (e.g. *sapiens* in *Homo sapiens*) should also be lowercased.

> The Evolution of HoxD-11 Expression in the Bird Wing: Insights from Alligator mississippiensis

3.7.17 Sentence Style

Sentence-style capitalization treats titles like sentences; that is, it capitalizes only the first word and any proper nouns. It is especially appropriate for titles or headings that are complete sentences, Latin titles, article titles in reference lists, and contexts where internal capitalization is important.

> Yamane et al. A neural code for three-dimensional object shape in macaque inferotemporal cortex.

3.7.18 Hyphenated Terms

In a headline-style title, the first word of a hyphenated term should always be capitalized. For other words in the term, capitalize according to the rules and exceptions in regular headline style (as if the term were not hyphenated), but leave common nouns that are objects of prefixes lowercased, as well as the second parts of spelled-out numbers. Terms that are connected by an en dash (*see* **Section 3.6.8, xvii**) should each be capitalized.

DNA-Based Vaccine Shows Promise against Avian Flu

Anti-inflammatory Properties of Solid-State Dihydrogen Monoxide

Thirty-one New Species of Spider Found in Amazon

Investigating Sulfuric-Acid–Water–Ammonia Particle Formation

In sentence-style capitalization, hyphenated terms appear as they would in a normal sentence. *See also* **Section 3.7.3** for guidelines on capitalizing proper nouns with prefixes.

3.7.19 Tables

For capitalization within a table or graph (along with general guidelines on table construction), *see* **Chapter 2, Section 2.2.1.6**.

3.7.20 Converting Titles

Titles of works, which often appear in all capitals on their actual works, may be changed to sentence style or headline style when made reference to. In either style, subtitles must always begin with capital letters. *See* **Sections 3.8** and **3.9** for specific titles.

3.7.21 Foreign Language

Foreign-language titles represented by roman characters should generally be capitalized sentence-style (*see* **Section 3.7.17**), even if they have been transliterated to English.

3.8 Names, Titles, Terms, and Organizations

3.8.1 Personal Names

i. Titles preceding. When a personal title is used in running text as a part of a name or in direct address, it should be capitalized; otherwise it should be lowercased. Be careful to distinguish a title that is part of a name from a title followed by a name used appositively (*see* **Section 3.6.4, xiv**).

> I told Dr. Espinosa about my headaches.

> Tell it to me straight, Doctor.

> The doctor will see you now.

> I'd like you to meet Monsignor Marco Battaglia.

> The monsignor Marco Battaglia does not agree.

In an index, such titles do not need to be included with personal names, except for clarification.

ii. Suffixes with. *Jr.* and *Sr.* may or may not be set off by commas, while roman numerals after a name never take them. Such suffixes should appear after full names but not after just a last name. In indexes, they should follow the first name and a comma. For word division of personal names, *see* **Section 3.3.4.**

iii. Compound surnames. Particles and hyphenated elements in names should be treated carefully. Capitalization should be maintained when full names are separated into given and surnames, but initial lowercase letters must be capitalized when at the beginning of a sentence. When in doubt as to how to spell or divide a name, check a biographical dictionary, a biographical appendix in a regular dictionary, or a respectable online reference.

> Oscar de la Hoya's last name is "de la Hoya."
> De la Hoya delivered a knock-out punch.

iv. Abbreviations and acronyms as. Initials of names should be capitalized and followed by a period and a space.

Mr. H. L. Mencken

But names represented entirely by two or more initials do not require periods or spaces. (Such initialisms should appear in indexes where they belong alphabetically, not where they would be placed if they were spelled out.)

FDR
JFK

v. Letters standing for. When letters are used to represent hypothetical people, they should be capitalized but not followed by periods.

A shakes hands with B and C, and B shakes hands with C.

vi. With laws and theories. Laws and theories do not need to be capitalized, except for proper names that happen to be a part of them.

theory of plate tectonics
Euclid's lemma

vii. Followed by place name. When a person's name is followed by the name of a place, the place name should be enclosed in commas only if it is optional information.

Dr. Anderson, of Milwaukee, treats many patients.

Dr. Anderson of Milwaukee is my doctor, not Dr. Anderson of Madison.

Helen of Troy was legendarily beautiful.

Helen White, of Trenton, is legendarily punctual.

See **Section 3.6.4, xii,** for more about restrictive versus nonrestrictive phrases, and **Section 3.6.4, xiv,** for names and appositives.

viii. In reference lists. In the reference lists of many publications, first names are only given as initials.

ix. Variants of a name. If variants of a name exist, use a biographical dictionary or reliable online reference to help decide which variant to use as the main entry and which to use as cross-references in an index.

x. In foreign languages. Many foreign languages, such as Spanish and Italian, do not capitalize titles before names.

el señor García

French titles *Monsieur*, *Messieurs*, *Madame*, and *Mademoiselle* may be abbreviated *M.*, *MM.*, *Mme*, and *Mlle*, respectively, when in front of a name, but should otherwise be spelled out.

3.8.2 Titles and Offices

i. Academic titles. Like most titles, academic titles should be lowercased when used generically or in front of a name used appositively (*see* **Section 3.6.4, xiv**), but capitalized when they are part of a name title or when they are used in place of a name to address someone directly.

the dean of the Department of Modern Languages

The provost, Ramona Ohmsford, will give a speech.

Did Professor Calavera seem pale today?

ii. Civil titles. Civil titles should be lowercased, except in cases described in **Section 3.8.1, i,** and the very few titles traditionally capitalized. They may only be abbreviated before a full name.

The chief justice banged her gavel.

We act under the authority of the mayor of Woodcrest.

I'd like you to meet Congressman Greene.

iii. Honorable titles. Titles used to show respect or honor are usually capitalized. Note that the words *honorable* and *reverend* should not be abbreviated after the word *the*, and usually appear only before a full name. *Sir* and *ma'am* are not capitalized.

> Your Majesty
>
> the Crown Prince
>
> the Reverend William Parks

iv. Organizational titles. Such titles are rarely used in direct address or as part of a name, so they have little occasion to be capitalized in running text. In formal lists, however, such titles often appear capitalized headline-style.

> Susan Villalobos was promoted to assistant director.
> Mortimer Namesman Jr., Custom Framing Manager

v. Professional titles. Some professional titles may be abbreviated after a name, as an academic degree would be. These should be capitalized when written out.

> Forsythe Beaumont, JP (Justice of the Peace)
> LPN (Licensed Practical Nurse)

vi. Social titles. Social titles should always be abbreviated with a period. They may be placed before a full name or just a last name, although they are not required before either. If they are used in direct address, they should be capitalized.

> Mrs. Thompson
> I need the number of a Dr. Richard Penbrook, please.

vii. Religious titles. Religious titles are capitalized in the same way as academic and civil titles: they remain lowercased unless they are part of a name or are used to address someone directly.

> Bishop Panucci; a/the bishop
> Imam Ahmad ibn Hanbal

3.8.3 Academic Degrees and Honors

i. Abbreviations of. Abbreviations of academic degrees do not require periods, but they should be set off by commas if they immediately follow a name.

Gregory Walter Graffin, PhD, is scheduled to give a lecture.
They both have MAs in film.

ii. Of author. The name of an author on a title page should not be abbreviated unless that author is best known by the abbreviated name. No author's academic degree should appear on the title page, unless the author is a doctor and the book health-related. To display information about the author of one chapter of an anthology, use an unnumbered footnote.

iii. Capitalization. Names of degrees and affiliations should be capitalized when a specific one is referred to, but otherwise left lowercase.

iv. Of contributors. Names of contributors to a multi-author work, often along with their academic affiliations and degrees, are usually listed before the index. They should be arranged alphabetically according to last name but should display the first name before the last.

Zelda W. Aaronson.
Douglas Bluff Jr.

v. Indexing of titles. Omit academic degrees and titles from index entries.

3.8.4 Fictitious Names

Characters in a fictitious work can be referred to by capitalized names, even if they are generic, descriptive titles.

The Clerk has only one spoken line in this play.

3.8.5 Geographic Names

i. Topographical names. Capitalize the name of a river, mountain, or other topographical feature, along with its accompanying generic term —but only when that term is a part of the name, not when it is simply a description or when it is used as an appositive.

Euphrates River; the river

Mount Ordeals; the mountain

Chimney Rock; the rock

ii. Parts of the world. Terms that appear on maps and words derived from them are capitalized, as are most regional terms based on compass directions. Do not capitalize generic terms or places used metaphorically.

The Strait of Gibraltar is a famous strait.
the Pacific Northwest
Mecca's role in Islam
the mecca of Elvis fans

iii. Political names. Names of political entities, such as countries and their subdivisions, are always capitalized. Generic terms included after these names are capitalized when they are part of the name. When the generic term comes before the name, it is usually only capitalized in the names of countries.

Salt Lake City
city of Venice
Kyoto Prefecture; prefecture of Kyoto

But the generic term may be capitalized when the name of a place represents a government rather than the place itself.

The event was paid for by the State of New Jersey

iv. Popular names. Popular but unofficial names or nicknames for geo-graphical locations and places are often capitalized.

the Florida Everglades
Philly's Chinatown
the Big Apple

v. Buildings, roads, and other structures. Structures such as buildings and roads should be capitalized when accompanied by generic terms. The generic terms should be lowercased when alone, whether the structure is famous or not.

Galloping Hill Road	the Metropolitan Museum of Art
the Sears Tower; the tower	the Great Wall of China; the wall

3.8.6 Nationalities and Other Groups

Nationalities and ethnicities are capitalized, but terms describing socioeconomic class, skin color, or disabilities should be lowercased. For appropriate terms to use, *see* **Chapter 1, Section 1.8**.

Tongan people	Jewish culture
middle class	white person

3.8.7 Names of Organizations

i. Scientific, business, and educational institutions and companies. Capitalize the names of such institutions, but not the word *the* before them, unless it is required syntactically. As with other titles, do not capitalize the generic terms that may accompany such names when they are used alone.

the University of Siena; the university
the Royal Astronomical Society
the James Randi Educational Foundation
Houghton Mifflin Company; the company
National Association of Science Writers

ii. Government and judicial bodies. Governmental bodies and their generic terms are capitalized, but the generic terms should be lowercased when used alone.

the Riverdale City Council; the council
the Supreme Court

iii. Associations and conferences. Titles of conferences, unions, associations, and the like should be capitalized according to rules for businesses and companies (*see* **3.8.7, i**).

National Conference for Academic Disciplines
Left-Handed Society of Australia

iv. Political organizations. Names of national political parties are capitalized, along with the word *party*, but nonspecific references to political philosophies are not.

the Democratic Party; the party
Socialist Party of 1968
communism's role in human history
libertarian viewpoints

3.8.8 Cultural and Historical Terms

i. Eras and periods. Names for time periods based on numbers should not be capitalized, except if they are proper names. *See* **Section 3.8.9, vii**, for the punctuation and abbreviation of names of decades.

the eighteen hundreds
the seventies
the Second Republic

Names of prehistoric ages are capitalized, but names given for modern ages are not.

the Stone Age
space age polymers

Descriptive names of time periods are often lowercased except for proper nouns they contain, but some periods are capitalized according to tradition.

the Era of Good Feelings
the Renaissance

ii. Events. Historical events are usually capitalized, while more recent events are usually lowercased. Major natural disasters are often capitalized, as are sporting events. Speeches are usually not, except for a few very famous ones.

the Great Leap Forward
the housing crisis
Hurricane Katrina
the Kentucky Derby
the Gettysburg Address

iii. Legislative terms. Legislative terms, such as laws, bills, and acts, are capitalized.

the Plant Variety Protection Act
the Tariff of 1832

iv. Military terms: Capitalize divisions of military forces when they are paired with proper nouns as parts of titles, but not when they are used generically.

United States Marine Corps
Spanish Armada
Royal Canadian Mounted Police
(unidentified) mounted police
16th century Spanish navy

v. Legal cases. Names of court cases are usually italicized.

vi. Awards. Award names are capitalized.

vii. Oaths and pledges. Most oaths and pledges are not capitalized.

viii. Academic subjects, courses of study, lectures. Academic subjects are not capitalized, except for any proper nouns they happen to contain. Course, lecture, and department titles, however, should be capitalized, headline style.

> a course on evolutionary biology
> women's studies major
> Shakespearean studies major
> Introduction to Logic 101
> Ichthyologic Research in the Twenty-first Century

ix. Religious names and terms. Names and epithets for divine beings and prophets of every sort are capitalized, but pronouns referring to them are not. The word *god* should be capitalized when used as a name, but not when used in generic reference to a deity.

> Odin   Jesus
> Yahweh   Horus
> Satan   the Lord
> the Buddha   God Almighty

Specific religions and denominations are capitalized, along with their adjective forms.

> Hinduism   Shiite
> Mormons   Roman Catholic

Titles of specific religious texts, including the Bible and its books, should be capitalized. Specific councils, places of worship, and major events are capitalized, except when used generically; but religious objects, doctrines, and services are not.

> the book of Psalms; a psalm
> the King James Bible
> Qur'an
> Nicene Council; the council
> The Exodus; an exodus
> the Great Synagogue of Rome
> chalice
> predestination
> liturgy

3.8.9 Time and Dates

i. Units and symbols. Abbreviate units for time as follows:

year:	yr.		hour:	h. *or* hr.
month:	mo.		minute:	min.
day:	d.		second:	sec.

ii. Clock time. When the time of day is expressed in numerals, the abbreviations *a.m.* and *p.m.* should be in roman type, or else in small capitals without the periods. While *12:00 a.m.* is often used for midnight and *12:00 p.m.* for noon, the words *midnight* and *noon* should be used instead, or a twenty-four hour system should be adopted (*see* **3.8.9, v**).

11:59 p.m.
12:00 midnight; midnight
2:30 AM

iii. Spelling out. Numerals are important when an exact time is necessary; otherwise, writing out the time is more appropriate.

Is it five o'clock yet?
Be here at one thirty sharp.
He was asleep by quarter of eight.

iv. Time zones. Abbreviations for time zones are often displayed in parentheses after a specified time. The abbreviations should be in all capital letters, but when time zones are written out, they are lowercased.

v. Twenty-four hour system. Twenty-four hour time is popular in Europe; in the United States it is known as *military time*. It may use four digits without punctuation, or four or six digits with colons in between the hours, minutes, and seconds. A decimal point may further divide seconds into tenths.

0800 hours
2205h (10:05 PM)
00:55:23.7 (12:55 AM and 23.7 seconds)

vi. Chronology systems. Year designations can be marked by CE and BCE (for *common era* and *before common era*), by AD and BC (for *anno Domini* and *before Christ*), or by less common systems of chronology—for instance, MYA or mya for *million years ago,* and BP or YBP for *years before the present*. They may be written in capital letters without periods, or they may be written in small capitals with periods optional.

65 MYA	*or*	65 MYA
12,000 B.C.	*or*	12,000 BC
600 CE	*or*	600 CE
1999 AD	*or*	1999 AD
AD 2300	*or*	AD2300 (*no space between* "AD" *and date*)
the first millennium AD	*or*	the first millennium AD

vii. Centuries and decades. If they are written out, names of centuries and decades should be lowercased. Decades can appear alone if it is clear which century is being talked about, and if the decade is not one of the first two in a century. Numbers are punctuated like normal words: they do not require apostrophes when plural, but there should be an apostrophe in any possessive form, as well as before the truncated, two-digit format.

the 1980s   the fifties
the seventeen hundreds   one of the 1830s' most prolific writers
'50s television (*note direction of punctuation*)

viii. Capitalization of. Capitalize names of months and days, but not names of seasons.

a Tuesday in spring

ix. As adjectives. Compound adjectives formed from a month and a year, or from a month and a day, do not require hyphens. Those formed from complete dates—containing a month, day, and year—can be unwieldy, and must take commas before and after the year if they are used at all.

a June 2007 study
the February 4th meeting
the September 3, 2008, conference

x. Commas with. When a date is expressed within a sentence, it only requires commas if it gives the month, the day, and the year, in that order. *See* **Sections 3.6.4, xx,** and **3.8.9, ix.**

xi. Alphabetizing. In an index, numerals should be alphabetized as if they were spelled out in letters.

Twigginsky, Ronald, 140–14
2001: A Space Odyssey 19–20, 305

xii. Holidays and holy days. Only specific holidays should be capitalized; their generic terms, when used alone, should remain lowercased.

New Year's Day; the day
Cinco de Mayo
Holy Week; the week
I wish you a happy anniversary.

xiii. Abbreviations for months. Months may be abbreviated according to one of the following three systems:

Jan.	Jan	Ja	July	Jul	Jl
Feb.	Feb	F	Aug.	Aug	Ag
Mar.	Mar	Mr	Sep.	Sep	S
Apr.	Apr	Ap	Oct.	Oct	O
May	May	My	Nov.	Nov	N
June	Jun	Je	Dec.	Dec	D

For abbreviating decades and centuries, *see* **Section 3.8.9, vii.**

xiv. With slashes. For usage rules regarding dates that contain slashes, *see* **Section 3.6.12, v.**

3.8.10 Vehicles

Names of ships are capitalized and italicized, except for the abbreviations USS and HMS. Names of cars, trains, aircraft, and spacecraft are capitalized but not italicized. Pronouns of neutral gender should be used to describe all vehicles, including ships.

USS *Swordfish* and its crew
2003 Hyundai Accent
Viking 1

3.8.11 Planets and Astral Bodies

Names of planets and other astral bodies should be capitalized, but only when used as names, and not in colloquial expressions.

My favorite planet is Earth.
How far is Haley's Comet from the Moon right now?
The object is probably a comet or a moon.
She lay on the earth and felt the sun warm her.
What on earth are you talking about?

3.8.12 Drugs and Reagents

Generic names for drugs should be lowercased. They are preferable to brand names, which must be capitalized and may be added within parentheses after the first instance of the generic name.

The drug rabeprazole sodium (Aciphex) is prescribed for acid-reflux disease.

3.8.13 Trademarks and Trade Names

Trademarked names should be referred to in their generic terms, if at all possible. Otherwise, they must be capitalized but do not legally require a superscript *TM* or any other symbol.

Post-it note → sticky note, repositionable note
Q-tips → cotton swabs

3.9 Titles of Works

3.9.1 Abbreviated

If an italicized title is abbreviated, that abbreviation should be italicized as well.

NYT article

In classical references, names of authors and works may be abbreviated, and no punctuation should appear between them.

3.9.2 Articles and Alphabetizing

The word *the*, when it begins the title of a newspaper or journal, does not need to be capitalized or italicized in running text. Foreign titles with an article in their original language may include it in the title's format as long as it is part of the official title.

She published an article in the American Journal of Psychiatry.

Any title forming a main entry in an index should have its initial article, if it has one, placed after a comma at its end or else removed entirely, and the title should be alphabetized in this format. It does not need to be inverted in a subentry.

amniocentesis, 262
Ancestor's Tale, The, 92–98
Anchorage Daily Tribune, 315

To the same end, the word *the* should be omitted from organization names in indexes.

Science Academy of South Texas, 53–57

Multiple titles by a single author in a bibliography should be arranged alphabetically, including but ignoring initial articles. Anonymous works should also be alphabetized according to their titles, again, disregarding initial articles. Entries beginning with numbers should be placed where they would be if the number were spelled out.

3.9.3 Newspaper Headlines

Within reference lists, newspaper headlines may be capitalized either headline- or sentence-style. In running text, small capitals are another option.

> "Flu shot in pregnancy protects newborns"
> BABOON ATTACKS PRIME MINISTER

3.9.4 Date in Title

When citing a title that ends in a date, include a comma before that date.

> *Fisticuffs: The History of American Boxing, 1900-1950*

3.9.5 Punctuation

For conflicting punctuation between a title and its surrounding text, *see* **Sections 3.6.7, xxiv** and **3.6.4, xxx.** For punctuating titles within other titles, *see* **Section 3.9.14.**

3.9.6 Capitalization

Titles in running text and in bibliographies are often capitalized according to headline style. Titles in Latin and those of many scientific works use sentence-style capitalization. *See* **Section 3.7, xvi-xvii.**

3.9.7 Italicized Terms

For italicized terms within a title, *see* **Section 3.4.7.**

3.9.8 Foreign Language

Non-English titles should be capitalized in sentence-style. For translations of documented titles, *see* **Section 3.5.7.**

3.9.9 Older Types

Old titles may retain their archaic spelling, punctuation, and capitalization (except for words in all capitals), but they may also be shortened—using ellipses—if overly long. The second part of a double-title joined by *or* should begin with a capital letter.

"A Modest Proposal For Preventing The Children of Poor People in Ireland From Being Aburden…to The Public"

3.9.10 Original Plus Translation

When a translated work and its foreign-language source both need to be put in the same bibliography, either list the original first, followed by the translation's information; or list the translation's information first, followed by *Originally published as* and the original's information.

3.9.11 Shortened

In notes, long or cumbersome titles may be shortened once they have already been cited in full, eliminating prepositional phrases or initial articles and keeping only a few important words. The shortened title should maintain the typographical formatting of the full title and be substantial enough to be easily identified with a bibliography or reference list entry.

"Fault Diagnosis Engineering of Digital Circuits Can Identify Vulnerable Molecules in Complex Cellular Pathways."   "Fault Diagnosis Engineering"

3.9.12 Permissible Changes

When citing a title, be faithful to the original style conventions, but change ampersands to the word *and*, and change fully capitalized text to headline or sentence style.

3.9.13 Author's Name in Title

A note citation may start with the title of a work instead of its author if the title already contains the author's name. Other reference lists, though, must begin with the full name of the author.

> 3. *Surely You're Joking, Mr. Feynman!* (New York: W. W. Norton & Company, 1997).

3.9.14 Titles within Titles

When an italicized title contains a title itself, put the internal title in quotation marks, whether it was originally in quotation marks or was also italicized.

> *Politics and "The Art of War"*
> *The Origin of "The Origin of Species"*

When a title in quotation marks contains an italicized title, leave that title italicized. If the title-within-a-title should be in quotes, use single quotes.

> "A Brief History of the USS *Constitution*"
> "Classical Allusions in Campbell's 'Letter to a Patriarchy'"

3.9.15 Very Long

Overly long titles, such as those from previous centuries that summarize their works, may be shortened with ellipses (*see* **Section 3.9.9**, for example). They should contain sufficient content to let readers locate the work if they need to.

3.9.16 Quotation in Title

Quotes used as titles or within titles do not require quotation marks, although the whole title might still need quotes if it is an article, short story, poem, etc. They are often capitalized sentence-style.

> "No man is an island: Sociology as the Study of Human Relationships"

3.9.17 Reference Lists

In reference lists, books and articles are usually capitalized sentence-style, while journal titles are capitalized headline-style (*see* **Section 3.7, xvi–xvii**). Articles and chapter titles, which are enclosed in quotation marks when referred to in running text or in a bibliography, are not enclosed in quotes in a reference list.

> Morris et al. Aggressive versus conservative phototherapy for infants with extremely low birth weight. New England Journal of Medicine, Oct. 30, 2008.

Journal names may be abbreviated, and, especially in scientific writing, do not necessarily require periods. See the appendices in Part Three for lists of scientific journals and their abbreviations.

J Syst Paleontol
Macromol Biosci

3.9.18 Containing Comma

Titles containing a single comma can seem awkward in the middle of a sentence, especially if not distinguished by italics or quotation marks. There is nothing technically ungrammatical about leaving it alone, but to get around the problem, the *Chicago Manual of Style* recommends rephrasing to alter or move the title, adding a comma at the end of it, or inserting a nonrestrictive phrase or clause after it (*see* **Section 3.6.4, xii**).

Below is an example of an awkwardly-phrased sentence that may be improved through rephrasing the sentence:

> *Awkward original:* He joined the Billiards Club of Springfield, Illinois to improve his game.

> *Rephrased:* To improve his game, he joined the Billiards Club of Springfield, Illinois.

> *Second comma inserted:* He joined the Billiards Club of Springfield, Illinois, to improve his game.

> *Phrase inserted:* He joined the Billiards Club of Springfield, Illinois, the largest of its kind in the area, to improve his game.

3.9.19 Classical References

Classical works should appear in italics and use sentence-style capitalization (*see* **Section 3.7.17**).

3.9.20 Specific Works

i. Articles in periodicals and parts of a book. Titles of book parts, such as chapters or sections, should be enclosed in quotation marks, along with articles in journals, newspapers, and magazines. Essays, short stories, and poems are treated the same way.

> Chapter 7 is called "Mathematics of Quantum Neutrino Fields."
> The most entertaining essay is Larra's "Vuelva usted mañana."

Generic parts of a book such as the preface, the introduction, and the glossary do not need to be capitalized, italicized, enclosed in quotation marks, or distinguished in any other way.

> The author explains in her introduction why she thinks her research is important.
> Refer to the glossary for a more complete definition.

A generic term that is a component of a numbered part of a book, such as *chapter* or *section* or *figure*, should also be written in lowercased roman type without quotation marks. When in parentheses, such a term may be abbreviated. Numbers should not be spelled out as words, but they may appear next to designative letters.

> Subjects in the control group displayed normal brainwave function (fig. 4A), but subjects in the test group...
> The most informative chapter was chapter 3: "Mole Physiology."

ii. Series and editions of books and periodicals. When referring to a book series or edition, capitalize but do not italicize it.

> Barnes & Noble Classics
> The Library of America series

iii. Unpublished works. Unpublished works should be capitalized, but have to be put in quotation marks instead of in italic type, even if they are books or plays. If a potential book is under contract to be published, it may be italicized with a qualifying *forthcoming* after the title in parentheses.

His controversial dissertation, "The Moral Imperative of Cloning"...
She is writing a book called "Psychological Egoism: The Selfless Myth."

iv. Electronic sources. Electronic sources, whether online or stored digitally on a disc, should be treated the same way as print media, where possible. Complete works should be italicized, while articles, essays, sections of works, etc. should be put in quotation marks. Website titles should be capitalized but not put in quotation marks.

v. Movies, television, and radio programs. Titles of movies, television shows, and radio programs are italicized, but scene titles of movies and episode titles of television shows should be enclosed in quotation marks.

Bill Murray in *Groundhog Day*
the radio series *Amos 'n' Andy*
the classic *Twilight Zone* episode "To Serve Man"

Television network and radio station abbreviations never take periods.

MSNBC
WNYC

vi. Paintings and sculptures. Titles of works of art are usually capitalized headline-style. Most are italicized, except for ancient, anonymous works, which are set in roman type, and photographs, which require quotation marks, not italics.

Salvador Dali's curious painting *Corpus Hypercubus*
the Venus of Willendorf
"Greenscape," a photograph of an alleyway in Siena

vii. Notices, mottoes, and inscriptions. To quote a short notice or motto, use headline-style capitalization. For a longer notice or motto, use quotation marks and capitalize it as it appears. For a motto in a foreign language, set in italics and use sentence-style capitalization.

> He saw the Bridge Out sign in the nick of time.
> Norway's royal motto is *Alt for Norge* ("Everything for Norway").
> "Warning," the sign read, "Radioactive materials within. Only
>   trained technicians may enter."

3.9.21 Subtitles

i. Format of. Titles and subtitles should be separated by a colon and a single space. If the title is italicized, the colon should be as well. A subtitle always begins with a capital letter.

> *A Cream Deferred: Reflections on American Cheesemaking*

ii. Of chapters. On a chapter title page, a chapter subtitle should appear left-justified and capitalized headline-style, below both the chapter number and the main chapter title.

iii. Headline style vs. sentence style. A title and its subtitle do not have to be capitalized according to the same style: if one element is a sentence or quote and the other is a more traditional title, both sentence- and headline-style (*see* **Section 3.7.16**) may be used.

iv. In indexes. A work's subtitle should not appear in its index entry unless it is required to identify the work.

v. Italics for. Subtitles generally match the typesetting of their main titles, so if a main title is in italics, the subtitle should be as well. *See* **Section 3.9.14** for titles within titles.

vi. Limited to one. Do not give more than one subtitle to a book. For treatment of older books with two subtitles, *see* **Section 3.9.9**.

vii. Location of. Subtitles should not appear on the "half title" page; that is, on the first page of a book.

3.10 Quotations

3.10.1 Format

There are two ways to represent quoted material: running it directly into the text with quotation marks around it; or setting it off on indented lines (called a block quotation). Run-in format, appropriate for short or multiple quotations, should match the surrounding text in typesetting and spacing. Block format, which works well for long quotations or those that have special formatting, is usually typographically distinct from its surrounding text—in size or font, for example.

> Organiser of the Turing Test, Professor Kevin Warwick from the University of Reading's School of Systems Engineering, said:
>
>> This has been a very exciting day with two of the machines getting very close to passing the Turing Test for the first time. In hosting the competition here, we wanted to raise the bar in Artificial Intelligence and although the machines aren't yet good enough to fool all of the people all of the time, they are certainly at the stage of fooling some of the people some of the time.
>
>> Professor Owen, who is a musician, confirms the foregoing statement, and remarks, though erroneously, that this gibbon "alone of brute mammals may be said to sing."

When quoting, make sure to keep track of tenses and pronouns outside the quoted text, and, if making changes to such elements within the quote itself, use brackets (*see* **Section 3.6.10, ii**).

> According to his journal, he was occupied at this time with "the curious puzzle of light and luminosity, of which [he had] been unable to explain either the composition or behaviour."

3.10.2 Block Quotations

i. Appropriate use. Block quotations should be used for quotes that are more than a few lines or more than one paragraph in length. They should also be used for text with unusual formatting, such as poetry, lists, or letters.

ii. Documentation following. A citation that immediately follows a block quotation should be in the same typeset and size as the quote. It should come after all other punctuation and be encased in parentheses, and it does not require a closing period itself.

> However, as the present pope emphasized…one must insist that God infused immortal souls into Adam and Eve—souls not possessed by their apelike ancestors….It forces the belief that the first humans… were reared and suckled by mothers who were soulless beasts. (Martin Gardner, *Did Adam and Eve Have Navels?* [New York: Norton, 2001], 14)

Abridged references, such as those to just the page or volume numbers, may consist of just the number or may begin with *p.*, *vol.*, etc. These abbreviations should be capitalized after a block quotation, but may remain lowercased for a run-in quote.

> Doyle wrote to Houdini in great excitement:

> > I have something…precious, two photos, one of a goblin, the other of four fairies in a Yorkshire wood. A fake! you will say. No, sir, I think not. (P. 118)

iii. Paragraphs in. Quotes that consist of multiple paragraphs should be set off as block quotations. The first paragraph does not need to be indented (and should not be, if the quoted material starts in the middle of its original paragraph), but subsequent paragraphs should be indented according to the original text.

When expository text interrupts a block quotation, that text may be inserted within brackets (*see* **Section 3.6.10.2**), or partially run into the preceding text, like so:

> "Just as there are DNA-DNA helicases," Kadonaga explains,

> > there are RNA-DNA helicases and RNA-RNA helicases. So it doesn't take a lot of imagination to foresee that there are probably going to be RNA–DNA annealing helicases and RNA–RNA annealing helicases. The field potentially can be fairly large…

When the text following a block quotation continues the paragraph that introduced it, it should not be indented as if it were a new paragraph.

iv. Manuscript preparation. When indenting block quotations, be sure to use the indentation features of your word processor, rather than inserting spaces. Do not insert hard line-breaks (with the Enter key) until the end of a paragraph.

v. Of letters (correspondence). Letters should be quoted as block quotations, with their formatting intact.

3.10.3 Run-in Quotations

i. Appropriate use. Run-in text is best for short quotations (those of less than a few lines), and for multiple quotations, especially when those vary in length or are syntactically integrated into the surrounding text. They are also useful in places where block quotations cannot easily go, such as within footnotes.

> Most generators operate by the "relative motion of conductors and fluxes." On the other hand, the Retro-Encabulator uses the "modial interaction of magneto-reluctance and capacitive directance." As plausible as this may sound to non-engineers, "modial" and "directance" are not even words, much less meaningful engineering terms.

ii. Text citation placement and. In running text, a parenthetical citation may precede a quotation instead of following it.

> Imagine Dalton's surprise when (1995, 104) "the entirety of the tub began to creak and rumble, and a great torrent of water issued forth into the main dining room."

iii. In notes. Quotations within notes should be run in and their sources should appear in the sentence or immediately following. (But *see* **Section 3.4.9** for endnotes arranged by quotes from their own work's text)

> 6. Woodrow Wilson, in his *Congressional Government* (1885), also commented on the difficulty of speaking amidst the "disorderly noises that buzz and rattle" through the vast House of Representatives chamber.

3.10.4 Capitalization

When quoted material plays a direct grammatical part in a sentence, rather than being set off from the sentence more distinctly as a quote, the quote's initial letter should be lowercased, whether or not it was capitalized in its original context.

> Can my client honestly have been expected to "show up on time every single day," given the nature of his impairment?

If a block quotation is introduced in such a way, its first word may also be changed to lowercase, as in the following example:

> According to Meyerson et al., such a methodology is critical in its field because
>
> > the new technique relies on the final product distribution from both reactions. Similar models have failed to incorporate the morphological and integrated procedural ramifications, a pitfall which is unfortunately common…

Conversely, a quotation of words that form a complete sentence may be changed to begin with a capital letter if the context demands it, whether or not it was a complete, capitalized sentence in its original context.

Care should be taken that such alterations do not change the implicit meaning of a quote. If precision is especially important, consider enclosing in brackets any letters whose capitalization have been changed.

3.10.5 Phrases Introducing

A colon is used for formal introductions of quotes, such as those using the words *the following* and *thus*. In more casual contexts, a comma is preferred. When quoted material plays a direct part in a sentence, rather than being set off from it more distinctly as a quotation, it does not require any introductory punctuation (beyond an opening quotation mark, if it is run in).

> The researchers made the following discovery: "Bacterial iron-oxide crystals once grew to eight times larger than the largest known of such crystals previously on record."

3.10.6 Interpolations

Anything replaced or inserted into a quote, whether to clarify or to adapt its grammar to that of the surrounding sentence, must be enclosed in brackets. *See also* **Sections 3.10.1** and **3.6.10, ii**.

3.10.7 Emphasis Added

For italicizing part of a quotation for emphasis, see **Section 3.4.2**.

3.10.8 Original Errors

When an author would like to quote a passage but it contains an error, three possible solutions are: 1) paraphrase the information rather than quoting it directly; 2) if the error is clearly a minor typographical one, simply correct it, or 3) add *sic*, in italics and within brackets, after the mistake. For a quote that contains many mistakes or uses archaic spelling, explain in the preceding text or in a note that the entire quotation has been left intact (or corrected).

3.10.9 Permissible Changes

i. Capital or lowercase letters. *See* **Sections 3.7.4** and **3.10.4** for changes of capitalization within quotes.

ii. Punctuation. Guillemets, punctuation marks that are sometimes used in foreign languages as quotation marks are in English, may be changed to quotation marks when used in an English context.

« Sacrebleu! » → "Sacrebleu!"

3.10.10 Original Spelling

Quoted material that uses British spelling should not be edited to conform to American spelling.

3.10.11 Note Numbers

Note numbers should come after quotations, including the closing quotation mark, if applied to a run-in quote.

> "…[I]t has been found that adult acclimation and developmental acclimation result in different hypotheses being supported."[4]

3.10.12 Page Numbers

When citing a quote with a note or a parenthetical reference, give only the page number(s) of the quote.

3.10.13 Translation

i. Translation following. After a quotation in a foreign language, an English translation may be added, in parentheses for run-in quotes or in brackets for block quotes. (After a translated quote, the original, foreign-language text may be added in the same way.) Quotation marks are not necessary, though they may be used as they appear within the quote; however, when putting a translation in a note instead of directly after its quotation, use quotation marks instead of parentheses or brackets.

> "Proprium humani ingenii est odisse quem laeseris." (It is human nature to hate a person whom you have injured.)

> Le langage est le seuil du silence que je ne puis franchir. Il est l'épreuve de l'infini. [Language is the threshold of silence that I cannot step across. It is the test of infinity.] (Brice Parain, *Recherches sur la nature et les fonctions du langage* [Editions Gallimard, 1942], conclusion)

ii. Crediting translation. Look for a published translation of a quotation in a foreign language before using one's own, and make sure to fully credit the original translator.

iii. Quoting in translation. Translated quotes don't necessarily require the original text after them, and, in certain cases, foreign-language quotes don't need to be followed by a translation. Either of these supplements may be acceptably placed in notes.

iv. Retranslation. Do not translate back into English a quote that has been translated from English into a foreign language. The original English quotation must be found in order to quote it, or else it must be paraphrased.

v. Typographic style: Foreign-language quotations should use the same punctuation as the original, with an exception: to avoid confusion, punctuation related to quotations should be matched to English style. This means that any guillemets (« ») should be changed to quotation marks, double and single quotes should be switched as necessary, and that ellipses should have spaces around them but should not be enclosed in brackets.

3.10.14 Proofreading

Quotations should be carefully proofread; block quotations for format (indentation, spacing, size, font) and all quotations for accuracy in relation to their source.

3.10.15 From Secondary Sources

If one source is only available through another source, both must be cited in the reference list entry.

3.10.16 Speech

i. Direct discourse. *See* **3.6.7, iv**.

ii. Faltering or interrupted speech. For ellipses used to indicate faltering speech, *see* **Section 3.6.1, xvi**. For em dashes used to indicate interrupted speech, *see* **Section 3.6.8, xii**.

iii. Of drama, discussion, and interviews. Quoted dialogue in a work of drama is not usually enclosed in quotation marks, but instead is preceded by a colon and distinguished by a different typeface or font. Indent any second and subsequent lines from a single speaker to match the beginning of the first, and italicize stage directions.

F. DOCTOR WEST: Good morning, madam.

HENRIETTA, *curtsying*: How do you do, Dr. West. May I take your hat?

Use a similar format for the transcription of an interview, but consider paragraph-style indentation. Actions should be italicized and put in brackets.

INTERVIEWER: Would you tell me a little bit about what happened?

SUBJECT A7: Yeah, well, for a minute I was dizzy, and then I felt this pain in my head and I was like [*gasps*] and I felt my heart pounding.

3.10.17 Epigraphs

Epigraphs are quotes used to introduce books or parts of texts. They do not require quotation marks, but should be followed, on the next line, by the quoted author's name and, if appropriate, by the title of the work. If a chapter epigraph needs a footnote or endnote, the reference number should follow the source, not the quote itself. Epigraphs fall under fair use and do not generally need permission to be used.

Science is the father of knowledge, but opinion breeds ignorance.
—Hippocrates

3.10.18 Common Facts, Proverbs

Common facts and proverbs do not need to be cited, except when the exact phrasing is quoted from a specific source.

Our solar system contains only one star.
Organisms evolve through mutation and natural selection.

3.10.19 Accuracy

Make sure that quotations accurately cite and reflect the content of their sources. *See also* **Sections 3.10.20–21**.

3.10.20 Attribution

Before using any copyrighted material outside of fair use, be sure to obtain permission from the owner—preferably written permission. This may apply even if the work was written by the same author for a different publication. Cite the original sources of borrowed material with footnotes. *See also* **Chapter 5, Sections 5.7–5.12**.

3.10.21 In Context of Original

Quotes should accurately reflect the spirit of their source. There should be no ambiguity about the meaning of the original material, whether it is conveyed through a misleading selection of words or through the implication of something contrary in the surrounding text.

3.10.22 Paraphrasing

Whether an idea is paraphrased or directly quoted, it should be cited in a note or within parentheses to give credit to the original author. In cases where a source is quoted extensively (more than a few paragraphs), permission may need to be obtained.

3.10.23 Editor's Responsibility

Editors should not correct any apparent mistakes in quoted material without making sure that the errors are transcription errors, not ones made by the original authors. Punctuation should be scrutinized, especially quotation marks and ellipses. All quotes should be checked for sources. *See* **Section 3.10.8** for [*sic*] and other ways around original errors in quotes.

3.11 Proofreading and Editing

3.11.1 Manuscript Editing

3.11.1.1 Online

i. Author's review of. Edited copies of an author's work should be sent to that author for review as a computer printout of a digitally corrected document. The editor may also wish to send a clean version (with changes made but not marked), and should certainly keep a marked version of the original file.

ii. Backup, conversion, and cleanup. An editor should keep separate copies of any files that an author sends, which should not be edited or converted themselves, but used instead as backups and for reference. If the files need to be converted to a format that works with the editor's software, note that the process may change the formatting or improperly translate certain characters.

iii. Changes in. Online editing offers the advantage that most software programs let one find and replace all instances of a specific text string. This allows an editor to quickly update every instance of an anomalous term or format in the complete text to reflect the correct version as written on a style sheet.

iv. Files for typesetter. Once an editor approves of the files returned from the author, a clean printout and copies of all files should be sent to the typesetter.

v. Formal markup languages. Manuscripts that are to be published online will often need to be converted to a formal markup language.

vi. Generic marking or coding. Parts of a work may be labeled by codes (often enclosed in angle brackets) that designate the intended typesetting format.

vii. Of notes. Many word processing programs keep track of notes automatically, renumbering and standardizing their typographical format as they are inserted, moved, or deleted.

viii. Proofreading type set from. Proofread type set from digital files against the last version of the manuscript returned from the author.

ix. Queries to author. When editing on a computer, take advantage of the feature that allows comments to be inserted separate from the author's footnotes. Other options for submitting queries to the author are to insert them in brackets directly in the text, or to list them in a separate file, their locations marked in the text with conspicuous symbols.

x. Redlining. Many word processors automatically have the option to keep track of changes as a work is being edited. The resulting document can be marked in various ways, depending on the software, but deleted text is usually struck through and inserted text is usually typographically distinguished by color, highlighting, underlining, etc.

3.11.1.2 Paper-Only

i. Author's review. Edited copies of an author's work should be sent to that author for review as a printed manuscript with hand-marked corrections in a distinct color. The editor may also wish to send a clean version (with changes made but not marked), and should certainly keep a photocopy of the hand-marked manuscript.

ii. Color of pencil or ink. Corrections to a manuscript should be made in a color that would show up easily if photocopied—typically red. If more than one person must mark the same pages, different colors should be used for each person.

iii. Page numbering. Manuscripts may be numbered from beginning to end, or each chapter may be numbered separately. In the latter case, every page should indicate which chapter it belongs to as well as the page number. Designate new pages inserted into a manuscript with the previous page number plus *a*, *b*, *c*, etc.

iv Preparing manuscript for. The author is responsible for delivering two copies of the manuscript to the publisher, double-spaced and with any corrections clearly written. Editors should keep one of these copies clean from their own marks, for backup and for reference.

v. Proofreading type set from. Proofreaders must check the retyped copy against the hand-marked original manuscript, making sure that all changes have been made correctly.

vi. Of quotations and previously published material. Quotations and material published previously should not be changed, except for conventions in marking quotes within that material, such as ellipses or quotation marks. Editors should take care to check the accuracy of quotes and sources, to make sure proper credit is given to the original authors and to eliminate any transcription errors.

3.11.1.3 Parts of Works

i. Bibliographies and reference lists. These should be checked for consistency, internal as well as with the main text of the work. Make sure they are properly alphabetized or chronologized.

ii. Captions and illustrations. Make sure illustrations and captions are in the right places. Check captions for consistency with the text in their corresponding illustrations and with the style of the whole work. In a list of illustrations for the front matter of a book, each caption should be condensed to a line or two.

iii. Cross-references within text. Cross-references should lead easily and reliably to the information that is promised, and superfluous cross-references should be removed.

iv. Indexes. To edit an index, do not refer to every page listed in every entry, but evaluate a sample of ten or so entries and extrapolate. Do check every entry for relevance to the work, correct format and punctuation, logical cross-referencing and subentry division, and consistency with the terms used in the text.

v. Mathematical copy. All mathematical characters that do not appear on the manuscript exactly as they should be set by the typesetter must be clearly written in by the author. Potentially ambiguous symbols should be clarified by the editor through special instructions or coding; these may include letters or numbers in unusual fonts, Greek letters, and subscripts and superscripts (especially those within other subscripts and superscripts). For information about the mathematical typesetting programs TeX and LaTeX, *see* **Section 7.1.4.**

vi. Notes. Editors should decide whether notes are in an appropriate location or whether they should be moved to an appendix, incorporated into the text, or deleted. They should be checked for consistency of style and, especially in hand-marked manuscripts, for numbering and format.

vii. Subheads. Check subheads for appropriate capitalization, for accurate correlation with the table of contents, and for logical division and subdivision.

viii. Tables. Tables should be scrutinized even more carefully than running text. They should be accurate, well organized, and consistent in format and style with other tables and text in the work.

ix. Titles. Titles should be checked for consistency with terms used elsewhere in the work, including the table of contents. They should also be checked for capitalization and, if applicable, for proper coding.

x. Translations. Unless a translation is the author's own, it must not be edited any more than a regular quote would be.

3.11.2 Mechanical Editing

Mechanical editing is the process of correcting typographical errors, inconsistencies in formatting, and simple language mistakes—those of spelling, capitalization, punctuation, grammar, etc. It is a part of copy-editing along with substantive editing, and may be done with the help of a computer program or entirely by hand. Computer programs can identify most typos and many grammar mistakes, but are not yet complex enough to be either reliable or authoritative, even in mechanical editing.

3.11.3 Substantive Editing

Substantive editing seeks to make a text clearer through rephrasing and reorganization of material. It identifies and fixes ambiguities, wordiness, awkward constructions, and content that may be inappropriate.

3.11.4 Editing for Style

Style, in an editorial sense, refers to a set of mechanical conventions of language and format. Journals or publishers often have their own house styles. Although different styles may have only minor differences between them, it is important that a writer, editor, and publisher all agree to use a specific style in order to maintain consistency within a work or group of works.

3.11.5 How to Mark a Manuscript

Although there are computer programs which keep track of editing automatically, many editors prefer to mark manuscripts by hand. Handwritten marks should be located directly around the parts of text to which they refer; only remarks that do not fit between the lines, such as queries to the author, should be in the margins. For a list of specific symbols, *see* **Section 3.11.6**.

Be aware that there is no universally agreed upon system for these marks; they vary from country to country, between publishing houses, and even between disciplines. Nevertheless, the system recommended in this text should be intelligible to anyone who has a general familiarity with proofreading marks.

i. Insertions and deletions. Material that is to be inserted into text should be written above where it belongs, with its exact location—whether between letters or between whole words—marked by a caret. Any editorial text that is *not* to be inserted into the manuscript should be circled.

Material in a manuscript that needs to be removed should be struck through. (To be certain such a marking is not confused with the mark for italics, the circled term "del" may appear in the margin, or the delete sign may be added to the cross-out mark.)

ii. Transpositions. Material that would be made correct by transposing two of its elements, whether two letters or two groups of letters or words, can be indicated with the transposition mark (*see* **Section 3.11.6**).

iii. Closing up and separating text. Text that needs to be closed up, usually an incorrectly open or hyphenated compound, can be marked with two curved lines. To separate text, insert a space with a vertical line and this symbol: #.

iv. Punctuation changes. Inserted periods should be circled; commas and other small symbols should always be accompanied by carets or inverted carets. For additional marks, *see* **Section 3.11.6**.

v. Dashes and hyphens. Distinguish carefully between hyphens, en dashes, and em dashes, as well as between soft and hard hyphens (soft hyphens exist solely to divide a word between lines), and mark those which are incorrect or ambiguous with the appropriate symbol from the list in the following section.

3.11.6 List of Proofreading Marks

On the following two pages are listed the standard proofreading marks, their appearance in the margins and in the body of the text, and indications of how these instructions will be understood and carried out by a typesetter or compositor. Because these are handwritten instructions subject to the variations in style and handwriting, it is important for an editor to verify that the markings have been correctly interpreted.

Proofreading Marks

Instruction	Symbol	Mark in Type	Corrected Text
Delete	*ɡ*	Now is̸ the time	Now is the time
Insert	∧	Now the time	Now is the time
Do not change	stet	Now is the ~~time~~	Now is the time
Make capital	cap	now is the time	Now is the time
Make lower case	lc	Now is the Ⳇime	Now is the time
Make small capitals	sc	Now is the time	Now is the time
Make italic	ital	Now is the time	*Now* is the time
Make bold	bold or bf	Now is the time	**Now** is the time
Move	tr	Now the time is	Now is the time
Transpose	tr	Now the is time	Now is the time
Close up space	⌒	Now i s the time	Now is the time
Delete and close up space	⌒	Now i̸s the time	Now is the time
Spell out	sp	for all ④ men	for all four men
Insert: space	#	Now is the time	Now is the time
– period	⊙	Now is the time	Now is the time.
– comma	⸳	Now we believe	Now, we believe,
– hyphen	=	seventy six	seventy-six
– semicolon	⸴	Now is the time this is the place	Now is the time; this is the place

Proofreading Marks, *continued*

3.11.7 Style Sheets

A style sheet is a list of words or mechanical conventions important to a specific manuscript that are not covered in—or are allowed to violate the rules of—the house style. If the manuscript is being marked by hand, each entry in the style sheet must note the page number of its first occurrence in the manuscript. The style sheet should accompany the manuscript when it is sent to the typesetter, and should also be on hand for proofreading.

3.11.8 Type and Typesetting

i. Description. Typesetting is the process by which a manuscript or marked proof is electronically updated (or retyped from a hard copy) to reflect editorial changes that have been made to it. In the past, professional typesetters were the only option for publishers, but today there are also computer programs that can help anyone accomplish the task. The typesetter's services are usually called for after the manuscript has been edited the first time, and again after that copy has been proofread and marked. The last version of the work that is typeset is the final version that is to be sent to the printer.

ii. Designer's layout: A document called a *designer's layout* should accompany a manuscript to the typesetter. On it are all the typesetting specifications needed, matched up with codes in the manuscript.

iii. Electronic files. Electronic files that may need to be sent to a typesetter include the separate elements of the manuscript (exactly as they appear in print copies), the designer's layout, the editor's style guide, and a list of all typesetting codes used.

iv. Errors. An error that occurs many times throughout a work, such as a formatting error introduced by converting a file, does not need to be marked every time. Instead, notify the typesetter of the situation with a single message.

v. Instructions. Proofreading marks will be understood by most typesetters. Written-out messages for the typesetter should be circled. If a table will not fit on one page, the typesetter should have instructions regarding where to place footnotes and *continued* lines.

vi. Master proofs. The final proof that the publisher sends to the typesetter is called the master proof. After it is returned with updated corrections from the typesetter, it should be kept by the publisher until the work is in publication.

vii. Sending files. Before a marked manuscript is sent to a typesetter, all queries to the author should be addressed and deleted.

viii. Components. Components of typesetting that need to be specified by a designer include the fonts and sizes of text in different parts of the work, the dimensions of those parts, and the dimensions of the pages themselves.

ix. Numbering pages. Many books place their page numbers, or folios, on the top-left of left-hand (*verso*) pages and on the top right of right-hand (*recto*) pages. Other books place page numbers on the bottoms of the pages, toward the outside edges or else centered on each page. Page numbers do not normally appear on the tops of the first pages of chapters, or at all on pages that consist only of tables, diagrams, or pictures.

x. Spine copy. The spine is the part of a book's cover that joins the front and back covers, and is usually the only part visible when the book is on a shelf. The spine of a hardcover book should have the author or editor's name, the book's title (but not its subtitle), and the publisher's name and/or logo.

A journal's spine should display its title, its volume number, and its exact date of publication, including the season. The range of page numbers the issue contains is optional.

xi. Running heads. Running heads are the lines of text that appear at the very top or, sometimes, the bottom of a page. They may consist of the title, chapter numbers, or other information. The title may need to be shortened in order to fit the running head on one line. If there is information in the running head which will reflect content of individual pages (rather than of chapters or sections), the copy for the headers should be written after page proofs have been typeset.

Running heads do not usually appear on pages that consist only of pictures or tables, unless there are several pages of them. Nor should they appear on top of a page that begins a chapter.

Most front matter (introduction, preface, etc.) should have running heads, but not usually on the first pages of sections or on display pages, such as the dedication.

In endnote sections, running heads should display the chapter or the pages to which those notes refer. If they display just the chapter, then the running heads within each chapter must have the chapter number.

Include running heads even in works that are only intended to be published electronically. For electronic journal articles, they should display the name of the article (shortened, if necessary) and of the author or authors.

3.12 Numbers

3.12.1 Expressing Numbers in Text

3.12.1.1 Words vs. Numerals

i. General Principals. In most scientific writing, single digit numbers are written out as words (*zero, one, two*, etc.) while longer numbers are kept in their numeral form. However, there are many exceptions to this rule.

Quantitative numbers or any number representing a measurement should be written in numeric form.

3 filters   9 times   2 tons

Single-digit numbers should not be spelled out when doing so will result in an obvious inconsistency.

Of the 71 subjects, only 6 improved on their previous scores.

On the other hand, if two separate categories of numbers are being talked about in the same context, distinguishing one by spelling it out may be the best option.

Seven of the monkeys were given 5 mg of the drug, twelve were given 10 mg, and six were given 50 mg.

ii. Technical vs. Nontechnical Copy. Where in nontechnical writing a popular rule is to spell out all numbers from zero to one hundred, in scientific writing it is preferable to write out only the numbers zero to nine. But if compound numbers are ever written out, a hyphen should be used, as in *twenty-one* and *ninety-nine*.

In nontechnical writing, zero and one are almost always spelled out.

on one hand
zero gravity
one example

But in mathematical writing, especially when zero and one are directly used with a unit of measurement or when they are used in a series of other numbers, they should be written as numerals.

1 hour   1 mm
0, 1, 2   1 out of 10

When numbers that are less than one are displayed as decimals, it is important to include a zero before the decimal point, so that the mark is not missed. For integers, a decimal point is not necessary unless it precedes zeros that are significant figures.

0.238   *not*   .238
$x = 0.93$ *not* $x = .93$
87 *or* 87.0 *not*   87.

When mathematical expressions are written out, use numerals.

x is multiplied by 6
the sum of 9 and *y*

In nontechnical writing, numbers are often used in figures of speech; these should always be spelled out. In technical writing, such figures of speech should be avoided.

Instead of: Wait an hour or two.
Use: Wait 1–2 hours.

Instead of: Compare one to the other.
Use: Compare the 2 cases.

3.12.1.2 First Word in a Sentence

A number at the beginning of a sentence must always be written out as a word. If this cannot be done, then the sentence must be reworded. This is done to avoid any confusion if there is a number or an equation at the end of a previous sentence. If spelling out such a number is too awkward, an inserted adverb or prepositional phrase can often solve the problem without the need for a complete recasting of the sentence.

Two hundred and fifty years ago, electricity and magnetism were
    viewed as separate phenomena.
In total, 184.43 mg of precipitant was collected.

If a number begins a sentence and is followed by another related number or a number in the same category, then both must be spelled out. If this seems awkward, the sentence should be reworded. For the use of inclusive numbers (e.g., 12–16) at the beginning of a sentence, *see* **Section 3.12.3.4**.

Thirty or forty scientists are employed by the laboratory.
The laboratory employs 30 to 40 scientists.

3.12.1.3 The Need for Consistency

As with all types of writing, it is important in scientific and mathematical writing to be consistent, especially when dealing with numerals written in their word form. Doing so will avoid any confusion on the reader's part, especially if two numbers describe measurements, or if two or more numbers or expressions are adjacent in a sentence.

3.12.2 Scientific Uses of Numbers

3.12.2.1 Quantities and Measurements

In scientific and mathematical writing, a measurement should always be written as a numeral followed by the unit, either written out or abbreviated. If the measurement and unit are used together as a modifier, join them with a hyphen.

2.5 cm	9 grams	12-km tunnel
3.0 gallons	275 kW	2.5-mg sample

If there is no numeral preceding the unit of measurement, then it must be written out, not abbreviated.

We measured the solution in liters rather than in quarts.

3.12.2.2 Fractions

In running text, fractions that are less than 1 can either be spelled out and hyphenated (*one-half* or *half, two-thirds*), written as numbers using a slash (3/4), or displayed as block fractions (¾). Larger fractions should only be written in numbers with a slash. If there are mixed numbers, use a block fraction (5¾). If this cannot be done, put a hyphen between the whole number and the fraction (5-3/4).

3.12.2.3 Percentages

Percentages should be written as numbers (35% or 35 percent). In scientific writing, use the percent symbol (%) especially when many percentages are given, while in the humanities the word *percent* should almost always be written out.

There is a 25% chance that the gene will be inherited.
The odds of success are less than 0.5%.

3.12.2.4 Scientific Notation

In technical writing, scientific notation is often used when writing either very large or very small numbers in order to simplify and save space. Scientific notation is expressed by multiplying a compact number by a power of ten. For example, 953 000 000 may be written as 9.53×10^8, and 0.000 000 107 as 1.07×10^{-8}.

3.12.3 General Usage with Numbers

3.12.3.1 Ordinals

In technical writing, single-digit ordinal numbers (those that describe rank or order) should be spelled out as words (*first* to *ninth*), while multiple-digit ordinals can remain as numerals (10th, 50th, 199th). If both single- and double-digit ordinal forms are included together in a series, use the number ("1st, 5th, and 15th trials"). If there are multiple single-digit ordinal forms used in one sentence, they may also be written as numerals.

3.12.3.2 Enumerations

When enumerating items in an outlined list, use arabic cardinal numerals; when running them directly into a sentence, use the written-out ordinal form, or pair the numerals with parentheses. *See also* **Section 3.6.9, xi.**

How to make brownies:
1. Combine butter, chocolate, sugar, eggs, vanilla, and flour.
2. Pour the mixture into a greased baking pan and bake for thirty minutes.
3. Let the confection cool and slice it into brownies.

To make brownies: first, combine chocolate, sugar, eggs, vanilla, and flour; second, pour the mixture into a greased baking pan and bake for thirty minutes; third, let the confection cool and slice it into brownies.

3.12.3.3 Plurals of Numbers

Numbers that are spelled out are pluralized as any noun would be, and numerals are made plural by adding an *s*; no apostrophe is needed.

Patients in their twenties and thirties may experience different side effects than patients in their forties and fifties.
Polio was prevalent in the 1950s.
Customers rated the service from the low 40s to the higher 90s.
three 7s, five 12s, and two 15s

3.12.3.4 Inclusive Numbers

When describing a range of numbers, use an en dash in between two numerals to represent *through* or *up to*. If the words *from* or *between* precede the numbers, or if the numbers are spelled out, forgo the en dash and separate the numbers with the words *to*, *until*, or *and*.

Please read chapters 3–7
There were between 20 and 30 patients waiting to be seen.
The package should arrive in three to four weeks.

Do not omit initial digits of the second number in a range to save space.

Incorrect: 1902–18
Correct: 1902–1918

When expressing ranges of quantities and measurements, the symbol or abbreviated unit of measurement can be written once after the second number in the range, or it can be written after both numbers. However, if the symbol is one that must appear without a space after a number, it should be written after both numbers in a range, unless they are joined by an en dash. Measurements of units with different prefixes should not be joined by an en dash.

20 to 40 mm *or* 20 mm to 40 mm
75% to 80% *or* 75–80%; *not* 75 to 90%
300 MHz to 50 GHz

When a range of numbers begins a sentence, write out the first number as a word and the second number as a numeral, or else reword the sentence. The unit of measurement should not be abbreviated unless it appears only after the second number.

Twenty to 40 mm of copper wire...
The range was from 20 to 40 mm.
Forty-five degrees to 90 degrees...
The angles ranged from 45° to 90°

To avoid any ambiguities, do not use the word *by* before a range of numbers. This may confuse the reader into thinking that there was a change from the first number to the second, rather than a range between the two.

Similarly, care should be taken when using the words *increase, decrease,* or *change* before giving a range. For example, if one were to state that "production increased from 25 to 30 units per day," it is unclear whether the production increased by anywhere between 25 to 30 units per day, or if production increased exactly 5 units per day, from 25 to 30. Specify clearly what is being stated: "production increased by a range of 25–30" or "production increased from an initial 25 to a final 30."

3.12.2.5 Commas in Numbers

If a number has more than four digits, use a comma to separate digits grouped in threes (1,500,000). In some countries, the comma is used as a decimal point (*see* **Section 3.6.4, xviii**). To avoid confusing foreign audiences, some journals use a thin space to separate groups of three digits (including those to the right of the decimal point) and use a period for the decimal place: (1 547 682.97).

Four-digit numbers do not require commas or any spacing to separate the last three digits from the first (3000). However, when four-digit numbers appear with a listing or a group of other large numbers that use commas or thin spaces, then they should be changed accordingly.

$$
\begin{array}{r}
3,000 \\
860,000 \\
+\quad 2,900 \\
\hline
915,900
\end{array}
$$

3.12.3.6 Arabic and Roman Numerals

Arabic numbers consist of the numerals 0, 1, 2, 3, 4, 5, 6, 7, 8, and 9; they are most commonly used in mathematical and scientific writing. Roman numerals consist of seven letters, each representing a specific value: I = 1, V = 5, X = 10, L = 50, C = 100, D = 500, and M = 1000. The letters are combined to create sums of the number that they represent. They are either all added together, or if a smaller letter precedes a larger one, then the smaller one is subtracted from the larger to create the total number. Roman numerals are mostly used in classification systems for tables, titles, headings, pagination, etc.; they should not be used when describing data or as units of measurement. They may also appear as both upper case (for main elements of an outline or list), or lower case (for lists within text or as subsidiary portions of an outline—*see* below).

3.12.3.7 Outline Style

In numbered lists or outlines, roman numerals, arabic numerals, and letters may all be used. Each system of numbering should be used for its specific subsection within the list. With outlines containing many subsections, the numbers and letters are usually organized in order, by roman numeral, capital letter, arabic numeral, lower-case letter, and lowercase roman numeral.

I. Volume Name
  A. Title
    1. Chapter
    a. paragraph
    b. paragraph
   i. table
    c. paragraph
    2. Chapter

3.12.3.8 Money

When large monetary units are expressed in writing, numerals or a combination of numerals and words may be used along with the appropriate currency symbol. For amounts in the millions, it is best to use only numerals, since different countries have different definitions for *million*. Do not use scientific notation to abbreviate monetary values.

$400 thousand
$400 000 000

If the number of a smaller sum of money is spelled out, *dollar*, *cent*, or any other monetary unit should also be spelled out. If the amount is written as a number, the currency symbol should be used.

five dollars   thirty-five cents $90

3.12.3.9 Parts of a Book

Any numerations used in books for headings, chapters, pages, tables, bibliographical information, references, etc. are written as numerals. In some cases the pagination of the front matter of the book is done in lowercase roman numerals, while the rest of the book is paginated with arabic numerals.

3.12.3.10 Names of People and Places.

i. Personal names. Royalty, monarchs, popes, etc. sometimes use roman numerals after their names to differentiate them from others who share their name and title.

>King Louis XIV
>Pope John Paul II

Roman numerals are also sometimes used with personal names; for this, however, using an ordinal number instead of a roman numeral is also correct.

>John J. Brown III
>John J. Brown 3rd

ii. Organizations. When ordinal numbers are used in the names or titles of an organization, numbers less than one hundred should be written as a word, and those over a hundred should be written as a number. It may be useful to refer to a website or pamphlet of the specific organization for their preferred usage.

>First Presbyterian Church
>Twenty-ninth Infantry

iii. Government designations. When ordinal numbers are used to describe specific governments, dynasties, or other designations of a period of rule, any number less than one hundred should be written out as a word.

>The pharaoh Akhenaten changed the face of the Eighteenth Dynasty by bringing monotheism to Egypt.
>The motto of the Second Republic was "liberty, equality, brotherhood."
>The 109th Congress was criticized as the "Do Nothing Congress."

iv. Addresses and road names. Roads that are named with an ordinal number should be spelled out as a word, while house, apartment, or building numbers that precede the name should be written out in their arabic form. If a street's name is a number over one hundred, a comma may be inserted in between the unit number and the street number in order to avoid confusion.

> The museum is located on 1000 Fifth Ave.
> We moved to 1771C, 190th St.

Numbered highways are written in arabic numerals.

> Route 22 is a dangerous road.
> We drove on Interstate 90 for three hours.

3.12.4 Mathematical Expressions in Text

3.12.4.1 General Usage

In mathematical and other technical writing, symbolic expressions and formulas are sometimes used to add emphasis, to clarify a point, or to convey a relationship. In general, only small or simple expressions should be written out in word form in the text, leaving larger expressions to be set in their symbolic form, either in displays or blocked examples.

3.12.4.2 Line Breaks in Text

When writing out expressions within the text, be wary of how line breaks may affect them, as a poorly placed break may confuse the reader or change the meaning of the expression. As a general rule, do not break an expression within fences ($\{\}$, [ ], ( ), $| \ |$, $\| \ \|$, etc.).

> $a(x + y)$ $\sum (a + b)$

Line breaks may occur after an operator that is a verb ($=, <, >, \neq, \equiv$, etc.) contained within fences.

> For this section, keep in mind that $a + b = x + y$.

When a collective sign ($\sum$, $\prod$, $\cap$, etc.) begins a mathematical expression, do not break until after an operator appears outside of fences.

> The trend can be represented by this formula: $\sum (x - y) +$
> $a(b - c)$.

When an expression begins with an integral symbol ($\int$), break after a d variable and its following punctuation or verb.

> $\int a\,(a + b)(x + y)\,dx$
> $= x(a + y)$

When there are two sets of fences adjacent in the text, break in between them and insert a multiplication symbol.

> This is a useful formula: $(a + b)$
> $\bullet(x + y)$.

3.12.4.3 Mathematical Expressions Set Off from Text

Writers may choose to set off long, complex, or important expressions within displays or blocked text in order to present them more clearly. The decision to place an expression within the text of a paragraph or to set it off separately on a new line must be made at the writer's discretion, keeping in mind the importance of the expression, its relationship to other expressions, and how readable it will make the text. Equations (containing an equals sign) are usually set off and labeled with numbers.

When there are multiple displays within a text, they may be numbered within the margins and the display is centered within the text.

> (97.2) $(a + b)(c + d)(e + f)$
> (97.3) $(u + v)(w + x)(y + z)$

For line breaks in a display, the rules of **Section 3.12.4.2** apply, except that, in displays, breaks should be made before an operator, not after.

If an expression is so long that it does not fit within the width of the page, the display may be set into a smaller font size to avoid numerous line breaks.

For more specific conventions of mathematical usage, *see* Chapter 6.

Chapter 4. Citations and References

Contents

Contents, 4.4, *continued*

Chapter 4. Citations and References

4.1 Citation and Reference Style

4.1.1 Standards of Clear and Proper Attribution

It is vital that projects appropriately credit information obtained from the works of other authors. In addition to the obvious and manifest reason of maintaining the highest standards of honesty and integrity in scientific research, there are several practical reasons for maintaining high standards in this area:

- It allows the reader to further his or her knowledge on the raised topic by being able to observe the research that led to the creation of a project.
- Citations bring concrete evidence to one's work and proves that what was written in one's project was not simply fabricated or invented.
- Conclusions are better supported when research of similar thinking colleagues are included and credited in one's project.
- Citations allow one to avoid plagiarizing someone else's work by correctly crediting the information gathered that was not one's own original research.

4.1.2 When to Reference and When Not to Reference

At times, one can be unsure as to what should and should not be cited in a project. Original research and ideas do not need to be cited, as they were conceived and gathered by the researcher.

But when ideas and research that are from another source are used, it is important to cite that source and to give credit to the original author. The following situations generally require some form of citation:

- Actual quotes from another person's work.
- Paraphrasing material from another work.
- Specific reference to another person's work.
- Using an idea or concept that has already been conceived.
- When someone else's work has been a critical element in the creation and development of your own ideas or assertions.

It is appropriate at times to use information from other sources without having to cite it in your own work. Commonly known facts, such as the name of the first president of the United States, and facts that can be found within multiple sources, such as Asia being the world's largest continent, fall into this category. Authors should use their judgment as to what information should and should not be cited, and be prepared to be queried by editors on this subject.

4.2 The Citation-Sequence System

The Citation-Sequence system is used when superscript numbers are attached to the end of a sentence within the text that the author believes has additional information to give. This is done as a way of giving more detailed information to the reader without cluttering the document. The number directs the reader to the endnotes, which are generally found at the end of the chapter, book, or manuscript, where a more detailed description of the original citation can be found. These references are numerically arranged in the endnotes according to the order in which they were cited in the text.

The following is an example of Citation-Sequence system:

- Studies have shown that a higher drinking age has led to fewer automobile fatalities.[1] But recent research suggests there isn't a connection between the two.[2]

Two ideas are referenced and cited numerically. The superscript numbers should always be placed outside of periods and commas and inside colons and semicolons.

When citing two or more references within the same sentence, a comma must separate the superscript numbers. If there are more than two references in a contiguous sequence, use a hyphen to indicate a closed series of numbers, and a comma to separate any additional numbers. The following is an example on how to cite multiple references within the same sentence.

- In a recent survey on people's view of psychiatry, many were found to disagree with the influence of the unconscious mind, which was originally conceived by Dr. Sigmund Freud.[2-4, 7,8]

Following is an example of how to cite material that is used multiple times within a document. In this case, the same reference number is used for the material.

Most elementary schools teach children that Columbus discovered America.[4] While it has been proven that he actually found islands in the Bahamas,[5] school children are still taught that the USA was discovered by Columbus.[4]

The final example in the Citation-Sequence system deals with secondary citations. This is when a writer is citing documents viewed by another source while being unable to look over the original material. An example on how to correctly cite this would be:

Patrick was driven to become a priest in order to view secret Vatican documents[13 (cited in 17)] that could answer his questions about religion.

When citing within a document one should avoid placing the citation directly after a number or a unit of measurement, as this could confuse the reader. For example, instead of "the answer was 13.[7]" use "13 was the answer.[7]"

4.3 The Author-Date (Name-Year) System (APA Style)

The Author-Date system refers to citations that immediately follow the text. Text citations are provided to identify information found within the text. A reference list, or cite list, at the end of the work is later included to provide a more detailed listing of the sources used.

Text citations contain, in parenthesis, the author's last name and the publication year of the work that is being cited, with the name and year being separated by a comma. Below is an example of a text citation for a book with a single author.

(Genings, 1969)

For two authors use:

(Genings & Jackson, 1983)

If one is citing multiple works by the same author that were published in different years, place the author's last name and then the publishing years in chronological order.

(Genings, 1973, 1976)

If one is citing multiple works by the same author that were published in the same year, add a letter to the end of each year to distinguish each work.

(Genings, 1984a, 1984b)

When the name of the author being cited is within the sentence itself there is no need to additionally cite the author. In this case one would cite the year after the name within the sentence. An example of this is:

Jackson (2006) was able to start the company after selling his previous business.

The reference list in the Author-Date system is organized differently then the information within the Citation-Sequence system. Unlike the Citation-Sequence system, which arranges the reference list numerically, the Author-Date system arranges its reference list alphabetically by the author's last name.

4.4 Citation-Sequence and Author-Date Citation Formats

This section examines the format used in the reference list for both the Citation-Sequence and the Author-Date citation systems.

4.4.1 General Format

This section will deal with how to properly cite work using both the Citation-Sequence system and the Author-Date system. Examples of both sequences will be provided.

For the Citation-Sequence, a reference list begins with the author's last name and the initial of the author's first name, with the entry followed by a period. An example of this is "Myers, T."

Following the author's name is the title of the book or of the publication. If the publication is a second or higher edition, include the edition number within the citation. The entry then has a period inserted at the end.

The city of publication comes after the title. The name of the city and state or country is followed by a colon, and then the name of the publisher. The whole citation ends with a semicolon, the publication year, and then a period.

An example of a Citation-Sequence reference is as follows:

Paige, C. The Complete History of Larping. New York, NY: Random House; 2001.

In a reference list for the **Author-Date system**, the last names of the authors cited are listed alphabetically.

Unlike the citation-system, the publication year follows the author's name. A period then follows.

Then, the title of the work is written in italics and is followed by a period.

After this comes the city and state of the publication, followed by a colon, the name of the publisher, and finally a period.

An example of a cited work using the Author-Date system is as follows:

Paige, C. 2007. *The Complete History of Larping*. New York, NY: Random House.

The rest of this chapter will deal with how to reference works in cite lists with more specific information than the example listed above. In each case, an example from the Citation-Sequence will be shown first, followed by an example of the Author-Date system.

4.4.2 Books with One Author

Citation-Sequence

Kayman, O. The Anatomy of Undersea Invertebrates. New York, NY: Scholastic; 1992.

Author-Date

Kayman, O. 1992. *The Anatomy of Undersea Invertebrates*. New York, NY: Scholastic.

4.4.3 Books with Two or More Authors

In the Citation-Sequence system and the Author-Date system a comma is used between the listed authors. However, in the Author-Date system use a comma and then an "and" after the last listed author. If there are more then six authors for the work, write the first three authors and then write the phrase "et al" after the third author.

Citation-Sequence

Karner, L., Lippy, D. The Secret Of Life. New Orleans, LA: Old World Press; 1999.

Author-Date

Karner, L., and Lippy, D. 1999. *The Secret Of Life.* New Orleans, LA: Old World Press.

4.4.4 Books with an Editor as the Author

Add "ed." if the work was edited by a single editor and "eds." if multiple editors are listed. The name or the group name of the editor or editors appears before the "ed." or "eds."

Citation-Sequence

Baker, B., Jade, T., eds. Soup and Those Who Enjoy It. New York, NY: Viking Press; 2002.

Author-Date

Baker, B., and Jade T., eds. 2002. *Soup and Those Who Enjoy It.* New York, NY: Viking Press.

4.4.5 More than One Work by the Same Author

The Citation-Sequence system references are listed numerically, so when referencing more than one work by the same author, no changes need to be made.

However, in the Author-Date system one has to separate references from the same author. If multiple references from the same author were also published within the same year, list the references alphabetically by the titles of the works, then add letters after the publication date. For example, use "a" and "b" for 2 works, "a", "b", "c" for 3 works, etc.

For the second and any other following listings, instead of writing the author's last name and first initial, use a triple em dash and follow with the publication year.

Author-Date

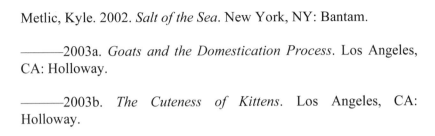

Metlic, Kyle. 2002. *Salt of the Sea.* New York, NY: Bantam.

———2003a. *Goats and the Domestication Process.* Los Angeles, CA: Holloway.

———2003b. *The Cuteness of Kittens.* Los Angeles, CA: Holloway.

4.4.6 Books with Authors and Editors

In both systems the name of the editor (or editors) is placed between the title of the book and the publishing location. The name of the editor or editors would come before the "ed." or "eds."

Citation-Sequence

Boatman, P., Bellock, M. The Downside of Sugar. Dawn, R., ed. New York, NY: Scholastic Publishing; 2001.

Author-Date

Boatman, P., and Bellock, M. 2001. *The Downside of Sugar.* Dawn, R., ed. New York, NY: Scholastic Publishing.

4.4.7 Books with Translators

The name of a translator would follow the title of the book, or the names of the editor or editors, if they are included in the reference. The word "translator" follows the actual name of the translator.

Citation-Sequence

Abney, L. Dissecting The Work Of Claude Monet. Fagan, E., ed. Quigly, H, translator. Paris, FR: Random House; 1996.

Author-Date

Abney, L. 1996. *Dissecting The Work Of Claude Monet*. Fagan, E., ed. Quigly, H., translator. Paris, FR: Random House.

4.4.8 Book Volumes with Separate Titles

If the title of a book is part of a set and contains a separate volume name, then the title of the volume will go after the title of the book, and before the name of any editors or translators. This is the same in both the Citation-Sequence system and the Author-Date System.

Citation-Sequence

Smith, C. The Blitz. Volume 4, December. Kracken K, ed. Buckinghamshire, UK: United Military Press; 1999.

Author-Date

Smith, C. 1999. *The Blitz*. Volume 4, December. Kracken, K, ed. Buckinghamshire, UK: United Military Press.

4.4.9 Chapters or Other Parts with Separate Titles, but the Same Author

In both systems the name of the chapter is written before the title of the book. The word "in" is then written between the chapter title and the title of the book, italicized in the Citation-Sequence and roman in the Author-Date sequence. In addition, the number of pages that references the material covered in the chapter are written after the title of the book, the names of the editor, and translator if such information is available.

Citation-Sequence

Mybis, M. The Failings of Mallrats, *in* The Work of Filmmaker Kevin Smith. Nathan K, ed. 103-111. Los Angeles, CA: Pacific Union; 2007.

Author-Date

> Mybis, M. 2007. The Failings of Mallrats, in *The Work of Filmmaker Kevin Smith*. Nathan, K., ed. 103-111. Los Angeles, CA: Pacific Union.

4.4.10 Chapters or Other Parts with Different Authors

There may be times when a person will write a book but have a specific chapter written by another author. In this instance, the author of the chapter is written at the beginning of the reference and the writer of the entire work is placed after the book title. Except for this, the reference is written as it would be in the examples of **Section 4.4.9**.

Citation-Sequence

> Affleck, H. How We Get Self-Sharpening Screws, *in* The How Do's and How Don'ts of Drilling. Deddie, N. 120-134. Philadelphia, PA: Pearl Publishing; 1994.

Author-Date

> Affleck, H. 1994. How We Get Self-Sharpening Screws, in *The How Do's and How Don'ts of Drilling*. Deddie, N. 120-134. Philadelphia, PA: Pearl Publishing.

4.4.11 Anonymous Author

If the author is unknown then place the word "Anonymous" in place of where the author would normally go. Write the rest of the reference as it would normally be written.

Citation System

> Anonymous. Go Ask Alice. New York, NY: Simon Pulse; 1971.

Author-Date System

> Anonymous. 1971. *Go Ask Alice*. New York, NY: Simon Pulse.

4.4.12 Place of Publication Clarified

If the location of a publication is not one that is immediately recognized to a reader, one should follow the name of the publishing location with the postal abbreviation of the state or country of the location. While this can be done at your discretion for most instances, it must be done if the name of the location is the same within two or more states or countries. This is so the reader will be able to know the exact location of the publication if they decide to look it up.

Citation-Sequence

Tammy, J. The Complete Guide To Babysitting. Bay City, TX: Open Range House; 2001.

Author-Date System

Tammy, J. 2001. *The Complete Guide To Babysitting*. Bay City, TX: Open Range House.

4.4.13 Article and Essay Types

When referencing an article, start out by writing the author's name and then the title of the article. This is followed by the publication in which the article appears, the volume and issue numbers of the publication, and the page numbers of the article. The year in which the article was printed is included at the end of the citation in the Citation-Sequence system and after the author in the Author-Date system

If the article is specialized then include this information in your citation. If the article cited is an editorial, obituary, letter to the editor, or something other than a regular article, it is specified in parentheses between the title of the article and the publication being cited.

Citation-Sequence

Bambing, R. The Truth Hurts: Why Tree Huggers Hate Guns. Shooting Times. 32 (2), pp. 12-17; 1997.

Author-Date

Bambing, R. 1997. The Truth Hurts: Why Tree Huggers Hate Guns. *Shooting Times*. 32 (2), pp. 12-17.

4.4.14 Organization as Author of Single Work or Series

If an article is credited to an organization rather than an individual author, follow the rules for articles written by an author and substitute the name of the organization for the name of the author. When alphabetizing the reference list in the Author-Date system, list this reference by the first word in the name of the organization.

Citation-Sequence

The History Channel. The Golden Age Of Piracy. New York, NY; 2007.

Author-Date

The History Channel. 2007. *The Golden Age Of Piracy*. New York, NY.

4.4.15 Articles in Journals Paginated by Issues

Citations for articles in journals operate differently when there is continuous pagination and when the pagination is not continuous in subsequent issues.

When pagination within an article is continuous, cite the author first, followed by the title of the article, the title of the journal, the volume number, and pages being referenced.

When pagination starts over at the beginning of each issue, the issue number must be put in parentheses after the volume number.

Citation-Sequence

Brodrick, B. A Flat Tire In The Fast Lane Of Life. Car Buying Journal, Vol 12, pp. 109-111; 2002.

Author-Date

Brodrick, B. 2002. A Flat Tire In The Fast Lane Of Life. *Car Buying Journal*, Vol 12, pp.109-111.

4.4.16 Newspaper Articles

First, write the name of the author of the article and then the title of the article. This is followed by the name of the newspaper and the specific date that the article was published. After this, write the location of the newspaper and the section and page number where the article is located.

Citation-Sequence

Mack, C. Swamp Monster Stalks Town: The Torch, February 13, 2003, Smallville, KS, A:03; 2003.

Author-Date

Mack, C. 2003. *Swamp Monster Stalks Town*: The Torch, February 13, 2003, Smallville, KS, A:03.

4.4.17 Magazine Articles

Citations for magazine articles are basically the same as citations for newspaper articles. Write the author's name and the name of the article, the name of the publication, the date of the issue, and the pages that are being referenced.

Citation-Sequence

Umor, D. Why Diet Soft Drinks Are Harmful. Living With Diabetes, November 2006, pp. 49-61; 2006.

Author-Date

Umor, D. 2006. *Why Diet Soft Drinks Are Harmful.* Living With Diabetes, November 2006, pp. 49-61.

4.4.18 Articles in Supplements to Issues or Volumes

Write the author of the article and then the title of the article. This is followed by the title of the magazine or journal, the word "Suppl.," the number of the supplement, and the page numbers of the article being cited.

Citation-Sequence

Babbles, G. European Photos From The First Half Of The 20th Century. Famous 20th Century Photos From Around The World. Suppl. No. 20. pp. 637-645; 2003.

Author-Date

Babbles, G. 2003. *European Photos From The 20th Century.* Famous 20th Century Photos From Around The World. Suppl. No. 20. pp. 637-645.

4.4.19 Electronic Publications

When writing a website citation, first put the author of the website. If no author can be found, write the name of the organization that created the site. Following that is the name of the website, the date the website was accessed, and the URL address in parenthesis. If the URL takes up more then one line, use a "/" at the end of the line and finish the address on the next line. When sending your reference list to someone electronically, it is preferable to remove the actual hyperlink and have it come up simply as text.

Citation System

Mayer, M. Let The Good Time's Role. Accessed April 20, 2005. (www.letthegoodtimesrole.com); 2005.

Author-Date System

Mayer, M. 2005. *Let The Good Time's Role.* Accessed April 20, 2005. (www.letthegoodtimesrole.com).

When citing from an online journal, newspaper article, or book, use the same format as a regular journal, article, or book, except include the access date and URL at the end of the reference.

4.4.20 Microform

After citing the author and the name of the work, write what medium was used in parentheses, followed by where the material was viewed and the section or collection where the microfilm was available. The reel and fiche numbers, which are separated by a colon, end the reference. One should also indicate if the microfilm can still be viewed. Write "Available" or "Not Available" after the medium being used.

Citation-Sequence

Nickels, B. Fall Play Loved By All. (Microfilm) Available: North Adams State Library, The Creative Arts, 665:2; 1932.

Author-Date

Nickels, B. 1932. *Fall Play Loved By All* (Microfilm) Available: North Adams State Library, The Creative Arts, 665:2.

4.4.21 Conference Presentations, Papers, and Abstracts

First, write the name of the author and the title of the paper. Write an italicized "in" followed by a colon and the title of the conference. Follow that information with the date that the work was presented and the location of the conference. Then write the city where the document was later published and the title of the document that the paper was published in. Finish with the page numbers that were used from the source. In the Citation-Sequence, the year the document was published follows the page numbers. In the Author-Date sequence this follows the author.

Citation-Sequence

Quick, S. The Increasing Dangers Of Marijuana. *In:* National Drug Convention, June 27, 1987, Babylon. Bay Shore: Police Guide On Illegal Drugs, pp. 17-26; 1986.

Author-Date

Quick, S. 1986. *The Increasing Dangers Of Marijuana. In:* National Drug Convention, June 27, 1987, Babylon. Bay Shore: Police Guide On Illegal Drugs, pp. 17-26.

4.4.22 Scientific and Technical Reports

For scientific and technical reports first write the author and the name of the report. Next, write the report number, the department of the company or university that sponsored the report, and the name of the company or university that the document came from.

Citation-Sequence

Strones, L. The Reduced Risks Of Genetically Enhanced Food. Report T-97, Department of Agriculture, Harvard University; 2006.

Author-Date

Strones, L. 2006. *The Reduced Risks Of Genetically Enhanced Food.* Report T-97, Department of Agriculture, Harvard University.

4.4.23 Dissertations and Theses

First write the name of the author and then the title of the dissertation or thesis being cited. Follow with the type of dissertation or thesis in brackets, the educational institution it was submitted to, and the location of the institution.

Citation-Sequence

Nett, S. The Impact Of Myths In The Modern World. [M.A. Thesis]. Georgia Tech, Atlanta, GA; 2002.

Author-Date

Nett, S. 2002. *The Impact Of Myths In The Modern World.* [M.A. Thesis]. Georgia Tech, Atlanta, GA.

4.4.24 Patents

When writing a patent reference, first write the name of the individual who filed the patent and then the name of the patent itself. This is followed by the number that was designated to the patent. After this comes the name and location of the particular national patent office that was used, the date the patent was filed on, and the date the patent was granted on. When writing the dates, indicate which is the filed date and which is the granted date. Try to get the most specific date possible, although if only the year can be found, that is acceptable.

Citation-Sequence

> Jones, N. Five Blade Razor. U.S. Patent No. 1,367,702. U.S. Patent and Trademark Office: Washington, D.C. Filed January 23, 2005, Granted April 1, 2008.

Author-Date

> Jones, N. 2008. *Five Blade Razor.* U.S. Patent No. 1,367,702. U.S. Patent and Trademark Office: Washington, D.C. Filed January 23, 2005, Granted April 1, 2008.

4.4.25 Maps, Legal Papers, Government, and Agency Documents

First, write the author of the map and then the map's title. Then write the city where the map was produced and the organization that produced it. In most cases this will be the map company, however, other organizations that could have also produced the map could be a legal body (NYC Water Works), a government institution (Department of Interior), or an agency (Yosemite National Park Commission).

Follow that information with the map's scale, the number of sheets of the physical map, and the year the map was published.

Citation-Sequence

> Stout, R. Map of Clumpsy Farm. Juneau: Alaskan Map Services, scale 1:8,000,000, 6 sheets; 2005.

Author-Date

> Stout, R. 2005. *Map of Clumpsy Farm.* Juneau: Alaskan Map Service, scale 1:8,000,000, 6 sheets.

In regards to a government reference, start with the author of the document. If the author is not available, substitute the name of the agency. Write the title of the document and, if available, a report, contract, or series number in parentheses. Follow with the location of the document, a colon, and the office from which the document came. Finish with the date that the document was written. If it is difficult to find the exact date, include at least the year.

Citation-Sequence

Piking, A. Maritime Law Handbook (FA-8083). Washington, DC: U.S. Government Maritime Office, 2004.

Author-Date

Piking, A. 2004. Maritime Law Handbook (FA-8083). Washington, DC: U.S Government Maritime Office.

4.4.26 Audiovisual Publications and Materials

In the case of audiovisual material, first determine the main director, writer, producer, and or editor of the work. If multiple individuals performed these jobs use the name of the director of the piece. If the director performed additional jobs on the piece (writer, producer) list all the functions he or she performed. Write the name and the job that was performed in the project. After this, write the title of the production and the medium in which the material is presented on, such as if it is a Motion Picture, a DVD, a VHS, a CD-Rom, etc. (If it is a motion picture that is released on DVD or VHS, write motion picture. However if it is a project that is made for a specific medium, such as a how to video, put the medium in which it was released.)

After this, write the main country of production, the distributor or publisher, and the original release date for the material.

Citation-Sequence

Smith, K. (Director/Writer/Producer). Clerks. (Motion Picture), USA: Miramax Films; 1994.

Author-Date

Smith, K. (Director/Writer/Producer). 1994. *Clerks* (Motion Picture), USA: Miramax Films.

4.4.27 Classical, Religious, and Secular Literature

When citing religious works such as the Bible, Old Testament, New Testament, Talmud, Koran, the Book of Mormon, etc., the names of the works are capitalized.

In these citations, use the "chapter and verse" or other division numbering instead of page numbers in the reference list.

For example, if one were citing a passage from the Book of Peter in the Bible it would be written "Peter, 3:16" instead of writing "Peter, p. 245."

This format of separating the larger section of the text from its subsection is common to many religious texts, although other religious texts may use terms other than chapter and verse.

This format may also be used in works from authors such as Dante and Shakespeare, where text is grouped in ways that differ from the regular paragraph format.

4.4.28 Unpublished Documents

When citing unpublished documents, write "Unpublished" after the title of the work, as there is no publishing house or location. With the date of publication try to cite an approximate date the document was created, although this might be difficult if sufficient records of the document were not kept.

Citation-Sequence

Brown, J. The Truth on the Iraq War. Unpublished. 2007.

Author-Date

Brown, J. 2007. *The Truth on the Iraq War*. Unpublished.

4.4.29 Forthcoming documents

For forthcoming documents, cite as one would with a regular document except write the word "forthcoming" in place of the date. Include page numbers within the citation if available.

Citation-Sequence

Maclay, A. Guide to Automobile Purchasing. New York, NY: Random House; Forthcoming.

Author-Date

Maclay, A. Forthcoming. *Guide to Automobile Purchasing.* New York, NY: Random House.

4.4.30 Bibliographies

A bibliography is a list of all the research material that was used in creating the work, even if the work was not directly cited in the notes. It is normally located after the reference list (which contains the works specifically cited in the work and the notes).

A bibliography is arranged alphabetically by the last name of the authors or editors. If no author or editor is given, the organization that sponsored the work is used. While most bibliographies are arranged as a single alphabetic list, there are times when the list is broken up into different subject sections. Each subject has an individual heading, with the material that falls under that subject arranged alphabetically. A source is not repeated after it is listed in a section, even if it falls under more then one subject heading. Generally however, it is best that a bibliography not be divided into sections. One may divide a bibliography by the nature of the sources when one uses material that would not fit into a normal alphabetic list, such as material from an archive collection, personal interviews, field observation, etc., or if the bibliography is intended to be primarily a guide for further reading.

Generally, bibliographies are categorized as either full bibliographies (which is assumed if no qualifier is used with the word "bibliography") or selected bibliographies. A full bibliography contains: (a) material that was cited in the author's work (both in the text itself and in notes); (b) material that was consulted but not referenced; and (c) works that were not used but would be of interest to any reader of the paper.

A selected bibliography is a bibliography in which only some of the materials an author used are listed. The author indicates that the bibliography is not complete by titling it "Selected Bibliography" or "Select Bibliography." Following the title the author may include a note explaining the criteria the author used in selecting the specified works.

At times an author may wish to provide additional information within a bibliography. This is achieved by creating an *annotated* bibliography or a bibliographic essay. An annotated bibliography includes a brief comment (between 100 to 200 words) following each (or many) of the listings. A bibliographic essay is an informal essay in which the bibliographic information in included within the text itself. The names of the authors and discussed works are written in the text and the additional publication information follows each discussed work in parentheses. Generally, a bibliographic essay is included in addition to a bibliography and reference list rather then instead of them. In some guides, lines after the first line of a bibliographic entry are indented; in others they are not.

In some scientific works, bibliographies use the following format: works in the notes (either in the text or in the endnotes) appear as the author's last name (or, on the case of an institution, an abbreviated form or acronym of the institution's name), followed by a date of publication in parentheses. For authors with more titles published in a single year, the works are identified by a letter following the year, in the order in which they appeared. (Three letters of the author's name are used if there are two authors whose name begin with the same two letters.)

Marion (1975)   Marion (1982a)
                     Marion (1982b)

In some instances, the reference will be noted by only the first two letters of the author's last name, followed by the year, both in parentheses. In this case, in text and in notes reference is made to the author by name and the reference appears in parentheses following the mention of the author's name.

A full presentation of the calculus of tensors is provided by Marion (Ma 1982), with many examples worked out.

In the bibliography of a work using this format, the notation would appear on the left, and is then followed by the full bibliographic reference. The bibliography entries are listed alphabetically by the

(Ma 1982) Marion, Jerry B. 1982. Classical Electromagnetic Radiation, 3rd Edition, New York, NY: Academic Press.

4.5 Further Reading and Resources

When citing references, it is important to bear in mind that different manuals advocate different styles of citation, as was amply demonstrated in the chapter. It is therefore important for an author to consult the editor of the publication to which one intends to contribute about the style used. (Often, of course, such information will be available in the publication's "Guidelines for Authors" publication or on its website.) As important as selecting the correct style for citations is being consistent, so that one should not use both the Citation-Sequence system and the Author-Date system within the same piece of work.

While each system is generally used for different material (the Citation-Sequence system is used for scholarly and literary works while the Author-Date system is used for research on social sciences), it is not unheard of for each system to be used for referencing different works. It is far more important to use the same system consistently within the work then to scrutinize over when to use what system.

There are several reference guides one can use to look up how to properly cite material within a paper and in a reference list. The premier guide to use is The *Chicago Manual Of Style*. This reference guide is precise in how to reference material in different mediums, although it favors the usage of the Author-Date system (which has become known as the APA style of citation because of its close association with the *Publications Manual of the American Psychological Association*).

The *AMA Manual Of Style* is another vital reference book for citation. Although not explicitly stated within the text, the book favors the usage of the Citation-Sequence system. The book gives examples of lesser-known citations, such as how to reference patents and microfilm.

Scientific Style and Format: The CSE Manual for Authors, Editors, and Publishers gives details on how to cite material in both the Citation-Sequence system and the Author-Date system.

The *MLA Style Manual* offers data on how to correctly cite references using the Citation-Sequence. It also offers brief descriptions on other citation systems, including the Author-Date system.

4.5.1 Sources Cited and Referenced

Council of Science Editors, Style Manual Committee. 2006. Scientific Style and Format: the CSE manual for authors, editors, and publishers. 7th ed. Reston, VA: The Council of Science Editors, Style Manual Committee.

Gibaldi, J. 1998. MLA Style Manual and Guide to Scholarly Publishing. 2nd edition. New York, NY: The Modern Language Association of America.

Oxford University Press. 2007. AMA Manual of Style: A Guide for Authors and Editors. 10th edition. New York, NY: Oxford University Press.

The University of Chicago Press. 2003. Chicago Manual of Style. 15th edition. Chicago, Illinois: The University of Chicago Press; 2003.

Chapter 5. Copyright and Permissions

Chapter 5. Copyright and Permissions

5.1 What is Copyright?

Copyright is a system used to ensure that an author's original works are protected from unauthorized use in the work of others. This protection is grounded in the U.S. Constitution and is granted by law. Both published and unpublished work is granted copyright protection.

When most people think of works that are protected by copyright, they think of artistic creations such as novels, music, multimedia expression, and works of art. However, copyright protection can also extend to nonfiction and technical writing, as well as to design work for computer software and architecture. Material that is posted on the Inter-net is also subject to copyright law. This includes original created material that others want to use and material that can be constituted as falling under fair use. Copying and pasting material from one web site to another without authorization can be viewed as copyright infringement.

There are several systems that are unprotected by copyright, but are protected by other designs. Ideas, concepts, discoveries, and methods of operation are protected under patent law rather than by copyright. Trademark laws are used to protect slogans, titles, names, and designs. Copyright is only applied to protect the medium in which such material is expressed.

5.2 "Poor Man's" Copyright

"Poor Man's" Copyright refers to an individual using registered dating in a attempt to establish a particular time and date that a work was created at. Most times this is done by an individual sending their created work to themselves by registered mail. The individual would then not open the sealed envelope and can use the postal date as an indicator to the creation of the work.

This practice is not encouraged, however, and is not a viable substitute for obtaining an actual copyright. Most times courts will not recognize this practice as open envelopes can be mailed to the person and material can then be added and sealed at a later date.

5.3 Work Made for Hire

The term "work made for hire" (or simply "work for hire") refers to work that is created by one individual or organization, but is in fact owned by another individual or organization. An example is a computer company hiring an outside programmer to write a new computer program. If both parties agree that the work in question is considered "work made for hire," the work becomes the property of the company and not of the individual. In such cases, both parties sign a contract stating this fact before the individual begins work.

5.3.1 Independent Contractors

Work that is created by an independent contractor can only become work made for hire if both parties agree beforehand that the work being created will be work made for hire. Copyright ownership of material created by contractors varies from country to country.

5.3.2 Contractors and Government Employees

Material that is solely created by federal employees of the US government is not protected by copyright law and falls into the public domain unless the government has reason to restrict the access of the work. However, work that is created by an outside contractor for the US government does fall under copyright law, with ownership of the work determined at the time the contract is drafted.

Work that is jointly created by federal employees and outside contractors falls under copyright law. Ownership of work created jointly by federal employees and outside contractors is often difficult to accurately determine. If select work was created independently with the intention of merging it with larger material, then creation and ownership of the work is fairly easy to determine. If the entire project was created by government employees and independent contractors then proper ownership is more difficult, if not impossible, to determine.

5.4 Rights of the Copyright Owner

The copyright owner possesses a number of rights pertaining to their work. These rights include:

- The right to have their work reproduced;
- The right to distribute copies of their work to others;
- The right to prepare future altered works of the original work;
- The right to have their work publicly displayed or performed.

The rights that a copyright holder possesses on a particular work are divided into two sections: economic rights and moral rights. Economic rights extend to the reproduction and distribution of a particular work. This includes the copyright owner deciding how they want their work reproduced, allowing their work to be performed, displayed or broadcast, and if they want to have copies of their work distributed. Moral rights pertain to the attribution (whether a copyright owner wants to be indicated as the creator of a particular work) and integrity of a work. Integrity of work includes interpreting and representing a copyright holder's work without betraying the beliefs and values of the copyright holder. Moral rights are more often enforced and recognized outside of the United States. Within the United States moral rights are only recognized with the representation of select visual artworks.

The limitation of the rights held by the copyright owner is a concept known as "fair use." This concept is further discussed in **Section 5.7** of this chapter.

5.5 Transferring Copyright

Transferring copyright is when copyright ownership is transferred, either in part or in full, between a publisher or agent acting on behalf of the original owner, or between the original owner and the new owner of the material. For legal reasons this transfer is done through a written agreement, although a written agreement is not required for transferring rights that are not on an exclusive basis. (A **sample copyright transfer agreement** appears on the next page.)

Copyright can also be transferred through legal decisions, or through a will or other estate law proceeding.

Copyright is a personal property right, and therefore its passage can be subjected to the state laws and regulations that govern the transfer, ownership, or inheritance of personal property. These laws vary from state to state.

5.5.1 Sample Copyright Transfer Agreement

AMERICAN SOCIETY OF CIVIL ENGINEERS
COPYRIGHT TRANSFER AGREEMENT

Publication Title: _____

Manuscript/Chapter Title: _____

Author(s) – Names and address of all authors _____

_____

The author(s) warrants that the above-cited manuscript is the original work of the author(s) and has never been published in its present form.

The undersigned, with the consent of all authors, hereby transfers, to the extent that there is copyright to be transferred, the exclusive copyright interest in the above-cited manuscript (subsequently called the "work") in this an all subsequent editions of the work, and in derivatives, translations, or ancillaries, in English and foreign translations, in all formats and media of expression now known or later developed, including electronic, to the American Society of Civil Engineers subject to the following.

- The undersigned author and all coauthors retain the right to revise, adapt, prepared derivative works, present orally, or distribute the work provided that all such use is for the personal noncommercial benefit of the author(s) and is consistent with any prior contractual agreement between the undersigned and/or coauthors and their employer(s).

- In all instance s where the work is prepared as a "work made for hire" for an employer, the employer(s) of the author(s) retain(s) the right to revise, adapt, prepare derivative works, publish, reprint, reproduce, and distribute the work provided that such use is for the promotion of its business enterprise and does not imply the endorsement of ASCE.

- No proprietary right other than copyright is claimed by ASCE.

- An author who is a US Government employee and prepared the above-cited work does not own copyright in it. If at least one of the authors is not in this category, that author should sign below. If all the authors are in this category, check this box: ☐ and sign here: _____. Please return this form by mail.

SIGN HERE FOR COPYRIGHT TRANSFER [Individual Author or Employer's Authorized Agent (work made for hire)]

_____   _____
Print Author's Name   Print Agent's Name and Title

_____   _____
Signature of Author (in ink)   Signature of Agency Rep (in ink)

Date: _____

Note: If the manuscript is not accepted by the ASCE or is withdrawn prior to acceptance by ASCE, this transfer will be null and void.

5.6 Length of Copyright Protection

Today the length of time that a work exists under copyright is determined by the copyright act of 1976. Through this act, works that are protected under copyright are generally divided into three groups:

- Works that were originally created on or after January 1, 1978.
- Works that were created before January 1, 1978, but not published or registered by that date.
- Works that were originally created and published or registered before January 1, 1978.

i. Works that were Originally Created on or After January 1, 1978

A work that is published on or after January 1, 1978 is automatically protected from the moment of its creation. In addition to copyright lasting throughout an author's life, the length of copyright is extended to last an additional 70 years after the death of the author.

In the case of work that is created by two or more authors on or after this date, the term of copyright lasts for 70 years after the death of the last surviving author. This does not apply to work made for hire.

Works made for hire created after 1978, as well as works that are anonymous, are granted a term of 95 years after publication, or 120 years after creation, whichever of the two periods is shorter.

ii. Works Originally Created Before January 1, 1978, But Not Published or Registered by That Date

Copyright length for work in these cases is generally the same as for work created after January 1, 1978. The copyright spans for the duration of the author's life plus 70 years, and in the case of multiple authors, the copyright spans for the duration of their lives plus 70 years after the death of the last surviving author.

Federal copyright law provides that the term of copyright for work in this category will not expire on or before December 31, 2002, and that for work published on or before December 31, 2002, the copyright term will not expire before December 31, 2047.

iii. Works Originally Created and Published, or Registered Before January 1, 1978

Before 1978, copyright was secured on a work's publication date, or on its registration date if the work was registered in unpublished form. In either case, the copyright lasted for a term of 28 years from the date it was secured, with the option of renewing the copyright for an additional 28 years during the final year of the original term.

Subsequent laws have extended the renewal period of copyrighted materials before 1978 by 67 years. This means that copyrighted materials in this category are protected for a total of 95 years under current law.

In 1992, a law was passed that made filing for renewal registration optional for copyrights in this category. This means that filing for renewal registration is no longer required to extend the original 28-year term of copyright protection to 95 years.

5.7 Fair Use

Fair use is the act of using material that is under copyright without obtaining permission from the original author. Some examples in which copyrighted material is used without permission are:

- Teaching
- Criticism and or Comment
- News Reporting
- Scholarship Writing
- Research

As can be expected, what is considered fair use can vary widely and be narrowly or widely interpreted. The use of some materials, such as whole passages, unpublished works, and photos are unavailable for usage even under the clause of fair use. For legal reasons it is better to seek permission when using copyrighted materials than to automatically assume one can use them according to fair use.

US federal copyright law establishes four factors in determining whether a particular use of copyrighted material falls under fair use:

- The purpose and character of the use, including whether such use is commercial or for nonprofit or educational purposes.
- The overall nature of the copyrighted work.
- The amount and substance of the portion used in relation to the copyrighted work as a whole.
- The effect of use upon the potential market for, or value of, the work in question.

5.8 Permissions

It can be difficult to accurately determine whether using copyrighted material falls into the category of fair use or copyright infringement. There is no specific number of words or sentences that can be safely used without permission. In addition, simple acknowledgment of the source of the copyrighted material is not a substitute for obtaining permission to use said material.

When one wishes to use copyrighted material they must obtain permission from the copyright owner beforehand. This is done by contacting the author and or publisher, and asking for permission to use his or her work. The way in which their work will be used also needs to be clearly stated. After this, one must obtain written permission to use the copyrighted work. (There may be a possible royalty fee for use of the copyrighted material.)

This may be difficult for several reasons:

- The licensing fees for use of the copyrighted material may be higher than one is willing or able to pay.
- The owner of the copyrighted might be difficult to identify and contact.
- The owner of the copyrighted material may refuse the use of their material, either for the use that you intend or for use altogether.

If permission for use cannot be obtained, one should avoid using copyrighted material unless the guidelines of "fair use" can clearly be applied. Use of copyrighted material without permission can lead to legal repercussions.

5.9 Changing Copyright Status

It is possible for copyright, either in full or in part, to be transferred from one party to another. Sometimes this is done simply as an option to gain benefits, and other times it is necessary to give up copyright in exchange for working with a publishing house. A publishing house may ask an author to give up copyright in exchange for royalties and or other benefits. The publisher may also ask for the copyright in exchange for the author profiting from the production, marketing, and advertising of the work.

This process is not limited to the publishing world. It also applies to filmmakers, musicians, architects, software writers, chorographers, and other persons in various creative fields.

It is possible for the original copyrighter to gain back copyright at a later date, most commonly by buying it back. It is also possible for copyright holders to surrender their copyright, but still legally use the work under certain conditions. Some copyright holders, for example, reserve the right to use their work in a different medium, time frame, or geographical area.

Transfer of copyright ownership should never be done orally, and instead should always be done with a written contract and recorded with the US Copyright Office.

5.10 Public Domain

A work that is no longer protected by current copyright law is considered to be "public domain." This means that anyone is free to use that work without legal repercussions. When using public domain work from other countries, one should keep in mind that public domain status can vary from country to country. Just because a work is in public domain in one country does not mean it's in the public domain of another. For legal reasons one should check the public domain status of a work if it originates from a foreign country.

Most countries allow the copyright of a published work to expire, or enter the public domain, when the following occurs:

- Work that was created and first published before January 1, 1923, or at least 95 years before January 1 of the current year, whichever of the two comes later.
- The author, or last surviving author in the case of multiple authors, died at least 70 years before January 1 of the current year.
- The creator of the work declared it able to be used in the public domain.
- The work is not renewed for copyright.

5.11 How Does Copyright Effect Scientists?

Most instances in which scientists use copyrighted material, such as in scholarly papers, research projects, and teaching, generally fall easily into the category of fair use. A few obvious exceptions are copying an entire paper word-for-word, or using copyrighted scientific material for profit without obtaining copyright approval. One should review the terms and obligations of fair use before using any copyrighted material.

Remember that even when using copyrighted material under fair use, one should still properly cite any outside contributions to scientific papers and projects (*See* **Chapter 4**).

Scientific inventions and processes fall under patent laws rather than copyright laws. Patents allow inventors to obtain the rights to their inventions. This prevents others from marketing, selling, making, or importing said invention without the inventor's permission. Generally a patent lasts for twenty years following the date the patent was filed on, although this may differ outside of the United States.

5.12 Copyright Conventions

There is no single copyright standard that will protect an author's creations throughout the world. Protection against copyright infringement can vary greatly from country to country.

This is not to say that any given country will not offer some form of protection to foreign works. Two international copyright conventions, The Berne Union for the Protection of Literary and Artistic Property (Berne Convention) and the Universal Copyright Convention (UCC), were set up to help break down the barriers in copyright law between foreign nations. An author from a country that is a member of either of these conventions can usually claim copyright protection under their terms. The United States has been a member of the UCC since September 16, 1955, and a member of the Berne Convention since March 1, 1989.

There are no formal requirements in the Berne Convention. Under the UCC, any problem in a national law may be overcome by the use of a copyright notice in the form and position specified by the UCC.

A UCC notice should consist of the symbol © (a C in a circle) accompanied by the year in which the work was first published and the name of the copyright owner. The phrase "All Rights Reserved" should follow. This notice should be clearly visible on the work.

- Example: © 2001 Lesley Hoffman. All Rights Reserved.

In 1989, the United States made the use of a copyright notice optional. However, US law can still provide advantages when a copyright notice is properly used. The use of a copyright notice can aid in confronting a defense of "innocent infringement," which is when an individual claims that they were unaware of copyright protection on the work they were using.

5.12.1 For Journalists

Journalistic copyrights are owned and maintained by a variety of different entities. This includes, but is not limited to, newspapers, networks, magazines, newspaper groups, websites, and the individual journalist. Examples of such are:

- © 2001 The New York Times. All Rights Reserved.
- © 2002 NBC. All Rights Reserved.
- © 2003 The New Yorker. All Rights Reserved.
- © 2004 Manic Madness Newspapers. All Rights Reserved.
- © 2005 generalnews.com. All Rights Reserved.
- © 2006 Mike Handle. All Rights Reserved.

In recent years, examples of organizations and individuals seeking copyright protection have grown and expanded with the development of online blogs and other new forums.

5.12.2 For Books

Copyright protection for books is usually granted to either the individual or to the publisher.

- © 2007 Jen Benson. All Rights Reserved.
- © 2005 Ducky Publishing. All Rights Reserved.

5.13 Author and Publisher Responsibilities

Most of the time, copyright infringement can be avoided or lessened by following certain rules and procedures:
- Only use original material in your work.
- Cite, document, reference, and footnote all material that is from other sources. This isn't a defense in and of itself, however, it can be helpful.
- Try to determine if the use of copyrighted material falls under the clause of fair use. Carefully look into the clauses and restrictions of fair use and have a copyright attorney review the work.
- Contact the owner of the copyright to find out if they will allow use of their work.
- Secure written permission from the owner of the copyright for your intended use of the material. This is the only way to fully rule out responsibility in a lawsuit.

5.14 Registration

In the United States, registering a copyright is encouraged rather than required. A registration creates a public record of the copyrighted material and helps vouch that a copyright is valid. It is also necessary for a copyright to be registered in order to file an infringement suit in court.

In order to register a copyrighted work, one must submit a copy of the registration form, copies of the materials that are being registered (the number of copies needed varies depending on what exactly is being registered), and a filing fee with the US Copyright Office. The material will become registered on the date all of the completed requirements have been received by the US Copyright Office.

5.15 Liability and Rights

When a copyright is infringed upon, the only person who can bring forth a lawsuit is the owner of the copyright.

Registration of a copyright is necessary before one proceeds with a lawsuit, although technically this is optional.

In a copyright lawsuit, a plaintive may bring charges not only to the person who used their work without authorization, but also to anyone who profited from using the work.

Willful infringement of copyright, as opposed to innocent infringement, is when a person or organization deliberately reproduces or manipulates copyrighted works with full knowledge of their protected status. In addition, willful infringement usually involves using the copyrighted material for profit.

Liability for a defendant can include tangible damages (through loss of profits or business opportunities), attorney's fees, court costs, and unspecified legal damages.

PART II.

Style and Usage for Specific Disciplines

Chapter 6: Style and Usage for Mathematics

Contents

Contents, 6.2, *continued*

[*See also* Chapter 3, Section 3.12, for related style issues.]

Chapter 6. Style and Usage for Mathematics

6.1 Manuscript Preparation

6.1.1 Structure of a Standard Mathematics Paper (in brief)

A standard mathematics paper contains many of the elements of a typical academic article, such as an abstract, an introduction, and reference section. (See **Part I, Chapter 2**: **Preparing the Manuscript.**) In addition to these elements, mathematics papers may contain theorems, mathematical definitions, and other similar constructs. While there are no specific guidelines for these later items, there are suggested formating styles.

It is recommended that the names of theorems, corollaries, lemmas, propositions, axioms, and definitions be placed in boldface type. The body of the text of these items may be placed in italics in order to distinguish these items from the bulk of the text of the article. The aforementioned items should be placed within their own paragraph.

Remarks, examples, cases, problems, and other mathematical constructs should also be placed in their own paragraphs. The headings of these items should be italicized, but the body of their text should be the same as the text in the main body of the article.

Leading spaces may be placed above and below theorems, corollaries, lemmas, and propositions. No leading spaces should be placed above or below axioms, definitions, examples, cases, problems, or other mathematical constructs.

6.1.2 Other Forms of Mathematics Manuscripts

There are numerous types of mathematical books including textbooks, handbooks of integration, books of tables of the values of mathematical functions, and other academic books.

6.1.3 Indexing—Mathematics Subject Classifications

6.1.3.1 The MSC

The Mathematics Subject Classification (MSC) is an indexing system designed to allow readers to easily find documents in their mathematical field of interest. The MSC is developed from the two reviewing databases *Mathematical Reviews* (*MR*) and *Zentralblatt MATH* (*Zbl*).

The MSC contains over 5000 classifications, categorized using a system of of between 2 and 5 digits. The most recent revision of the MSC was re-leased in the year 2000. Several publications, including all of those published by the AMS, use the MSC to classify mathematical works. Each publication is assigned a primary classification number and may also be assigned one or more secondary classification numbers.

6.1.3.2 Obtaining the MSC

A hard copy version of the MSC may be found within the annual index of *Mathematical Reviews*, published with the December issue, or ordered through the AMS Customer Services Department. An electronic version of the MSC is available on the Internet at:

http://www.ams.org/msc/ or via telnet at e-math.ams.org.

(The login and password, for the telnet site, are both *e-math*.)

6.1.4 Typesetting Mathematical Text

6.1.4.1 TEX and LATEX

TEX (often appearing as "TeX"; the same is true of similarly-named programs such as LATEX, which appears often as "LaTex"), a typesetting program designed to enable extensive manipulation of mathematical text, has become the standard typesetting system in mathematics publishing. Available in various versions both commercially and free, TEX is fairly easy to use and allows the author a large amount of control over the appearance of text and equations. It can be easily converted by a publisher into a finished product, minimizing the introduction of errors, reducing costs, and allowing the work to be processed more quickly.

LATEX incorporates a number of macros that enable the author to more easily format complicated documents, but loses some of the versatility of TEX.

6.1.4.2 AMS-TEX and AMS-LATEX

The American Mathematical Society (AMS) offers AMS-TEX and AMS-LATEX, which contain style files, fonts, and instructions. The AMS also offers publication-specific packages and instructions. The webpage http://www.ams.org/tex/ contains links to other webpages from which AMS-TEX and AMS-LATEX can be downloaded, as well as links to other resources related to TEX and LATEX. The AMS website contains material on commands and shortcuts to facilitate the use of these programs by authors of mathematics papers.

6.1.4.3 Other Programs Used to Typeset Mathematics

Commercial word processors, such as Microsoft Word and Word-Perfect, often come with the ability to typeset mathematics. Additional commercial plug-ins are available to enhance the abilities of these word processors to format various characteristics of equations such as font and color.

Programs such as Mathematica are able to produce TEX and LATEX typesetting code.

Some versions of TEX, such as HyperTEX and AMS-TEX, can be used to create and format equations for use on the Internet. Alternatively, a technology known as MathML can be used to produce webpages containing equations. The MathML standards can be found at : http://www.w3.org/Math/.

Amaya is a freely available W3C MathML compliant editor, produced by W3C in collaboration with INRIA, and can be downloaded at: http://www.w3.org/Math/Software/mathml_software_cat_browsers.html #Iamaya.

Lists of other MathML editors, plug-ins, and other software can be found at: http://www.w3.org/Math/Software/.

6.2 Usage

6.2.1 Alphabets Used in Mathematical Expressions

6.2.1.1 Latin Characters and Their Uses

i. Roman. Abbreviations of several common mathematical functions and operators are placed in roman font.

ii. Italic. Names of common functions, numbers, the imaginary number i (sometimes referred to as j), and mathematical operations should not be italicized, whether they appear within an equation, subscript, or superscript. For example, trigonometric, logarithmic, and exponential functions should not be italicized.

Names of general functions not representing any specific operation should be written using italics.

$$\text{K.E.} = \tfrac{1}{2}\,(m/g_c)\,v^2 \qquad\qquad \log k = \log k_{\text{H+}} + \text{Log}\,[\text{H}_3\text{O}^+]$$

A variable that needs to be assigned a value, in order to evaluate a mathematical expression, should be in italics. A symbol whose purpose is to signify the nature of a quantity (e.g., a variable, function, tensor, vector, etc.) should not be italicized. For example, let p_s represent the price of one metric ton of steel. The letter p is italicized because it is to be replaced by a number while the subscript s is not italicized because it symbolizes the fact that this is the price of steel.

Indices appearing within the elements of sets, sums and products as well as within the components of vectors, tensors, and n forms should be italicized.

$$\sum_{i=0}^{5} b_i$$

Matrices, but not their components, should be italicized.

iii. Cursive, german, and sans serif. Cursive, german, and sans serif characters are seldom used within mathematics.

iv. Bold. Titles of definitions, theorems, corollaries, lemmas, propositions, and axioms appearing prior to the definition, theorem, corollary, lemma, proposition, or axiom should be placed in bold font.

Matrices, vectors, tensors, and n-forms (but not the components of these objects) should be bold, though other signs—signs of aggregation (e.g., brackets), subscripts and superscripts—should be in lightface.

6.2.1.2 Greek Letters

Greek letters are often used to denote variables appearing within the arguments of trigonometric functions as well as to denote some statistical parameters.

A table of the conventions regarding the use of letters for functions, constants, and mathematical expression may be found in **Appendix A** in **Part III**.

6.2.2 Mathematical Expressions

6.2.2.1 When to Run and When to Display

Those short mathematical expressions that will not be referred to by other portions of the work should be placed within the body of the text. Longer mathematical expressions or mathematical expressions that will be referred to elsewhere within the work should be centered on a separate line.

6.2.2.2 Punctuation

Mathematical expressions appearing within the body of the text follow traditional grammatical rules, where symbols function as parts of speech. For example, the equation $y = mx + b$ could be thought of as a sentence with a noun (y), a verb ($=$), a noun phrase (mx), a conjunction ($+$), and another noun (b). Punctuating a run-in expression this way reduces ambiguity when a work is typeset, relating it more clearly to both other mathematical expressions and to the surrounding prose.

Punctuation marks should not appear immediately prior to, within the body of, or immediately following a displayed equation. This rule applies even if the displayed equation ends a sentence and a new sentence begins right after the displayed equation.

i. Ellipsis. Ellipses (... or $\cdots$) are not an officially defined mathematical symbol, but their use in mathematics is common. They usually signify *and so forth*, especially when a pattern has clearly been established and when it would be inconvenient to write out the terms they replace.

Ellipses have many uses in set theory, and are also used in matrices to indicate missing numbers. When they occur between commas in a list, they appear as they would in nonmathematical contexts.

$$x_1, x_2, x_3, \ldots, x_{100}$$

They can also be used after irrational numbers, as in

$$\sqrt{3} = 1.73205\ldots$$

When they appear between operators or brackets, they are usually raised.

$$1 + \frac{1}{2} + \frac{1}{3} + \frac{1}{4} + \frac{1}{5} + \cdots = \sum_{n=1}^{\infty} \frac{1}{n}$$

For general usage of ellipses, *see* **Chapter 3, Section 3.6.1, xvi**.

ii. Prime symbol. The prime ($'$) and double prime ($''$) symbols should not be confused or replaced with the apostrophe ($'$), or with single or double quotation marks ($''$). In mathematics, they often replace subscripts as a way to distinguish a variable from an alternate but similar variable, especially in transformations, set complements, and derived functions.

$$f(x) = x^3$$
$$f'(x) = 3x$$
$$f''(x) = 3$$

6.2.2.3 Spacing Between Symbols and Operators

Most modern mathematical typesetting programs automatically set the spacing between symbols and operators. However, these same typesetting programs also offer the ability to manually set the spacing between individual symbols and operators.

The following set of commands can be used to control the spacing within LATEX:

Table 6.1 LATEX Spacing Commands

Size of Spacing	LATEX Command
thin space	\,
thick space	\;
quad space	\quad
double quad space	\qquad
negative thin space	\!

The standard conventions regarding the amount of space to place between operators and symbols are itemized below.

No space:

No space is placed prior to or following superscripts, subscripts, fences (i.e., parentheses, brackets, braces, and vertical bars), arrows (e.g., the arrow appearing within the argument of a limit), or between a number or variable multiplying another variable.

$$2x, 3xy, |z|, f(u)$$

Thin space:

A thin space (1/6 em quad) should be used within the context of the following circumstances:

• On both sides of the symbols given by $=, \neq, <, >, >>, <<, \leq, \leq, \approx, \equiv, \subseteq, \supseteq, \subset, \supset, \in, \notin, \times$ (multiplication), $\cup, \cap$.

• On both sides of the symbols given by $+, -, \pm$ and $\mp$ if the symbols are used to explicitly represent the operations of addition and/or subtraction (as opposed to representing the signs of a number or variable).

• Prior to, but not following, the symbols given by $+, -, \pm$ and $\mp$ if the symbols are used to explicitly represent the signs of a number or variable (as opposed to representing the explicit operations of addition and/or subtraction).

• After the commas appearing within a set or sequence.

• On both sides of integration, summation and product symbols.

• On both sides of the abbreviations of common functions.

• On both sides of a single vertical line or colon that is used as a mathematical symbol.

• On both sides of an integration measure or n-form.

$$x \approx y$$
$$x \pm y$$
$$(x, y)$$
$$\sum_i x_i$$
$$AB : UV$$
$$\int f(x)\,dx$$

Thick space:

A thick space is 1/3 em quad in width. Thick spaces are seldom used. The single exception occurs when a constraint is explicitly mentioned.

$$\{x_n\}(n \text{ is real number}).$$

Em quad space:

The width of a person under em quad space is equal to one em. Em quad spaces are used whenever a verbal phrase or word occurs within a line containing one or more mathematical expressions. An em quad space is placed between each verbal word and mathematical expression.

$$x = x_0 + vt \text{ where the speed } v \text{ is constant}$$

Two-em quad space:

The width of a two-em quad space is equal to two ems. Two-em quad spaces are used within display mathematical expressions appearing on the same line; or between a display mathematical relation and a condition or constraint placed on the same line as the mathematical relation.

$$y = cx^2 \text{ and } y = a + bx$$

$$w = h - kz \text{ (where } z \text{ is a rational number)}$$

6.2.2.4 Breaking Mathematical Expressions

i. When to break a mathematical expression. Mathematical expressions appearing within the body of the text as running equations should, ideally, never be broken. If breaking is necessary, then the break should come immediately before or after an operator sign or a relation sign, not in the middle of a fenced term or after an integral or summation sign.

A display mathematical expression should only be broken if it consists of one or more equalities, inequalities or proportionalities; or if it is too long to fit within a display line (as used here, a display line includes equation numbering and margins). Expressions in mathematical logic are more likely to be misinterpreted as a result of breaks, so extra caution should be observed with such material.

ii. Appropriate Places to Break

a. Running

Symbols within a mathematical expression appearing within the body of the text should all appear on the same line.

b. Display

The following criteria list those locations at which a display mathematical expression may or may not be broken.

• A display mathematical expression may be broken either prior to or after one of the following symbols: $=, \neq, <, >, >>, <<, \leq, \leq, \approx, \equiv, \subseteq, \supseteq,$ $\subset, \supset, \in, \notin, +, -, \times$ (multiplication), $\pm, \cup, \cap$.

• A display mathematical expression may be broken at any thick space (please see section 2.3).

• If it can be avoided, display mathematical expression should not be broken at a location within a set of nested fences (i.e., parentheses, brackets, braces and vertical bars).

• A break should not occur following an integral, a summation, or a product symbol.

• If the break precedes one of the symbols $+, -, \times$ (multiplication), $\pm,$ $\cup, \cap$ then the symbol following the break should be aligned with the first mathematical symbol appearing to the right of relational symbol (i.e., $=, \neq, <, >, \leq, \leq, \approx, \equiv, \subseteq, \supseteq, \subset, \supset, \in, \notin$) appearing on the preceding line.

6.2.2.5 Numbering and Arrangement of Displayed Equations

Mathematical expressions appearing on a separate line should be numbered with Arabic numerals placed within parentheses that are aligned with the right margin.

A series of several equations displayed on separate lines may each receive a number, but if only one is significant to the text, a single equation may be numbered.

The following conventions apply to numbering equations displayed on multiple lines:

i. When a display equation requires two or more lines, the equation number should appear on the last line of the equation.

$$
\begin{aligned}
P(x, y, t + \Delta t) = {} & s\,\Delta t\ P(x - 1, y, t) \\
& + \{[(y + 1)\,P(x - 1, y + 1, t)\,v\alpha\,\Delta t]\,/\,h\,\} \\
& \quad + \{[(y + 1)\,P(x, y + 1, t)\,v\alpha\,\Delta t]\,/\,h\,\} \\
& \qquad + \{[(x + 1)\,P(x + 1, y, t)\,v\,\alpha\,\Delta t]\,/\,h\,\} \\
& \qquad\quad + \{1 - s\,\Delta t - [y(\mu\alpha + v\alpha)\,\Delta t\,/\,h\,]\}P(x,y,t) \qquad (15)
\end{aligned}
$$

The same would apply to an equation in series, such as:

$$
\begin{aligned}
Q &= 2\int (c^3 p_0\,/\,3\mu\,L)\cos^2 \alpha\,dx \\
&= 2\,(c^3 p_0\,/\,3\mu\,L)\int [L^2\,/\,(L^2 + 4x^2)]\,dx \\
&= (c^3 p_0\,/\,3\mu)\ \tan^{-1}(2\pi\,r)/\,L \qquad (22)
\end{aligned}
$$

In the above example, the equation is deemed to begin with "$Q =$" and continues to the end of the third line.

ii. When a single number is used to identify two or three equations, the equation number may be placed in the center of the group, as in:

$$
e(t) = E_m \sin(\omega t + \theta)
$$
$$
(17)
$$
$$
A(t) = (1 - \varepsilon^{-Rt/L})\,/\,R
$$

or in:

$$
\begin{aligned}
M_{12} &= M_{21} = k_{12}\,(L_1 L_2)^{1/2} \\
M_{23} &= M_{32} = k_{23}\,(L_2 L_3)^{1/2} \qquad (23) \\
M_{31} &= M_{13} = k_{31}\,(L_3 L_1)^{1/2}
\end{aligned}
$$

iii. Two equations may appear on the same display line and may be separated by only a (2 em) space or by an "and" or an "or," and be numbered by a single number, as in:

$$
c_n = \sqrt{(a_n^2 + b_n^2)} \qquad\qquad \theta_n = \tan^{-1}(b_n/a_n) \qquad (8)
$$

$$
\alpha_1 = \Delta P_1\,/\,(T_\alpha P_o)_1 \quad and \quad \alpha_2 = \Delta P_2\,/\,(T_\alpha P_o)_2 \qquad (12)
$$

6.2.3 Abbreviations

6.2.3.1 Of Functions

Abbreviations of common mathematical functions and operators are set in roman font.

6.2.3.2 Using e and exp

The notation *exp* should be used when referring to the exponential function within the body of the text.

The notation *exp* should be used within display mathematical expressions whenever superscripts, fractions, or awkward notations (e.g., summation sign, product sign, and integral) appear within the argument of the exponential function.

- *use* $\exp(-x^2)$ *instead of* e^{-x^2}

- *use* $\exp\left(\dfrac{4}{7-2x}\right)$ *instead of* $e^{\left(\frac{4}{7-2x}\right)}$

- *use* $\exp\left(\int_1^3 x\,dx\right)$ *instead of* $e^{\left(\int_1^3 x\,dx\right)}$

6.2.4 Superscripts and Subscripts

Superscripts and subscripts should be placed flush against the relevant symbol. The only exception is tensors containing both covariant and contravariant components. Thus, $x_\mu{}^\nu$ has different meaning than $x^\nu{}_\mu$ Either the covariant (subscript) or the contravariant (superscript) indices should be placed flush against the symbol representing the tensor.

If the covariant indices are placed flush against the symbol of the tensor, then a space (equal to the width of the space occupied by the covariant indices) should be placed between the tensor's symbol and the contravariant indices. If the contravariant indices are placed flush against the symbol of the tensor, then a space (equal to the width of the space occupied by the contravariant indices) should be placed between the tensor's symbol and the covariant indices.

Following are several rules regarding subscript and superscripts:

i. Except for tensor notation (where the order is critical, as indicated above), subscripts general precede superscripts. This is especially true when the superscript is a power (indicating that the entire term as defined or characterized by the subscript is to be operated upon). The exceptions are superscripts of primes, asterisks, and degree symbols.

$$D'_m \qquad\qquad D^*_m \qquad\qquad D^\circ_m$$

ii. When a superscript is used in a term that has three or more subscripts, the superscript is placed adjacent to the main term. (The subscript is sometimes placed after the superscript and sometimes right below if the usage is clear and unequivocal.)

$$R^2{}_{opt} \qquad\qquad or \qquad\qquad R^2_{opt}$$

iii. Indices that indicate limits (as with integrals), subscripts and superscripts are aligned.

$$\sum_0^\infty \qquad\qquad \int_0^\infty \qquad\qquad C_0^\infty$$

iv. For certain scientific expressions, subscripts and superscripts are conventionally aligned:

Refractive index: $\quad n_D^{20}$

Optical rotation: $\quad [\alpha]_D^{20}$

Density: $\qquad\qquad d^{20}_{20}$

v. Symbols that have two subscript or two superscripts that are unrelated to each other (e.g., one describes placement in an array while the other indicates the value is a maximum), the subscripts are separated by a comma with no space following the comma:

$$R_{1,max} \qquad\qquad B^{R,min}$$

6.2.5 Bracketing ("Fences")

The term "bracket" is used to describe both the general practice of aggregating terms and the specific "square-cornered" mark ("[" and "]"). A general term that avoids confusion is "fences," though the verb form "bracketing" is widely used to describe the form. Parentheses—"(" and ")"—brackets, and braces ("{}" and "}") must appear in pairs in order to indicate the order of operations within a mathematical expression. Open parentheses, brackets, or braces that are not closed lead to much confusion in reading mathematical or scientific material. (Exceptions are when a specific kind of product is indicated by mixed used of fences, as in:

(a,b]   and   [a,b).

Also, the "less than" and "greater than" signs ("<"; ">") have been adapted for use as brackets in certain areas of computer science (in SGML and its applications, as in "<div>"; as well as for URLs, as in "<www.websitename.com>") and in physics (to denote time averages and, with a vertical line, a system state vector, as in: " $| z >$ ").

6.2.5.1 Sequence of Nested Brackets

The order of fences—from the first, innermost aggregation to the last, outermost use—is: parenthesis, bracket, brace. (A visual mnemonic is: "{[( )]}.")

If there is only a single type of fence (i.e., parentheses, brackets, or braces), then only pairs of parentheses are to be used.

$$1 + 2(x - (4(y + 11)))$$

If two types of fences are to be used to indicate the order of operations within a mathematical expression, then only pairs of parentheses are to be used as the innermost pairs of fences while only pairs of brackets are to be used for the outermost pairs of fences.

$$1 + 2[x - (4(y + 11))] \quad or \quad 1 + 2[x - [4(y + 11)]]$$

If all three types of fences are to be used to indicate the order of operations within a mathematical expression, then only pairs of parentheses are to be used for the innermost pairs of fences, only pairs of brackets are to be used for the intermediate pairs of fences, and only pairs of braces are to be used for the outermost pairs of fences.

$$1 + 2\{x - [4(y + 11)]\}$$

6.2.5.2 Changes in Size of Brackets

If nested fences are used to control the flow of operations within a mathematical expression, then they should increase in size from innermost to outermost. This does not apply to brackets that are used in exponents, where brackets are not considered part of the main sequence of fences for the expression as a whole. Thus:

correct:　　$m\left(e^{-[p/(p+1)]N} + g\right)$

incorrect:　$m\left\{e^{-[p/(p+1)]N} + g\right\}$

6.2.5.3 Special Bracket Notation for Specific Operators

• In interval notation, only parentheses are to be used to denote an open boundary while only brackets are to be used to denote a closed boundary.

• Parentheses should be used for functions and functional expressions: $f(x)$; $\rho_v(\rho_v(x))$.

• Braces or brackets should be used for Laplace transforms: $\mathcal{L}\{f(t)\}$.

• Only braces should be used to enclose sets.

• Square brackets or a pair of vertical lines (in preference to large parentheses) can be used as the delimiters of a matrix.

• Brackets only should be used to indicate chemical concentrations: $[S^{--}]$; $[Na^+]$; and isotopic prefixes: $[^{32}P]AMP$.

• Fences are also used in narrative text (*see* **Chapter 3**, **Section 000**, above, for relevant usage rules in narrative text), and care must be taken that the two uses are not confused, particularly when single letters are enclosed by braces or parentheses (and can easily be mistaken for mathematical notation).

Note that the sequence in prose is reversed—braces are innermost, followed by brackets, followed by parentheses (and the visual mnemonic is "([{ }])").

6.2.6 Limits

Limits, and related operations, are placed in lowercase, non-italicized roman characters. The arguments are placed below their abbreviation if they are part of a display mathematical expression and to the lower right if they are part of the body of the text.

$$Display:\ \lim_{x \to 5}$$

$$Running:\ \lim_{x \to 5}$$

6.2.7 Fractions

6.2.7.1 General Use

Fractions occurring within the body of the text of a work should not be upright ("stacked"), but written using a slash ("shilling"). If the numerator or the denominator contain two or more terms, they should be placed in parentheses.

$$(1 + a)/(3 - 2a)$$

Fractions within display mathematical expressions should be upright and placed within a set of parentheses if they are multiplied by a mathematical expression or object (e.g., a variable, function, another fraction, etc.).

$$\left(\frac{4x^2}{x-5} \right)(2x+5) - 3x^4$$

Complex fractions should always be written in display form.

$$\cfrac{x}{1 + \cfrac{x}{1 + \cfrac{x}{1 + \ldots}}}$$

6.2.7.2 Fractions in Matrix Notation

Simple fractions within matrices should be written using a slash whenever possible. If upright fractions need to be included within one or more elements of the matrix, then both the vertical and horizontal spacing between the elements of the matrix should be increased as appropriate.

6.2.7.3 Equations Containing Multiple Simple Fractions

Equations containing two or more simple fractions should be written in display form. The fractions should be upright and not written using a symbol character code (which are available in many word processors).

$$Write \quad \frac{1}{2} + \frac{1}{4} \quad and\ not \quad \frac{1}{2} + \frac{1}{4}.$$

6.2.7.4 Fractions in Superscripts, Subscripts, and Limits

Fractions appearing in superscripts, subscripts, and the number or expression which is being approached by a variable within a limit should not be upright, but written using a slash.

$$x^{1/3}$$

$$S_{2/5}$$

$$\lim_{x \to 5/3} f(x)$$

6.2.7.5 Fractions in Exponential Expressions

Fractions in exponents should always be set as running and not as stacked or built up, to facilitate reading (particularly on computer screens).

$$e^{(a+b)/(c+d)} \qquad (not: \quad e^{\frac{(a+b)}{(c+d)}} \quad)$$

6.2.8 Multiplication

6.2.8.1 Multiplication Sign

i. Standard multiplication. If a multiplication sign is to be explicitly included to signify the product of any mathematical object with a number, variable, function, or any mathematical expression containing only numbers, variables, and functions, then the explicit multiplication sign is to be denoted by either a raised dot or an asterisk. (The multiplication sign used routinely in arithmetic—"×"—should be used with extreme care because of the ease with which it may be confused with the classic denotation of an unknown variable—"x"—and because of its specialized use in denoting products in vector algebra. Therefore, five multiplied by twelve can be written as: $5 \cdot 12$ or $5*12$, and not as:

$$5 \times 12.$$

ii. Vector products. Vector products, which are often referred to as "cross products," are represented by the x-shaped multiplication sign (×) placed midway between the two vectors:

$$\vec{u} \times \vec{v}$$

iii. Other Forms of Multiplication

Two other forms of product are used widely in advanced mathematical literature:

a. Wedge product. Multiplication within a group or ring can be indicated implicitly or explicitly using the group operation of the group or ring. If, for example, g_1 and g_2 are elements of a group G having a group operation *, then the group product of g_1 and g_2 can be written as either: $g_1 g_2$ or $g_1 * g_2$.

An upright wedge, $\wedge$, is used to denote the "wedge product" (also known as the "exterior product") of n-forms. The wedge product of the two one forms $d\mu$ and $d\eta$ is written as: $d\mu \wedge d\eta$.

b. Tensor product. Also called the "outer product" and symbolized by "⊗" (as in "$a \otimes b$"), this product is the most general bilinear operation between two tensors that yields another tensor in the same vector space.

6.2.8.2 Raised Dot

i. Standard multiplication. As mentioned in **Section 6.2.8.1, i**, standard multiplication can be indicated using either a raised dot or asterisk. In this context, many instances where a raised dot is used can be just as effectively described by simply placing multiplicands next to each other or through the use of parentheses.

ii. Vectors and dyadics. Journals in mathematical fields and in physics (and in different countries) follow various conventions regarding the display of vectors and dyadics. It is important to check with the publication to which material is being sent to determine its accepted practices. (See the following section for some notational conventions for vectors.)

Following are some general conventions regarding products between vectors and between dyadics (tensors of second rank) and vectors:

• A raised dot is used to represent the inner product between two vectors. A raised dot is also used to denote the "dot product" between a dyad and a vector. (This product will also often be signified by angled brackets, as in: $\langle x,x \rangle$.)

• The inner product between the vectors **A** and **B** is written as **A • B**. The result is often set in an open face font (e.g., $\mathbb{R}$).

• The "left" dot product of the dyad **CD** with the vector **A** is the vector defined as **(A•C)D** while the "right" dot product of the dyad **CD** with the vector **A** is the vector defined as **C(A•D)**.

• The "colon product" of two dyads is denoted by a colon between the dyads. The colon product of the two dyads **AB** and **CD** is defined as:

AB:CD $\equiv$ **(A•C)(B•D)**.

6.2.9 Vectors, Tensors, and n-forms

Table 6.2 Notational Conventions for Vectors

Convention	Example
Set vectors in bold type	**n** ; **r**
Place arrow above vector	$\vec{a}$
Underline the vector	k̲
Shown vector components	$(x_i, x_j, x_k, \ldots)$

6.2.9.2 Components

The following rules apply to the subscripts of vectors and tensors:

• Contravariant components of vectors and tensors should be written as superscripts.

• Covariant components of vectors and tensors should be written as subscripts.

• Components of n-forms should be written as subscripts.

• Contravariant components should not be placed directly adjacent to the covariant components of a tensor, but should be adjacent to one another and placed either to the left or right of the set of covariant components, which should also be placed adjacent to one another.

• The components of a tensor having three contravariant indices and two covariant indices should be written as:

$$T_{\alpha\beta}{}^{\chi\delta\phi} \text{ or } T^{\chi\delta\phi}{}_{\alpha\beta}$$

6.2.9.3 Summation Convention

An index repeated within the contravariant and covariant positions of a tensor; or when two or more vectors are multiplied; or in a product of tensor and vector components—signifies a sum over that index. (This convention is due to Einstein, after whom it is sometimes named.)

Thus, the following expressions are to be summed over the index k:

$$T_{mk}{}^{p} v^{k} v^{p}$$

$$T^{k}{}_{k}$$

$$v_{k} v^{k}$$

6.2.10 Summations, Products, Unions, and Integrals

6.2.10.1 Size of Summation and Product Signs

Summation and product signs are to be placed in a larger font to distinguish them from the capital Greek letters sigma and pi, respectively.

6.2.10.2 Placement of Arguments within Summations, Products, Integrals, and Unions

The upper and lower arguments of a summation, product, or union are to be placed to the right side of the symbol when they appear within the text of the body.

$$\sum_{i=1}^{5} , \prod_{k=4}^{12} , \bigcup_{m=2}^{20}$$

The upper and lower arguments of a summation, product or union are to be placed directly above and below the symbol when they appear within a display mathematical expression.

$$\sum_{i=1}^{5} , \prod_{k=4}^{12} , \bigcup_{m=2}^{20}$$

6.2.10.3 Integration Limits

The upper and lower limits of a single integral are always placed to the right of the integral whether the integral is within the body of the text or within a display mathematical expression.

$$\int_{-3}^{0}$$

The upper and lower limits of multiple integrals may be placed directly above and below the integrals if the limits are the same for each of the integrals.

$$\iiint_{0}^{\infty}$$

6.2.10.4 Integrals over Open and Closed Manifolds

A symbol representing the manifold is to be placed on the lower right hand side of the integral over a manifold. A circle is placed over the center of the integral if the manifold is closed.

Open manifold: $\int_S$

Closed manifold: $\oint_C$

Table 6.3 The Greek Alphabet

Uppercase	Lowercase	Name
A	α	alpha
B	β	beta
Δ	δ	delta
Γ	γ	gamma
Δ	δ	delta
E	ε	epsilon
Z	ζ	zeta
H	η	eta
Θ	θ	theta
I	ι	iota
K	κ	kappa
Λ	λ	lamda
M	μ	mu
N	ν	nu
Ξ	ξ	xi
O	o	omicron
Π	π	pi
P	ρ	rho
Σ	σ, ς	sigma
T	τ	tau
Y	υ	upsilon
Φ	φ	phi
X	χ	chi
Ψ	ψ	psi
Ω	ω	omega

For meanings of Greek letters in mathematics, *see* **AppendixA, A5**.

6.3. Lists of Tables Related to Mathematics

6.3.1 List of Tables in This Chapter

6.3.2 Contents of Appendix A in Part III

Chapter 7. Style and Usage for Physics

Contents

Contents, 7.2.2, *continued*

Chapter 7. Style and Usage for Physics

7.1 Format and Indexing

7.1.1 Format

The information provided in the Physics sections (in this chapter and in **Appendix B in Part III**) is based on several sources, none more important than the instructions provided to journal contributors by the American Institute of Physics, available online at:

http://www.aip.org/pubservs/style/4thed/toc.html

Other sources of information specific to physics (and also available online) are the following:

- The Physical Review Style and Notation Guide, compiled and edited by Anne Waldron, Peggy, Judd, and Valerie Miller, and published by the American Physical Society, revised in 2005 (available at http://authors.aps.org/STYLE/)

- Reviews of Modern Physics Style Guide, edited by Karie Friedman, published by the American Physical Society, third edition, 1998 (available at: http://web.njit.edu/~sirenko/Phys450/style/style.pdf).

It is important to note, however, that individual journals may have different or additional style guidelines. Further, it is often difficult to provide direction that will guide an author through any and all writing contingencies. Authors should therefore consult the "Information for Contributors" publication or webpage of any journal to which they intend to contribute, and become familiar its style and format.

7.1.2 The Physics and Astronomy Classification Scheme (*PACS*)

The Physics and Astronomy Classification Scheme (*PACS*) codes are prepared by the America Institute of Physics, in collaboration with the International Council on Scientific and Technical Information (ICSTI). *PACS* is a subject classification system created to order and categorize the literature of physics and astronomy (and related research areas) for journal indexes and other databases. The scheme is updated regularly and is available (in print or as a download) online at:

http://www.aip.org/pacs/pacs08/pacs08-toc.html.

PACS has served a useful purpose in the physics community by providing the following:

- allowing researchers and authors to identify the journals publishing actively in areas of physics.

- allowing indexing services and their users to identify the papers of interest in the many diverse areas of physic research and publication.

- allowing journals to identify the areas in which faculty and researchers in the physics community who are capable of serving as referees and reviewers in specific and specialized areas of physics..

7.1.3 Using *PACS*

The *PACS* system assigns numbers to areas of physics according to a hierarchical structure that starts with ten broad subject categories and then divides to more specialized categories at each level. The hierarchy includes five levels of depth, with the narrowest term giving the most detailed characterization.

PACS also includes detailed appendices for acoustics and geophysics, a nanoscale science and technology supplement, and a topical alphabetical index with corresponding *PACS* codes. *PACS* is continually reviewed and assessed, and amended significantly every two to three years, so authors are advised not to rely on older classification codes.

i. *PACS* code

The most detailed *PACS* code may be found at the third, fourth, or fifth hierarchical levels. At these three levels, each *PACS* code consists of six alphanumeric characters divided into three pairs.

The American Physical Society offers the following recommendations for choosing PACS numbers:

- Choose no more than four index number codes.
- Place your principal index code first.
- Always choose the lowest-level code available.
- Always include the check characters.
-

The *PACS* code should follow the abstract on any submitted manuscript on a line set apart from the abstract itself.

7.2 Usage

7.2.1 Grammar and Notation Rules Specific to Physics

7.2.1.1 Capitalization

Most of the rules regarding capitalization can be found in **Part I, Chapter 3**, above. Included here are only the capitalization rules specific to physics writing.

i.. Adjectives and Nouns Derived from Names

Adjectives or nouns formed from proper names should be capitalized:

Poisson   Ohmic   Cauchy-Lorentz

Exceptions: In the following four categories, nouns derived from proper names are not capitalized:

(1) units of measure (newton, pascal),
(2) particles (fermion, boson),
(3) elements (mendelevium, curium),
(4) minerals (bertrandite, garnierite).

ii. Names used with common nouns. When a name is used with a common noun, only the name is capitalized.

Avogadro's number   Bohr radius   Ohm's law

iii. Starting a sentence with a lowercase symbol. Symbols and abbreviations that are lower case—such as *ac Stark effects*—are never capitalized when starting sentences or in titles or headings. It is preferable, however, to reword the sentence so that the lowercase term is not

iv. Reference to elements in the paper. Items that are part of the paper referred to in the course of the narrative are not capitalized, so that "curve A," "sample 4," and "column 2" are lowercase. But it is permissible to capitalize major portions of the paper itself, as in "Appendix A." "Theorem 4," and "Table 2."

v. Laws, Theories, and Hypotheses Associated with a Person's Name

There are many principles, theories, laws, phenomena, and constants in physics that are associated with the name of their discoverer. In these cases, only the proper name should be capitalized:

Heisenberg's uncertainty principle
Newton's second law of motion
Kepler's first law of planetary motion
Hall effect

Principles, theories, laws, phenomena, and constants that are unaffiliated or not mentioned in conjunction with a person's name should be written in lowercase.

the second law of thermodynamics
the principle of the conservation of energy
the theory of general relativity

vi. Small capitals. Note that small capital letters ("small caps") are frequently used for computer programs and for ionization states in atomic spectroscopy. He I denotes an isotope of Helium; He III denotes an ionization state of the element.

7.2.1.2 Abbreviations and Acronyms

Included here are rules concerning abbreviation and acronym specific to physics writing.

i. Defining acronyms and abbreviations. An abbreviation or acronym should be defined the first time it occurs if it is either:
(1) newly invented, or
(2) unfamiliar to those outside the author's area of expertise. Unfamiliar abbreviations should be used as infrequently as possible.
If an abbreviation or acronym is defined first in the abstract or in a caption or table, it should be redefined when it initially appears in the body of the paper. In longer papers, the abbreviations or acronyms should be redefined periodically.
Abbreviations invented by an author should be rarely used, and then only to emphasize a significant concept central to the thesis of the work.

ii. Word versus phrase abbreviations. An abbreviation for a word is generally a shortened version of the word and should be set in lowercase without punctuation; an abbreviation for a phrase is generally an acronym, and should be set in all capitals and unpunctuated.

av for "average"   MO for "molecular orbital"
const for "constant"   BCS for "Bardeen–Cooper–Schrieffer"

iii. Notation for Isotopes and Nuclides

The American Institute of Physics requires that authors follow the notation prescribed by the Symbols, Units, and Nomenclature (S.U.N.) Commission of the International Union of Pure and Applied Physics for nuclides and their states.

A nuclide is notated with its mass number as an anterior postscript:

$$^{12}C$$

A posterior subscript can be used to indicate the total number of atoms in a molecule:

$$^{16}O_2$$

The use of a posterior superscript can be used to indicate the ionization state:

$$He^{2+} \quad \textit{or an excited state:} \quad ^{12}C* \textit{ or } ^{110}Ag^m$$

iv. Abbreviations in Mathematical Expressions

Multiletter abbreviations should not be used in mathematical expressions, particularly not as variables, nor when computations or derivations are to be performed. Use the conventional symbol instead. For example, for kinetic energy, use E_k instead of *KE*. (*See* **Part III, Appendix B, B3: Symbols Commonly Used in Physics.**)

Exception: The abbreviation for the Reynolds number is conventionally written as "Re"—the *e* following the *R* is lowercase and not a subscript.

v. Abbreviations in subscripts. When a proper name is used as either a subscript of a superscript, the capitalization of the name is retained.

E_C or E_{Coul} *for* "Coulomb Energy" (*Note the subscript is roman in the second example.*)

An extensive table of standard abbreviations accepted across a wide spectrum of science journals appears in **Appendix F, F8**, beginning on page 791, below. A table of abbreviations used in physics appears in **Appendix B, B5**. *See* the contents page of the various appendices in Part III for lists and tables of abbreviations used in other disciplines.

7.2.1.3 Characters and Fonts in Physics Publications

Physics publications generally follow the practice of most scientific journals around the world in using serif typefaces, and the one most frequently used is Times New Roman.

Italics should be used for all mathematical symbols, particle symbols, symbols for quantum states, and group-theoretic designations. The following table contains the general rules for symbol font styles:

Table 7.1 Characters and Fonts Used in Physics

Font Style	Symbol	Example
lower case, italic	constants, variables, and ordinary functions	x, y, β
upper case	matrices and functions	S, F
script upper case	operators	$\mathcal{H}$
bold lower case, roman	three- vectors	**r**
bold upper case, roman	matrices three-vectors	**J, B**

i. Other fonts. In addition to Latin and Greek characters (in roman and italic styles), other fonts are occasionally used for special purposes. These include the following:
  a. Script font: *see* the table above.
  b. Old German (Fraktur): used in some areas of mathematics, such as Lie algebra (e.g., $\mathfrak{R}, \mathfrak{g}$)
  c. Sans serif: used to describe shapes and as operators in quantum theory (e.g., S, H)

7.2.2 Symbols for Subatomic and Atomic Physics

7.2.2.1 Subatomic Particles

Greek and Roman letters are used to represent particles that are involved in nuclear reactions, either as projectiles or as products. Symbols for these particles are indicated in **Table 7.2, a-c**, below, and should always be italicized.

Table 7.2a Elementary Bosons

Name	Symbol	Antiparticle	Charge (e)	Spin	Mass (GeV/c^2)	Force mediated	Confirmation
Photon	Γ	Photon	0	1	0	Electromagnetism	Observed
W Boson	W^-	W^+	−1	1	80.4	Weak	Observed
Z Boson	Z	Z Boson	0	1	91.2	Weak	Observed
Gluon	G	Gluon	0	1	0	Strong	Observed
Graviton	G	Graviton	0	2	0	Gravity	Theorized
Higgs boson	H^0	Unknown	0	0	> 112	Unknown	Theorized

Particles that are more radiation than substance (mass-bearing) matter

Table 7.2b Fermions (Quarks and Antiquarks)

Name	Symbol	Antiparticle	Charge (e)	Mass (MeV/c^2)
Bottom	b	b	$-\frac{1}{3}$	4,130–4,370
Charm	c	c	$+\frac{2}{3}$	1,160–1,340
Down	d	d	$-\frac{1}{3}$	3.5–6.0
Strange	s	s	$-\frac{1}{3}$	70–130
Top	t	t	$+\frac{2}{3}$	169,100–173,300
Up	u	u	$+\frac{2}{3}$	1.5–3.3

Particles that combine to make up the particles of the atomic nucleus.

Table 7.2c Leptons

Name	Symbol	Antiparticle	Charge (e)	Mass (MeV/c^2)
Electron	e^-	e^+	−1	~ 0.511
Electron neutrino	v_e	v_e	0	$< 2.2\ eV/c^2$
Muon	μ^-	μ^+	−1	~ 105.6
Muon neutrino	v_μ	v_μ	0	< 0.170
Tauon	τ^-	τ^+	−1	$\sim 1,776.8$
Tauon neutrino	v_τ	v_τ	0	< 15.5

Particles that "orbit" the nucleus and flit around the universe in great waves. The three classes of particles described above make up all the matter thus far known.

7.2.2.2 Indicating Particle Charge

The charge of each particle should be indicated by including a + superscript for a positive charge, a − superscript for a negative charge, or a zero (0) superscript for a neutral particle.

$$e^+ \qquad t^+ \qquad e^- \qquad \mu^+$$

Exceptions: p and e without a charge superscript represent these particles in their natural form (i.e., a positive proton—p^+—and a negative electron—e^-).

7.2.2.3 Antimatter Symbols

An antiparticle can be indicated either by putting a bar above the particle or by placing the antiparticle's charge as a superscript to the particle symbol.

$$e^+ \ or \ \bar{e} \qquad p^+ \ or \ \bar{p}$$

7.2.2.4 Notations for Nuclear Reactions

The following is the formula and an example for denoting a nuclear reaction is the following order of symbols, with spaces surrounding or in between the parentheses:

- The original nuclide (^{6}Li in this example) followed in parentheses by the incoming particle or photon (d in this this example), then a comma, then the outgoing particle(s) or photon(s) (α—a Helium–4— in this example), and, following a closed parentheses, the resultant nuclide (α in this example), to wit:

$$^6\text{Li} \ (d,\alpha) \ \alpha \qquad or \qquad ^6\text{Li} \ (d,\alpha) \ ^4\text{He}$$

Note that the particles listed above represent only a small percentage of the total number of subatomic particles that have been detected and measured. The current state of this field is summarized and updated continually by the **Particle Data Group** (http://pdg.lbl.gov/), a US government-sponsored program, and from the **Elsevier**-operated site: http://www.sciencedirect.com/science/issue/5539-2008-993329998-695406

334

7.2.2.5 Atomic Particles

The subatomic particles combine in a myriad of ways to create the particles that make up the nucleus of the atom. Hadrons come in two varieties: **baryons**, which are comprised of three quarks and include the familiar particles of chemistry, the proton and the neuron; and **mesons**, which are comprised of one quark and another particle, an antiquark.

Below is a brief, simplified outline of the particles involved and a guide to their symbols. Refer to the two sources listed on the previous page for further detail (which, thanks to the phenomenal growth of particle physics during the twentieth century, is prodigious).

Although the particles of the atomic nucleus are quite small, they are themselves believed to be, according to current physical theory, comprised of yet smaller particles in a complex scheme involving the particles of **Table 7.2b**, above. The class of particles that make up the bulk of matter is called the **hadrons**; the other class, lighter and less substantial (but no less important), are the **leptons**. Add the **bosons**, and one has accounted for the known material universe, albeit in an extraordinarily complex scheme. The theory that gave rise to the quark model is known as the **Standard Model**. This formulation focuses on the forces that at work at the atomic and subatomic level, which are greater at those microscopically short distances than the two forces that dominate the macroworld—gravity and electromagnetism.

There are two main varieties of hadrons: baryons and mesons. Ordinary experience is dictated largely by the baryons, the two best known of which are protons and neutrons. Protons are observed to be very stable; whether protons ultimately decay after billions of years is an open question in contemporary physics. A surprising discovery is that, while neutrons, the uncharged particle that shares the nucleus of the atom with the proton and is of comparable size, seems very stable within the atomic nucleus, are much less stable once freed from the atom. On the following pages are tables of some members of these particle groups.

i. Baryons. In the table below, note the preponderance of quarks giving most of the members of this group a "Bottomness" (a measure of the quark-ladenness of the particle), and an electric charge.

ii. Mesons. The members of this group have to deal with a combination of quarks and antiquarks in their makeup, which makes them all very unstable, and incapable of partaking in the atomic life of the macroworld like baryons. One important member of the group, the pion, decays into an electron, possibly serving as a continuing source for what would otherwise be a fleeting nuclear phenomenon.

Table 7.3 Baryons—Symbols and Properties

Particle	Symbol	Makeup	Rest mass MeV/c²	Spin	B (Bottomness)	Q (e) (charge)	Lifetime (seconds)	Decay Modes
Proton	p	uud	938.3	1/2	+1	+1	Stable	...
Neutron	n	ddu	939.6	1/2	+1	0	920	$pe^-\bar{\nu}_e$
Lambda	Λ^0	uds	1115.6	1/2	+1	0	2.6×10^{-10}	$p\pi^-, n\pi^0$
Sigma	Σ^+	uus	1189.4	1/2	+1	+1	0.8×10^{-10}	$p\pi^0, n\pi^+$
Sigma	Σ^0	uds	1192.5	1/2	+1	0	6×10^{-20}	$\Lambda^0\gamma$
Sigma	Σ^-	dds	1197.3	1/2	+1	-1	1.5×10^{-10}	$n\pi^-$
Delta	Δ^{++}	uuu	1232	3/2	+1	+2	0.6×10^{-23}	$p\pi^+$
Delta	Δ^+	uud	1232	3/2	+1	+1	0.6×10^{-23}	$p\pi^0$
Delta	Δ^0	udd	1232	3/2	+1	0	0.6×10^{-23}	$n\pi^0$
Delta	Δ^-	ddd	1232	3/2	+1	-1	0.6×10^{-23}	$n\pi^-$
Xi Cascade	Ξ^0	uss	1315	1/2	+1	0	2.9×10^{-10}	$\Lambda^0\pi^0$
Xi Cascade	Ξ^-	dss	1321	1/2	+1	-1	1.64×10^{-10}	$\Lambda^0\pi^-$
Omega	Ω^-	sss	1672	3/2	+1	-1	0.82×10^{-10}	$\Xi^0\pi^-, \Lambda^0K^-$
Lambda	Λ^+_c	udc	2281	1/2	+1	+1	2×10^{-13}	...

Some baryons have mass, charge, and two of them have long lifetimes, thanks to the dominance of quarks in their makeup.

Table 7.4 Mesons—Symbols and Properties

Particle	Symbol	Anti-particle	Makeup	Rest mass—MeV/c²	S	C	B	Lifetime	Decay Modes
Pion	π^+	π^-	$u\underline{d}$	139.6	0	0	0	2.60×10^{-8}	$\mu^+\nu_\mu$
Pion	π^0	Self		135.0	0	0	0	0.83×10^{-16}	2γ
Kaon	K^+	K^-	$u\underline{s}$	493.7	+1	0	0	1.24×10^{-8}	$\mu^+\nu_\mu$, $\pi^+\pi^0$
Kaon	K^0_s	$\overline{K}^0_s$	1*	497.7	+1	0	0	0.89×10^{-10}	$\pi^+\pi^-$, $2\pi^0$
Kaon	K^0_L	$\overline{K}^0_L$	1*	497.7	+1	0	0	5.2×10^{-8}	$\pi^+e^-\underline{\nu}_e$
Eta	η^0	Self	2*	548.8	0	0	0	$<10^{-18}$	2γ, 3μ
Eta prime	η^0	Self	2*	958	0	0	0	...	$\pi^+\pi^-\eta$
Rho	ϱ^+	ϱ^-	$u\underline{d}$	770	0	0	0	0.4×10^{-23}	$\pi^+\pi^-$
Rho	ϱ^0	Self	$u\underline{u}$, $d\underline{d}$	770	0	0	0	0.4×10^{-23}	$\pi^+\pi^-$
Omega	ω^0	Self	$u\underline{u}$, $d\underline{d}$	782	0	0	0	0.8×10^{-22}	$\pi^+\pi^-\pi^0$
Phi	ϕ	Self	$s\underline{s}$	1020	0	0	0	20×10^{-23}	K^+K^-, $K^0\underline{K}^0$
D	D^+	D^-	$c\underline{d}$	1869.4	0	+1	0	10.6×10^{-13}	$K + \_$, $e + \_$
D	D^0	$\overline{D}^0$	$c\underline{u}$	1864.6	0	+1	0	4.2×10^{-13}	$[K,\mu,e] + \_$
D	D^+_s	$\overline{D}^-_s$	$c\underline{s}$	1969	+1	+1	0	4.7×10^{-13}	$K + \_$
J/Psi	J/ψ	Self	$c\underline{c}$	3096.9	0	0	0	0.8×10^{-20}	e^+e^-, $\mu^+\mu^-\dots$
B	B^-	B^+	$b\underline{u}$	5279	0	0	-1	1.5×10^{-12}	$D^0 + \_$
B	B^0	$\underline{B}^0$	$d\underline{b}$	5279	0	0	-1	1.5×10^{-12}	$D^0 + \_$
B_s	B^0_s	$\overline{B}^0_s$	$s\underline{b}$	5370	0	0	-1	...	$B^-_s + \_$
Upsilon	Υ	Self	$b\underline{b}$	9460.4	0	0	0	1.3×10^{-20}	e^+e^-, $\mu^+\mu^-\dots$

S = Strangenes; C = Charm; B = Bottomness—all the legacy of the quark-like constituents of each particle.

7.2.3 Notation for Physical Quantities and Units

i. The SI System

For over three centuries, physics has been a discipline that has involved an international array of great theoreticians and experimentalists. As a result, there has been a great need for uniformity in the system of units used by physicists in reporting their results and in making their predictions. This problem was exacerbated by the fact that British-American measurement systems in use in everyday life were very different from the metric system used in Europe since the days of the French Revolution. (Physicists have in the past traditionally used two systems—the MKS system and cgs system. Both were metric, but the cgs system was used to formulate the basic theory of electromagnetism, which is why it is still used in some circles and in some areas of physics.).

The SI system begins with seven basic quantities and seeks to define or "derive" all other units and measures in terms of these units. There will be cases where the phenomena observed will give rise to measures that are not reducible to these base units, but may still be couched in compatable terms These are the units that may still be used with SI units.

Finally, there are units that are derived as coefficients of equations that appear frequently in the course of research. These are called "dimensionless" quantities, because they serve only to balance an equation.

Table 7.5 (= Table B1.1) SI Base Units

Quantity	Name	Symbol
length	meter	m
mass	kilogram	kg
time	second	s
electric current	ampere	A
thermodynamic temperature	kelvin	K
amount of substance	mole	mol
luminous intensity	candela	cd

SI and the International Organization for Standardization (ISO) give the following recommendations for notation of these units. If there are alternatives in notation, authors must select one system, make clear which system is being used, and apply it uniformly throughout the paper.

ii. When to Use SI

Only SI units and those units recognized for use with the SI should be used to express the values of quantities. Equivalent values in other units may be given in parentheses following values in acceptable units only when deemed necessary for the intended audience.

As mentioned previously, the edifice of electromagnetic theory developed in the 19th and early twentieth century took advantage of features of the cgs system that served important pedagogical interests. Today, however, conversion from one system to another is easily accomplished. It is useful for students and researchers to periodically carry out these conversions in order to become "hands-on" familiar with the SI set of parameters and values in real laboratory terms.

iii. Abbreviations

Abbreviations such as sec, cc, or mps should be avoided and only standard unit symbols, prefix symbols, unit names, and prefix names are used.

correct: s or second; cm^3 or cubic centimeter; m/s or meter per second

incorrect: sec; cc; mps

iv. Plurals

Unit symbols should be unaltered when described in the plural.

correct: $l = 75$ cm

incorrect: $l = 75$ cms

v. Punctuation

Unit symbols should not be followed by a period unless they appear at the end of a sentence.

correct: The length of the bar is 75 cm.
               The bar is 75 cm long.

incorrect: The bar is 75 cm. long.

vi. Multiplication and Division

A space or half-high dot should be used to signify the multiplication of units. A solidus (or slash), horizontal line, or negative exponent is used to signify the division of units. The solidus must not be repeated on the same line unless parentheses are used.

correct: The speed of sound is about 344 m·s^{-1} (meters per second)
The decay rate of ^{113}Cs is 21 ms^{-1} (reciprocal milliseconds)
m/s, m·s^{-2}, m·kg/(s^3·A), m·kg·s^{-3}·A^{-1}
m/s, m s^{-2}, m kg/(s^3 A), m kg s^{-3} A^{-1}

incorrect: The speed of sound is about 344 ms^{-1} (reciprocal milliseconds)
The decay rate of 1^{13}Cs is 21 m·s^{-1} (meters per second)
m ÷ s, m/s/s, m·kg/s^3/A

vii. Typeface for Variables

Variables and quantity symbols should be in italic type; unit symbols in roman type. Numbers should generally be written in roman type. These rules apply irrespective of the typeface used in surrounding text.

correct: She said, "That dog weighs 10 kg!"
$t = 3$ s, where t is time and s is second
$T = 22$ K, where T is thermodynamic temperature, and K is Kelvin

incorrect: He said, "That dog weighs 10 kg."
$t = 3$ s, where t is time and s is second
$T = 22$ K, where T is thermodynamic temperature, and K is kelvin

viii. Typeface for Subscripts and Superscripts

Subscripts and superscripts should be in italic type if they represent variables, quantities, or running numbers; in roman type if descriptive.

Table 7.6 Typefaces for SI Subscripts and Superscripts

subscript category	typeface	proper usage
quantity	italic	c_p, specific heat capacity at constant pressure
descriptive	roman	m_p, mass of a proton
running number	italic	$$\bar{x} = X = 1/n \sum_{i=1}^{n} X_i$$

ix. Abbreviations for "Parts Per" Phrases

The combinations of letters "ppm," "ppb," and "ppt," and the terms part per million, part per billion, and part per trillion, and the like, are not used to express the values of quantities.

correct: 2.0 μL/L; 2.0 x 10^{-6} V;
4.3 nm/m; 4.3 x 10^{-9} l;
7 ps/s; 7 x 10-12 t, where V, l, and t are the quantity symbols for volume, length, and time.

incorrect: "ppm," "ppb," and "ppt," and the terms part per million, part per billion, and part per trillion, etc.

x. Unit Modifications

Unit symbols (or names) are not modified by the addition of subscripts or other information. The following forms, for example, are used instead.

correct: $V_{max} = 1000$ V
a mass fraction of 10 %

incorrect: $V = 1000$ V_{max}
10 % (*m/m*) or 10 % (by weight)

xi. Percent Symbol

The symbol % is used to represent simply the number 0.01.

proper: $l_1 = l_2(1 + 0.2$ %),
    or: $D = 0.2$ %, where D is defined by the relation $D = (l_1 - l_2)/l_2$.

improper: the length l_1 exceeds the length l_2 by 0.2 %

xii. Information and Units

Information is not mixed with unit symbols or names.

correct: the water content is 20 mL/kg

incorrect: 20 mL H_2O/ kg
20 mL of water/ kg

xiii. Mathematical Notation

It is important that it be clear to which unit symbol a numerical value belongs and which mathematical operation applies to the value of a quantity.

correct   35 cm x 48 cm
1 MHz to 10 MHz or (1 to 10) MHz
20 °C to 30 °C or (20 to 30) °C
123 g ± 2 g or (123 ± 2) g
70 % ± 5 % or (70 ± 5) %
240 x (1 ± 10 %) V

incorrect: 35 x 48 cm
1 MHz-10 MHz or 1 to 10 MHz
20 °C-30 °C or 20 to 30 °C
123 ± 2 g
70 ± 5 %
240 V ± 10 % (one cannot add 240 V and 10 %)

xiv. Unit Symbols and Names

Unit symbols and unit names are not mixed and mathematical operations are not applied to unit names.

correct: kg/m^3, $kg \cdot m^{-3}$, or kilogram per cubic meter

incorrect: $kilogram/m^3$, kg/cubic meter, kilogram/cubic meter, kg per m^3, or kilogram per $meter^3$.

xv. Numerals and Unit Symbols

Values of quantities are expressed in acceptable units using Arabic numerals and symbols for units.

correct   m = 5 kg
the current was 15 A

incorrect: m = five kilograms
m = five kg
the current was 15 amperes

xvi. Unit Spacing

There is a space between the numerical value and unit symbol, even when the value is used in an adjectival sense, except in the case of superscript units for plane angle.

correct: a 25 kg sphere
          an angle of 2° 3' 4"

If the spelled-out name of a unit is used, the normal rules of English apply: "a roll of 35-millimeter film."

incorrect: a 25-kg sphere
          an angle of 2 ° 3 ' 4 " (*note additional space after numbers*)

xvii. Digit Spacing

The digits of numerical values having more than four digits on either side of the decimal marker are separated into groups of three using a thin, fixed space counting from both the left and right of the decimal marker. Commas are not used to separate digits into groups of three.

correct: 15 739.012 53

incorrect: 15739.01253
          15,739.012 53

xviii. Quantity Equations

Equations between quantities should be used in preference to equations between numerical values, and symbols representing numerical values are different from symbols representing the corresponding quantities.

When a numerical-value equation is used, it is properly written and the corresponding quantity equation is given where possible.

correct: $(l/\text{m}) = 3.6^{-1} [v/(\text{km/h})](t/s)$

incorrect: $l = 3.6^{-1} vt,$ [*accompanied by text saying*:
          "where l is in meters, v is in kilometers per hour,
          and t is in seconds"]

xix. Standard Symbols

Standardized quantity symbols should be used. Similarly, standardized mathematical signs and symbols are used. More specifically, the base of "log" in equations should be specified when required by writing $\log_a x$ (meaning log to the base a of x), lb x (meaning $\log_2 x$), ln x (meaning $\log_e$ x), or lg x (meaning $\log_{10}$ x).

correct: tan x
           R for resistance
           A_r for relative atomic mass

incorrect: tg x for tangent of x
            words, acronyms, or improvised groups of letters

xx. Weight vs. Mass

When the word "weight" is used, the intended meaning should be clear. In science and technology, weight is a force, for which the SI unit is the newton; in commerce and everyday use, weight is often used as a synonym for mass, for which the SI unit is the kilogram. Though this is technically incorrect, it is so widely practiced colloquially that it has become accepted even in academic, though not scientific, writing.

xxi. Quotient Quantity

A quotient quantity should be written explicitly.

correct: mass divided by volume

incorrect: mass per unit volume

xxii. Object and Quantity

An object and any quantity describing the object should be distinguished. (Note the difference between "surface" and "area," "body" and "mass," "resistor" and "resistance," "coil" and "inductance.")

correct: A body of mass 5 g

incorrect: A mass of 5 g

xxiii. Obsolete Terms

The obsolete terms normality, molarity, and molal and their symbols N, M, and m are not used.

correct: amount-of-substance concentration of B (more commonly called concentration of B), and its symbol cB and SI unit mol/m3 (or a related acceptable unit)

molality of solute B, and its symbol bB or mB and SI unit mol/kg (or a related unit of the SI)

incorrect: normality and the symbol N, molarity and the symbol M molal and the symbol m

xxiv. Consistency of Abbreviation

Never mix names and abbreviations in a unit expression.

correct: W/A   *or*   Watt/Ampere

incorrect: Watt/A

xxv. Compounding SI Prefixes

Do not use compound prefixes (prefixes created by combining two or more SI prefixes).

correct: m

incorrect: 1 mnm

xxvi. Use of the Solidus (Slash)

A solidus (or slash or oblique stroke, "/") should not be used more than once in a given expression.

correct: $kg \cdot m^{-1} \cdot s^{-2}$
$kg / m \cdot s^{2}$

incorrect: $kg/m/s^{2}$

7.3 Symbols and Constants Commonly Used in Physics

For nearly a century, no laboratory in any of the physical sciences, particularly in chemistry and physics, could consider itself a serious research facility without a copy of "the CRC"—*The CRC Handbook of Chemistry and Physics*—on hand. Now approaching its 90th edition, the CRC provides careful and painstakingly assembled information on all manner of constants and scientific notation. Several other useful guides are listed in **Appendix I**.

The tables below list (in summary) the symbols and constants approved for use in physics by the International Organization for Standardization (ISO), and by the International Unions of Pure and Applied Chemistry (IUPAC) and of Pure and Applied Physics (IUPAP).

Table 7.7a Symbols Commonly Used in Physics (Classical)

Symbol	Name	Definition	SI Unit	Symbol	Name	Definition	SI Unit
	Space and Time			P	momentum	$= mv$	$kgms^{-1}$
l	length		m	F	force	$dp/dt=ma$	N
$x, y, z /$ r, θ, ϕ	cartesian/ spherical coordinates		m	ρ ρ_A , ρ_S	density surface density	m /V m /A	kgm^{-3}
d	distance		m	$T, (M)$	tourque	$T = r \times F$	N m
h b	height breadth		m	E	energy		J
				E_P, V, Φ	potential energy	$=\int F \cdot ds$	J
s	path length			E_K, T, K	kinetic energy	$= \frac{1}{2} mv^2$	J
A, A_s	area		m^2	W, w	work	$=\int F \cdot ds$	J
$V, (v)$	volume		m^3	P	power	dW/dt	W
α, β, θ	angle	$\alpha = s/r$	rad., 1	η, μ	viscosity	$R_{x,z}=\eta(dv_x/dz)$	Pa s
$\omega, \Omega...$	solid angle	$\omega = A/r^2$	Sr, 1	$\sigma /$ τ	stress / shear stress	$\sigma = F/A /$ $\tau = F/A$	Pa Pa
t	time		s	E	Young's modulus	σ/ε	Pa
T	period	$T=t/N$	s	H	Hamiltonian	$= T(q, p)+V(q)$	J
v, f	frequency	$v= 1/T$	Hz	L	Lagrangian	$=T(q, \dot{q})-V(q)$	J
	Classical Mechanics				**Electricity and Magetism**		
r	position vector	$r = xi +$ $yj + zk$	m	Q, q	electric charge		C
$v, u, \dot{r}$	velocity	$= dr/dt$	m s^{-1}	C	capacitance	$=Q/U$	F, CV$^-$
ω	angular velocity	$= d\phi/dt$	rad s^{-1}	I	electrical current	$=dQ/dt$	A
$a, (g), \ddot{r}$	acceleration	$= dv/dt$	m s^{-1}	ρ	charge density	$=Q/V$	Cm^{-3}
m	mass		kg	V, ϕ	electric potential	dW/dq	JC$^-$
μ	reduced mass	$\dfrac{m_1m_2}{(m_1+m_2)}$	kg	$\varepsilon /$ ε_0	permittivity/ p. of vacuum	$D = \varepsilon E /$ $1/(\mu_0 c_0^2)$	F m^{-1}

Table 7.7a, *continued*

Symbol	Name	Definition	SI Unit	Symbol	Name	Definition	SI Unit
Electricity and Magnetism, continued				*Thermal Physics*			
j, J	current density	$\int j \cdot dA$	m	q, Q	heat		J
E	electric field	$F/Q = -grad\ V$	M	w, W	work		J
H	magnetic field	$B = \mu H$	A m⁻²	U	internal energy	$\Delta U = q+w$	J
φ / B	magnetic flux/m.f. density	$\phi = \int B \cdot dA$ / $F=Qv\times B$	T / A m⁻²	T	thermody-namic temp.		K
μ_0, μ_r	permability	$B=\mu H$	H m⁻¹	H	enthalpy	$= U + pV$	J
R	resistance	$R = U/I$	Ω	S	entropy	$dS \geq dq/T$	JK⁻¹
G	conductance	$= 1/R$	S	C_p / C_V	heat capacity	$(\partial H/\partial T)_p$ / $(\partial H/\partial T)_V$	JK⁻¹
X	reactance	$= (U/I)\sin\delta$	Ω	Z	compresibility	pV_m/RT	
Z	impedance	$Z= R+iX$ $Y = 1/Z$	S		expansion coefficient		K⁻¹
κ, γ, σ	conductivity	$\kappa = 1/\rho$	S m⁻¹	μ_0, μ_{JT}	Joule-Thompson coefficient	$(\partial T/p)_N$	KPa⁻¹
M, L₁₂	inductance	$E=-L(dI/dt)$	H	$< x >$	average x		
A	magnetic vector potential	$B = \nabla\times A$	Wbm⁻¹	A	Helmholtz energy	$= U - TS$	J
S	Poynting vector	$S = E \times H$	W m⁻²	G	Gibbs energy	$= H - TS$	

Table 7.7b. Symbols Commonly Used in Physics (Modern)

Symbol	Name	Definition	SI Unit	Symbol	Name	Definition	SI Unit
Quantum Theory				*Atoms and Molecules*			
$\hat{A}$	operator			$[\hat{A}, \hat{B}]$	commutator of A, B	$= \hat{A}\hat{B} - \hat{B}\hat{A}$	
$\dot{p}$	momentum operator			n	principal quantum no.	$E= \dfrac{-hcR}{n^2}$	1
$\dot{T}$	kinetic energy operator	$-(h^2/2m)\nabla$	J	*M, μ*	magnetic dipole	$E_p=-m\cdot B$	JT⁻¹
$\hat{H}$	Hamiltonian operator			ν_L	Larmour frequency	$\nu_L=\omega_L/2\pi$ $\omega_L=e/2m)B$	Hz s⁻¹
Ψ, ψ, φ,...	wave (state) function	$H\Psi = E\Psi$		A	radioactivity	$= - dN_B/dt$	Bq
P	probability density			λ	radioactive decay	$A = \gamma N_B$	s⁻¹
$\hat{A}^\dagger$	Hermitian conjugate	$(\hat{A}^\dagger)_{ij} = (A_{ij})^*$		t½, T½	half life		s

347

Table 7.8 Constants Commonly Used in Physics

Symbol	Name	Value	Symbol	Name	Value
c	speed of light in a vacuum	$2.99\ 792\ 458 \times 10^8$ m s^{-1}	N_A, L	Avogadro constant	$6.022\ 1415 \times 10^{23}$ mol^{-1}
g_n	acceleration of gravity	$9.806\ 65$ m s^{-2}	F	Faraday constant	$96\ 485.3383$ C mol^{-1}
h	Planck constant	$6.626\ 0693(11) \times 10^{-34}$ J s	G	Gravitational constant	$6.6742(10) \times 10^{-11}$ m^{3}kg^{-1}s^{-2}
$\hbar$	("h-bar") reduced Planck constant	$1.054\ 571\ 68(18) \times 10^{-34}$ J s	R_∞	Rydberg constant	$10\ 973\ 731.568\ 525(73)$ m^{-1}
e	electronic charge	$1.602\ 176\ 53(14)$ x 10^{-19} C	eV	electron volt	$1.602\ 176\ 53(14) \times 10^{-19}$ J
k or k_b	Boltzmann constant	$1.380\ 6505(24) \times 10^{-23}$ J K^{-1}	σ	Stefan-Boltzmann constant	$5.670\ 400(40) \times 10^{-8}$ W m^{-2} K^{-4}
m_e	electron mass	$9.109\ 3826(16) \times 10^{-31}$ kg	m_p/m_e	proton-electron mass ratio	$1836.152\ 672\ 61(85)$
m_p	proton mass	$1.672\ 623\ 1 \times 10^{-24}$ grams $= 1.007\ 276\ 470$ amu	u $= m_u$ (amu)	atomic mass unit	$1.660\ 538\ 86(28) \times 10^{-27}$ kg
m_n	neutron mass	$1.674\ 928\ 6 \times 10^{-24}$ grams $= 1.008\ 664\ 904$ amu	R	molar gas constant	$8.314\ 472(15)$ J mol^{-1} K^{-1}
m_d	deuterium nucleus mass	$3.343\ 586\ 0 \times 10^{-24}$ grams $= 2.013\ 553\ 214$ amu	G_F	Fermi coupling constant	$1.166\ 39(1) \times 10^{-5}$ GeV
m_α	alpha particle mass	$6.644\ 6565(11 \times 10^{-24}$ grams	E_h	Hartree energy	$4.359\ 744\ 17(75) \times 10^{-17}$ J
a_0	Bohr radius	$0.529\ 177\ 210\ 8(18) \times 10^{-10}$ m	μ_B	Bohr magneton	$972.400\ 949(86) \times 10^{-26}$ Ω
α	magnetic constant	$4\pi\ (=12.566\ 370\ 614...) \times 10^7$ NA^{-2}	α	fine structure constant	$7.297\ 352\ 568(24) \times 10^{-3}$
ε_0	electric constant	$8.854\ 187\ 817... \times 10^{-12}$ F m^{-1}	v_S	speed of sound in air (at STP)	3.31×10^2 m s^{-1}

7.4 Lists of Tables Relevant to Physics

7.4.1 List of Tables in This Chapter

7.4.2 Contents of Appendix B of Part III • 599:

B1 International System of Units (SI) Tables

B1.1 SI Base Units
B1.2 Standard SI Prefixes
B1.3 SI-Derived Units with Special Names
B1.4 Units in Use with SI
B1.5 Units Approved for Use with SI
B1.6 SI to Non-SI Conversion Factors

B2 Electromagnetic Spectrum

B3 Style Guide to Physical Effects and Phenomena

B4 Coordinate Systems

B5 Abbreviations Commonly Used in Physics

Chapter 8. Style and Usage for Astronomy

Contents

Contents, *continued*

Chapter 8. Style and Usage for Astronomy

8.1 Preparing the Manuscript

8.1.1 The Changing Face of Astronomy as a Science

The modern view of astronomy sees pre-twentieth century astronomical observation as purely observational—largely a pastime undertaken by aristocratic amateurs. Once serious observational tools became available, and the foundations of classical physics were laid, astronomy could be-come worthy of scientific notice. The discipline could not be considered a science, however, until the space-time structure of the universe could be understood (through general relativity) and cosmic processes could be described (through atomic physics). In this view, astronomy became a sub-division of the general discipline of physics, but because of the scope of the subject and the special tools employed, has become a discipline operating independently. (A *post-modern* view, still in formation, would see the work of the pre-modern period as relevant and even inspirational of the modern discipline.)

8.1.2 Publishing Guidelines

Some journals may require special instructions for categorization, such as this list of approved keywords for astronomy articles in *The Monthly Notices of the Royal Astronomy Society, The Astrophysical Journal,* and *Astronomy and Astrophysics*:

http://www.blackwellpublishing.com/pdf/mnraskey.pdf.

Articles that contain many equations may be better suited as TEX or LATEX files than as Microsoft Word documents or file formats of other word-processing programs (*see* **Chapter 7**, **Section 7.1.4**). Some journals' webpages supply downloadable templates that are already in the format preferred by the specific journal.

Most journals will require that all tables and figures be referenced in the main text of the work. Many will not charge extra for a color illustration in an article that is to be published electronically, but there usually is a fee for such items in a printed publication.

Journals may elect to receive manuscript files via an online service such as Manuscript Central, which organizes authors' submissions in an online database before they are published.

For more general guidelines about preparing a manuscript, *see* **Chapter 2**. For more specific guidelines, see the style or submission guide section of the website of an individual journal.

8.2 Usage

8.2.1 Formal Astronomical Name

There are more than a billion identified astronomical objects, and all of them need unique, systematic designations. The authority on astronomical nomenclature is the International Astronomical Union (IAU) located in Paris, France. Some objects have proper names, especially planets and naked-eye visible stars, but most are more reliably given a catalog name.

8.2.1.2 Astronomical Catalogs

An astronomical catalog (British spelling *catalogue*) is a database of similar astronomical objects with systematic designations. Catalogs generally have acronyms or initialisms that precede numerical identifiers for specific objects within them. For recommended formatting conventions, see **Section 8.2.4**. Some important catalogues follow:

Hipparcos Catalogue (HIP) – contains more than 118,000 stars, and is complete to magnitude 7.3.

Washington Double Star Catalog (WDS) – maintained by the United States Naval Observatory, with over 100,000 systems.

United States Naval Observatory B1.0 (USNO B1.0) – documents over 1,000,000,000 objects, including proper motions.

Tycho-2 Catalogue – lists over 1,050,000 stars.

Second U.S. Naval Observatory CCD Astrograph Catalog (UCAC2) – over 48,000,000 stars; designed for R magnitudes of about 7.5 to 16.

Yale Bright Star Catalog (BSC) - one of the most widely used star catalogues; provides detailed basic astronomical and astrophysical data for 9110 of the brightest stars.

Guide Star Catalog II (GSC-II) - positions, classifications, and magnitudes for 998,402,801 objects

8.2.2 Standard Capitalization Rules

Generic terms and general descriptive terms, including celestial, atmospheric, and meteorological phenomena, should be lowercased:

black hole	solar system
cepheid variable	universe
galaxy	white dwarf

Exception: Generic terms with proper names should only have the proper name capitalized:

Van Allen radiation belt

Adjectival forms of proper names are lowercased:

the mercurial craters	the saturnian rings
the venerian clouds	the martian volcanoes

Capitalize the names of the nine planets of the solar system and all planetary features:

Mercury	Amazonis Planitia
Venus	Maxwell Montes
Earth	Arecibo Vallis

Note: The term Earth may be written occasionally lowercased because it has become generic, however when it is used in an astronomical context, it should always be capitalized. In cases where Earth is capitalized, the definite article "the" is typically omitted.

The tunnel was dug deep into the earth.
The two planets neighboring Earth are Venus and Mars.

Capitalize the names of planetary satellites and their features:

Callisto (Jupiter)	Valhalla
Triton (Neptune)	Abatos Planum

A moon is a type of planetary satellite. Earth's satellite is called the Moon. Similar to the capitalization of *earth*, *moon* should only be capitalized when it is used in an astronomical context.

The moon shone over the ocean.
The Apollo project first sent astronauts to the Moon.

Capitalize the names of stars, asteroids (minor planets), and other unique celestial objects (nebulae, galaxies, etc.):

the Orion Nebula	the Andromeda Galaxy
Polaris	Comet Halley
1 Ceres (asteroid)	Virgo Cluster
NGC 2023	M97

Note: Similar to the capitalization of *earth* and *moon*, the word *sun* should only be capitalized when it is used in an astronomical context.

The sun rose early that morning.
The scientists detected neutrinos from the Sun.

8.2.3 Celestial Coordinate Notation

Right ascension (abbreviated as RA, signified by α) and declination (abbreviated as dec, signified by δ) are a system of coordinates for locating celestial objects similar to the system of longitude and latitude used on the Earth. Right ascension is a measurement along the celestial equator (see Figure 3.1) and is denoted in hours (h), minutes (m), and seconds (s), or sidereal time.

There should be no spaces between numbers and units.

$$6^h45^m10^s$$

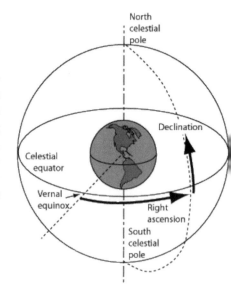

Figure 3.1 Equatorial Coordinate System

Declination is denoted in degrees, arc-minutes, and arc-seconds of arc north (using a plus sign or no marking) or south (using a minus sign) of the celestial equator. There are 60 arc-minutes in a degree, and 60 arc-seconds in an arc-minute. Declination measures how far overhead an object will rise in the sky, and is measured as 0° at the equator, +90° at the North Pole and -90° at the South Pole.

$$-16° \ 43' \ 22''$$

Should decimals be used in the coordinate, the decimal point is placed between the unit symbol and the decimal.

$$6^{h}45^{m}10^{s}.7 \qquad -16° \ 43'.32$$

8.2.4 Format for the Creation of a New Designation

The following specifications were created by the International Astronomical Union (IAU) to minimize confusion concerning the creation of designations for newly discovered astronomical sources of radiation. In source listings, positional information and/or a second designation must be included next to a primary designation so as to avoid uncertainties.

The designation of an astronomical source of radiation takes the following format, with the underscores taking the place of actual spaces for illustrative purposes:

Acronym_Sequence_(Specifier)

The specifier is optional, but it must be enclosed in parentheses if one is included. The following examples are correctly formatted:

NGC 201   PKS 1209-52   R 136:a3 (30 Dor)

8.2.4.1 Creating Acronyms for the New Designation

The acronym is a code that indicates the catalogue or collection of sources. It can be created out of catalogue names (e.g. NGC, BD), instruments or observatories used for large surveys (VLA, IRAS, 3C, 51W), or the names of authors (RCW). For a list of existing astronomical acronyms, *see* **Appendix C2**.

In creating new acronyms, the following rules apply:

(1) Acronyms should consist of at least three characters, since two-letter acronyms are almost all already in use.

(2) Acronyms should consist of numbers and letters only—avoid the use of subscripts, superscripts, and other special characters.

(3) Each new acronym should be unique—check all appropriate sources to avoid duplication with existing catalogue designations.

(4) The title of a catalogue should include the acronym by which it is known [e.g., Fifth Fundamental Catalogue (FK5)].

(5) Excessively long acronyms should be avoided.

(6) Never abbreviate an acronym.

8.2.4.2 Sequence

The sequence for an astronomical designation is an alphanumeric string of characters, typically consisting only of numbers, that uniquely defines the source within a catalogue or collection. It can be a sequence number within a catalogue, based on coordinates, or a combination of fields.

8.2.4.3 Using Coordinates

i. Truncating coordinates. Astronomical coordinates should be truncated, not rounded. This way, the position of an object will always be correctly restricted to a small area of the sky.

Original:	QSO 004848-4242.8
Correctly truncated:	QSO 00488-4242
Incorrectly truncated:	QSO 00484-4242
	QSO 00488-4243

ii. Prefix (flag) letters. The flag letters B, G, and J may be placed before the equatorial coordinates, and directly after the space that follows the acronym.

8.2.4.4 Specifiers

The inclusion of a specifier is optional and allows for the indication of other source parameters. A specifier is not required syntax and should be enclosed in parentheses.

8.2.4.5 Subcomponents

Subcomponents of a source should never be named without their parent sources unless it is clear to which parent source they apply. For example, do not refer to the subcomponent IRS 3 alone when you might more clearly refer to W 3 IRS 3 or NGC 7538 IRS 3.

8.2.4.6 Descriptive Designations

Descriptive designations are letters or numbers that may help identify a subcomponent of an astronomical object. A colon may be used to separate the parent source from its subcomponent's descriptive designation and to indicate the hierarchical relationship. Below are a few examples.

Parent source	With subcomponent
Sgr B	Sgr B2
W 49	W 49A
DR 21	DR 21:CO 1
W 3	W 3 IRS 5

8.2.4.7 Punctuation and Special Characters

The following rules apply for astronomical designations:
(1) [ ] (blank character) should be used to separate characters or phrases instead of [.] or [/].

(2) [_] (underscore) may be used as a separator instead of [ ] when necessary, such as entries in electronic catalogue.

(3) [-] (hyphen) should be restricted to use as a minus sign. It is occasionally used as a separator, but such use is discouraged, especially when it might potentially be confused for a minus sign.

(4) [.] (period) should be limited to use as a decimal point.

(5) [/] (slash) should be used for linking of quoted sources.

(6) [:] (colon) should be used to indicate subdivision or subcomponent.

8.2.5 Existing Designations

Existing designations should never be altered. This includes truncation, rounding, and shortening, among other possibilities. When citing a designation in a publication, the bibliographical reference should be given.

8.2.6 Astronomy Notation: Near Earth Celestial Bodies

8.2.6.1 Planets, Satellites, and Rings

Responsibility for the names of newly discovered planets, satellites, and planetary features falls to the Working Group for Planetary System Nomenclature of the International Astronomical Union (IAU). The determinations of this group are published in the Gazetteer of Planetary Nomenclature and posted on the Gazetteer's website at:

http://planetarynames.wr.usgs.gov.

In 2006, the IAU redefined a planet in this solar system as a celestial body that:

- is in orbit around the Sun,

- has sufficient mass so that it assumes a hydrostatic equilibrium (nearly round) shape, and

- has "cleared the neighborhood" around its orbit.

As of the publication of this book, the IAU has not yet come to a precise definition for extrasolar planets (also called exoplanets), as the description would be much more complex and will require more time and experience to decide.

When listing satellites, order them by increasing mean distance from the planet.

A non-satellite body fulfilling only the first two of these criteria is classified as a *dwarf planet*, whilst a non-satellite body fulfilling only the first criterion is termed a *small solar system body* (SSSB).

> **Table 8.6, found at the end of this chapter, contains the designations of astronomical bodies frequently cited in the scientific literature, with their IAU-approved alphabetical and numerical designations.**

8.2.6.2 Minor Planets (Asteroids) and Comets

Minor planets are identified by using a combination of serial numbers and proper names, following the definition of reliable orbital information (*see below,* page 362).

1501 Baede   71445 Marc

Since 1995, newly discovered comets have been given a designation that references the date of their first observation. The name consists of a capital letter that indicates in which half-month the comet was observed, the comet's year of discovery, and a serial number indicating the order in which comets were discovered during that half-month. A prefix is added to indicate the type of comet:

P/ for a periodic comet
C/ for a comet that is not periodic
X/ for a comet whose orbit cannot be computed
D/ for a periodic comet that no longer exists

Examples of these designations are:

Comet Pojmański (C/2006 A1)
(discovered by Pojmański, on January 2, 2006)

Comet Zhu-Balam (C/1997 L1)
(discovered by Zhu on June 3 and Balam on June 8, 1997)

Prior to 1995, a comet was designated by the name of its discoverer, the year of its perihelion, and a number in capitalized roman numerals that indicated its order of perihelion passage that year.

Examples of these designations are:

Comet Kohoutek 1973 XII
(discovered by Kohoutek on March 7, 1973)

Comet Brooks 1911 V
(discovered by Brooks on July 21, 1911)

8.2.6.3 Meteor Showers, Meteors, and Meteorites

The same particle from a comet or asteroid orbiting the sun goes through a series of names as it passes from space down to Earth. A *meteoroid* is the particle that has separated from a comet or asteroid in space. A *meteor* is the light phenomenon that occurs as the meteoroid enters the atmosphere of Earth and begins to vaporize. A *meteor shower* results from the passage of a comet near the sun, which causes the production of large quantities of meteoroids. A *meteorite* is a meteor that lands on Earth, not entirely disintegrating as it passes through the atmosphere.

Meteor showers are typically named after the constellation in which their radiant points appear or the comet with which they are associated. Examples include:

the Perseids   the Leonids

Large meteorites are typically named according to the location where they are discovered. A fireball is identified by the date on which it was observed. Examples include:

the Willamette meteorite (Willamette Valley, Oregon)
the Manson crater (near Manson, Iowa)

The main responsibility for designating and naming minor celestial bodies has been given by the IAU to the Committee for Small Body Nomenclature (CSBN). Initially, provisional names were given to asteroids discovered in the solar system by the discoverer, and these names were either made permanent or the discoverer was given an opportunity to provide a permanent name once the discovery was confirmed and an orbit was determined.

In recent years, however, the discovery of asteroids has accelerated due to the application of computer-assisted techniques (LINEAR and LONEOS, among others), which has resulted in the confirmed discovery of many thousands of asteroids. The CSBN has therefore placed a limitation on the naming process: a maximum of two names may be designated per discoverer per month. This means that there will be many thousands of asteroids that will be discovered in the course of the coming decades that will never be named officially. The IAU is currently considering solutions to this problem.

8.2.7 Cosmology Notation: Far Celestial Bodies

The nomenclature system for far celestial bodies varies by category. Their respective designation rules are detailed below.

8.2.7.1 Bright Stars

i. Bayer designation. The IAU is the body with the internationally recognized authority to assign star designations. The primary designation system for bright stars, called Bayer designations (after Johann Bayer, who first introduced the system in 1603), consists of a Greek letter followed by the name of the constellation in which it appears. The name of the constellation is sometimes abbreviated with a standard 3-letter abbreviation. The Greek letters are assigned in order (α, β, γ, δ, etc.) according to brightness. In the case of the roughly 1,000 stars that have proper names, the Bayer designation should be used as well.

> α Ori (proper name: Betelgeuse)
> α Aql (proper name: Altair)

Variable stars use a naming system that is a variation of the Bayer system. *See* **Section 8.2.7.2**, below for these conventions.

ii. Flamsteed system. The Flamsteed system, an alternative designation system, is similar to the Bayer system except that the Greek letter is replaced by a number and the constellation's name is used in the genitive form. The Flamsteed designation for a star is generally used only if there is no Bayer desig-nation.

> 79 Cnc (in the constellation Cancer)
>
> 41 Capricorni A and 41 Capricorni B
>   (binary star system in Capricorn)

iii. Spectral designation. Stars may also be categorized by spectral temperature. There are nine groups represented by the Latin letters (in order from hottest to coolest) O, B, A, F, G, K, M, L, and T. These groups can be further divided into groups represented by the digits 0 to 9 linked to the letters listed above (e.g. F3, T6). (There are no K8 or K9 stars.)

iv. Luminosity classification. Another classification scheme designates stars according to their luminosity with roman numerals, which may be subdivided with lower-case letters.

Table 8.1 Classes of Stars

Class	Luminosity type
Ia and Ib	Supergiants
II	Bright giants
III	Giants
IV	Subgiants
V	Main-sequence (dwarfs)

There are about 900 bright stars listed in The Bright Star Catalogue and about 1,500 bright stars listed in the annual Astronomical Almanac.

8.2.7.2 Variable Stars

Variable stars are either stars whose luminosity actually changes—for example, because the star periodically swells and shrinks—or stars with eclipsing and rotating variables, where the changes in brightness are only a perspective effect.

In 1862, astronomer Friedrich Wilhelm Argelander suggested that the letters R through Z be reserved for variable stars. At that time, this scheme provided ample possible star names, but as the number of variables increased, the naming system had to be expanded.

Designations of variable stars use the name of the constellation preceded by either a one- or two-letter (capitalized) code or the capital letter V and a number. If the star has a proper name, which can be found in the General Catalogue of Variable Stars or the SIMBAD database, use it as part of the title.

Below are two examples of these designations:

V1500 Cyg   YZ Cet

8.2.7.3 Novae or Cataclysmic Variables

A nova is designated by the name or abbreviation of the constellation preceded by the word Nova and followed by the year. If more than one nova is discovered in a single constellation in one year, the year is followed by a number. A nova may also be given a designation as a variable star.

Nova Cyg 1975 Nova Aql 1943

8.2.7.4 Supernovae

A supernova is designated by the capital letters *SN* in front of the year and a letter or combination of letters. The letter portion of the designation is assigned alphabetically—first as a single capital letter, then two lowercase letters (aa, ab, ac, etc. to zz). No punctuation should be used.

SN 1995B   SN 2002cn

8.2.7.5 Nebulae, Galaxies, Clusters, Star Clusters

There are two catalogues of nebulae, galaxies, and clusters that are commonly used. The catalogue created by Charles Messier in the late 18th century currently lists 110 objects. Objects in this catalogue are designated by the letter M and a serial number without a space between them.

M45 (The Pleiades)   M57 (The Ring Nebula)

The other catalogue, called the New General Catalogue (NGC), was originally compiled by J.L.E. Dreyer in the late 19th century. Subsequently, two Index Catalogue (IC) supplements have been added to the NGC. Objects in this catalogue and its supplements are designated by the letters *NGC*, a space, and the object's catalogue number. Objects may have both a Messier and NGC designation.

NGC 1313   NGC 1976 (The Orion Nebula, also M42)

8.2.7.6 Pulsars

Pulsars are named by the prefix *PSR* (for *Pulsating Source of Radio*) followed by a *J* and the declination, including minutes. Older pulsars are conventionally preceded by a *B* instead of a *J*..

PSR B1919+21   PSR J1921+2153

8.2.7.7 Quasi-stellar Objects (QSOs) or Quasars

It is only in the last 40 years that quasi-stellar objects (or "quasars"—abbreviated as QSO) have been identified as immensely powerful radiation sources that exist in the heart of galaxies extremely far away (on the order of a significant fraction of the extent of the universe itself).

One difficulty in observing (and thus naming) quasars derives from the fact that the great distance radiation must travel from the source to Earth makes it possible for the radiation to undergo so-called gravitational (or Einstein) lensing, where an intervening galaxy bends the light from a quasar and directs two beams to Earth from the same object. (Or else, light from a QSO is redirected to Earth and observed at a location other than its actual location.)

Older quasars have designations from earlier observations (when they were thought to be simply extremely bright stars). Thus, 3C 273 was among the first quasars identified and its name was derived from the order in which it was listed in the *Third Cambridge Catalogue of Radio Sources*, published in 1959. Recently discovered QSOs are designated by the letter Q and a designation that indicates its location n the night sky.

8.2.7.8 Other Radiation Sources (DRAGNs)

The increased sensitivity of earth-based and space-based observational tools has provided an exponential increase in the radiation sources that are being recorded. Among the most studied are the so-called *d*ouble *r*adiosource *a*ssociated with a galactic *n*ucleus, given the acronym DRAGN. These radiational sources are created from plasmas in the heart of galaxies or from the collision of (very) large clouds of intergalactic gas. (Currently, the designation is given to many sources of radio waves unassociated with identifiable cosmic structures. The use of the term "double" is an accident of early observations, but has been retained.)

As yet, no internationally accepted convention has emerged for the naming of these sources, now numbering in the thousands.

8.2.8 Notation for Astronomical Date and Time Systems

According to the IAU Style Manual (available online at: http://www.iau.org/static/publications/stylemanual1989.pdf):

> It should be noted that sidereal, solar and universal time are best regarded as measures of hour angle expressed in time measure; they can be used to identify instants of time, but they are not suitable for use as precise measures of intervals of time since the rate of rotation of Earth, on which they depend, is variable with respect to the SI second.

8.2.8.1 International Atomic Time (TAI)

Beginning in 1967, the definition of a second by the International System of Units (SI) has been the period equal to 9,192,631,770 cycles of the radiation corresponding to the transition between the two ground state energy levels of the Cesium-133 atom. A cesium oscillator, or atomic clock as it is more commonly known, is therefore the primary standard for time measurements.

International Atomic Time (TAI) is calculated by the International Bureau of Weights and Measures (BIPM) in France from a collection of over 200 clocks in 50 national laboratories around the world. TAI is broadcast from BIPM for use by scientists and engineers around the world.

8.2.8.2 Coordinated Universal Time (UTC)

Coordinated Universal Time was created to account for the slightly irregular rotation of the earth. Since TAI is a stable and uniform scale, it does not track the earth's rotation. The rate of UTC and TAI are equal, except adjustments known as "leap seconds" are inserted from UTC either in June or December to ensure that the sun crosses the Greenwich prime meridian at noon UTC to within 0.9 seconds.

8.2.8.3 Ephemeris or Dynamical Time (ET or TDB)

Ephemeris time is an obsolete time standard based on the ephemeris second, which was a fraction of the tropical year.

8.2.8.4 Coordinated Barycentric Time (TCB)

Coordinated Barycentric time is based on Dynamical time, but includes relativistic corrections that move the origin to the barycenter, or the center of the solar system. It is used especially for the calculation of ephemerides for use in the solar system as a whole.

8.2.8.5 Greenwich Mean Sidereal Time (GMST)

Sidereal time is a method of timekeeping that predates international atomic time and is based on the rotation of the earth with respect to fixed distant stars. The measurement of sidereal time is the hour angle of the vernal equinox. GMST is the hour angle measurement made with respect to the Greenwich meridian.

8.2.8.6 Julian Dates

The date of an astronomical event is typically referenced by the Julian date (JD) jointly with an appropriate conventional form.

The Julian calendar begins with 0 at noon on January 1 in the year 4713 BCE. The Julian date, therefore, is the number of days (including fractions of the day expressed with a decimal) that have passed since the beginning of the Julian calendar. It is important to note that a new day in the Julian system begins at 12:00 Universal Time (UT) (noon on the Greenwich meridian) and not midnight, like our common calendar. For example, the Julian date for midnight on January 1, 2007 is 2454101.50. The Julian date for 12:00 UT on January 1, 2007 is 2454102.00.

The modified Julian date (MJD) is a shorter version of the Julian date sometimes used by astronomers. The modified Julian calendar begins with 0 at midnight on November 17, 1858. The formula for finding the MJD is:

$$MJD = JD - 2400000.5$$

8.2.8.7 Besselian and Julian Epochs

The Julian year and century are the standard format for referencing astronomical epochs. The Julian year is 365.25 days, and, correspondingly, the Julian century is 36,525 days.

8.2.9 Field Specific Units and Measurements

8.2.9.1 International System of Units (SI) Units

SI units are the current prevailing system of measurement in science today. For general information on the system, see **Chapter 7, Section 7.2.3.** For SI tables—including prefixes, derived units, and conversion factors—see **Appendix B1.** The following table lists units that are not a part of the SI system, but that are accepted for use in astronomy by the IAU.

Table 8.2 Non-SI Units Recognized for Use in Astronomy by the IAU

Quantity	Unit	Symbol	Value
time	minute	min or "	60 s
time	hour	h	3600 s
time	day	d	86 400 s
time	year (Julian)	a	31.5576 Ms = 365.25 d
angle	second of arc	"	$(\pi/648\ 000)$ rad
angle	minute of arc	'	$(\pi/\ 10\ 8000)$ rad
angle	degree	°	$(\pi/180)$ rad
angle	revolution (cycle)	c	$2\ \pi$ rad
length	astronomical unit	au	0.149 598 Tm
length	parsec	pc	30.857 Pm
mass	solar mass	Mo	1.9891×10^{30} kg
mass	atomic mass unit	u	$1.660\ 540 \times 10^{-27}$ kg
energy	electron volt	eV	0.160 2177 aJ
flux density	jansky	Jy	$10^{-26} \text{W} \cdot \text{m}^{-2} \cdot \text{Hz}^{-1}$

8.2.10 Commonly Used Constants in Astronomy

Table 8.3 Astronomical Constants

Quantity	Value
astronomical unit (A.U.)	149,597,870.691 kilometers
light year (ly)	$9.460536207 \times 10^{12}$ km = 63,240 A.U.
parsec (pc)	$3.08567802 \times 10^{13}$ km = 206,265 A.U.
sidereal year	365.2564 days
tropical year	365.2422 days
Gregorian year	365.2425 days
Earth mass	5.9736×10^{24} kilograms
Sun mass	1.9891×10^{30} kg = 332,980 × Earth
mean Earth radius	6371 kilometers
Sun radius	6.96265×10^{5} km = 109 × Earth
Sun luminosity	3.827×10^{26} watts

Table 8.4 Physical Constants

Quantity	Symbol	Value
speed of light	c	299,792.458 kilometers/second
gravitational constant	G	6.6726×10^{-11} m^3 /(kg sec^2)
Boltzmann constant	k	1.380658×10^{-23} Joules/Kelvin
Stefan-Boltzmann constant	s	5.67051×10^{-8} J/(m^2 K^4 s)
Wien's law constant		2.897756×10^6 nanometers Kelvin
Planck constant	h	$6.6260755 \times 10^{-34}$ Joules second
Dirac's constant (reduced Planck's constant; "h-bar")	ℏ	$1.05457168 \times 10^{-34}$ Joules second
electron mass	m$_e$	$9.1093898 \times 10^{-28}$ grams $= 5.48579903 \times 10^{-4}$ amu
proton mass		$1.6726231 \times 10^{-24}$ grams $= 1.007276470$ amu
neutron mass		$1.6749286 \times 10^{-24}$ grams $= 1.008664904$ amu
deuterium nucleus mass		$3.3435860 \times 10^{-24}$ grams $= 2.013553214$ amu

8.2.11 Abbreviations of Common Astronomical Journals

A more complete list of astronomy journals is found in **Appendix C**. In the table below are listed several prominent journals in the field.

Table 8.5 Abbreviations of Common Astronomical Journals

Abbreviation	Journal Title
A&A	Astronomy and Astrophysics
A&AS	Astronomy and Astrophysics Supplement Series
AJ	Astronomical Journal
ApJ	Astrophysical Journal
Ap&SS	Astrophysics and Space Science
ARA&A	Annual Review of Astronomy and Astrophysics
AZh	Astronomicheskij Zhurnal
BAAS	Bulletin of the American Astronomical Society
MNRAS	Monthly Notices of the Royal Astronomical Society
MmRAS	Memoirs of the Royal Astronomical Society
PASJ	Publications of the Astronomical Society of Japan
PASP	Publications of the Astronomical Society of the Pacific
QJRAS	Quarterly Journal of the Royal Astronomical Society
S&T	Sky and Telescope

8.3 Lists of Astronomy Tables and Appended Material

8.3.1 Tables in This Chapter

8.3.2 Contents of Appendix C in Part III

C1 Astronomy Units and Symbols

C2 Astronomy Acronyms

C3 Hertzsprung–Russell Diagram

C4 Glossary of Astronomy and Physics Terms

C5 Astronomical Catalogues and Their Abbreviations

C6 Astronomy and Physics Journals and Their Abbreviations

Table 8.6 IAU-Approved Designations of Frequently Cited Astronomical Objects in the Solar System

Body Type	Common Name	Scientific Name	Body Type	Common Name	Scientific Name
Star	Sun		Satellite	Thyone	Jupiter XXIX
Planet	Mercury		Satellite	Hermippe	Jupiter XXX
Planet	Venus		Satellite	Aitne	Jupiter XXXI
Planet	Earth		Satellite	Eurydome	Jupiter XXXII
Satellite	Moon	Earth I	Satellite	Euanthe	Jupiter XXXIII
Planet	Mars		Satellite	Euporie	Jupiter XXXIV
Satellite	Phobos	Mars I	Satellite	Orthosie	Jupiter XXXV
Satellite	Deimos	Mars II	Satellite	Sponde	Jupiter XXXVI
Asteroid	Eros	433 Eros	Satellite	Kale	Jupiter XXXVII
Asteroid	Gaspra	951 Gaspra	Satellite	Pasithee	Jupiter XXXVIII
Asteroid	Ida	243 Ida	Satellite	Hegemone	Jupiter XXXIX
Asteroid	Kalliope	22 Kalliope	Satellite	Mneme	Jupiter XL
Asteroid satellite	Dactyl	243 Ida I	Satellite	Aoede	Jupiter XLI
Asteroid	Mathilde	253 Mathilde	Satellite	Thelxione	Jupiter XLII
Asteroid satellite	Linus	22 Kalliope I	Satellite	Arche	Jupiter XLIII
Asteroid satellite	Petit-Prince	45 Eugenia I	Satellite	Kallichore	Jupiter XLIV
Planet	Jupiter		Satellite	Helike	Jupiter XLV
Satellite	Io	Jupiter I	Satellite	Carpo	Jupiter XLVI
Satellite	Europa	Jupiter II	Satellite	Eukelade	Jupiter XLVII
Satellite	Ganymede	Jupiter III	Satellite	Cyllene	Jupiter XLVIII
Satellite	Callisto	Jupiter IV	Satellite	Kore	Jupiter XLIX
Satellite	Amalthea	Jupiter V	Planet	Saturn	
Satellite	Himalia	Jupiter VI	Satellite	Mimas	Saturn I
Satellite	Elara	Jupiter VII	Satellite	Enciladus	Saturn II
Satellite	Pasiphaë	Jupiter VIII	Satellite	Tethys	Saturn III
Satellite	Sinope	Jupiter IX	Satellite	Dione	Saturn IV
Satellite	Lysithea	Jupiter X	Satellite	Rhea	Saturn V
Satellite	Carme	Jupiter XI	Satellite	Titan	Saturn VI
Satellite	Ananke	Jupiter XII	Satellite	Hyperion	Saturn VII
Satellite	Leda	Jupiter XIII	Satellite	Iapetus	Saturn VIII
Satellite	Thebe	Jupiter XIV	Satellite	Pheobe	Saturn IX
Satellite	Adrastea	Jupiter XV	Satellite	Janus	Saturn X
Satellite	Metis	Jupiter XVI	Satellite	Epimetheus	Saturn XI
Satellite	Callirrhoe	Jupiter XVII	Satellite	Helene	Saturn XII
Satellite	Themisto	Jupiter XVIII	Satellite	Telesto	Saturn XIII
Satellite	Megaclite	Jupiter XIX	Satellite	Calypso	Saturn XIV
Satellite	Taygete	Jupiter XX	Satellite	Atlas	Saturn XV
Satellite	Chaldene	Jupiter XXI	Satellite	Prometheus	Saturn XVI
Satellite	Harpalyke	Jupiter XXII	Satellite	Pandora	Saturn XVII
Satellite	Kalyke	Jupiter XXIII	Satellite	Pan	Saturn XVIII
Satellite	Iocaste	Jupiter XXIV	Satellite	Ymir	Saturn XIX
Satellite	Erinome	Jupiter XXV	Satellite	Paaliaq	Saturn XX
Satellite	Isonoe	Jupiter XXVI	Satellite	Tarvos	Saturn XXI
Satellite	Praxidike	Jupiter XXVII	Satellite	Ijiraq	Saturn XXII
Satellite	Autonoe	Jupiter XXVIII	Satellite	Suttungr	Saturn XXIII

Table 8.6, *continued*

Body Type	Common Name	Scientific Name	Body Type	Common Name	Scientific Name
Satellite	Kiviuq	Saturn XXIV	Satellite	Prospero	Uranus XVIII
Satellite	Mundilfari	Saturn XXV	Satellite	Setebos	Uranus XIX
Satellite	Albiorix	Saturn XXVI	Satellite	Stephano	Uranus XX
Satellite	Skathi	Saturn XXVII	Satellite	Trinculo	Uranus XXI
Satellite	Erriapus	Saturn XXVIII	Satellite	Francisco	Uranus XXII
Satellite	Siarnaq	Saturn XXIX	Satellite	Margaret	Uranus XXIII
Satellite	Thrymr	Saturn XXX	Satellite	Ferdinand	Uranus XXIV
Satellite	Narvi	Saturn XXXI	Satellite	Perdita	Uranus XXV
Satellite	Methone	Saturn XXXII	Satellite	Mab	Uranus XXVI
Satellite	Pallene	Saturn XXXIII	Satellite	Cupid	Uranis XXVII
Satellite	Polydeuces	Saturn XXXIV	**Planet**	**Neptune**	
Satellite	Daphnis	Saturn XXXV	Satellite	Triton	Neptune I
Satellite	Aegir	Satu‚n XXXVI	Satellite	Nereid	Neptune II
Satellite	Bebhionn	Saturn XXXVII	Satellite	Naiad	Neptune III
Satellite	Bergelmir	Saturn XXXVIII	Satellite	Thalassa	Neptune IV
Satellite	Bestla	Saturn XXXIX	Satellite	Despina	Neptune V
Satellite	Farbauti	Saturn XL	Satellite	Galatea	Neptune VI
Satellite	Fenrir	Saturn XLI	Satellite	Larissa	Neptune VII
Satellite	Fornjot	Saturn XLII	Satellite	Proteus	Neptune VIII
Satellite	Hati	Saturn XLIII	Satellite	Halimede	Neptune IX
Satellite	Hyrrokkin	Saturn XLIV	Satellite	Psamathe	Neptune X
Satellite	Kari	Saturn XLV	Satellite	Sao	Neptune XI
Satellite	Loge	Saturn XLVI	Satellite	Laomedeia	Neptune XII
Satellite	Skoll	Saturn XLVII	Satellite	Neso	Neptune XIII
Satellite	Surtur	Saturn XLVIII	Dwarf Planet	Ceres	1 Ceres
Satellite	Anthe	Saturn XLIX	Dwarf Planet	Pluto	134340 Pluto
Satellite	Jarnsaxa	Saturn L	Satellite	Charon	Pluto I
Satellite	Greip	Saturn LI	Satellite	Nix	Pluto II
Satellite	Tarqeq	Saturn LII	Satellite	Hydra	Pluto III
Planet	**Uranus**		Dwarf Planet	Haumea	136108 Haumea
Satellite	Ariel	Uranus I	Satellite	Hi'iaka	136108 Haumea
Satellite	Umbriel	Uranus II	Satellite	Namaka	136108 Haumea
Satellite	Titania	Uranus III	Dwarf Planet	Eris	136199 Eris
Satellite	Oberon	Uranus IV	Satellite	Dysnomia	136199 Eris I
Satellite	Miranda	Uranus V	Dwarf Planet	Makemake	136472 Makema
Satellite	Cordelia	Uranus VI	Periodic Comet	Halley	1p/Halley
Satellite	Ophelia	Uranus VII	Periodic Comet	Encke	2P/Encke
Satellite	Bianca	Uranus VIII	Periodic Comet	Faye	4P/Faye
Satellite	Cressida	Uranus IX	Periodic Comet	Tuttle	8P/Tuttle
Satellite	Desdemona	Uranus X	Periodic Comet	Borrelly	19P/Borrelly
Satellite	Juliet	Uranus XI	Periodic Comet	Schwassmann-Wachmann 1	29P/Schwassmann Wachmann
Satellite	Portia	Uranus XII			
Satellite	Rosalind	Uranus XIII	Periodic Comet	Hale-Bopp	C1995 01
Satellite	Belinda	Uranus XIV	Periodic Comet	Bennett	C 1965 Y1
Satellite	Puck	Uranus XV	Periodic Comet	Hyakutake	C1996 B2
Satellite	Caliban	Uranus XVI	Periodic Comet	Ikeya-Seki	C 1965 S1
Satellite	Sycorax	Uranus XVII	Periodic Comet	Shoemaker-Levy 9	D 1993/F2

Chapter 9. Style and Usage for Chemistry

Contents

Contents, 9.2.5 Atoms, Molecules, and Isotopes, *continued*

Chapter 9. Style and Usage for Chemistry

9.1 Preparing the Manuscript

9.1.1 Publishing Guidelines

The literature of chemistry, like most scientific writing, can be published in monographs, which examine a single subject in depth; in multi-authored handbooks, which treat a broader scientific field in detail; in proceedings volumes, which correspond to meetings or symposia; but most often as articles in peer-reviewed scientific journals. In fact, most articles in chemistry journals represent original research, and most original research must be published in a peer-reviewed journal in order to have credibility in the international scientific community.

Some chemistry journals, such as the American Chemical Society's *Accounts of Chemical Research*, treat topics relevant to the entire field of chemistry, but most journals are more specialized; for example, the Royal Society of Chemistry's journal *Photochemical & Photobiological Sciences*.

Chemistry journals generally follow the standard scientific article format of abstract, introduction, methods, etc., as described in **Chapter 2**, above, but individual journals retain the right to expand upon, relax, or otherwise alter the conventions of this format. It is advisable to (a) become familiar with the style and content of a journal to which one wishes to submit a paper; and (b) contact the journal editors to assess their interest in the subject matter of the paper one wishes them to consider publishing.

Electronic publishing. Journals will often have webpages that contain very specific submission guidelines. These will not only detail the journal's house style but may also give information about topics appropriate for the journal, prices of publication (per page or as a whole), their review process with a general time frame, ethical and legal policies, and complete instructions for electronic and nonelectronic submission, including tables, illustrations, and other multimedia.

To find a chemistry journal's webpage, refer to a list of hyperlinked chemistry journals such as this one provided by the University of Cambridge: http://www.ch.cam.ac.uk/c2k/cj/. Alternately, try typing the journal's name into a search engine (within quotation marks if necessary). For a list of over sixteen hundred chemistry journals—which is by no means exhaustive—*see* **Appendix D5**, in **Part III**, below.

9.1.2 Typesetting Chemical Formulas and Structures

Most line formulas, including those with subscripts, superscripts, braces, and basic formatting conventions like italicization, can be easily produced with modern word-processing software such as Microsoft Word. If a chemical or chemicals are better displayed as illustrations, connected by lines or simulating a three-dimensional image, there are many programs, both commercially available and downloadable for free, that are designed to help create them.

Journals often request that images produced by these programs be sent separately as digital files. Many programs even supply templates that correspond to illustration preferences of specific journals. Some popular programs are ISIS/Draw, ChemSketch, ChemDraw, Chemistry 4-D Draw, and DrawIt. For a detailed and reliable comparison of some of the major options, visit:

http://dragon.klte.hu/~gundat/rajzprogramok/dprog.html.

The typesetting systems TEX and LATEX are useful for typesetting mathematical equations and other technical information. For more information about these programs, *see* **Chapter 6**, **Section 6.1.4, i**.

9.2 Usage

9.2.1 Subatomic Particles

9.2.1.1 Symbols for Subatomic Particles

Subatomic particles are represented by lowercase Latin or Greek letters. Commonly used symbols for subatomic particles are listed in **Part Three, Section D1.6**.

9.2.1.2 Electric Charge Notation

Electric charge for a subatomic particle is indicated by a superscript of the corresponding charge. A *p* or *e* used without charge notation refers to a positive proton and a negative electron.

$$p^- \qquad\qquad e^+ \qquad\qquad \mu^-$$

9.2.2 Electronic Configuration

Electron shells are signified by the uppercase letters K, L, M, and N.

Electron subshells and atomic orbitals are signified by the lowercase roman letters s, p, d, and f. Principal energy levels 1 through 7 are written to the left of the letter. The orbital axes are specified with italic subscripts, and the number of electrons in each orbital is written next to the letter as a superscript.

7p electron

The electron configuration of iron is $1s^2\,2s^2\,2p^6\,3s^2\,3p^6\,4s^2\,3d^6$.

Greek letters are used to indicate some bonding orbitals and the bonds they create.

π bond　　　　　　σ orbital

The electronic state of the atom, representing the angular momentum quantum number L, is denoted with the uppercase roman letters S, P, D, F, G, H, I, and K. The letters represent the L values 0 through 7, respectively. The same letters in lowercase are used to denote the orbital angular momentum of a specific electron. A superscript to the left of the letter represents the spin multiplicity, and a subscript to the right of the letter indicates the total angular momentum quantum number J.

$^7D_1$　　　　　　$^7d_1$　　　　　　$^2P_{3/2}$　　　　　　$^2p_{3/2}$

The electronic state of the molecule is denoted with the uppercase roman letters A, B, E, and T, with the ground state X. The same letters in lowercase are used to denote one-electron orbitals. A subscript to the right of the letter is used to express the symmetry of the orbital. A tilde (~) over the letter is used to indicate polyatomic molecules.

$^2A_{2g}$　　　　　　$\tilde{E}$　　　　　　$^3b_1$　　　　　　T_{2g}

9.2.3 Names and Symbols of Chemical Elements

(The complete list of chemical elements and their symbols can be found in the periodic table, explained beginning on the next page and appearing on pages 000–000.)

The names of chemical elements and chemical compounds are written in roman type and treated as common nouns, whether they are named after proper nouns or not.

carbon	californium
mercury	einsteinium
sodium chloride	mercury sulfide
hydrogen bromide	

The symbols of chemical elements are written in roman type, and the first letter of the symbol is capitalized.

C   Pb   Hg   H

When using the symbol for a chemical element, the name of the element is pronounced. Therefore, when writing using chemical symbols, use the article that is compatible with the pronunciation of the chemical element name.

a Ag metal ("a silver metal")

an Fe ion ("an iron ion")

When referencing a chemical compound, use either the complete symbol or the complete name of the compound; do not mix symbols and names.

HCl   *or*   hydrogen chloride

not

H chloride   *or*   hydrogen Cl

9.2.3.1 Periodic Table of the Elements

The periodic table is an arrangement of the known chemical elements according to their atomic properties. The elements are listed by increasing atomic number from left to right and top to bottom, with new periods (horizontal rows) beginning as electrons of those elements begin to occupy new valence shells and chemical trends of those elements repeat. As a result, groups of elements (those in the same vertical columns) share similar properties, and the table as a whole shows certain trends as well. Metallic elements tend to occupy the left side of the table, semimetallic elements the middle, and nonmetals the right side. Elements of like physical states tend to be grouped together. Electropositivity and ionization energy increase from left to right across the table, while atomic radius decreases.

Periods

The periods of the periodic table represent elements whose electrons fill the same valence shells. The first period contains only hydrogen and helium, whose atoms contain electrons that occupy only the lowest energy levels. The next period contains elements of atomic numbers 3–10, filling the second energy level with electrons and ending with neon. The third period fills up another energy level and ends with argon, of atomic number 18. Elements in the fourth and fifth periods are mostly transition elements, which have outer electrons that occupy the next subshell. The sixth and seventh periods are the last rows of the periodic table, and comprise the two series of elements known as the lanthanides and the actinides, respectively. These are elements that draw on the higher-energy f shell orbitals.

Groups

Groups of elements are vertical columns in the periodic table, once numbered with roman numerals but now conventionally referred to by the arabic numerals 1–18. They can be divided into metals, semimetals, and nonmetals. Group 1 elements, with the exception of hydrogen, are often referred to as the alkali metals. In general, they exhibit properties associated with metals: ductility, malleability, high luster, good electrical conductivity, and low electronegativity and ionization energy.

Periodic Table of

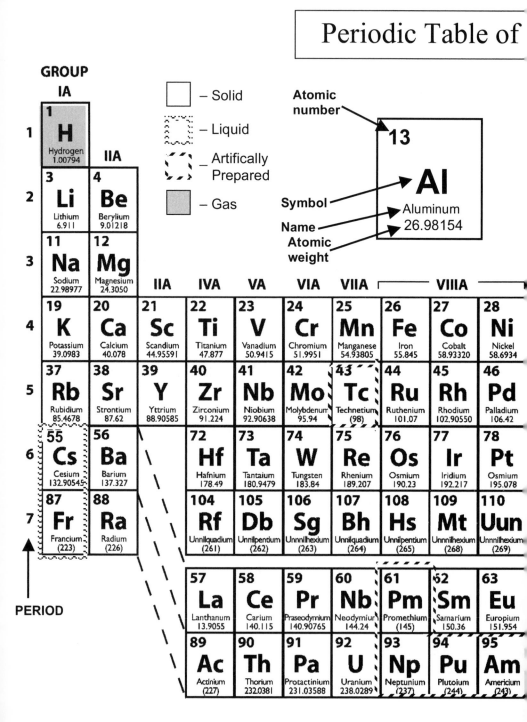

the Elements

							VIII
							2 **He** Helium 4.00260

		IIIB	IVB	VB	VIB	VIIB	
		5 **B** Boron 10.811	6 **C** Carbon 12.0107	7 **N** Nitrogen 14.00674	8 **O** Oxygen 15.9994	9 **F** Fluorine 18.99840	10 **Ne** Neon 20.1797
IB	IIB	13 **Al** Aluminum 26.98154	14 **Si** Silcon 28.0855	15 **P** Phosphorus 30.97376	16 **S** Sulfur 32.066	17 **Cl** Chlorine 35.4527	18 **Ar** Argon 39.948
29 **Cu** Copper 63.546	30 **Zn** Zinc 65.39	31 **Ga** Gallium 69.723	32 **Ge** Germanium 72.61	33 **As** Arsenic 74.92160	34 **Se** Selenium 76.95	35 **Br** Bromine 79.904	36 **Kr** Krypton 83.80
47 **Ag** Silver 107.8682	48 **Cd** Cadmium 112.411	49 **In** Indium 114.818	50 **Sn** Tin 118.710	51 **Sb** Antimony 121.760	52 **Te** Tellurium 127.60	53 **I** Iodine 126.90447	54 **Xe** Xenon 131.29
79 **Au** Iridium 192.96655	80 **Hg** Mercury 200.59	81 **Tl** Thallium 204.3833	82 **Pd** Lead 207.2	83 **Bi** Bismuth 208.98038	84 **Po** Polonium (209)	85 **At** Astatine (210)	86 **Rn** Polonium (222)
111 **Uuu** Unnnilhexium (272)	112 **Uub** Unnnilhexium						

64 **Gd** Gadolinium 157.25	65 **Tb** Terbium 158.92534	66 **Dy** Dysprosium 162.50	67 **Ho** Holmium 164.93032	68 **Er** Erbium 167.26	69 **Tm** Thulium 168.93	70 **Yb** Ytterbium 173.04	71 **Lu** Lutetium 174.957
96 **Cm** Curium (247)	97 **Bk** Berkelium (247)	98 **Cf** Californium (251)	99 **Es** Einsteinium (252)	100 **Fm** Fermium (257)	101 **Md** Mendelevium (258)	102 **No** Nobelium (259)	103 **Lr** Lawrencium (260)

Groups, *continued*

Group 2 elements are called alkaline earth metals. They share similar metallic properties, and interact with other groups of elements in similar ways. From left to right through the transition elements, metals give way to metalloids and finally to nonmetals, which include the highly reactive halogens (group 17), and the noble gases (group 18), the rightmost group of elements which are characterized by their low chemical reactivity.

9.2.3.2 Referencing the Periodic Table

The phrase *periodic table* should be treated as a common noun and written in lowercase.

When referencing the periodic table, the word *group* should always be in lowercase, with or without a specific number. Groups are sometimes seen designated by roman numerals, but this guide, along with the International Union of Pure and Applied Chemistry, recommends use of the arabic numerals 1–18.

group 17 elements   group 8 metals

9.2.4 Chemical Formulas

9.2.4.1 Empirical Formulas

An empirical formula is the simplest possible means of expressing chemical composition. It is frequently used in indexes and other instances when exact molecular composition of the compound is not necessary.

SiO_2 LiOH $Ni(OH)_2$

9.2.4.2 Molecular Formulas

The molecular formula of a compound represents its atomic composition. The order of the formula is determined by relative electronegativities, with the first component having the greatest electropositivity.

$C_6H_{12}O_6$ (*not* CH_2O)
H_2O_2 (*not* HO)
Hg_2Cl_2 (*not* HgCl)

9.2.4.3 Structural Formulas

A structural formula provides details about how the components of a molecule are connected. A stereoformula gives information about the relative locations of the components in a 3-dimensional space. A line formula provides other types of structural information such as single, double, or triple bond locations.

i. Representing Bonds. In a line formula, a single bond need not be represented, but may be represented by an en dash (–) if necessary. Double bonds are represented by an equals sign (=) and triple bonds by a triple line (≡), which is available in most word processing programs.

$$CH_3CH=CHCH_3$$

$$CH_3CH_2CH_2CH_2CH_2CH_3$$

<pre>
    H H
    | |
H - C - C - O -H
    | |
    H H
</pre>

ii. Representing Stereoformulas. In order to represent a molecule in 3-dimensional space, the IUPAC recommends using a thick line or a solid wedge for bonds reaching above the plane and a dashed thick line or a dashed wedge for bonds receding behind the plane. For information about software for drawing chemical structures, *see* **Section 9.1** in this chapter.

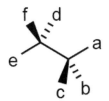

9.2.4.4 Abbreviations

Abbreviations for organic groups may be used in formulas and structures, but not in written text. Abbreviations other than the ones listed below must be defined when used.

Table 9.1 Abbreviations for Organic Group Names

Abbreviation	Group	Abbreviation	Group	Abbreviation	Group
Ac	acetyl	*sec*-Bu	*sec*-butyl	Ph	phenyl
Ar	aryl	*t*-Bu	*tert*-butyl	Pr	propyl
Bu	butyl	Bz	Benzoyl	*i*-Pr	isopropyl
i-Bu	isobutyl	Et	Ethyl	R or R'	alkyl
n-Bu	*n*-butyl	Me	Methyl		

9.2.4.5 Unique Cases

Coordination entities are enclosed in square brackets.

$$[Cr(C_6H_6)_2] \qquad\qquad K[PtCl_3(C_2H_4)]$$

Addition compounds are represented with a solid dot in the formula with no spaces on either side. Water of hydration is represented in the same manner, with the H_2O following the dot.

$$Me_3Al{\cdot}NMe_3 \qquad\qquad ZnSO_4{\cdot}7\,H_2O$$

To indicate the components of a mixture, use either a slash (/) or an en dash (–). Do not use a colon.

or:
  dissolved in a 4:1 methanol–water mixture

  dissolved in a 4:1 methanol/water mixture
not:
  dissolved in a 4:1 methanol:water mixture

9.2.5 Atoms, Molecules, and Isotopes

9.2.5.1 Mass Number

The mass number is indicated by a superscript on the left side of the chemical symbol. Mass numbers are typically only shown for isotopes or in isotope-related discussions.

^{14}C $\quad\quad\quad$ ^{12}C $\quad\quad\quad$ ^{238}U $\quad\quad\quad$ ^{235}U

9.2.5.2 Atomic Number

The atomic number, or number of protons in an atom, is indicated by a subscript on the left side of the chemical symbol. Atomic numbers are typically only referenced during discussions of nuclear chemistry.

$_{84}Po$ $\quad\quad\quad$ $_{88}Ra$ $\quad\quad\quad$ $_{90}Th$ $\quad\quad\quad$ $_{93}Np$

9.2.5.3 Number of Atoms

The number of atoms in a formula is indicated by a subscript on the right side of the chemical symbol.

H_2O $\quad\quad\quad$ MnO_2

9.2.5.4 Ionic Charge

The ionic charge is indicated by a superscript on the right side of the chemical symbol. If the charge is positive or negative 1, use only a plus sign $(+)$ or a minus sign $(-)$.

Cl^- $\quad\quad\quad$ K^+ $\quad\quad\quad$ Be^{2+}

In case there are both and a charge superscript, the two should not be aligned. The subscript should be written first.

NO_3^- $\quad\quad\quad$ SO_4^{2-}

9.2.5.5 Excited State

An excited electronic state is indicated by an asterisk on the right side of the chemical symbol.

$H*$ $\quad\quad\quad$ $NO*$

9.2.5.6 Oxidation Number

The oxidation number of an atom is indicated by a superscript in roman numerals on the right side of the chemical symbol. Roman numerals should never be used on the line next to the chemical symbol due to possible confusion with chemical symbols of iodine (I) and vanadium (V).

$$Al^{III} \qquad\qquad Ba^{II} \qquad\qquad O^{-II}$$

In the case of an oxidation number superscript and a subscript, the two should not be aligned; the oxidation number superscript should be written first.

$$Br^{-I}{}_{2}$$

9.2.5.7 Referencing an Isotope

An isotope may be referenced either by writing the chemical symbol with its mass number as described in **Section 9.2.5.1,** or by writing the full name of the element with a hyphen connecting the name to its mass number.

$$^{15}N \qquad or \qquad nitrogen\text{-}15$$

When pronounced, however, the element name always comes first. When writing the isotope name in text, therefore, the article chosen (*a* or *an*) should always correspond to how the element's full name will be pronounced.

9.2.5.8 Deuterium and Tritium

If no other nuclides are present, the symbols D and T may be used to represent deuterium and tritium respectively.

$$D_2O \qquad\qquad T_2S$$

9.2.5.9 Isotopically Modified Compounds

An isotopically modified compound is one that is comprised of nuclides not commonly found in nature. An isotopically unmodified compound is comprised of nuclides in proportion to their natural occurrence.

9.2.5.10 Isotopically Substituted Compounds

An isotopically substituted compound is a compound in which each molecule has only specific nuclides in their designated positions. When a compound is isotopically substituted, the nuclide symbol is included in the compound formula.

If the name of the compound is written out, the number, symbol, and locants (if necessary) are included in parentheses with no spaces between the parentheses and the compound name.

$^{238}UI_3$ fluoro(^{3}H)benzene

9.2.5.11 Isotopically Labeled Compounds

An isotopically labeled compound is a mixture of an isotopically unmodified compound with one or more similar isotopically substituted compounds. In order to indicate an isotopically labeled compound, the number and symbol of the isotope are enclosed in brackets next to or within the compound name or formula.

$[^{15}N]NF_2$ $CH_2[^2H_2]$

9.2.5.12 Radicals

Indicate a free radical by placing either a centered dot or a superscript dot, signifying the unshared electron, next to the chemical symbol or compound formula.

Cl·   HO$^{\cdot}$

Radical cations and anions are referenced by a superscript on the right side of the chemical symbol or compound formula of a dot followed by the positive or negative charge symbol. When writing in terms of mass spectroscopy, however, it is customary to reverse the order of the dot and the charge symbol.

Cl$^{\cdot\,-}$ R$^{\cdot\,+}$

9.2.5.13 Bonds

When writing linear formulas in text, single bonds are not shown unless there is a specific reason to highlight them.

NaCl CH_3COOCH_3

If they are being discussed, indicate single bonds with en dashes.

$$Na–Cl \qquad the \ C–C \ bond$$

If it is necessary, show double and triple bonds within the linear formula. Make every effort to keep the formula on a single line.

$$O=C=OH_2C=CH_2 \ CH_3$$

9.2.6 Crystallography

9.2.6.1 Crystal Planes and Directions

The various types of indices of a crystal face are referenced in different ways, as show in the following chart:

Table 9.2 Crystal Face Index Format

Indices	How Referenced	Example
Miller indices (of a crystal face or a single net plane)	enclosed in parentheses	(xyz) $(x_1x_2x_3)$
Laue indices	not enclosed	xyz
indices of a set of symmetrically equivalent crystal faces or net planes	enclosed in braces	$\{xyz\}$ $\{x_1x_2x_3\}$
indices of a set of symmetrically equivalent lattice directions	enclosed in angle brackets	$<abc>$
indices of a zone axis or lattice direction	enclosed in square brackets	[789] [xyz]

If an index is written together with a spelled-out chemical element name, the index is separated from the element name by a space. If the index is used with the chemical symbol, on the other hand, the index is placed next to the symbol with no space.

9.2.6.2 Types of Crystal Lattices

There are four main types of crystal lattices.

Table 9.3 Crystal Lattices and Their Symbols

Type of crystal lattice	Symbol
face-centered cubic	fcc
body-centered cubic	bcc
cubic close-packed	ccp
hexagonal close-packed	hcp

9.2.7 Chirality

Chirality symbols and symmetry terms are written in uppercase italics. See **Appendix D1.7** for a list of those symbols.

9.2.8 Concentration

In reactions or equations, a concentration is indicated by enclosing the chemical formula in square brackets. In text, the concept must be written out.

$[H_2SO_4] = 0.4$ M   or   The concentration of H_2SO_4 was 0.4 molar.

9.2.9 Chemical Reactions

There are two ways to treat chemical reactions in text. If the reaction is relatively short, it may be included in the line of text or set apart and numbered, if necessary. Longer chemical reactions should be set apart from the text. Sequences of reactions may be numbered or lettered.

The beaker contained materials undergoing the reaction $N_2 + 3H_2 \rightarrow 2NH_3$.

or:

$$NaCl + KNO_3 \rightarrow NaNO_3 + KCl \qquad (1)$$

The number of each reactant and product should be placed before its symbol in the same typeface, and its state (solid, liquid, gaseous, or aqueous) may also be indicated after the formula—without a space—by (s), (l), (g), and (aq), respectively.

$$SnO_2(s) + 2H_2(g) \rightarrow Sn(s) + 2H_2O(g)$$

Different types of arrows signify different properties of a reaction. A single arrow pointing right ($\rightarrow$) should be used to indicate a one-way, non-reversible reaction. Two half-headed arrows pointing both ways ($\rightleftharpoons$) signifies a reversible reaction in equilibrium. Catalysts and conditions such as heat or light may be placed as symbols over the arrow in a smaller font size.

$$2SO_2 + O_2 \rightleftharpoons 2SO_3$$

$$C_3H_8(g) + 5O_2(g) \xrightarrow{\Delta} 3CO_2(g) + 2H_2O(g)$$

9.2.10 Spectroscopy

A chemical element and a spectral line associated with it should be separated by a single space.

Cu Kα   Cr K

A list of common spectroscopic symbols and abbreviations can be found in **Appendix D1.8**.

9.2.11 Nomenclature Conventions

Names of chemical compounds are determined by the International Union of Pure and Applied Chemistry (IUPAC), the International Union of Biochemistry and Molecular Biology (IUBMB), the Committee on Nomenclature of the American Chemical Society, and other authorities when appropriate.

Many websites offer databases of chemical substances that are searchable by formal and trivial names, registry identification numbers, and chemical properties. Some offer a search based on chemical structure, via a browser-based molecule-drawing program. One database is kept and updated daily by the American Chemical Society's CAS Registry, including structural information and known names for over forty million organic and inorganic compounds, polymers, metals, elements, alloys, isotopes, proteins, etc.

For rules on where to divide the name of a chemical between two lines on a page, *see* **Appendix D2**.

9.2.11.1 Case and Type Style

Chemical names in text should be treated as common nouns. They should not be capitalized unless they are found at the start of a sentence or in a title. Every name should be written in roman type unless there is a specific rule to set it in italic or Greek letters.

9.2.11.2 Locants and Descriptors

Numerals or letters used as locants or descriptors in a chemical name are placed at the beginning or inside of the name. The locants should be set off from the rest of the name with hyphens. When a chemical symbol is used to signify attachment to a particular atom or a site of ligation, it should be italicized.

2-amino-*N*-methyl-*N*-phenylbenzene sulfonamide
2-amino-4-methylthiazole

However, if chemical symbols are used as descriptors for types of reactions, roman type should be used for the symbol instead of italics and a hyphen should set it off from the rest of the word.

O-methylation   N-carbamylation   N-substituted

When using a Greek letter to denote stereochemistry, use a hyphen to connect it to the word that follows and use only the symbol form of the letter.

α-(ω-cyanoethyl)-β-naphthol   α-amino-β-mercapto-n-valeric acid hydrochloride

• Prefixes that denote position, stereochemistry, configuration, or structure should always be italicized when used in conjunction with the chemical name. Follow any prefix with a hyphen.

• Prefixes that are not capitalized should remain in lowercase, even in a title or at the beginning of a sentence, and those that are capitalized should remain capitalized. The prefixes R, R*, S, S*, E, and Z should be enclosed with parentheses.

• IUPAC accepted prefixes are listed in the table below.

Table 9.4 Common Chemical Prefixes

abeo	facgem	rel
ac	hexahedro	retro
altro	hexaprismo	ribo
amphi	hypho	s
anti	icosahedro	S
antiprismo	klado	S*
ar	l	sec
arachno	m	sn
as	M	sym
asym	mer	syn
c	meso	t
cis	nido	tert
cisoid	o	threo
closo	octahedron	trans
cyclo	p	transoid
d	P	triangulo
dodecahedro	pentaprismo	triprismo
E	quadro	uns
endo	r	vic
erythro	R	xylo
exo	R*	Z

9.2.12 Types of Nomenclature

There are three types of nomenclature for inorganic compounds. The different types of nomenclature are applied to specific chemical entities.

9.2.12.1 Binary Nomenclature

Binary nomenclature is applied in the case of salt-like ionic species of compounds. The name *binary* applies to its most common use in naming simple salts comprised of cation and anion; however, it is sometimes used for naming more complicated compounds. In this style of nomenclature, the electropositive component is listed first, followed by the electronegative component.

Do not use hybrids of chemical symbols and spelled out names.

KCl   *or*   potassium chloride   *not*   K chloride

The binary system is also used for simple organometallic species derived from inorganic salts.

diethylaluminum bromide

9.2.12.2 Substitutive Nomenclature

Substitutive nomenclature, which was originally used with organic compounds, is also used for the naming of some organometallic compounds. The system is applicable for compounds in groups 13, 14, 15, and 16, and is based on a parent hydride with its hydrogen in part or entirely replaced by organic groups called substituents. The parent hydride structure is identified and constitutes the main part of the compound name. Atoms or groups that attach to the parent hydride are then cited as prefixes or suffixes. Remaining parts of the parent hydride are cited as substituents as well.

Parent Hydride: SiH_4 (silane)
Substituted Parent Hydride: $Si(CH_3)_3Cl$
                            (chlorotrimethylsilane)

Table 9.5 Common Parent Hydrides

Formula	Name	Formula	Name
BH_3	borane	SbH_5	λ^5-stabane
CH_4	methane	BiH_3	bismuthane
SiH_4	silane	SH_4	λ^4-sulfane
GeH_4	germane	SH_6	λ^6-sulfane
SnH_4	stannane	SeH_2	selane
PbH_4	plumbane	TeH_2	tellane
NH_3	azane	PoH_2	polane
PH_3	phosphane	FH	fluorane
PH_5	λ^5-phosphane	ClH	chlorane
AsH	arsane	BrH	bromine
SbH_3	stibane	IH	iodane

Table 9.6 Common Characteristic Groups Cited Only as Prefixes

Characteristic Group	Prefix	Characteristic Group	Prefix
—Br	bromo	—IO_2	iodyl
—Cl	chloro	—$I(OH)_2$	dihydroxyiodo
—ClO	chlorosyl	=N_2	diazo
—ClO_2	chloryl	—N_3	azido
—ClO_3	perchloryl	—NO	nitroso
—F	fluoro	—NO_2	nitro
—I	iodo	=N(O)OH	*aci*-nitro
—IO	iodosyl	—OR	R-oxy

Table 9.7 Common Characteristic Groups Cited as Prefixes or Suffixes

Characteristic Group	Prefix	Suffix
—COOH	carboxy-	-carboxylic acid
—COO^-	carboxylato-	-carboxylate
—(C)OOH	----	-oic acid
—$(C)OO^-$	----	-oate
—SO_2OH	sulfo-	-sulfonic acid
—SO_2O^-	sulfonato-	-sulfonate
—COX^1	halocarbonyl- carbonohalidoyl-	-carbonyl halide
—$CONH_2$	carbomoyl- aminocarbonyl-	-carboxamide
—$(C)ONH_2$	----	-amide
—$C(=NH)NH_2$	carbamimidoyl- amidino-	-carboximidamide -carboxamidine
—CN	cyano-	-carbonitrile
—(C)N	----	-nitrile
—CHO	formyl-	-carbaldehyde -carboxaldehyde
—(C)HO	----	-al
—OH	hydroxyl-	-ol
—O^-	oxido-	-olate
—SH	sulfanyl- mercapto-	-thiol
—S^-	sulfido- sulfanidyl-	-thiolate
—NH_2	amino-	-amine
=NH	imino-	-imine

1. Where X represents F, Cl, Br, I, N_3, CN, NC, NCO, NCS, NCSe, or NCTe.

9.2.12.3 Coordination Nomenclature

The coordination nomenclature system for inorganic chemistry is an additive one. It is based on the concept of identifying and naming a central atom and then naming the ions, atoms, or groups (called ligands) that are attached to it. In a coordination name, the ligands are named first in alphabetical order, followed by the name of the central atom. The full formula is enclosed in square brackets. If the coordination unit is negatively charged, it is preceded by a cation formula.

$[Co(en)_3]Cl_3$

tris(ethane-1,2-diamine)cobalt trichloride

$[IrClH_2(CO)\{P(CH_3)_3\}_2]$
carbonylchlorodihydridobis(trimethylphosphane)iridium

For more information on nomenclature, see "Nomenclature of Organometallic Compounds of the Transition Elements" at: http://www.iupac.org/publications/pac/1999/71_08_pdf/7108salzer_1557 .pdf.

9.2.13 Functional Replacement Nomenclature

The functional nomenclature system utilizes prefixes and infixes to denote the alterations to parent compounds by replacement of hydroxy groups or oxygen atoms with other atoms or functional groups. The chemical symbols for the atoms or groups being substituted should be placed in italics.

Parent compound:
acetic acid

Compound with functional replacement:

thio-acetic *S*-acid

9.3 Lists of Chemistry Tables and Appended Material

9.3.1 Tables in this Chapter

Table 9.1 Abbreviations for Organic Group Names
Table 9.2 Crystal Face Index Format
Table 9.3 Crystal Lattices and Their Symbols
Table 9.4 Common Chemical Prefixes
Table 9.5 Common Parent Hydrides
Table 9.6 Common Characteristic Groups Cited
     Only as Prefixes
Table 9.7 Common Characteristic Groups Cited
     as Prefixes or Suffixes

9.3.2 Contents of Appendix D of Part III

D1 Chemistry Symbols and Abbreviations (tables)

    D1.1 General Chemistry
    D1.2 Melting and Boiling Point Abbreviations
    D1.3 Chemical Kinetics
    D1.4 Polymer Chemistry
    D1.5 Electrochemistry
    D1.6 Subatomic Particles
    D1.7 Chirality
    D1.8 Spectroscopy
        D1.8.1 NMR Spectroscopy
        D1.8.2 IR Spectroscopy
        D1.8.3 Mass Spectroscopy
        D1.8.4 UV-Visible Spectroscopy

D2 Word Division of Chemical Names

D3 Atomic Weights of the Elements (table)

D4 Chemistry Journals and Their Abbreviations (table)

*Note: For a glossary of chemistry terms, see the Glossary in the Organic Chemistry chapter—***Appendix E2***. That glossary covers both organic and inorganic chemical science.*

Chapter 10. Style and Usage for Organic Chemistry

[*Note: For general conventions of writing manuscripts in the field of chemistry, see Section 9.1 of the previous chapter.*]

Contents

Contents, *continued*

Chapter 10. Style and Usage for Organic Chemistry

10.1 Usage

10.1.1 Structural Formulas

A simple molecular formula that indicates only the number of atoms of each element in a compound (such as $C_4H_{10}O$) is often insufficient in organic chemistry. In addition to the conventions described in **Section 2.4** of **Chapter 9**, complex organic molecules may also be displayed as condensed structural formulas, with distinct groups arranged and numbered separately—as in $CH_3(CH_2)_3OH$. In organic chemistry, complete stereo-formulas are often simplified to skeletal formulas (*pictured below*), which omit the direct symbols C and H for carbon and hydrogen. In a skeletal formula, carbon atoms are understood to exist at the vertexes and unlabeled ends of each line, and hydrogen atoms are assumed to be bonded to carbon atoms to such a degree that each carbon atom is left with exactly four bonds.

For information on creating and typesetting chemical structures, *see* **Chapter 9, Section 1.2**.

Complete and Skeletal Formulas of Acetone:

Skeletal Formula of Vitamin C:

10.1.2 Common Conventions and Rules for Naming Organic Compounds

10.1.2.1 Alkanes

Alkanes are saturated hydrocarbons with the general formula C_nH_{2n+2}. They may be unbranched (normal) or branched.

i. Number of carbon atoms. Alkane names are based on the number of carbon atoms in the longest continuous chain in the compound. The names of alkanes containing up to twelve carbon atoms per chain are as follows:

Table 10.1 Naming Alkanes

Number of Carbon Atoms in Longest Continuous Chain	Name of Alkane	Molecular Formula
1	methane	CH_4
2	ethane	C_2H_6
3	propane	C_3H_8
4	butane	C_4H_{10}
5	pentane	C_5H_{12}
6	hexane	C_6H_{14}
7	heptane	C_7H_{16}
8	octane	C_8H_{18}
9	nonane	C_9H_{20}
10	decane	$C_{10}H_{22}$
11	undecane	$C_{11}H_{24}$
12	dodecane	$C_{12}H_{26}$

ii. Isomers. Branched alkanes containing three or more carbon atoms in their longest chain have isomers with unique names.

iii. IUPAC naming rules. A systematic method for the naming of many organic compounds is contained in the IUPAC publication: *Nomenclature of Organic Chemistry* (1979) and clarified in *A Guide to IUPAC Nomenclature of Organic Compounds* (Recommendations 1993).

Branched alkanes contain a continuous carbon chain skeleton with one or more side branches composed of alkyl groups. A list of the simplest alkyl groups follows:

Table 10.2 Simple Alkyl Groups

Alkyl Group	Formula
methyl	CH_3-
ethyl	CH_3CH_2-
propyl	$CH_3CH_2CH_2-$
butyl	$CH_3CH_2CH_2CH_2-$
pentyl	$CH_3CH_2CH_2CH_2CH_2-$

Branched alkanes are named using these IUPAC conventions:

(1) Identify the parent alkane and number the atoms in the chain so that branches are attached at the lowest carbon numbers possible.
(2) Identify attached radicals. For multiples of the same simple radical, use a multiplying prefix before the group name.

Table 10.3 Prefixes for Simple Radicals

Multiplying Prefix	Number of Like Unsubstituted Radicals
di-	2
tri-	3
tetra-	4
penta-	5
hexa-	6

Other multiplying prefixes are used to indicate the presence of identical complex radicals.

Table 10.4 Prefixes for Complex Radicals

Multiplying Prefix	Number of Like Radicals Substituted in the Same Manner
bis-	2
tris-	3
tetrakis-	4
pentakis-	5
hexakis-	6

Radical names are then put in alphabetical order before the name of the parent alkane.

(3) Use carbon numbers (if needed) as locants in the name to precisely identify the locations of the alkyl group(s) along the carbon chain. Multiple locants are separated by commas (e.g., 2,3,5-trimethylhexane).

The IUPAC allows usage of some common names for certain branched alkanes and radicals. These names and additional information about the naming of complex branched alkanes are included in Rule A-2 of *Nomenclature of Organic Chemistry* (1979).

iv. Cycloalkanes. Cycloalkanes are alkanes arranged in a cyclic structure.

a. IUPAC rules for cycloalkane nomenclature. Monocyclic alkanes with no side chains are named similarly to acyclic alkanes, but with the prefix *cyclo-* attached to the parent hydride name (e.g., cyclohexane).

10.1.2.2 Alkenes

Alkenes are unsaturated hydrocarbons containing at least one $C=C$ bond. They have the general formula C_nH_{2n}, and are named similarly to alkanes (*see* **Section 10.1.2.1**) with the following changes:

(1) The longest chain selected as the parent hydride must include the double bond.

(2) The suffix *-ane* in the corresponding alkane parent name is replaced with the suffix *-ene* if the compound contains only one double bond, *-adiene* for two double bonds, *-atriene* for three double bonds, etc.

(3) Carbons in the parent chain are numbered beginning with the chain end nearest a double bond.

(4) The stem name (e.g., pentene or hexene) is affixed with a numerical locant identifying the lowest carbon number associated with the double bond (e.g., pent-2-ene for $CH_3CH=CHCH_2CH_3$).

Some trivial names for alkenes are retained by the IUPAC. They are listed along with more detailed instructions for naming alkenes in Rule A-3 of *Nomenclature of Organic Chemistry* (1979) and Section R-3.3.1 of the 1993 Recommendations.

i. Alkenes with more than one double bond. In alkenes with more than one double bond, the suffix *-ane* in the corresponding alkane parent name is replaced with the suffix *-adiene* if there are two double bonds, and *-atriene* if there are three double bonds. Positions of the double bonds are further specified using a numerical locant identifying the lowest carbon number associated with the double bond (e.g., 1,4-hexadiene for $CH_2=CHCH_2CH=CHCH_3$).

ii. Cycloalkenes. Cycloalkenes are alkenes arranged in a cyclic structure. Monocyclic alkenes with no side chains are named similarly to acyclic alkenes with the prefix *cyclo-* attached to the parent hydride name (e.g., cyclohexene).

10.1.2.3 Alkynes

Alkynes are unsaturated hydrocarbons containing at least one $C\equiv C$ bond. They have the general formula C_nH_{2n-2}. They are named similarly to alkanes with the following changes:

(1) The longest chain selected as the parent hydride must include the triple bond.

(2) The suffix *-ane* in the corresponding alkane parent name is replaced with the suffix *-yne* if the compound contains only one triple bond.

(3) Carbons in the parent chain are numbered beginning with the chain end nearest a triple bond.

(4) The stem name (e.g., pentyne or hexyne) is affixed with a numerical locant identifying the lowest carbon number associated with the triple bond (e.g., pent-2-yne for $CH_3C\equiv CCH_2CH_3$).

Some trivial names for alkynes are retained by the IUPAC. They are listed along with more detailed instructions for alkyne naming in Rule A-3 of *Nomenclature of Organic Chemistry* (1979) and Section R-3.3.1 of the 1993 Recommendations.

i. Alkynes with more Than One Triple Bond. The suffix *-ane* in the corresponding alkane parent name is replaced with the suffix *-adiyne* if the compound contains two triple bonds, *-atriyne* for three triple bonds, etc.

ii. Cycloalkynes. Cycloalkynes are alkynes arranged in a cyclic structure. Monocyclic alkynes with no side chains are named similarly to acyclic alkynes with the prefix *cyclo-* attached to the parent hydride name (e.g., cyclooctyne).

10.1.2.4 Molecules with Substituent Groups

Hydrocarbons with substituent groups are often named using systematic substitutive nomenclature in accordance with IUPAC guidelines. Other types of naming systems are allowed in specific cases by the IUPAC. In addition, some traditional (trivial, common, or semi-systematic) names are retained because of their long-standing usage in the discipline.

Systematic substitutive nomenclature relies on the following general process:

(1) Identify the parent hydride and number its carbon atoms.

(2) Identify the substituents and prioritize them by importance.

(3) Supplement the name of the parent hydride with affixes (prefixes, infixes, and/or suffixes) that precisely identify the locations and types of the substituents.

Nomenclature for alkyl groups substituted on the parent chain was described in **Section 10.1.2.1, iii**. Other substituents have unique names based on their chemical compositions.

i. Alcohols. Alcohols contain an –OH functional group. Aliphatic alcohols have the general formula R–OH. Aromatic alcohols (phenols) have the general formula Ar–OH. Two-word common or trivial names are widely used for the simpler alcohols and are based on parent hydride structure (e.g., hexyl alcohol or benyzl alcohol).

a. Primary, secondary, tertiary. Alcohols are classified as primary when the carbon atom including the –OH group is attached to only one other carbon atom. Secondary alcohols are those in which the carbon atom including the –OH group is attached to two other carbon atoms. Tertiary alcohols feature a more complex structure in which the carbon atom including the –OH group is attached to three other carbon atoms.

b. IUPAC rules for naming alcohols. Most alcohols are named using an *-ol* suffix with the parent chain or ring name (e.g., ethanol). A locant is used (if needed) to identify the location of the hydroxyl group (e.g., 4-heptanol).

ii. Diols. Diols (alcohols containing two hydroxyl groups) use a *-diol* suffix with the parent chain or ring name. Locants identify the positions of the hydroxyl groups (e.g., 2-ethyl-2-methyl-1-butanol). Diols are also called *glycols*, but the IUPAC only accepts two glycol names: *ethylene glycol* for 1,2-ethanediol and *propylene glycol* for 1,2-propanediol.

iii. Thiols. Thiols contain an –SH group and have the general formula R–SH. In IUPAC substitutive nomenclature most thiols are named similarly to alcohols using the suffix *-thiol* with the name of the parent chain or ring (e.g., CH_3CH_2SH is called ethanethiol).

iv. Carboxylic acids. Carboxylic acids contain a carboxyl group (–COOH) and are oxoacids with the general formula R–COOH or Ar–COOH.

a. Systematic naming. Systematic name designation depends on whether the carbon atom in the carboxyl group is counted (or not counted) as one of the carbon atoms in the parent hydride.

For simple carboxylic acids—acyclic hydrocarbon chains including only one or two carboxyl groups at the chain end(s)—the carbon atom in the carboxyl group is counted as one of the carbon atoms in the parent hydride and the suffix form "-oic acid" (indicating one carboxyl group) or "-dioic acid" (indicating two carboxyl groups) is used with the name of the parent hydride.

$CH_3CH_2CH_2COOH$
butanoic acid

$COOHCH_2CH_2COOH$
butanedioic acid

More complicated carboxylic acids are typically named using the suffix form "-carboxylic acid" with the name of the parent hydride (e.g., benzenecarboxylic acid).

Exceptions to this naming convention are described by the IUPAC in Section R-5.7.1 of the 1993 Recommendations.

Many simple carboxylic acids are known by trivial names:

Table 10.5 Common Names of Carboxylic Acids

Structural Formula	Structural Name	Common Name
HCO_2H	Methanoic Acid	Formic Acid
CH_3CO_2H	Ethanoic Acid	Acetic Acid
$CH_3(CH_2)_{16}CO_2H$	Octadecanoic Acid	Stearic Acid
$CH_3CHOHCO_2H$	2-Hydroxypropanoic Acid	Lactic Acid
$CH_2=CHCO_2H$	Propenoic Acid	Acrylic Acid
C_6H_5COOH	Benzenecarboxylic Acid	Benzoic Acid
$CH_2(COOH)_2$	Propanedioic Acid	Malonic Acid
$CH_3CH_2CH_2\text{-}COOH$	Butanoic Acid	Butyric Acid

The presence of one or more carboxyl groups in an organic compound in which another group takes priority is typically indicated using the prefix *carboxy-* with the appropriate locant(s).

v. Acid anhydrides. Acid anhydrides contain two acyl groups and have the general formula $RC(=O)OC(=O)R$ or acyl–O–acyl. The acyl groups may be the same (symmetric) or different (asymmetric or mixed).

The simplest and most common acid anhydrides are symmetrical derivatives of monobasic acids. Whether substituted or unsubstituted, these compounds are named per IUPAC guidelines by replacing the word *acid* in the parent acid name with the word *anhydride* (e.g., ethanoic acid becomes ethanoic anhydride). Note that the parent acid name may be either the systematic or trivial name (e.g., ethanoic anhydride is also known as acetic anhydride).

IUPAC rules for more complicated anhydrides are specified in Section 5.7.7 of the 1993 Recommendations.

vi. Acid or acyl halides. Acyl halides contain an acyl group ($RC(=O)$) bonded to one or more halogens and have the general formula acyl–X. Their names consist of two or more words including the acyl group name followed by the name(s) of the halides. Multiple halides are listed alphabetically with multiplicative prefixes, if needed.

$CH_3C(=O)Br$
Acetyl bromide

$C_6H_5C(=O)Cl$
Benzoyl chloride

$BrC(=O)CH_2C(=O)Cl$
Malonyl bromide chloride

$ClC(=O)CH_2C(=O)Cl$
Malonyl dichloride

vii. Nitriles. Nitriles (or cyanides) contain a triply bound nitrogen atom and have the general formula RC≡N. In substitutive nomenclature nitrile names are typically based on the names used for the associated carboxylic acids. Nitriles of acids ending in *-oic acid* or *-dioic acid* use the suffixes *-nitrile* and *-dinitrile*, respectively. Nitriles of acids ending in *carboxylic acid* use the suffix *–carbonitrile*. Nitriles of acids called by trivial names ending in *-ic acid* use the suffix *-onitrile*.

In addition, the IUPAC allows the use of functional class names for nitrile compounds with the general structures of RC≡N, RCOC≡N, and RSO$_2$C≡N. These names are based on the R group name followed by the word *cyanide* (e.g., $CH_3CH_2C≡N$ is called ethyl cyanide. Note that the same compound is called propanenitrile when using substitutive nomenclature).

The presence of one or more nitrile groups in an organic compound in which another group takes priority is typically indicated using the prefix *cyano-* with the appropriate locant(s).

IUPAC guidelines for isocyanides and related compound are included in Section R-5.7.9 of the 1993 Recommendations.

viii. Aldehydes. Aldehydes contain a carbonyl group (C=O) with at least one attached hydrogen atom and have the general formula RC(=O)H.

a. IUPAC rules for naming aldehydes. Aldehydes of carboxylic acids that have trivial names ending in *-ic acid* or *-oic acid* use the suffix *-aldehyde* (e.g., the name *acetic acid* becomes *acetaldehyde*). Acyclic mono- and dialdehydes are also named using the parent hydride name and the suffix *-al* or *-c-dial*, respectively (e.g., acetaldehyde, which has the formula $CH_3C(=O)H$, is also called ethanal). More complex aldehydes are named using the parent hydride name and the suffix *-carbaldehyde*.
The presence of one or more aldehyde groups in an organic compound in which another group takes priority is typically indicated using the prefix *formyl-* with the appropriate locant(s).

b. Ring aldehydes. The names of cyclic aldehydes are typically formed from the parent hydride name added to *-carbaldehyde* (e.g., cyclohexanecar-baldehyde).

ix. Ketones. Ketones contain a carbonyl group (C=O) bonded to two carbon atoms and have the general formula RC(=O)R' where neither R is a hydrogen atom.

a. Rules for naming ketones. Ketones are named similarly to aldehydes: the name of the parent hydride with the suffix -*one* or -*dione* indicates the presence of one or two carbonyl groups, respectively, and locants identify carbonyl group placement (e.g., $CH_3CH_2C(=O)CH_2CH_2CH_2CH_3$ is hepton-3-one). Functional class names are formed by listing the names of the attached radical(s) in alphabetical order followed by the word *ketone* (e.g., butyl ethyl ketone).

IUPAC rules for more complicated ketones, thioketones, and their analogues are included in Section R-5.6.2 of the 1993 Recommendations.

The presence of one or more ketone groups in an organic compound in which another group takes priority is typically indicated using the prefix *oxo-* with the appropriate locant(s).

x. Amines. *Amine* is a generic term for organic compounds with the general formulas of NH_2R, $NHRR'$, and $NRR'R''$. These are classified as primary, secondary, and tertiary amines, respectively. Amine nomenclature is complicated. There are several different naming conventions in use, and many trivial or common names exist for amine compounds.

a. IUPAC rules for naming amines. The IUPAC preferred meth-od for naming primary monoamines is as follows: simple structures are named using the radical name with the suffix -*amine* (e.g., $CH_3CH_2NH_2$ is called ethylamine) and complex cyclic structures utilize the same suffix but with the name of the parent compound (RH).

b. Precedence. The presence of one or more $-NH_2$ groups in an organic compound in which another group takes priority is typically indicated using the prefix *amino-* with the appropriate locant(s).

c. Primary, secondary, tertiary. Primary diamines and polyamines are named similarly, using the suffixes -*diamine*, -*triamine*, etc., but only when all of the amino groups are bonded to an aliphatic chain or bonded directly to a cyclic nucleus.

Additional IUPAC conventions for primary amines in which R is a nitrogen-containing heterocyclic nucleus and for cyclic groups with side chains are described in Sections C-812 and C-813 of *Nomenclature of Organic Chemistry*. Sections C-813 and C-814 each describe nomenclature for different types of secondary and tertiary amines. Section C-815 contains naming rules for ammonium compounds.

xi. Aniline

Aniline ($C_6H_5NH_2$) is the trivial name for the simplest compound of the aromatic amines. Aromatic amines contain a phenyl group (C_6H_5) and have the general formula Ph–NH_2. Aromatic amines are named using the conventions described above for amines.

10.1.2.5 Compounds with Heteroatoms in Their Carbon Skeletons

i. Ethers. Ethers are compounds with the general formula ROR'. In systematic nomenclature they are named using the name of group R'O– followed by the name of the parent hydride associated with R (e.g., $CH_3OCH_2CH_3$ is called methoxyethane). In functional class nomenclature, ether names are constructed by listing the names of R and R' in alphabetical order followed by the word *ether* (e.g., $CH_3OCH_2CH_3$ is called ethyl methyl ether in this system).

ii. Cyclic ethers (epoxides). Cyclic ethers can be named using substitutive or additive methods. Three-member cyclic ethers are known as epoxides. The prefix *epoxy-* is typically used in naming conventions for cyclic ethers (e.g., CH_2CH_2O is called epoxyethane). Use of the word *oxide* after the name of the associated parent alkene is also acceptable. In this convention, CH_2CH_2O is called ethylene oxide. More complicated cyclic ethers are commonly named using the conventions for heterocyclic compounds.

iii. Alkylthio groups. Alkylthio groups are components of sulfides, which have the general formula RSR'. Sulfides are named similarly to simple ethers with the word *sulfide* replacing the words *ether* or *oxide* in the name (e.g., $CH_3CH_2SCH_3$ is called ethyl methyl sulfide). The prefix *thio-* is used for structures containing identical units.

IUPAC naming conventions for more complicated sulfides are included in Section C-514 of *Nomenclature of Organic Chemistry* (1979).

iv. Esters. Esters are most commonly derived from carboxylic acids by replacing the H in the carboxyl group with an alkyl group or more complicated hydrocarbon. Esters of this type have the general formula RC(=O)OR'. They are named by replacing the suffixes *-oic acid* or *-carboxylic acid* in the parent acid name with *-ate* and prefacing the resulting word with the name of the alkyl or aryl group(s) comprising R. Thus, the compound $CH_3CH_2C(=O)OH$, which is named propanoic acid, becomes $CH_3CH_2C(=O)OCH_3$ and is named methyl propanoate. Likewise, benzenecarboxylic acid becomes the ester named ethyl benzoate. Naming conventions for more complicated esters and those derived from other acids are described in Section R-5.7.4.2 of the 1993 Recommendations.

v. Anhydrides. The nomenclature conventions for simple acid anhydrides are described in **Section 10.1.2.4, v.** IUPAC rules for more complicated anhydrides are specified in Section 5.7.7 of the 1993 Recommendations.

vi. Amides. *Amide* is the generic term for compounds derived from oxoacids by replacement of an acidic hydroxy group with an amino group or substituted amino group. Compounds containing up to three acyl groups (RC(=O)C) bonded to a single nitrogen atom are included in the generic class. Primary, secondary, and tertiary amides contain one, two, and three acyl groups, respectively. Amides are classified based on the types of oxoacids from which they are derived (e.g., carboxylic acids become carboxamides, sulfonic acids become sulfonamides, etc.). The simplest carboxamides are primary amides containing an unsubstituted NH_2 group. These compounds are named by replacing the acid name suffixes *-oic acid* and *-ic acid* with *-amide* or by replacing the suffix form *-carboxylic acid* with *-carboxamide*. Thus propanoic acid becomes propanamide ($CH_3CH_2C(=O)NH_2$) and benzoic acid becomes benzamide ($C_6H_5C(=O)NH_2$).

IUPAC naming conventions for more complicated carboxamides and amides based on other acid classes are described in Section R-5.7.8 of the 1993 Recommendations and Rules C-821 through C-825 of *Nomenclature of Organic Chemistry* (1979).

10.1.2.6 Prefixes for Atom Substituents

i. Heterocycles. Heterocycles are cyclic compounds in which ring member atoms include at least two different elements (e.g., carbon and oxygen). The simplest heterocycles are monocyclic and contain no more than ten ring members. These compounds are most often named using the extended Hantzsch–Widman system. This system uses prefixes to indicate the non-carbon ring members with stems that indicate ring size and whether the compound is unsaturated or saturated. The prefixes are known as the "a" prefixes and include *oxa-* for oxygen, *thia-* for sulfur, and *aza-* for nitrogen. The terminal letter *a* is omitted when combined with a stem.

Name stems for three-, four- and five-membered unsaturated rings are *-irene*, *-ete*, and *-ole*, respectively. Name stems for like-sized saturated rings are *-irane*, *-etane*, and *-olane*, respectively. However, some traditional stem names are retained for heterocyclic compounds in which nitrogen is a ring member (viz., *-irine* for a 3-membered unsatu-rated ring and *-iridine*, *-etidine*, and *-olidine* for three-, four-, and five-membered saturated rings, respectively).

Replacement nomenclature is commonly used for monoheterocyclic compounds containing greater than ten ring members. In addition, a number of trivial names are retained for monoheterocyclic compounds.

a. Additional names for 3-membered rings. The heterocycle CH_2CH_2O, known as oxirane in the extended Hantzsch–Widman system, has the trivial name *ethylene oxide*. Aziridine (CH_2CH_2NH) has the trivial name *ethylenimine*.

b. Additional names for 4-membered rings. The heterocycle oxetane ($CH_2CH_2CH_2O$) has the trivial name *trimethylene oxide*.

c. Additional names for 5-membered rings. The heterocycle oxole ($CH_2CH_2CH_2CH_2O$) has the trivial name *furan*. The heterocycle thiole ($CH_2CH_2CH_2CH_2S$) has the trivial name *thiophene*. Additional commonly used trivial names for five-membered rings are listed in Appendix R-9.1 of the 1993 Recommendations.

d. Additional names for 6-membered rings. The six-membered heterocycle containing one oxygen member and two double bonds has the trivial name *pyran*. Additional commonly used trivial names for six-membered rings are listed in Appendix R-9.1 of the 1993 Recommendations.

ii. Some rules for heterocycles. IUPAC naming conventions for heterocycles containing more than one non-carbon atom stipulate that extended Hantzsch–Widman prefixes be listed in decreasing order of priority as specified in Section R-2.3.3.1.3, Table 3 in the 1993 Recommendations. The three most common non-carbon atoms in heterocycles in decreasing order of priority are oxygen, sulfur, and nitrogen. Thus, a heterocycle containing both nitrogen and sulfur would have the complex prefix *thiaza-* (note that the terminal *a* on *thia-* was omitted).

iii. Double and triple bonds in heterocycles. Monoheterocycles containing no more than ten ring members and named using the extended Hantzsch–Widman system rely on the use of stems that indicate whether the compound is unsaturated or saturated.

iv. Numbering the atoms of heterocycles. The ring members of monoheterocyclic compounds are numbered so that a heteroatom is in first position. In rings containing only one heteroatom, the heteroatom is considered locant 1. The remaining members are numbered in a clockwise fashion. In rings containing two or more of the same heteroatom, numbering is performed so as to give the lowest locants possible to the heteroatoms. In rings containing different heteroatoms, the lowest locant is assigned to the heteroatom ranked highest in priority in Section R-2.3.3.1.3, Table 3 in the 1993 Recommendations.

v. Numbering by ring size. Guidelines for numbering the ring members of monoheterocyclic compounds are included in **Section 10.1.2.6, iv**.

vi. Two or more rings attached. IUPAC naming conventions for fused heterocyclic systems are described in Section 2.2 of *Nomenclature of Fused and Bridged Fused Ring Systems* (IUPAC Recommendations 1998). Some polyheterocyclic ring systems have trivial or semi-systematic names with accepted usage by the IUPAC (e.g., purine and indole). Those that do not are often given a name based on the components (individual rings or ring systems) making up the entire compound. One component is chosen as the principal component; the remaining components are considered attached components. Prefixes are selected for the attached component(s) using either trivial names or the Hantzsch–Widman system. Locants (numbers and/or letters) follow the prefix(es) in brackets and indicate the points and sides of the components involved in fusion. These are followed by the base name (the name of the principal component) which may be a trivial name (e.g., furan) or an extended Hantzsch–Widman name.

10.1.2.7 Benzene and Its Derivatives

i. Substituents

a. Organic. The IUPAC retains trivial names for certain benzene derivatives with organic substituents. Examples include toluene ($C_6H_5CH_3$) and styrene ($C_6H_5CH=CH_2$). Additional organic substituents are indicated by pre-facing the trivial name with the appropriate prefix (*ethyl-*, *methyl-*, etc.) unless the additional organic substituent is identical to one already present on the compound with the trivial name. In that case the compound is named using the appropriate suffix(es) with the stem *benzene* (e.g., toluene with a second methyl group is not methyltoluene, but dimethylbenzene). Organic-substituted benzene derivatives lacking accepted trivial names are named using the appropriate prefix with the stem *benzene* (e.g., ethylbenzene).

b. Inorganic. Inorganic substituents in benzene derivatives are indicated by prefacing the stem benzene with the appropriate prefix for the characteristic group. Common prefixes include *bromo-* for bromine, *chloro-* for chlorine, *fluoro-* for fluorine, and *iodo-* for iodine. Thus, C_6H_5Cl is named chlorobenzene.

c. Order of appearance. When more than one substituent is attached to a benzene ring, the substituents are numbered so as to give the lowest numbers possible to the substituents. In derivatives with accepted trivial names (such as styrene) the carbon including the functional group inherent to the derivative is assigned the lowest number.

d. Ortho, meta, and para. Benzene derivatives including only two substituents can be assigned non-numerical locants that indicate specific numerical positions as follows: *ortho* is indicated by the italic prefix *o-* for substituents at the number 1 and 2 positions, *meta* is indicated by the italic prefix *m-* for substituents at the number 1 and 3 positions, and *para* is indicated by the italic prefix *p-* for substituents at the number 1 and 4 positions (e.g., *p*-cresol).

e. Benzene as a substituent group. The presence of benzene as a substituent group in another compound (e.g., an alkane or alkene) is indicated using the prefix phenyl- with the base name of the compound.

f. Polycyclic aromatic compounds. The IUPAC retains trivial names for many polycyclic aromatic compounds. Common examples include *naphthalene* for two fused benzene rings and *anthracene* for three fused benzene rings arranged in a straight-line fashion. Naming conventions for more complicated polycyclic aromatic compounds are described in *Nomenclature of Fused and Bridged Fused Ring Systems* (IUPAC Recommendations 1998).

10.1.2.8 Monosaccharides or Simple Sugars

Monosaccharide is a generic term that includes compounds with the general formula $(CH_2O)_n$, where $n = 3$ to 8. Both aldoses and ketoses are included in this group and are distinguished by the presence of an aldehyde $(RC(=O)H)$ functional group or ketone $(RC(=O)R')$ functional group, respectively, in the compound.

Monosaccharides with $n > 3$ exhibit chirality (the existence of non-superimposable mirror images known as enantiomers). Enantiomers are distinguished by the configurational symbols D and L in their names. Note that these letters are typed as small capital letters. D or L configuration is determined by the orientation of the H–C–OH group farthest from the functional group. The number of stereoisomers for a given sugar is equal to 2n where n is the number of asymmetric carbons (i.e., carbons comprising H–C–OH groups).

The IUPAC naming conventions for monosaccharides are described in *Nomenclature of Carbohydrates* (Recommendations 1996). In general, monosaccharide names carry the suffix -*ose*, and the numbering of carbon atoms is performed so that the carbonyl carbon in the aldehyde is locant 1 and the carbonyl carbon in the ketone is locant 2.

i. Tetroses. Tetroses are monosaccharides containing four carbon atoms. There are two D configurations and two L configurations.

ii. Pentoses. Pentoses are monosaccharides containing five carbon atoms. There are four D configurations and four L configurations.

iii.. Hexoses. Hexoses are monosaccharides containing six carbon atoms. There are eight D configurations and eight L configurations.

10.1.2.9 Organometallics

Organometallics are organic compounds featuring metal bound to carbon. For compounds containing antimony, bismuth, germanium, tin, or lead, the names are derived using substitutive nomenclature. The parent names corresponding to the metals are included in Table 2 of Section R-2.1 in the 1993 Recommendations. The parent hydrides are prefaced with prefixes distinguishing the organic groups. Naming conventions for organometallics containing metal bound only to carbon atoms of organic groups and hydrogen and for organometallic compounds with anionic ligands are described in Sections 5.2.2 and R-5.2.3, respectively, of the 1993 Recommendations.

10.1.2.10 Fat

Fat is a generic term for organic esters formed from fatty acids and glycerol. IUPAC naming conventions for these compounds are described in *Nomenclature of Lipids* (Recommendations, 1976). In general, fatty acids and their acyl radicals are named similarly to carboxylic acids. The carbon atom in the carboxyl group is considered the first carbon in the carbon skeleton. A number of trivial names are also in accepted use for higher fatty acids. Common glycerol derivatives are named in accordance with conventions for carbohydrates.

10.1.2.11 Amino Acids

Amino acids are biomolecules and important structural components of proteins. The twenty amino acids commonly found in proteins are called α-amino acids. Their nomenclature is described in *Nomenclature and Symbolism for Amino Acids and Peptides* (Recommendations 1983).

In general, substituted α-amino acids are named in a semi-systematic fashion by prefacing the trivial name of the amino acid with the name of the substituent group. Individual amino acids should not be abbreviated in running text. For a list of α-amino acids and their abbreviations, see **Chapter 12, Table 12.4.**

10.1.2.12 Lignans

Lignan is the generic name for natural plant products derived from derivatives of cinnamic acid. They are characterized by the coupling of two C_6C_3 units, each of which is considered a propylbenzene for the purposes of nomenclature. Numbering of the carbons in the cyclic component begins with the carbon to which the propyl group is attached. The carbons in the propyl groups are subsequently numbered 7 through 9, beginning with the carbon closest to the cyclic structure. The carbon numbers in the second C_6C_3 unit are followed by a prime mark (e.g., 3'). IUPAC naming conventions for lignans are described in detail in *Nomenclature of Lignans and Neolignans* (Recommendations 2000).

10.1.2.13 Carotenoids

Carotenoids include hydrocarbons and their oxygenated derivatives containing eight isoprenoid units in a particular structural arrangement. The parent structure is acyclic, includes conjugated double bonds, and has the general formula $C_{40}H_{56}$. Names use the stem *carotene* prefaced by alphabetically listed Greek-letter prefixes indicating particular structural arrangements. Numerical locants are also used. IUPAC naming conventions are described in detail in *Nomenclature of Carotenoids* (Rules Approved 1974).

10.1.2.14 Steroids

Steroids are complex organic compounds containing multiple rings and side chains. They are based on a skeleton or derivative of cyclopenta[α]phenanthrene. Locants include numerals and Greek letters. A number of trivial names are retained. IUPAC naming conventions are described in detail in *The Nomenclature of Steroids* (Recommendations 1989). Testosterone is the trivial name for 17^β-hydroxyandrost-4-en-3-one.

10.1.2.15 Vitamins

According to the IUPAC, generic terms such as *vitamin A* and *vitamin B-12* should only be used to refer to derived terms such as *vitamin A deficiency.*

i. Vitamin E or tocopherols. *Vitamin E* is the generic term for certain tocol and tocotrienol derivatives that exhibit the biological activity of natural α-tocopherol. *Tocol* is the trivial name for a specific compound. *Tocopherol* is a generic term for compounds including all mono-, di-, and trimethyl-tocols; it is not considered synonymous with the term *vitamin E*. IUPAC naming conventions are described in detail in *Nomenclature of Tocopherols and Related Compounds* (Recommendations 1981).

ii. Vitamin A or retinoids. *Vitamin A* is a generic term for retinoids that exhibit the biological activity of retinol. Retinoids contain four isoprenoid units joined in particular structural arrangements. Although no parent hydrocarbon is named for retinoids, three stereoparents are identified: retinol, retinal, and retinoic acid. IUPAC naming conventions are described in detail in *Nomenclature of Retinoids* (Recommendations 1981).

iii. Vitamin B-6 and pyridoxal. *Vitamin B-6* is the generic term for all 3-hydroxy-2-methylpyridine derivatives that exhibit the biological activity of pyridoxine. IUPAC naming conventions are described in detail in *Nomenclature for Vitamin B-6 and Related Compounds* (Recommendations 1973). *Pyridoxine* is the trivial name for a particular compound and is not considered a synonym for vitamin B-6.

iv. Vitamin B-12. *Vitamin B-12* is the generic term for certain chemicals that contain a macrocyclic ring including four reduced pyrrole rings. These chemicals are part of a broader class known as the corrinoids that are based structurally on the skeleton of corrin ($C_{19}H_{22}N_4$). The naming conventions for the corrinoids are quite complex and are described in detail in *The Nomenclature of Corrinoids* (Recommendations 1975).

10.1.2.16 Folic Acid

Folic acid is the trivial name for pteroylglutamic acid. The term *folates* is a generic term for certain pteroylglutamates or their mixtures. Specifically they are heterocyclic compounds based on the skeleton of 4-[(pteridin-6-ylmethyl)amino]benzoic acid confugated with at least one L-glutamate unit. The singular term *folate* is the trivial name for pteroylglutamate. IUPAC naming conventions for the folates are described in detail in *Nomenclature and Symbols for Folic Acid and Related Compounds* (Recommendations 1986).

10.1.2.17 Prenol

The term *prenol* describes a particular chain-based structure including repeating C_5H_8 units and a terminal hydroxyl group. Numbering of the carbons begins with the carbon attached to the hydroxyl group. The prenols consist of compounds based on the prenol skeleton, including polyprenols and derived esters and their derivatives. IUPAC naming conventions for the prenols are described in detail in *Prenol Nomenclature* (Recommendations 1986).

10.1.2.18 Enzymes

Enzyme is the generic term encompassing numerous macromolecules that act as biocatalysts for certain reactions. The IUPAC has categorized enzymes and lists them by an alphanumeric identifier called an Enzyme Commission (EC) number. The list includes the common names for the enzymes. The EC categories are as follows:

Table 10.6 EC Number Categories

EC Number	Category
1	oxidoreductases
2	transferases
3	hydrolases
4	lyases
5	isomerases
6	ligases

IUPAC naming conventions for the enzymes are described in detail in *Enzyme Nomenclature* (1992).

10.1.2.19 Stereochemistry

Stereochemistry involves the structural arrangement of chemical compounds in three-dimensional space. Isomers are compounds with identical molecular formulas, but which contain atoms attached to one another in different ways. Thus, they are different compounds. Stereoisomers are isomers that differ in the way the atoms are oriented in three-dimensional space. In systematic nomenclature affixes called stereodescriptors are used to indicate stereochemical information about compounds. Stereodescriptors include letters, symbols, and Latin words.

i. Enantiomers. Enantiomers are stereoisomers that are nonsuperimposable mirror images of each other.

ii. Diastereomers. Diastereomers are stereoisomers that are not enantiomers. Diastereomerism is indicated in nomenclature through the roman-type prefixes *erythro-* and *threo-*, and through the italicized prefixes *l, u, R*, S** and *rel-*.

iii. *R* or *S*. The prefixes *R* and *S* are used to designate particular configurations of stereoisomers. The prefix *R* is from the Lain word *rectus*, meaning "right." The prefix *S* is from the Latin word *sinister*, meaning "left." The prefixes are determined based on a set of sequence rules that prioritize atoms or groups of atoms based mainly on their atomic numbers. The prefixes are typically enclosed in parentheses. Note that they are in italic type, but the parentheses are not.

(*R*)-glyceraldehyde   (*S*)-glyceraldehyde

iv. + or −. The symbols + and − are used to designate differences in optical activity between stereoisomers. The prefixes are determined experimen-tally. They are typically enclosed in parentheses.

(+)-glucose   (−)-glucose

v. D or L. The prefixes D and L are used to designate configurations of stereoisomers. In monosaccharides the D or L configuration is determined by the orientation of the H–C–OH group farthest from the functional group. The letters are typeset as small capitals.

L-erythrose   D-threose

vi. *E* or *Z*. The prefixes *E* and *Z* are used to describe configurations at double bonds. The italic letters *E* and *Z* are enclosed in nonitalic parentheses and placed in front of the name, separated by a hyphen. The prefix *E* is from the German word *entgegen*, meaning "opposite," and describes preferred groups in a double bond attached in an opposite configuration. The prefix *Z* is from the German word *zusammen*, meaning "together," and describes the attachment of preferred groups to the same side of each atom in a double bond. The prefixes *trans*- and *cis*- can be used interchangeably with *E* and *Z*, respectively.

(*E*)-2-butene   *trans*-2-butene

(*Z*)-1,2-dichloroethane   *cis*-1,2-dichloroethane

10.2 Lists of Organic Chemistry Tables and Appended Material

10.2.1 Tables in this Chapter

Table 10.1 Naming Alkanes
Table 10.2 Simple Alkyl Groups
Table 10.3 Prefixes for Simple Radicals
Table 10.4 Prefixes for Complex Radicals
Table 10.5 Common Names of Carboxylic Acids
Table 10.6 EC Number Categories

10.2.2 Contents of Appendix E in Part III

E1 List of Organic Reactions
E2 Glossary of Chemistry Terms

Note: For a list of organic chemistry journals, see **List of Chemistry Journals and Their Abbreviations** *in* **Appendix D4**.

Chapter 11. Style and Usage in Earth Science and Environmental Science

Contents

Contents, 11.2 Usage: A. Geology and Earth Science, *continued*

<u>**Contents, 11.2 Usage,** *continued*</u>

Chapter 11: Style and Usage in Earth and Environmental Sciences

11.1 Manuscript Preparation

As with all scientific writing, when preparing a manuscript, it is best to become familiar with the journal to which one is sending material. Read the journal's recent issues and their mission statement to see if the subject of the paper will make a good fit, and whether that journal is likely to publish it. This will also determine the type of readership for whom to write. For example, if a journal's audience is made up of a preponderance of college students, it is important to keep the writing on a level that collegians can understand. Lists of journals and their abbreviations can be found in **Part III, Appendix F:** *see* **Table F4** for earth science journals; **Table F10** for environmental science journals.

11.1.1 Submitting Text

To submit a text to a journal publication, one needs a written and signed statement of copyright transferal from any authors, illustrators, photographers, cartographers, and other people involved in the writing and research process. The journal must also be informed if the manuscript has been published elsewhere or in another language, or if it is under consideration for publication by any other journal. *See* **Part I, Chapter 5** for more regarding copyright transfer.

The length, spacing, font and margin size, and citation style in the manuscript should be determined by the specific preferences of the journal. Many journals also ask for a specific number of copies of the manuscript and photocopies of any illustrations. For example, the American Meteorological Society requires five copies of the text and photocopies of large drawings along with the complete set of originals.

Some journals require a publication charge for authors; these may be based on an amount per page or for the entire manuscript, depending on the journal. Charges may also be higher if there is a need for the reproduction of color images in print.

Manuscripts should include a cover letter indicating the subject of the paper, along with the number of words or pages in the text and the abstract, separately; a list of other published papers on related topics; a statement on the roles of the authors and others involved in the paper; a list of multimedia files used within the paper; and a statement declaring that all material is new and original and that the author agrees to any publishing charges. This helps editors determine what section of the journal the manuscript may be published in. Some journals also require that a query letter for the intent of submission be sent to the journal before the actual manuscript.

11.1.2 Images, Illustrations, Maps, and Tables

In geological manuscripts, many authors chose to include photos, drawings, tables, figures, and maps to supplement their writing. When this is done, the same care must be taken with illustrations as is taken with research and writing. It may sometimes be necessary to invest in a professional photographer, illustrator, graphic designer, or cartographer. Many journals now also require that authors send their images as a hard copy, along with photocopies, and in an electronic format. Consult with the specific journal to find out which format they prefer. Some journals may also limit the amount of illustrations you may use. For example, the *Journal of Geocryology* only allows three pieces of artwork, while *Geotimes* asks that authors send as many images as possible. Generally, images are sent separate from the text and should include any numbering, titles, and captions.

i. Illustrations. Photographs should be of high quality and give a clear and accurate portrayal of the subject. Photos that are close up rather than far away will communicate an intimacy with your topic of research to the reader. Take the photos in the best, most natural light possible. A poorly lit photo may not come out clearly in print. To determine the size of the illustration, it is best to consult with the specific journal publishing the paper. To ensure detail quality, many journal editors prefer the image to be slightly larger than what will be printed. Captions should be included with all images; however, do not use any text on the actual image, as it may become distorted in print. If scales are included in the illustration (for example, a 20-meter tree), be sure that they are identified in a caption. Many editors prefer graphic scales to mathematical ones, as they are less likely to be miscalculated.

ii. Maps. Maps are extremely important in geographical writing, because they can present millions of bits of information in an effective and efficient manner. As with illustrations, before making a map, it is best to consult with the journal editor to learn their specific preferences for size, paper type, scale, printing limitations, and other details. Often, a geologist may consult with a cartographic supervisor at the site before any fieldwork has even begun. To choose what symbols or colors to use within a geological map, it is best to research other similar maps or review the U.S. Geological Survey for geologic map symbols, as there is no universally recognized standard for map colors and symbols. Make sure to always include a legend defining all symbols, coloring, lettering, and other important terminology used within the map.

Symbols for map units consist of capital and lowercase letters. The capitalized first letter stands for one or several of the chronostratigraphic, geochronologic, and geochronometric units (for example, *Qa* in which the *Q* stands for Holocene). The lowercase second letter stands for the rock unit name (*Qa* in which the *a* stands for alluvium). When referring to map symbols within the running text, do not use the abbreviation; spell out the word and the use the symbol within parenthesis.

iii. Tables. Tables should be used only when necessary; they should be used in place of text and not as a repetition of the text. Many journals prefer that no vertical lines be used, and that the number of horizontal lines should be limited. Consult with the journal to learn their specific preferences for the number and size of tables, as well as the line thickness. Many journals also have a preference as to how many columns should be used. For example the *Journal of Paleoliminology* asks that tables fit into one column, allowing tables to run over two columns only when absolutely necessary. Tables should be numbered consecutively as they are mentioned in the text. Many journals will not use a table unless it is referred to within the text.

11.1.3 A Note on Style

As in most scientific writing, one should use the active voice whenever possible. However, exceptions may be made when writing about, for example, the structure and texture of diatomite, where it may be clearer and more logical to use the passive voice.

i. Commonly Misspelled Geological Terms

consistent, persistent
desiccate
discernible
eutasy, isostasy
fluorite, fluorspar
liquid, liquefy
Mohs

occurred, occurrence
permeable, permeability
phosphorus
predominant, resistant
soluble
symmetrical

11.2 Usage

A. Geology and Earth Science

Earth science, in its simplest terms, is the study of all of the physical aspects of the planet Earth. Major disciplines include geology, mineralogy, hydrology, and agronomy. For a glossary of earth science terms, *see* **Part III Appendix F3.**

11.2.1 Units of Measurement in Geologic Time

11.2.1.1 Geologic Time

The following units should be used to express geologic dates or durations of time:

Table 11.1 Absolute Dates in Geologic Time

Unit	Symbol		Number of Years Ago
	Date	Duration	
kiloannum (*no hyphen*)	ka	ky	10^3
mega-annum	Ma	My	10^6
giga-annum	Ga	Gy	10^9

The following are the most common units used to discuss temporal issues of geologic history:

Table 11.2. Geologic Time Units

Classification Systems	Definition	Units (in order of decreasing rank)
chronostratigraphy	specifies the position of the material in geologic time	eonothems, erathems, systems, sub-systems, series, stages, chronozones
geochronology	specifies the age of the material within the geologic time	eons, eras, periods, subperiods, epochs, ages, chrons
lithostratigraphic	specifies the sequence of rock strata, or the locality of the unit	supergroups, groups, subgroups, formations members, beds

11.2.1.2 Capitalization of Units

i. All formally named chronostratigraphic and geochronologic units are capitalized.

Phanerozoic Eon   Paleozoic Era   Jurassic System

ii. Modifiers *Early, Middle,* or *Late* are used solely in conjunction with geochronologic units (time). Modifiers *Upper, Middle,* or *Lower* are used in combination with chronostratigraphic units (rocks). When such modifyers are a component of the formal name, they should be capitalized. With geochronological units, *Medial* is sometimes used in place of *Middle* to avoid confusion.

Upper Cretaceous   Late Jurassic   Middle Cambrian

iii. Terms such as *Eon*, *Series*, and *Era* should be capitalized when they are used as part of a formal name, but remain lowercased when used generically.

Cenozoic Era   Precambrian Eon   Triassic Period

iv. When a modifier is used informally, it should not be capitalized.

late Jurassic Period   middle Cambrian Era

11.2.1.3 Formal versus Informal Names

Capitalization rules are different for formal and informal names of chronostratigraphic and geochronologic units. Formal and informal names are classified by The International Union of Geological Sciences (IUGS) in the *International Stratigraphic Guide,* which may be found online at: http://www.stratigraphy.org/guide.htm.

Informal names are used for stratigraphic units that were not defined according to the standards at the time of their original publication. Informal units may have names representing their specific color, position, lithology, type of deposit, letter, number, or locality. For informal units, the first letter of the first word may be capitalized, but the first letter of the second word should be lowercase.

The glacial-climate classification scheme is informal, and the terms *interglaciation* and *glaciation* should be written in lowercase. The word *facies* and *sequence* are not considered to be a part of the formal stratigraphic unit system and should not be capitalized.

Devensian glaciation   Whiteface facies

11.2.1.4 Time and Place

As stated above in **Section 11.1.2, i**, *early*, *middle* or *medial*, and *late* are used to describe units of time; while *upper*, *middle*, and *lower* are used with units of place. Some exceptions to this rule include: lithodemic units, or rocks that do not conform to the law of superposition; terraces, which would use the time modifiers instead of place; and *Precambrian*, which is used mostly with a time modifier, but is also correct with a place modifier.

Below are further examples of informal time and place modifiers:

Table 11.3 Time and Place Modifiers

Time	Place
late	upper
middle (medial)	middle
early	lower
young(er)	high(er)
old(er)	low
post-	super-
pre-	sub-
after	above
before	below
when	where
while	whereas
sometime(s)	someplace(s)
often, frequent	abundant, common
occasionally	locally
during	in

11.2.1.5 Summary of Geochronologic and Chronostratigraphic Nomenclature

In the following table (as well as in geologic literature), geochronologic modifiers are set in roman type, while chronostratigraphic modifiers are in italics.

Table 11.4 Geochronological/Chronostratigraphic Nomenclature

Formal	Informal
Cenozoic	early, *lower;* middle, *middle;* late, *upper*
Quaternary	early, *lower;* late, *upper*
Holocene (Recent)	early, *lower;* late, *upper*
Pleistocene	early, *lower;* late, *upper*
Tertiary (Neogene + Paleogene)	early, *lower;* late, *upper*
Pliocene	early, *lower;* late, *upper*
Miocene	early, *lower;* middle, *middle;* late, *upper*
Oligocene	early, *lower;* late, *upper*
Eocene	early, *lower;* middle, *middle;* late, *upper*
Paleocene	early, *lower;* late, *upper*
Mesozoic	early, *lower;* middle, *middle;* late, *upper*
Cretaceous	middle, *middle*
Late, *Upper*	
Early, *Lower*	
Jurassic	
Late, *Upper*	
Middle, *Middle*	
Early, *Lower*	
Triassic	
Late, *Upper*	
Middle, *Middle*	
Early, *Lower*	
Paleozoic	early, *lower;* middle, *middle;* late, *upper*
Permian	middle, *middle*
Late, *Upper*	
Early, *Lower*	
Carboniferous	middle, *middle*
Pennsylvanian or Late, *Upper* Carboniferous	

**Table 11.4 Geochronological/Chronostratigraphic Nomenclature
Paleozoic, *continued***

Formal	Informal
Mississippian or Early, *Lower* Carboniferous	
Devonian	
Late, Upper	
Middle, Middle	
Early, Lower	
Silurian	middle, *middle***
Late, *Upper*	
Early, *Lower*	
Ordovician	
Late, *Upper*	
Middle, *Middle*	
Early, *Lower*	
Cambrian	
Late, *Upper*	
Middle, *Middle*	
Early, *Lower*	
Precambrian	early, *lower;* middle, *middle;* late, *upper*
Proterozoic	
Late	*upper*; Precambrian Z and VI
Middle	*middle;* Precambrian Y and V
Early	*lower;* Precambrian X and IV
Archean	
Late	*upper;* Precambrian W and III
Middle	*middle;* Precambrian V and II
Early	*lower;* Precambrian U and I***

[1] *International Stratigraphic Guide* (http://www.stratigraphy.org/chron.htm)

* For tables of generally accepted, formal age/stage names, see the references cited at the ISG website. Those references include isotopic ages of age/stage boundaries; however, they do not agree on all age/stage names or numerical ages of boundaries.

** The references cited by ISG formally recognize a two-fold subdivision of the Silurian; the U.S.G.S. formally recognizes three (Luttrell et al. 1986).

*** Roman numerals are widely used for Precambrian subdivisions in Russian scientific literature.

The following table displays the geological time scale into which the history of the Earth is divided. The table starts with the current era, Cenozoic, and ends with the earliest era, Precambrian. The table is based on the presence of discernibly distinct layers in the geological record.

Table 11.5 Division of Earth's History in Geologic Time

CENOZOIC ERA					MESOZOIC ERA				
Millions of Years Ago	Duration (in millions of Years)	Period	Epoch	Age	Millions of Years Ago	Duration (in millions of Years)	Period	Epoch	Age
NOW—			HOLOCENE		70—				MAASTRICHTIAN
			PLEISTOCENE	CALAMBRIAN				LATE	CAMPANIAN
5—		NEOGENE	PLIOCENE L / E	PIACENZIAN / ZANCLEAN	80—		CRETACEOUS		(several)
			MIOCENE L	MESSINIAN	90—	80			CENOMANIAN
10—				TORTONIAN	100—				ALBIAN
15—			MIOCENE M	SERRAVALLIAN	110—			EARLY	APTIAN
				LANGHIAN	120—				
20—				BURDIGALIAN					BARREMIAN
		TERTIARY	MIOCENE E	AQUITANIAN	130—			NEOCOMIAN	HAUTERIVIAN
25—			OLIGOCENE L	CHATTIAN					VALANGINIAN
30—					140—				DERRIASIA
				RUPELIAN	150—			LATE	TITHONIAN
35—			OLIGOCENE E		160—				KIMMERIDGIAN
			EOCENE L	PRIASONIAN					OXFORDIAN
40—					170—	65	JURASSIC	MIDDLE	CALLOVIAN
				BARTONIAN	180—				BATHONIAN
45—		PALOGENE	EOCENE M	LUTETIAN					BAJOCIAN
					190—				AALENIAN
50—					200—			EARLY	TOARCIAN
				YPRESIAN					SINEMURIAN
55—			EOCENE E		210—				HETANGIAN
			PALEOCENE L	SELANDIAN	220—			LATE	NORIAN
60—					230—	40	TRIASSIC		CARNIAN
65—			PALEOCENE E	DANIAN	240—			MIDDLE	LANDINIAN
									ANISIAN
								EARLY	SCYTHIAN

439

Table 11.5, continued

| PALEOZOIC ERA | | | | | PRECAMBRIAN ERA | | | |
Millions of Years Ago	Duration (in millions of Years)	Period	Epoch	Age	Millions of Years Ago	Duration (in millions of Years)	Eon	Era
260	40	PERMIAN	LATE	TATARIAN / KAZANIAN	750	2,000	PROTERZOIC	LATE
280			EARLY	KUNGURIAN / ARTINIKIAN / SAKMARIAN	1000			
300	80	CARBONIFEROUS — PENNSYLANIAN	LATE	KASIMOVIAN / MOSCOVIAN / BASHKIRIAN	1250			MIDDLE
320								
340		CARBONIFEROUS — MISSISSIPPIAN	EARLY	SERPUKHOVIAN / VISEAN / TOURNAISIAN	1500			
360					1750			
380	45	DEVONIAN	LATE / MIDDLE	PAMENNIAN / FRASNIAN / GIVETIAN / EIFELIAN	2000			EARLY
400			EARLY	EMSIAN / SIEGENIAN	2250			
420	40	SILURIAN	LATE	PRIDOLIAN / LUDLOVIAN	2500	1,500+	ARCHEAN	
440			EARLY	WENDLOCKIAN / LLANDOVERIAN	2750			LATE
460	65	ORDOVICIAN	LATE	ASHGILLIAN / CARADOCIAN	3000			
480			MIDDLE	LLANDEILIAN / LLANVIRNIAN	3250			MIDDLE
500			EARLY	ARENGIAN / TREMADOCIAN	3500			
520	80	CAMBRIAN	LATE	TREMPEALEAUAN / FRANCONIAN / ORESBACHIAN	3750			EARLY
540			MIDDLE		4000			
560			EARLY					
580								

11.2.1.6 Abbreviations

To avoid ambiguity, the full formal names should always be used in narrative text.

Permian and Carboniferous   *not*   Permo-Carboniferous

11.2.1.7 Format for Statements Regarding Accuracy

In order to express uncertainty about the accuracy of the assignation of a geochronologic or chronostratigraphic unit, a question mark can be placed after the part of the temporary name that is in doubt.

Middle? Cambrian [connotes doubt regarding the accuracy of the epoch]

Middle Cambrian? [connotes doubt regarding the accuracy of the period]

To avoid any confusion, do not place a question mark that represents doubt of a period of time at the end of a sentence.

11.2.2 Rocks and Minerals

11.2.2.1 Rock Names

Formally named rock units should be capitalized. The formal name of a rock unit is a compound of the geographic name and the stratigraphic rank term (*bed, group, family, formation, member*) or lithic term (*limestone, shale, sandstone, coal*). All formally named stratigraphic units should be capitalized, with the exception of the species name in a biostratigraphic unit.

Lava Creek Tuff   Blisworth Limestone   Old Red Sandstone

In the case of informally named stratigraphic units, capitalize only the name of the place, not the stratigraphic rank term or lithic term. The first time an informal term is introduced; the term should be explicitly indicated as informal.

gneiss of Baltimore   tuff of Lava Creek

When a rank term is used alone in the text, or as a reference to a specific rock unit, only capitalize the unit, not the rank.

The formation for Madeira Canyon Formation

Rank terms may be abbreviated within maps and tables, but should be written out in running text. After its first use in the text, the rank and lithic term may be omitted.

11.2.2.2 Mineral Names

The complete collection of mineral names, chemical formulas, crystal systems, and other characteristics can be found in *Fleischer's Glossary of Mineral Species*. The names of rocks and minerals should not be abbreviated in narrative text. They are only appropriate on maps or in tables. Avoid using colloquial, outdated, nonspecific, and varietal names without referring to the parent mineral.

For a list of common mineral names and abbreviations *see* **Appendix F2** in **Part III**.

11.2.2.3 Prefixes, Suffixes, and Adjectival Modifiers

The International Mineralogical Association (IMA) considers the addition of a prefix or suffix to an already existing mineral name to be the creation of a new mineral name. The use of adjectival modifiers, however, does not constitute a new name. New names are subject to approval by the IMA.

i. Prefixes. Prefixes can be connected to the mineral name or separated from it with a hyphen, and should be used only when the element named is dominant in an isomorphous series.

Table 11.6 Use of Prefixes in Geology

Purpose of Prefix	Example
to indicate crystallographic information	clinoenstatite
to indicate chemistry	ferroglaucophane
no unique implication	parachrysotile

After *Suggestions to Authors of the Reports of the United States Geological Survey* by Wallace Hansen (http://cuadra.cr.usgs.gov).

ii. Suffixes. Suffixes to mineral names are either a chemical or crystallographic symbol and should be attached to the mineral name with a hyphen. The suffixes are used to give some additional information about the mineral. The most common use of a chemical suffix is to denote rare earth minerals.

> Monazite-(La) is a monazite in which La is the dominant rare earth element.

iii. Adjectival Modifiers. Adjectival modifiers are added to mineral names to indicate non-dominant chemical substitution in an isomorphous series. Adjectival modifiers should be separated from the mineral name by a space. Do not use nouns as adjectival modifiers if possible.

> ferroan glaucophane

11.2.3 Crystals

11.2.3.1 Symbols for Basic Symmetry Operations of a Space Group

Unlike crystallography in physics, crystallography in earth science deals mainly with the study of crystals that can be found in nature. A crystal is formed from periodic repetitions of atoms, molecules, or ions at equal intervals throughout the volume of a specimen. Crystals are distinguished by class, family, and system. A crystal class designates 1 of 32 categories determined by operations such as rotations, reflections, and inversions. For examples, *see* **Table 11.7**.

Table 11.7 Symbols for Basic Symmetry Operations

Symbol	Definition
$\{E/0\}$	identity operation
$\{Cn/0\}$	*n*-fold rotation
$\{\sigma/0\}$	reflection
$\{I/0\}$	inversion
$\{Sn/0\}$	*n*-fold rotation followed by a fractional translation parallel to the plane

443

Table 11.7, continued

{E/τ}	translation
{$\sigma/\tau/m$}	reflection followed by a fractional translation parallel to the plane
{$Cn/\tau/m$}	rotation followed by a fractional translation parallel to the rotation axis

11.2.3.2 Symbols in Crystallography

A crystal system refers to a category, in which 1 of 7 space groups characterize the main symmetry of structures and the unit-cell shape of the crystal's lattice.

Table 11.8 Symbols in Crystallography

Symbol	Definition
1	not a diad
2	Diad
A	centered in *yz* plane
A	glide reflection in the *x* direction
B	centered in *zx* plane
B	glide reflection in the *y* direction
C	centered in *xy* plane
C	glide reflection in the *z* direction
D	diamond glide rotation
F	centered in all 3 planes *(xy, yz, zx)*
g	glide reflection for plane groups
I	body-centered
i or I	Inversion
m	reflection lines
n	diagonal glide reflection
O	center of inversion
P	primitive plane
p	primitive net
R	Row
σ	Reflection
T	Translation

11.2.3.3 International (Hermann–Mauguin) Symbols

The Hermann–Mauguin symbol has both a short and a full version. The symbol is formed from two parts: a letter indicating the centering type of the conventional cell and a set of characters indicating symmetry elements of the space group (modified point-group symbol). Use lowercase italic letters for two dimensions (nets) and capital letters for three dimensions (lattices).

Full	**Short**
$C1m1$	Cm
$P2_1/n2_1/m2_1/a$	$Pnma$
$P6_3/m\ 2m\ 2/c$	$P6_3/mmc$

11.2.4 Sediment

Sediment may consist of organic material, rock or mineral fragments, precipitates from water, loess, desert sand, volcanic ash, settled dust, and small atmospheric or cosmic particles. Sediment is material deposited by water, wind, or glaciers; it can provide information about the origin, age, and transport of material in rivers, lakebeds, and ocean basins. Sediments can be classified according to their color, texture, and source.

11.2.4.1 Source

The sources of original material that make up a sediment defines what type of sediment it is. Lithogenous and terrigenous sediments consist of material from weathered rocks; cosmogenous sediments contain materials deposited from the atmosphere or space; hydrogenous sediments come from chemical reactions in water; and biogenous sediments come from plant and animal remains. If a sediment consists of more than 30% biogenous materials it is considered an ooze.

11.2.4.2 Texture

Grain size and sorting are used to classify sediment. The Wentworth and Phi scales, as shown in **Table 11.9**, are used to define limits of class for particles based on the maximum particle diameter.

Table 11.9 Classifying Texture of Sediments

Size Class	Diameter (mm)	Phi (Φ)
Boulder	> 256	< -8
Cobble	64 to 256	-8 to -6
Pebble	4 to 64	-6 to -2
Gravel	2 to 4	-2 to -1
Sand		
– very coarse sand	1 to 2	-1 to 0
– coarse sand	0.5 to 1	0 to 1
– medium sand	0.25 to 0.5	1 to 2
– fine sand	0.125 to 0.25	2 to 3
– very fine sand	0.0625 to 0.125	3 to 4
Silt		
– coarse silt	0.0310 to 0.0625	4 to 5
– medium silt	0.0156 to 0.0310	5
– fine silt	0.0078 tp 0.0156	6
– very fine silt	0.0039 to 0.0078	7
Clay	< 0.0039	>8

11.2.5 Agronomy and Crop Science

Agronomy describes the study of plant and soil sciences and how they are applied for usage in and the creation of food, fuel, fiber, feed, and pharmaceuticals.

The American Society of Agronomy (ASA), Crop Science Society of America (CSSA), and Soil Science Society of America (SSSA) issue style conventions for scientific publications in the fields of agronomy and crop science. The guidelines can be accessed online at:

https://www.agronomy.org/publications/style/.

11.2.5.1 Crop Growth Staging Scales

The ASA, CSSA, and SSSA recommend the use of specific growth staging scales for describing the phenological development of crops and weeds. These scales feature descriptors and associated numerical index values to indicate growth stages.

The following is a reference list for crop-specific scales recommended by the ASA, CSSA, and SSSA style guide.

Table 11.10 Recommended Staging Scales and Sources

Crop	Citation
Alfalfa	Kalu, B.A., and G.W. Fick. 1981. Quantifying morphological development of alfalfa for studies of herbage quality. Crop Sci. 21: 267–271 Fick, G.W., and S.C. Mueller. 1989. Alfalfa: Quality, maturity, and mean stage of development. Cornell Coop. Ext. Inf. Bull. 217. Cornell Univ., Ithaca, NY.
Corn	Ritchie, S.W., J.J. Hanway, H.E. Thompson, and G.O. Benson. 1996. How a corn plant develops. Spec. Rep. 48. Rev. ed Iowa State Univ. Coop. Ext. Serv., Ames.
Cool-season forage grasses	Haun, J.R. 1973. Visual quantification of wheat development. Agron. J. 90:235–238. Moore, K.J., L.E. Moser, K.P. Vogel, S.S. Waller, B.E. Johnson, and J.F. Pedersen. 1991. Describing and qualifying growth stages of perennial forage grasses. Agron. J. 83:1073–1077.
Cotton	Elsner, J.E., C.W. Smith, and O.F. Owen. 1979. Uniform stage descriptions in upland cotton. Crop Sci. 19:361–363
Red clover	Ohlsson, C., and W.F. Wedin 1989 Phenological staging schemes for predicting red clover quality. Crop Sci. 29:416–420.
Small-grain cereals	Haun, J.R. 1973. Visual quantification of wheat development. Agron. J. 90:235–238. Zadoks, J.C., T.T. Chang, and C.F. Konzak. 1974. A decimal code for growth stages of cereals. Weed Res. 14:415–421. Tottman, D.R. 1987. The decimal code for the growth stages of cereals, with illustrations. Ann. Appl. Biol. 110:441–454.
Sorghum	Vanderlip, R.L., and H.E. Reeves. 1972. Growth stages of sorghum [*Sorghum bicolor* (L.) Moench]. Agron. J. 64:13–16.

Table 11.10, *Continued*

Soybean	Fehr, W.R. and C.E. Caviness. 1977. Stages of soybean development. Spec. Rep. 80. Iowa Agric. Home Econ. Exp. Stn., Iowa State Univ., Ames. Ritchie, S.W., J.J. Hanway, H.E. Thompson, and G.O. Benson. 1994. How a soybean plant develops. Spec. Rep. 53. Rev. ed. Iowa State Univ. Coop. Ext. Serv., Ames.
Stoloniferous grasses	West, C.P. 1990. A proposed growth stage system for bermudagrass. p. 38–42. *In* Proc. Am.. Forage and Grassl. Counc., Blacksburg, VA. 6–9 June 1990. Blacksburg, VA. AFGC, Georgetown, TX.
Sunflower	Schneiter, A.A., and J.F. Miller. 1981. Description of sunflower growth stages. Crop Sci. 21:901–903.
Warm-season forage grasses	Moore, K.J., L.E. Moser, K.P. Vogel, S.S. Waller, B.E. Johnson, and J.F. Pederson. 1991. Describing and quantifying growth stages of perennial forage grasses. Agron. J. 83:1073–1077. Sanderson, M.A. 1992. Morphological development of switchgrass and kleingrass. Agron. J. 84:415–419.
All crops and weeds	Lancashire, P.D., H. Bleiholder, T. van den Boom, P. Langelüddeke, R. Stauss, E. Webber, and A. Witzenberger. 1991. A uniform decimal code for growth stages of crops and weeds. Ann. Appl. Biol. 119:561–601.

11.2.5.2 Light Measurements and Photosynthesis

The ASA, CSSA, and SSSA advocate the use of SI radiometric units (e.g., joule and watt) rather than photometric units (e.g., lumen) for expression of the energy or quantum content of radiation used by plants during growth. The following is a list of terms and definitions as recommended by the Committee on Crop Terminology for the expression of photosynthetic energy and photosynthetic capacity.

- *Photosyntheticaly active radiation* (PAR): radiation in the 400- to 700-nm waveband.
- *Photosynthetic photon flux density* (PPFD): the number of photons in the 400- to 700-nm waveband incident per unit time on a unit surface. Suggested units: $\mu mol\ cm^{-2}\ s^{-1}$.

- *Photosynthetic irradiance* (PI): the radiant energy in the 400- to 700-nm waveband incident per unit on a unit surface. Suggested units: W m^{-2}.

- *Apparent photosynthesis* (AP): photosynthesis estimated indirectly and uncorrected for respiratory activity. The term *apparent photosynthesis* is preferred to *net photosynthesis* or *net assimilation*, because the latter terms imply measurement of a photosynthetic product.

- *CO_2 exchange rate* (CER): The net rate of carbon dioxide diffusion from (–) or to (+) an entity, such as a plant tissue; organ or canopy; a soil surface, etc. Suggested units: μmol cm^{-2} s^{-1}. (Use this term instead of *net CO_2 exchange*—except in the rare instances when the measurement does not involve a rate.)

11.2.5.3 Special Usage in Chemical Names of Organic Substance Used for Pesticides

The ASA, CSSA, and SSSA recommend adherence to American Chemical Society style conventions for the chemical names of organic substances used for pesticides. The following is a selection of common usages for writing chemical names of organic substances in pesticides.

- Use italics for the prefixes *anti, asym, c, cis, cyclo, d, endo, exo, l, m, n, o, p, r, s, sec, t, tert*, and *trans*. Do not capitalize these prefixes, even at the beginning of a sentence or in a title.

- Use italics for the capitalized prefixes *R, R*, S, S*, E*, and *Z* and enclose them in parentheses.

- Use italics for symbols of chemical elements indicating ligation or attachment to an atom (e.g., *O, P, N, S*) or when indicating added hydrogen (*H*).

- Use Greek letters to denote position or stereochemistry (e.g., α-amino acids).

- Enclose the stereochemistry prefixes for plus and minus in parenthesis: (+), (–), and (±).

- Use roman type for multiplying prefixes (e.g., hemo-, mono-, di-, tri-, deca-, semi-, uni-, sesqui-, bi-, ter-, deci-, bis-, tris-, decakis-).

11.2.5.4 Field-Specific Units and Measurements

Many of the units used in general environmental science are also applicable in agronomy and crop science. In addition, the latter disciplines feature units for field-specific parameters such as agricultural yield, soil and plant properties, growth and fertilization rates, etc. Many of these units are compound terms (e.g., $\text{mmol m}^{-2}\text{ s}^{-1}$) derived from SI units. Following are some commonly used SI-based units.

Table 11.11 Commonly Used SI-Based Units for Agronomy

Unit	Common Usage
g m^{-2}	fertilizer application rate
$\text{m}^2\text{ kg}^{-1}$	surface area of soil
Mg m^{-3}	soil bulk density
$\text{kg m}^{-2}\text{ s}^{-1}$	water flow off land surface
$\text{g m}^{-2}\text{ d}^{-1}$	plant growth rate
g kg^{-1}	plant water content
kg kg^{-1}	soil water content
g m^{-2}	yield
kg ha^{-1}	yield
Mg ha^{-1}	yield or application rate
$\text{mg m}^{-2}\text{ s}^{-1}$	transpiration rate
$\text{kg ha}^{-1}\text{ y}^{-1}$	element transfer rate

In addition, many non-SI units are used in agronomy and crop science (viz., pounds, feet, acres, and combinations thereof). General convention calls for SI units to be used with numerical quantities whenever possible. Values with non-SI units are often presented in parentheses after values in SI units. For a complete listing of units and their symbols *see* **Appendix F8**. For SI-to-non-SI conversion factors, *see* **Appendix B1.5**.

11.2.5.5 Non-SI Units

i. Exchange Composition and Capacity. SI base or derived units are preferred for expression of exchange composition and capacity. The ASA, CSSA, and SSSA recommend units of moles of charge per kilogram of soil (e.g., $4\text{ cmol}_c/\text{kg}$). The sign of the charge is not included, but should be evident from the descriptive text. If the cation exchange capacity is determined experimentally by replacement of cations with a single cation, that cation should be identified in parentheses and affixed to the unit expression (e.g., 6 cmol$_c$ (cation)kg^{-1}). For a list of common non-SI units and their abbreviations used in Earth science, *see* **Appendixes F5–F8**.

ii. Use of Percentage versus SI units. SI base or derived units are preferred for expression of concentrations or compositional quantities (e.g., mmol/m^3). The ASA, CSSA, and SSSA deem the use of percentage acceptable only when the use of SI base or derived units is not possible or when describing specific parameters commonly cited in percentage (e.g., yield).

iii. Cotton Fiber. The ASA, CSSA, and SSSA style guide recommends the following for measuring cotton fiber:

> Official standards for cotton staple length are given in terms of inches and fractions of an inch, generally in gradations of thirty-seconds of an inch. Stapling is done by a classer in comparison with staple standards. Measurement by instrument has shown unequal increments between consecutive staples in these standards. Because the classer is the authority on length, these unequal increments have been maintained. When staple length is determined by a classer, it may be reported as a code number, with the code being the number of thirty-seconds of inch called by the classer.

For more information on units of applied chemistry used in soil and agronomic analysis, *see* The *ASA–CSSA–SSSA Publication Handbook and Style Guide,* Chapter 3.

11.2.5.6 Field-Specific Abbreviations

The ASA, CSSA, and SSSA style guidelines list abbreviations that are so commonly known within the field that they do not have to be defined when they are first used in print. Field-specific abbreviations are listed in the following table. The list also contains abbreviations that may or may not have to be defined on first use, depending on whether they are used in the text, tables and figures, or with numerical values. Definitions that are marked with a (T) may be restricted to use in tables and figures, an (N) with numeric values, and an (A) with addresses.

Table 11.12 Common Abbreviations in Agronomy

Abbreviation	Meaning (restriction)
a.i.	active ingredient
Agric.	Agriculture, Agricultural (A)
ARS	Agricultural Research Service
ASA	American Society of Agronomy
avg.	average(T)
CI	Cereal Investigation [number][1]
coef.	coefficient (T)
conc.	concentration (T)
CSREES	Cooperative State Research, Education, and Extension Service (A)
CSSA	Crop Science Society of America
cv.	Cultivar[2]
d	day (N)
Dep.	Department (A)
diam.	diameter (T, N)
dry wt.	dry weight (N, T)
EC	Enzyme Commission [number]
SCS	Soil Conservation Service
ELISA	enzyme-linked immunosorbent assay
Eq.	Equation, Equations (N)
Exp.	Experiment (A, N)
Fig.	Figure [number], Figures [range of numbers]
fresh wt.	fresh weight (N,T)
g	gravity constant
h	hour (N)
i.d.	inside diameter (N)
Inst.	Institute, Institution (A)
Int.	International (A)
max.	maximum (T)
min	minute (T)
min.	minimum (T)
mo	month (N)[3]
Natl.	National (A)
no.	number[4]
NRCS	Natural Resources Conservation Service (A)
o.d.	outside diameter
PI	Plant Introduction, Plant Identification [number]
Res.	Research (A)

Table 11.12, continued

Abbreviation	Meaning (restriction)
s	second (N)
sp., spp.	species[5]
SSSA	Soil Science Society of America
stn.	Station (A)
TVA	Tennessee Valley Authority
Univ.	University (A)
USA	United States of America
USDA	United States Department of Agriculture
USEPA	United States Environmental Protection Agency
vs.	versus
wk	week (N)
yr	year (N)

[1] The CI must be followed by a two-letter abbreviation for the applicable cereal genus: CI*av* for oat, CI*ho* for barley (Hordeum), CI*tr* for wheat (Triticum), etc.

[2] Use cv. Only before a cultivar name, and preferably only if also after a scien-tific name.

[3] Abbreviate only with values ≥ 6; otherwise, spell out both number and month, with some indication that the value is approximate.

[4] Despite the strictures of the CBE style manual (CBE, 1994. P. 187), do not use "nr" as an abbreviation for number; do not end this abbreviation with a period ("no.").

[5] Use only after a genus name.

For more specific abbreviations (including many used in a wide variety of scientific disciplines), *see* **Appendix F1**.

The Following is a list of common statistical abbreviations. All abbreviations requiring a definition on first use should be assembled into an alphabetically arranged list within the publication.[1]

Table 11.13 Common Statistical Abbreviations

Statistic	Preferred symbol	Acceptable symbol	Symbol for population[2]
arithmetic mean	$\bar{x}$		μ
regression coefficient	b		β
sample size	n		N
standard error of mean	SE	$s_{\bar{x}}$	$\sigma_{\bar{x}}$
variance	s^2		σ^2
chi-square	x^2		
correlation coefficient	r		
coefficient of multiple determination	R^2		
coefficient of simple determination	r^2		
coefficient of variation	CV		
degrees of freedom	df	DF	
least significant difference	LSD		
multiple correlation coefficient	R		
not significant	NS		
probability of a Type I error	α		
probability of a Type II error	β		
Student's t	t		
variance ratio	F		

[1] In addition, the symbols *, **, and *** are used to show significance at the P = 0.05, 0.01, and 0.001 levels, respectively. Significance at other levels is designated by additional footnotes, using the next available symbol from the standard sequence (†, ‡, §, ¶, #, ††, ‡‡, etc.).

[2] Where a symbol for population is given, the other form or forms are for the sample.

11.2.5.7 Standard Symbols and Diagrams

Standard symbols used in agronomy and crop science include the general symbols of environmental science and field-specific symbols (e.g., soil horizons O, A, B, etc.). The ASA, CSSA, and SSSA style guidelines recommend that symbols be identified at first use unless they are extremely common (e.g., % for percent, n for sample size in statistics, and the chemical element symbols). Single letter symbols (excluding Greek letters) used in mathematical and statistical expressions are printed in italics (e.g., g for gravity constant).

Diagrams commonly encountered in the fields of agronomy and crop science include soil maps, hierarchical arrangements, and morphological structures. These and other types of diagrams should adhere to the style recommendations provided by the ASA, CSSA, and SSSA in Chapter 5 at https://www.agronomy.org/publications/style. Any units used in diagrams should be consistent with units used in the accompanying text.

11.2.5.8 Sources for Specialized Terminologies

Authoritative sources for specialized terminologies within agronomy and crop science include professional organizations, government agencies, and universities. The following are glossaries of specialized terminologies for these disciplines:

(1) Glossary of Crop Science terms (Barnes and Beard, 1992), available online at: www.crops.org/cropgloss/.

(2) Crop Science (Leonard et al. 1968; Shibles, 1976).

(3) Glossary of Soil Science Terms (SSSA, 1997), available online at: www.soils.org/sssagloss.

11.2.6 Soils

11.2.6.1 Soil Classification System

The Food and Agriculture Organization of the United States (FAO) soil classification system (also called World Soil Classification) is the most commonly used. Soil taxonomy includes orders, suborders, great groups, subgroups, families, and series. These groups are capitalized and do not need to be italicized.

Table 11.14 Examples of Orders

Orders	Description
Alfisol	Subsoil accumulation of clay; not very leached.
Andisol	Mineral soil with volcanic ash makeup
Aridisol	Soil from hot and dry areas
Entisol	Simple soil without subsoil diagnostic horizons
Histosol	Dark organic soil with little mineral matter
Inceptisol	Minimal development; little subsoil or clay
Mollisol	Thick, soft, dark
Oxisol	Leaching with iron and aluminum oxides and quartz
Spodosol	Subsoil accumulation of iron
Ultisol	Subsoil accumulation of clay; very leached
Vertisol	Clay and few subsoil horizontal layers

Series classification takes into account climate, thickness of profile, acidity, mineralogy, and other soil and environmental properties. If there are horizons, for example, they are classified at this level, specifically color, kinds, numbers, and thickness.

Suborder names contain letters from their orders. Some examples:

Suborders of Entisols: Arquents and Arents
Suborders of Inceptisols: Andepts and Ochrepts
Suborders of Mollisols: Albolls and Borolls
Suborder of Vertisols: Torrerts and Xererts

11.2.6.2 Pedon Descriptions

A pedon is a 3-dimensional sample of a body of soil that is 1 m^2 at the surface and extends to the bottom of the soil. Pedon descriptions are used to classify soil, to divide pedons into units for analysis, and to define map units for a soil survey. In soil surveys, pedon descriptions should be written in narrative form within a series of short phrases separated by semicolons; while in journal articles, pedons are displayed within tables, usually containing less information than there would be in a narrative text.

Use shorthand labels to indicate the properties and relations of the horizons of the pedon sample. These may consist of up to four parts: a numeral prefix representing the geological origin of the soil; a capital letter representing the most obvious horizons in the sample or the master horizon; a lowercase letter representing the processes or properties involved in creating the sample; and a numeric suffix representing a minor subdivision.

11.6.2.3 Morphological Properties

i. Color. When describing the color of soil, the properties of the Munsell System are used (describing color by hue, value, and chroma). For example, 5YR 6/8 indicate that 5YR is the hue, 6 is the value, and 8 is the chroma. In soil surveys a description of the color is written, with the Munsell designation following in parentheses.

Reddish yellow (5YR 6/8)

In scientific journals, the Munsell designation is written first, with the color description following in parentheses.

5YR 6/8 (reddish yellow)

If there are multiple colors being described, the dominant color should be written first. The abundance of mottles within a sample can be designated by a lowercase letter (*c* for *common*, *f* for *few*, and *m* for *many*).

Reddish yellow (5YR 6/8) (2.5YR 3/6 m)

or:

5YR 6/8 (reddish yellow) 2.5 YR 3/6 m

ii. Texture. Use a texture modifier to describe the presence of a condition or component other than sand, silt, or clay. These include terms such as ashy, bouldery, cobbly, gravelly, hydrous, mucky, permanently frozen, stony, and woody. Texture modifiers should only be used when the component's content comprises more than 15% of the sample.

iii. Structure. Structure describes the grouping of soil particles to form larger units or peds, or it can describe how these peds fall apart. Structure features are represented in three ways: a number indicates the grade or distinctness; a lowercase letter indicates the size; and two lowercase letters indicate the type or shape.

11.2.6.4 Tracts of Land

U.S. public land tract locations are designated by rows of townships (each six miles square) and columns of ranges. These rows and columns form the grid system that is used to survey all land west of the Ohio River (excluding Texas) in relation to various east-west baselines and named north-south meridians. Format as follows:

SE1/4 NW1/4 sec 4, T 12 S, R 15 E, of the Boise Meridian
[designates the southeast quarter of the northwest quarter of
section 4, which is 12 townships south and 15 ranges east of
the Boise Meridian]

N1/2 sec 20, T7N, R2W, sixth principal meridian
[designates the northern half of section 20, which is 7 townships north and 2 ranges west of the sixth principal meridian]

Periods should be omitted from abbreviations. Do not use a space between the compass direction and the fraction. If fractions are spelled out in land descriptions, use *half* or *quarter* in place of *one-half* or *one-quarter.*

south half of T 47 N, R 64 E

The plural of T (for *township*) is "Tps", and the plural of R (for *range*) is "Rs".

Tps 9, 10, 11, and 12 S, Rs 12 and 13 W

If breaking a land-description symbol group at the end of a line is unavoidable, break it after a fraction and without a hyphen. For example, break the land description NE1/4SE1/4 sec 4 at the end of a line as follows:

NE1/4SE1/4
sec 4

B. Environmental Science

Environmental science consists of a wide range of scientific disciplines dealing with the natural environment and its interactions within physics, chemistry, and biology. Some examples of disciplines classified strictly as environmental science are meteorology, oceanography, and ecology. Although the environment has been studied for centuries, environmental science is a rather recent term, which came about due to the growing need for a vast multi-disciplined team to study complex environmental issues. For a glossary of Earth science terms see **F9** in the Appendix.

11.2.7 Aquifers

The term *aquifer* can have many different definitions. For the most part, an aquifer can be described as an underground geological formation capable of receiving, storing, and transmitting large quantities of water.

Do not capitalize the following terms even when named: *aquifer, aquifer system, zone,* and *confining unit.* Terms like *sand and gravel aquifer* and *limestone aquifer* are neither capitalized nor hyphenated. Adjectival modifiers and relative-position terms are not capitalized unless they are part of the formal geographic name.

Ogallala aquifer   Guarani aquifer   Yarkon-Taninim aquifer

To avoid any confusion, do not use words that are meant to be synonyms for *aquifer*. For example, terms like *hydrofer*, *aquiformation*, and *aquigroup* have different definitions or degrees of precision than *aquifer*.

11.2.8 Water

11.2.8.1 Descriptive Terms

Terms that are used to describe water are usually made up of a combination of a modifier and the word *water*. These terms can be used as both nouns and adjectives, however the construction of the term may vary, using either a separating space or hyphen accordingly.

> saltwater (noun or adjective)
> deep water (noun) *but* deepwater (adjective)
> open water (noun) *but* open-water (adjective)

Since there is not a consistent set of rules for formatting these terms, it is best to consult a dictionary for correct usage.

11.2.8.2 Currents

Currents are horizontal movements of water. Currents are described by the velocity and direction of water flow, and current flow is expressed as a rate.

i. Names of Currents. Currents that are permanent or large in scale are usually named for the area on the globe where they occur. The names of these currents should be capitalized.

> Antarctic Circumpolar Current   East Australian Current

Currents that occur off the coast of Japan may use their Japanese name either with or without the word *current* as a modifier.

> Kuroshio   *or*   Kuroshio Current   *or*   Japan Current
> Tsugaru   *or*   Tsugaru Current
> Oyashio   *or*   Oyashio Current

Other well-known currents may also be used with or without the word *current* as a modifier.

| Gulf Stream | *or* | Gulf Stream Current |
| Humboldt | *or* | Humboldt Current |

ii. Current Direction. Current direction is determined by which direction the water is flowing.

A southerly ocean current flows *to* the South but a southerly wind blows *from* the South.

Currents are defined as cyclonic when they circulate counterclockwise in the northern hemisphere and clockwise in the southern hemisphere. While, anticylconic currents circulate clockwise in the northern hemisphere and counterclockwise in the southern hemisphere,

11.2.8.3 Tides

i. Datum. A tidal datum is a base elevation from which to measure heights and depths. Abbreviations for tidal datum are composed of the first letter of each word in the datum name in uppercase letters. When describing more than one base elevation, use the plural *datums* rather than *data,* which is used to define multiple units of information.

Table 11.15 Tidal Datum Abbreviations

Term	Abbreviation
mean lower low water	MLLW
mean low water	MLW
mean higher high water	MHHW
mean sea level	MSL

ii. Tidal height. Tidal height is represented in meters relative to the datum for an area. If the tidal height is less than the datum, it should be reported as a negative number. In scientific writing, tidal heights should be reported in meters.

Yesterday's low tide of −2.5 m brought many tourists to the beach.

When tides exceed 1 m, the streets of Venice may flood.

11.2.8.4 Seafloor Features

Ocean floor features and other descriptive terms, such as *ridge, trench*, or *bank*, should be capitalized.

Great Barrier Reef   Mid-Oceanic Ridge   Bear Seamount

General features of the sea floor remain lowercased.

deep-sea trench guyot   abyssal plain

11.2.9 Meteorology

11.2.9.1 Wind

Describe winds in terms of the direction from which they are blowing. Use the suffix *erly* to convey the same sense.

north wind *or* northerly wind [a wind blowing from the north]
south wind *or* southerly wind [a wind blowing from the south]

Also appropriate in non-technical writing:

northward [a wind blowing toward the north]
southward [a wind blowing toward the south]

11.2.9.2 Specialized Terminologies

i. Skill. The term *skill* is used in meteorology to convey a statistical evaluation of a forecast or class of forecasts in comparison with some other forecast or climatological average.

Varying samples can lead to overestimated forecast skills

ii. Satellite names. During design and construction of satellites used for meteorology, a letter designating the specific satellite is added to its name; after successful launch and operation implementation, the satellite name becomes official and the letter is changed to a number.

When referring to a satellite that has not yet become operational, set the name in roman type; after the name becomes official, italicize the name (as with a ship's name).

GOES-K   [during design and construction]
GOES-10   [after christening]

11.2.9.3 Units in Meteorology

The use of the International System of Units (SI) is standard in all American Meteorological Society (AMS) publications. However, some exceptions are made with the usage of non-SI units on the basis of communicating clearly and its universal usage within a particular discipline.

The following practices are recommended by the AMS for writing units:

i. Celsius: Celsius temperature is often used in place of the standard kelvin (K) for observational, synoptic, and climatic work within meteorology and oceanography. The thermodynamic temperature in kelvin minus 273.15 K is equal to Celsius temperature. In writing it can be expressed as *degrees Celsius* or °C.

For more on SI units *see* **Appendix B1** in **Part III**.

ii. Millibar and decibar: Although the pascal (Pa), the standard SI unit for pressure, is the preferred unit of measurement, meteorologists will commonly use the millibar (mb or mbar), while oceanographers will use the decibar (dbar).

iii. Logarithmic measures: The use of logarithmic measures and units are all accepted within AMS publications (pH, dB, dBZ, Np, etc.).

iv. Units of time: When *day* and *month* are used as a unit of measurement, they should be spelled out. The AMS allows for the abbreviations *d* and *mo* when they are used in a table or chart with restricted space.

v. Liters: The International Committee on Weights and Measures has accepted the use of a roman capital *L* as the SI unit symbol for liter. This replaces the former use of a lowercase *l* to avoid confusion with the arabic numeral *one*.

vi. Metric ton: The AMS accepts the use of the *metric ton* over the preferred SI unit, *megagram* (Mg). However, *ton* must be written with the prefix *metric* in order to avoid confusion with the US *short ton* and the British *long ton*. The term *tonne* should not be used in its place.

vii. Nautical Mile: Although it is not recognized by the International Committee on Weights and Measures, the AMS accepts the use of *n mi* to represent the nautical mile.

viii. Knot: The unit of speed for nautical miles, or knots, may be represented with the abbreviation *kt*; however, the corresponding unit of speed, expressed in SI units as $m\ s^{-1}$, must be indicated as well.

11.2.9.4 Writing in Meteorology

The American Meteorological Society (AMS) publishes several different journals on the many disciplines within meteorology. However, it is the decision of the editorial board to choose which journal a specific article will be published in. For this reason, it is important to become familiar with the many different journals, to format the article in the style that the journal prefers, increasing the chances of publication.

AMS journals stress the need for brevity and clarity within their articles. They also prefer writing that uses impersonal construction and the passive voice. (*See also* **Section 11.1** of this chapter and **Chapter 2**.) For a listing of journals in environmental science, *see* **Appendix F10**.

**Table 11.16 Sources of Specialized Terminology
in Environmental Science**

Discipline	Source	Website for Information
Atmospheric Chemistry	Glossary of Atmospheric Chemistry Terms (IUPAC, 2000)	http://www.iupac.org/reports/ 1990/6211calvert/glossary.ht ml
Chemistry	Glossary for Chemists of Terms Used in Toxicology (IUPAC Recommendations 1993)	http://sis.nlm.nih.gov/enviro/ glossarymain.html
Coastal Science	Glossary of Coastal Terminology (Washington State Department of Ecology, March 1998)	http://www.csc.noaa.gov/text/ glossary.html
Ecology	UCMP Glossary – Ecology (University of California Museum of Paleontology)	http://www.ucmp.berkeley .edu/glossary/gloss5ecol.html
Ecology and Restoration	Glossary of Terms in Ecology and Restoration (National Park Service)	http://www.nps.gov/plants/ restore/library/glossary.htm
Ecosystems	Glossary of Ecosystem-Related Terms (U.S. Fish and Wildlife Service)	http://www.nps.gov/plants/ restore/library/glossary.htm
Energy	Glossary (Energy Information Administration)	http://www.eia.doe.gov/ glossary/
Environmental Science	Terminology Reference System (U.S. Environmental Protection Agency)	http://epa.gov/trs/

Table 11.6, *continued*

Geology	Glossary of Geology (American Geological Institute)	http://www.agiweb.org/pubs/glossary
Geology	Geologic Glossary (U.S. Geological Survey)	http://wrgis.wr.usgs.gov/parks/misc/glossarya.html
Hydrology	Glossary of Hydrologic Terms (National Weather Service)	http://www.srh.noaa.gov/wgrfc/resources/glossary/default.html
Meteorology	American Meteorological Society Glossary of Meteorology (1999)	http://amsglossary.allenpress.com/glossary
Toxicology	IUPAC Glossary of Terms Used in Toxicology, 2nd Edition (IUPAC Recommendations 2007)	http://sis.nlm.nih.gov/enviro/iupacglossary/frontmatter.html
Water Science	Water Science Glossary of Terms (U.S. Geological Survey)	http://ga.water.usgs.gov/edu/dictionary.html

11.3 Lists of Earth Science and Environmental Science Tables and Appended Material

11.3.1 Tables in This Chapter

Table 11.1 Absolute Dates in Geologic Time

Table 11.2 Geologic Time Units

Table 11.3 Time and Place Modifiers

Table 11.4 Geochronological/Chronostratigraphic Nomenclature

Table 11.5 Division of Earth's History in Geologic Time

Table 11.6 Use of Prefixes in Geology

Table 11.7 Symbols for Basic Symmetry Operations

Table 11.8 Symbols in Crystallography

Table 11.9 Classifying Texture of Sediments

Table 11.10 Recommended Staging Scales and Sources

Table 11.11 Commonly Used SI-Based Units for Agronomy

Table 11.12 Common Abbreviations in Agronomy

Table 11.13 Common Statistical Abbreviations

Table 11.14 Examples of Orders

Table 11.15 Tidal Datum Abbreviations

Table 11.16 Sources of Specialized Terminology in Environmental Science

11.3.2 Contents of Appendix F in Part III

Earth Science

> **F1 Field-Specific Abbreviations**
>
> **F2 Common Mineral Abbreviations**
>
> **F3 Glossary of Earth Science Terms**
>
> **F4 Earth Science Journals and Their Abbreviations**

Environmental Science

> **F5 Common Units in Environmental Science**
>
> **F6 Common Alphabetic Symbols for Variables**
>
> **F7 Units in Environmental Science Diagrams**
>
> **F8 Abbreviations, Signs, and Symbols for Scientific and Engineering Terms**
>
> **F9 Glossary of Environmental Science Terms**
>
> **F10 Environmental Science Journals and Their Abbreviations**

Chapter 12. Style and Usage for Life Science

Contents

Contents: *continued*

Contents, *continued*

Contents: Taxonomy and Nomenclature *continued*

Chapter 12. Style and Usage for Life Science

12.1 Manuscript Preparation

The rules for writing in life science are consistent with those in earth and environmental sciences and medical science (*see* **Chapters 11** and **13**, respectively, as well as **Chapter 2** for basic information on manuscript preparation). As with all scientific writing, it is important to research the specific needs and preferences of the journal to which the manuscript is being sent. This will not only ease the writing process, but will also increase the chances of publication.

12.1.1 Writing for Life Science

In scientific writing, especially in the United States, articles dealing with life science issues are appearing much less frequently than those in other scientific disciplines. Because of this, it is important that writing in this field should be as concise and readable as possible. Articles written in this way, will not only have a higher chance of publication, but will also be more accessible and understandable to readers based in other scientific fields of study.

Life science journal articles will usually contain a summary (in addition to an abstract) or a basic introduction to the piece, separate from the actual text, Summaries, along with the actual title of the article, are scanned for keywords for the purposing of indexing. Many articles (even in professional journals) are made available to the general public. Therefore, titles and summaries should be free of any abbreviations, acronyms, and measures with which those not within the field would be unfamiliar.

A section describing research methods is usually included in a separate section at the end of the manuscript; in life science, however, if this section is brief, it may be omitted and described within the actual text. Descriptions of methods that have already been published may also be omitted and, a reference may be cited in its place.

12.1.2 Taxonomy and Nomenclature

In life science writing, it is important to use the correct taxonomic terms and nomenclature within a document. This is especially true, when these terms are used within the title and the abstract of the manuscript. As previously stated, keywords in the title and abstract are weighed more heavily than those in the rest of the text. These keywords are then used to index the article for its journal. If incorrect terms or unfamiliar or improper abbreviations are used in the title or abstract the article may not be properly indexed, or it may not be indexed at all. This will cause serious obstacles in your audience obtaining, reading, and learning from your document.

If papers contain taxonomy and nomenclature of newly discovered species, authors should be aware that it is possible for third parties to exploit the prior publication of nomenclature at any time between an online posting and the print publication date within a journal. Journals will not take responsibility for assertions of priorities in the cases of manuscripts it publishes, if they have previously appeared in the public domain as online posts or preprints. It is therefore particularly important in this field that the first publication of original research is in a peer-reviewed journal.

When creating any new nomenclatures or taxonomic ranks, one must adhere to the rules and principles established by the governing code in their specific field. For example, when naming plant life and fungi, authors should follow the International Code of Botanical Nomenclature (ICBN); for animals, authors should follow the International Commission on Zoological Nomenclature (ICZN); for bacteria, follow the International Code of Nomenclature of Bacteria (ICNB); and for viruses, the International Committee on Taxonomy for Viruses (ICTV).

These codes are set forth in order to promote uniformity, accuracy, and stability in the nomenclature of new species. Since codes are updated every few years, authors should be aware of any new or revised guidelines before publishing new nomenclature. In order for new nomenclature or taxonomic ranks to be valid, they must be approved and published with their specific code.

12.2 Usage

12.2.1 Biochemical Nomenclature and Abbreviations

12.2.1.1 Nucleic Acids

i. Abbreviations for bases and nucleosides.

Table 12.1 Abbreviations for Bases and Nucleosides

Nitroge-nous Base	Nucleoside	Abbrevi-ation	Deoxynucleoside	Abbrevi-ation
Adenine	adenosine	A	deoxyadenosine	dA
Guanine	guanosine	G	deoxyguanosine	dG
cytosine	cytidine	C	deoxycytidine	dC
Uracil	uridine	U	deoxyuridine	dU
Thymine	thymidine	T	deoxythymidine	dT
Unknown purine	unknown purine nucleoside	R		
Unknown pyrimidine	unknown pyrimidine nucleoside	Y		

ii. Conventions for naming bases and nucleosides. The names of bases and nucleosides are treated in written text as common nouns in lower-case.

> Concentration of uridine in plasma was elevated.

iii. Designating nucleotide sequences. Nitrogenous bases and nucleosides can be designated by one-letter abbreviations when indicating the sequences of bases.

> The telomere DNA sequence in humans is GGGTTA.

By convention, nucleotide sequences start (on the left) at the 5′ end and finish (on the right) at the 3′ end.

> 5′-ATGGCTATGGCTTACCCAGTGC-3′
> ATGGCTATGGCTTACCCAGTGC

Codon triplets can be indicated by placing a space every 3 base pairs.

GCA TTA ACC GGT AGA TAC GCA

iv. Designations for nucleotide length. DNA sequence length is designated in base pairs, which can be abbreviated as *bp* when used as a unit in written text. Abbreviate kilobases as kb and megabases as Mb.

The enzyme recognizes a 6 bp sequence.
The plasmid is 3.4 kb in length.
C. elegans has a genome of 97 Mb.

Nucleotide length can be indicated by using the suffix *-mer*.

The primer consisting of 22 nucleotides is a 22mer.

v. Designating nucleotide mutations.

Table 12.2 Meaning of Mutation Symbols

Type of Mutation	Symbol	Meaning
substitution	T152C	substitution of T to C at nucleotide 152
deletion	29delCG	deletion of C and G at position 29
	164del20	deletion of 20 bp at nucleotide 164
insertion	774insA	insertion of A at nucleotide 774
	58ins12	insertion of 12 bp at nucleotide 58

Examples:

• Substitution: **TA<u>G</u>** → **TA<u>A</u>**

• Deletion: **T<u>T</u>GACT** → **TGACT**

• Insertion: **ACG** → **A<u>T</u>CG**

476

vi. Nucleic acid abbreviations.

Table 12.3 Nucleic Acid Abbreviation

Deoxyribonucleic Acids		Ribonucleic Acids	
deoxyribonucleic acid	DNA	ribonucleic acid	RNA
complementary DNA	cDNA	complementary RNA	cRNA
single-stranded DNA	ssDNA	single-stranded RNA	ssRNA
Double-stranded DNA	dsDNA	double-stranded RNA	dsRNA
Nuclear DNA	nDNA	nuclear RNA	nRNA
ribosomal DNA	rDNA	ribosomal RNA	rRNA
heterogeneous nuclear cDNA	hn-cDNA	heterogeneous nuclear RNA	hnRNA
mitochondrial DNA	mtDNA	messenger RNA	mRNA
		microRNA	miRNA
		RNA interference	RNAi
		small interfering RNA	siRNA
		small nuclear RNA	snRNA
		small nucleolar RNA	snoRNA
		transfer RNA	tRNA

12.2.1.2 Amino Acids and Proteins

i. Designating amino acids. The names of amino acids are treated as common nouns in written text.

    alanine
    glycine
    tryptophan

ii. Amino acid abbreviations. Amino acids have three-letter and one-letter abbreviations. Abbreviations for amino acids are not used in written text.

Table 12.4 Amino Acid Abbreviations

Amino Acid	Three-Letter Abbreviation	One-Letter Abbrev.	Systematic Name
alanine	Ala	A	2-aminopropanoic acid
arginine	Arg	R	2-amino-5-guanidinopentanoic acid
asparagine	Asn	N	2-amino-3-carbamoylpropanoic acid
aspartic acid	Asp	D	2-aminobutanedioic acid
cysteine	Cys	C	2-amino-3-mercaptopropanoic acid
glutamic acid	Glu	E	2-aminopentanedioic acid
glutamine	Gln	Q	2-amino-4-carbamoylbutanoic acid
glycine	Gly	G	aminoethanoic acid
histidine	His	H	2-amino-3-(1H-imidazol-4-yl)propanoic acid
isoleucine	Ile	I	2-amino-3-methylpentanoic acid
leucine	Leu	L	2-amino-4-methylpentanoic acid
lysine	Lys	K	2,6-diaminohexanoic acid
methionine	Met	M	2-amino-4-(methylthio)butanoic acid
phenylalanine	Phe	F	2-amino-3-phenylpropanoic acid
proline	Pro	P	Pyrrolidine-2-carboxylic acid
serine	Ser	S	2-amino-3-hydroxypropanoic acid
threonine	Thr	T	2-amino-3-hydroxybutanoic acid
tryptophan	Trp	W	2-amino-3-(1H-indol-3-yl)propanoic acid
tyrosine	Tyr	Y	2-amino-3-(4-hydroxyphenyl)propanoic acid
valine	Val	V	2-amino-3-methylbutanoic acid

iii. Designating protein sequences. When presenting protein sequences, use either the three-letter or the one-letter abbreviations for amino acids.

When three-letter symbols are used to represent polypeptides, a hyphen between amino acids indicates a peptide bond. When one-letter symbols are used, a hyphen between amino acids is not necessary.

Met-Glu-Ala-Thr-Arg-Arg-Arg-Gln-His-Leu-Gly-Ala-Thr
MEATRRRQHLGAT

iv. Naming proteins. The word *protein*, as well as names of proteins, should be written as common nouns in lowercase.

motor protein
hemoglobin
actin

Abbreviations for proteins must be defined when used.

multidrug resistance protein (MDR)
Krüppel-like factor 6 (KLF6)

v. Amino acid mutations. Amino acid sequence mutations may be represented in shorthand form in either 3-letter or 1-letter notation.

A histidine to glutamine amino acid mutation would be designated as follows:

His527Gln
H527Q

vi. Transfer RNAs. The type of transfer RNA (tRNA) may be specified based on its amino acid attachment.

nonacylated tRNA: $tRNA^{Val}$
aminoacylated tRNA: $Val-tRNA^{Val}$

12.2.1.3 Sugars and Carbohydrates

Abbreviations are often used when naming sugars and carbohydrates, especially those that are lengthy with substituted carbons. An example of three-letter abbreviations for parent aldoses is given below.

O-α-D-Glc*p*-(1→4)-D-Glc*p*   maltose
O-β-D-Man*p*-(1→4)-D-Man*p*   mannobiose

Table 12.5 Abbreviations for Parent Aldoses

Number of Carbon Atoms	Parent Name	Three-Letter Abbreviation
5	ribose	Rib
	arabinose	Ara
	xylose	Xyl
	lyxose	Lyx
6	allose	All
	altrose	Alt
	glucose	Glc
	mannose	Man
	gulose	Gul
	idose	Ido
	galactose	Gal
	talose	Tal

12.2.1.4 Fatty Acids

Fatty acids can be designated in shorthand by their number of carbons and their number of double bonds, separated by a colon.

18:1   oleic acid

If the number of carbons and double bonds is the same between two or more different fatty acids, the positions of the double bonds is added in parentheses.

20:4(8,11,14,17)   eicosatetraenoic acid
20:4(5,8,11,14)   arachidonic acid

A fatty acid radical is indicated by adding "acyl" in parentheses.

18:2(acyl)   acyl radical of linoleic acid

12.2.2 Genetics

12.2.2.1 Chromosomes and Chromosomal Components

i. Standard Chromosome Nomenclature

(1) Human chromosomes are autosomal chromosomes numbered 1 through 22, or sex chromosomes designated X or Y.
(2) The chromosome number is followed by the arm designation: p for the short arm of the chromosome, or q for the long arm.
(3) The arm designation is followed by the region number (1 to 4), which specifies a region on the chromosomal arm.
(4) The region number is followed by the band number, a period, the subband number, and, when applicable, the sub-subband number.

Example: 7q32.31

> 7 = chromosome number
> q = long arm
> 32 = region 3, band 2
> 31 = subband 3, sub-subband 1

ii. Dimensional units for chromosomes. The unit for distances between genetic loci on a chromosome is the centimorgan (cM).

iii. Designations for anonymous DNA sequences. Anonymous DNA sequences are designated by D-number nomenclature and are named according to their chromosomal location, the sequence type, and the site.

Examples:

D4S7E	DXF12S1
D = anonymous DNA sequence	D = anonymous DNA sequence
4 = chromosome 4	X = X chromosome
S = unique sequence*	F = family sequence*
7 = sequence number	12 = sequence number
E = expressed sequence*	S1 = site number

Table 12.6 Conventions for Sequence Type Abbreviation

Abbreviation	*Sequence Type	Conventions
S	unique DNA sequence	S is followed by a sequence number
Z	repetitive DNA sequence	Z is followed by a sequence number
F	family DNA sequence	F is followed by a sequence number and then S for site number
E	expressed DNA sequence	E is added at the end of the symbol

iv. Designating karyotypes. The karyotype designation begins with the autosomal chromosome number separated by a comma from the sex chromosomes.

47,XY,+13 (male with trisomy 13)

When describing chromosomal abnormalities, abnormal sex chromosomes are designated first, followed by abnormal autosomal chromosomes listed in numerical order.

Karyotype designations may also indicate structural alterations in chromosomes. A single chromosomal rearrangement is indicated in the karyotype using a symbol that identifies the type of chromosomal alteration, followed by the chromosome number in parenthesis. If there has been a rearrangement of more than one chromosome, a semicolon separates the designations. For abbreviations used to indicate chromosomal rearrangements, *see* **Section 12.2.2.5**.

Examples:

46,XX,dup(3)(q32q33)   indicates duplication of the region between bands 3q32 and 3q33.

46,XY,del(19)(q23q25)   indicates deletion in chromosome 19 with rejoining of 19q23 and 19q25.

12.2.2.2 Human Gene Nomenclature

i. Conventions: symbols for genes and phenotypes. Among the nomenclature committees involved in establishing a system for standardizing human gene nomenclature, are the International Committee on Gene Symbols and Nomenclature, and the HUGO Gene Nomenclature Committee.

Gene names most commonly describe the mutant phenotype or the protein encoded. Gene symbols are usually derived by shortening the original name or by using the initials of a multiword name. They are almost always italicized. Gene symbols followed by an additional letter or Arabic numeral indicate genes with different loci but similar phenotypes.

BRCA1, BRCA2 (breast cancer genes)
PFN1, PFN2 (profilin genes)

ii. Alleles. Alleles are alternative forms of genes. They are often designated by the gene symbol followed by an asterisk, followed by the italicized allele designation.

CFTR*N   *CFTR*R*

12.2.2.3 Common Genes and Phenotypes

i. HLA nomenclature. Nomenclature guidelines for the human major histocompatibility complex are established by the World Health Organization Committee for Factors of the HLA System. Human major histocompatibility complex gene names begin with HLA-, followed by a locus symbol and designations for subregions or chains.

HLA-DQA1   *HLA-DRB4*

ii. Symbols representing human retroviral genes. Human retroviral genes are italicized and have a variety of alphanumeric designations.

HIV genes: *gag, pol, env*
HTLV genes: *p19, p24, gp68*

iii. Symbols for oncogenes. Human oncogene sequences are three letters long, written in lowercase, and italicized. For example:

src
myc
ras
erb

To further specify the location or source of the gene, non-italicized prefixes may be used.

c-*myc*	c- for cellular
v-*abl*	v- for virus
H-*ras*	H- for Harvey rat sarcoma
B-*lym*	B- for B-cell lymphoma

iv. Transspecies gene families: the P450 supergene family. Cytochrome P450 genes are italicized and abbreviated *CYP*, followed by an arabic numeral designating the gene family, a capital letter indicating the subfamily, and another numeral for the individual gene.

CYP2A6
CYP3A43
CYP8A1

v. Conventions for representing bacteriophage genes. Bacteriophage genes are designated by a prefix for the phage, either spaced or unspaced from the name of the gene.

T4 *soc*
λ int
ϕX174

12.2.2.4 Genes: Related Functional Elements

i. Abbreviated Prefixes for Initiation and Elongation Factors.

Initiation factors are abbreviated IF, followed by a hyphen and alphanumeric designations. Bacterial initiation factors have no prefix, while eukaryotic initiation factors are designated by the lowercase prefix e.

IF-1, IF-2   eIF-4B , eIF-4G

Bacterial elongation factors are designated EF-Tu and EF-G.
Eukaryotic elongation factors are designated EF-1 and EF-2.

ii. Designations for Probes

Table 12.7 Abbreviations for Plasmids

Vector type	Abbreviation
plasmid	p
cosmid	c
lambda phage	l
yeast	y

iii. Plasmids notation. Plasmids are designated by a lowercase p for plasmid, followed by a variety of alphanumeric designations. Deletions are conventionally identified using a Greek delta sign. Insertions, transpositions, and translocations are designated using a Greek omega sign.

pMG101
pI258Δ7
pLF273Ω7

iv. Enzymes. Specify the type of enzyme being used, as different enzymes may come from the same organism and have similar names.

Taq DNA ligase
Taq DNA polymerase
Dam DNA ligase
Dam methylase

v. Abbreviations for restriction endonucleases. Restriction endonucleases have a standard three-letter italicized abbreviation (for the source organism) with the first letter capitalized. This is followed by a non-italicized strain designation including arabic or roman numerals.

A *Bam*H1 and *Sma*I double digest was performed.
The vector includes four *Eco*RI restriction sites.

vi. Abbreviations for transposons. Transposons in bacteria are designated by Tn followed by italicized alphanumeric designations.

Tn*A*   Tn*10*

Transposons in eukaryotes are usually not italicized.

P element (*Drosophilia melanogaster*)

12.2.2.5 Field-Specific Abbreviations

Table 12.8 Field-Specific Abbreviations

Abbreviation	Term
ace	acentric fragment
add	additional material of unknown origin
b	break
c	constitutional anomaly
cen	centromere
chr	chromosome
cht	chromatid
del	deletion
dic	dicentric
dup	duplication
e	exchange
fra	fragile site
g	gap
h	heterochromatin
i	isochromosome
ins	insertion
inv	inversion
mar	marker chromosome
mos	mosaic
rea	rearrangement
rec	recombinant chromosome
s	satellite
t	translocation
tan	tandem
tel	telomere
tri	tricentric chromosome
v	variable region

12.2.2.6 Standard Symbols

Table 12.9 Symbols for Terms in Genetics

Symbol	Term
superscript plus sign ($^+$)	wild-type allele
superscript minus sign ($^-$)	mutant allele
Δ	deletion
IN	inversion
double colon (::)	insertion
Φ	fused genes
p	promoter site
t	terminator site
o	operator site
a	attenuator site
p	short arm of chromosome
q	long arm of chromosome

12.2.2.7 Sources for Specialized Terminologies

• Home Page of the Human Cytochrome P450 (CYP) Allele Nomenclature Committee. Available from: http://www.cypalleles.ki.se/

• HUGO Gene Nomenclature Committee. London (UK): HUGO Nomenclature Committee. Available from: http://www.gene.ucl.ac.uk/nomenclature

• IMGT/HLA Sequence Database. Cambridge (UK): European Bioinformatics Institute; 2005. Available from: http://www.ebi.ac.uk/imgt/hla/

• International Committee on Genetic Symbols and Nomenclature. Report of the International Committee on Genetic Symbols and Nomenclature. Union Int Sci Biol, Ser B. 1957; (30):1-6.

• ISCN 1995. An International System for Human Cytogenetic Nomenclature. Mitelman F, editor. Basel (Switzerland): Karger, 1995.

- Marsh SGE, Albert ED, Bodmer WF, Bontrop RE, Dupont B, Erlich HA, Geraghty DE, Hansen JA, Mach B, Mayr WR, Parham P, Petersdorf EW, Sasazuki T, Schreuder GMT, Strominger JL, Svejgaard A, Terasaki PI. Nomenclature for factors of the HLA system. Tissue Antigens 2002;60:407-464. Available from: http://www.anthonynolan.com /HIG/lists/nomenc.html

- Novick R, Clowes RC, Cohen SN, Curtiss III R, Datta N, Falkow S. Uniform Nomenclature for Bacterial Plasmids: a Proposal. Bacteriological Reviews. 1976;40(1):168-189.

- Wain HM, Bruford EA, Lovering RC, Lush MJ, Wright MW, Povey S. Guidelines for human gene nomenclature. Genomics. 2002;79(4):464-470. Available from: http://www.gene.ucl.ac.uk/nomenclature/guidelines.html

12.2.3 Other Abbreviations and Nomenclature Conventions

12.2.3.1 Format for the Description of Cell Lines

Cell lines are commonly referred to by acronyms and should be defined at first mention.

> HUVEC (human umbilical vein endothelial cells)
> CHO (Chinese hamster ovary cells)

12.2.3.2 Isotopes

Abbreviations for radioactive isotopes should be defined when used for the first time. To indicate that the nonradioactive isotope is normally part of the compound, use brackets around the isotope symbol.

> The phospholipid was labeled with radioactive phosphate $[^{32}P]$

> $[^{32}P]$phospholipid

When the isotope is not normally part of the compound, do not use brackets. Instead, separate the element from the compound with a hyphen.

> ^{131}I-human growth hormone

For uniformly labeled compounds, write the abbreviation *ul* in parenthesis following the compound name.

$[^{14}C]$glucose (ul)

12.2.4 Taxonomy and Nomenclature

See **Appendix G.000, for a complete list of taxonomy ranks and endings**.

12.2.4.1 General Rules

i. Standard format for naming taxa. There are seven basic taxa: kingdom, phylum (or division), class, order, family, genus, and species. Any of these taxa can be prefixed by either *sub-* or *super-* to further extend the taxonomic categories.

The names of taxa at the rank of family and above are written in plain roman type; names of taxa for genus and below are italicized. Names of taxa at the level of genus and above have an initial capital letter.

When the taxonomic term precedes the name, the taxonomic term is written in lowercase, as in "kingdom Bacteria."

Names of taxa at the rank of family and above are treated as plural, while names of taxa for genus and below are treated as singular.

The family Micrococcaceae are…
The genus *Python* is…

ii. Standard format for scientific names. Only the names of taxa at the rank of genus and above may stand alone as monomials. Names of taxa at the level of species and below cannot stand alone and must be preceded by the genus name. The first letter of the genus name is capitalized, the species name is written in all lowercase letters, and the entire name is italicized.

The genus *Homo*…
The species *Homo sapiens*…

A species name may be abbreviated if the full name is used at first mention. Thus, *Staphylococcus aureus* becomes *S. aureus*.

12.2.4.2 Bacteria

i. Format for descriptions of taxa. Bacterial nomenclature is defined by the International Committee on Systematic Bacteriology in the *International Code of Nomenclature of Bacteria*. Typical taxonomic endings are shown for examples of bacterial species. *See* **Section 12.2.4.1**, above.

Table 12.10 Bacterial Nomenclature

Taxon	Name		
Kingdom	Archaea	Bacteria	Eubacteria
Phylum	Euryarchaeota	Proteobacteria	Firmicutes
Class	Halobacteria	Gamma Proteobacteria	Bacilli
Order	Halobacteriales	Enterobacteriales	Lactobacillales
Family	Halobacteriaceae	Enterobacteriaceae	Streptococcaceae
Genus	*Halobacterium*	*Escherichia*	*Streptococcus*
Species	*Halobacterium salinarum*	*Escherichia coli*	*Streptococcus pneumoniae*

ii. Standards for scientific names. Using the name of a genus on its own suggests the genus as a whole.

Streptococcus is comprised of gram-positive bacteria.

Using the term species after the genus name implies that the genus is certain but the species is not.

Streptococcus species are part of the normal flora of the mouth.

iii. Designations for infrasubspecific taxa and strains. Infrasubspecific subdivisions, or subdivisions below the subspecies level, are not included in the Bacteriological Code but are useful for practical purposes. These designations include:

biovar or biotype (bv.)
serovar or serotype (sv.)
pathovar or pathotype (pv.)

Infrasubspecific subdivisions, biovars, and biotypes are designated with letters or numbers.

Agrobacterium vitis biovar III
Fusobacterium necrophorum biovar A

Infrasubspecific subdivisions serotype, serovar, and type are useful for designating strains. For instance, *Escherichia coli* strains are designated by the O:K:H serotype profile.

Escherichia coli O1:K1:H7
Escherichia coli O126:H27

Haemophilus influenzae strains are designated by types a through f.

Haemophilus influenzae type a

Salmonella strains are designated by the infrasubspecific divisions serotype/serovar.

Salmonella serotype Enteritidis, serovar Enteritidis
Salmonella serotype Muenchen, serovar Muenchen

Furthermore, *Salmonella* serotypes are expressed as O, Vi, and H antigen types. The letters O, Vi, and H are not included in the serotype, which is composed of alphanumeric designations separated by colons.

Salmonella serovar Typhi 9:d:k

iv. Vernacular names and adjectival forms. Vernacular names for bacteria are written in lowercase roman letters.

rhizobia   cholera   typhoid

Adjectival forms derived from scientific names usually end in *-al*, but the noun form may also serve as the adjective. A genus name in lowercase roman letters may be used as a vernacular adjective.

staphylococcal infection
staphylococcus infection

Traditional plural designations can be used for vernacular plurals. If the generic plural is unknown, add the word "organisms" to the genus name.

staphylococci
Escherichia organisms

v. Field-specific abbreviations. Bacteria that are well characterized but cannot be maintained or isolated in culture are given *Canddatus* status. These terms should be written in quotation marks, with the word *Candidatus* italicized and the taxon name in roman: "*Candidatus* Phytoplasma allocasuarinae." *Candidatus* can subsequently be abbreviated *Ca.* Names of bacteria used in laboratory media are written in lowercase roman letters, as in "salmonella agar."

vi. Standard Symbols

Table 12.11 Standard Symbols for Taxonomy and Nomenclature for Bacteria

Symbol	Term
p	promoter site
t	terminator site
o	operator site
a	attenuator site
superscript s, as in Kan^{s}	drug sensitive
superscript r, as in Amp^{r}	drug resistant

vii. Sources for Specialized Terminologies

- Euzeby JP. List of bacterial names with standing in nomenclature [Internet]. Societe de Bacteriologie Systematique et Veterinaire; 1997. Available from: http://www.bacterio.cict.fr/

- Holt JG, editor-in-chief. Bergey's manual of determinative bacteriology. 9th ed. Baltimore: Williams & Wilkins; 1994.

- International Committee on Systematic Bacteriology. International code of nomenclature of bacteria: bacteriological code. Washington (DC): American Society for Microbiology; 1992.

- Krieg NR, Holt JF, eds. Bergey's Manual of Systematic Bacteriology. Baltimore, Md: Williams & Wilkins; 1984.

- Murray RGE, Stackebrandt E. Taxonomic Note: implementation of the provisional status Candidatus for incompletely described prokaryotes. Int J Syst Bacteriol. 1995;45(1):186-187.

- Skerman VBD, McGowan V, Sneath PHA, editors. Approved lists of bacterial names. Amended ed. Washington (DC): American Society for Microbiology; 1989.

12.2.4.3 Viruses

i. Format for descriptions of taxa. A system for classifying viruses was established by the International Committee on Taxonomy of Viruses (ICTV). In viral classification, there are no ranks above the level of order. Specialist groups (not the ICTV) deal with taxa below the level of species. Names of orders are not italicized, while names of families, genera, and species are italicized. *See* **Section 12.2.4.1.1.**

Table 12.12 Suffixes for Virus Taxonomy

Taxon	Suffix
Order	-virales
Family	*-viridae*
Subfamily	*-virinae*
Genus	*-virus*
Species	*-virus*

ii. Standards for scientific names. Although virus species do not have Latin names, the name of the virus is formally written in italics, with the first word of the species name capitalized. Other words in the virus name are capitalized only if they are proper nouns.

Ebola virus
Epstein-Barr virus

iii. Acronyms for viruses. Acronyms can be used to designate viruses when the full virus name is used at first mention. Abbreviations for some common virus species are listed below.

human papillomavirus (HPV)
rabies virus (RABV)

Table 12.13 Abbreviations for Common Virus Species

Virus Species	Abbreviation
California encephalitis virus	CEV
Ebola virus	EBOV
hepatitis A virus	HAV
hepatitis B virus	HBV
hepatitis C virus	HCV
human adenoviruses 1 to 47	HAdV-1 to 47
human coronavirus	HCV
human herpesvirus 1 to 6	HHV-1 to 6
human papillomavirus	HPV
human parainfluenza virus	HPIV
human rhinovirus	HRV
human T-lymphotropic virus	HTLV
measles virus	MeV
rabies virus	RABV
rotavirus	ROTAV
rubella virus	RUBV
variola (smallpox) virus	VARV

iv. Strain designations. Strain designations are regulated by international specialist groups, not the ICTV. Information about the virus strain may be included in the virus name. The strain designation may be separated from the virus name by a dash.

HaCPV-B (*Heliothis armigera cypovirus*, strain B)

v. Designations for vernacular names of viruses. Family, genus, and species classifications can all be given in the vernacular. Vernacular terms can be ambiguous because the same name can be used for more than one taxonomic level. Therefore, vernacular names should be used with the specific taxonomic rank term.

topovirus genus   topovirus species

vi. Sources for Specialized Terminologies

• The Universal Virus Database of the International Committee on Taxonomy of Viruses. Available from: http://www.ncbi.nlm.nih.gov/ICTVdb/index.htm

• Regenmortel MHV van, Fauquet CM, Bishop DHL, Carstens EB, Estes MK, Lemon SM, Maniloff J, Mayo MA, McGeoch DJ, Pringle CR, Wickner RB. Virus taxonomy: classification and nomenclature of viruses. 7th report of the International Committee on Taxonomy of Viruses. New York (NY): Academic Press; 2000.

• Sander DM. All the virology on the WWW [Internet]. D Sander; 1995 May. Available from: http://www.virology.net/garryfavweb.html

12.2.4.4 Plants, Fungi, Lichens, and Algae

i. Plants

a. Rules for Stylistic Treatment of Nomenclature

1. Format for descriptions of taxa. The ultimate authority on the *International Code of Botanical Nomenclature* is the International Botanical Congress. Typical endings for the taxonomic ranks are shown below. *See* **Appendix G, Table H.000** for a full table on ranks in plants taxonomy.

Table 12.14 Suffixes for Taxonomic Ranks for Plants

Taxon	Suffix
Division	–phyta
Subdivision	–phytina
Class	–opsida
Subclass	–idae
Order	–ales
Family	–aceae
Subfamily	–oidene
Tribe	–eae
Subtribe	–inae
Genus	–us

2. Standards for scientific names. Scientific names for plants are written using the standard Latin binomial system of nomenclature. The first name represents the genus and is capitalized, the second name identifies the species and is written in lowercase, and the entire name is italicized. *See* **Section 12.2.4.1: General Rules**.

3. Format for the publication of names. Requirements for publishing plant names of any taxonomic rank are specified in the *International Code of Botanical Nomenclature*. A person who publishes a scientific name is considered the author of that name, though citation of the author's name is optional. The author's name is not italicized, and follows the italicized scientific name. For names published by Linnaeus that are still valid, the author is designated as "L."

4. Vernacular names for plants. Vernacular names may be genus or family names and should be written in all lowercase letters. Vernacular names are not capitalized unless they are proper nouns.

> pine
> Canary Island pine

When the vernacular name does not reflect the correct taxonomic position of the plant, join the terms with a hyphen. However, it should be noted that this rule is practiced inconsistently.

> poison-oak (belongs to a different genus from true oak)
> white oak (a true oak)

5. Synonyms and homonyms. Valid scientific names should not be synonyms or homonyms. However, synonyms and homonyms do exist for scientific and vernacular names. Multiple names used for the same species (synonyms) or names that sound the same but represent different species (homonyms) pose problems in consistency and in retrieval of data and should be avoided if possible.

b. Cultivar notation. Cultivar notation is defined by the *International Code of Nomenclature for Cultivated Plants* (ICNCP). Cultivar names are written in roman type with an initial capital letter and are placed within single quotation marks when they come after the scientific name. Single quotation marks are not needed when the cultivar name is used alone.

Miscanthus sinesis 'Adagio'
Adagio

c. Designating hybrids. Hybrids are designated by a formula in which the names of the parents are separated by a multiplication symbol. If no other convention is specified, list the names in alphabetical order.

Magnolia × soulangeana
Fragaria × ananassa

d. Orchid nomenclature. Orchid nomenclature is governed both by the *International Code of Botanical Nomenclature* and the *International Code of Nomenclature for Cultivated Plants*. All described orchid species are named using the standard Latin binomial system.

e. Construction of common names for plant diseases. Plant disease names are usually based on the major disease symptom or on the pathogen responsible for the disease. A recommended list of common names for plant diseases has been published by the American Phytopathological Society. When the Latin name of the pathogen is part of the disease name, the Latin word should be italicized and its first letter should be capitalized.

Aphanomyces root-rot

The same disease name may be used even when the disease is caused by different pathogens in different host species. In order to avoid confusion, include the host name as part of the disease name.

southern corn leaf blight (caused by a fungus)
leaf blight of rice (caused by a bacterium)

Plant diseases caused by nematodes are described using the common name of the nematode pathogen.

root-gall nematode disease
root-knot nematode disease

ii. Fungi

a. Format for descriptions of taxa. Fungal nomenclature is defined by the *International Code of Botanical Nomenclature*. Typical endings for the taxonomic ranks are shown below.

Table 12.15 Suffixes for Taxonomic Ranks for Fungi

Taxon	Suffix
Division	-mycota
Subdivision	-mycotina
Class	-mycetes
Subclass	-mycetidae

b. Standards for scientific names. The scientific name of a fungus is written with the first letter of the genus name capitalized, the species name written in all lowercase letters, and the entire name italicized. *See* **Section 4.1: General Rules**.

Glugea heraldi
Vairimorpha plodiae

Fungal genera are not written in plural form. To refer to a group of species in a genus, write the abbreviation for *species* after the genus name. The abbreviation for species is either sp. (for a group composed of one species) or spp. (for a group consisting of two or more different species).

Pleistosporidium sp.
Bacillidium spp.

c. Designations for infrasubspecific taxa and strains. Infrageneric or infraspecific rank names are preceded by an abbreviation indicating the taxonomic rank. This abbreviation is not capitalized or italicized.

Banksia subg. *Isostylis*
Erigonum longifolium subsp. *Diffusum*

Below the rank of species, the scientific name may be written in shortened form.

E. longifolium var. *plantagineum*

d. Yeasts and slime molds. Rules for naming both yeast and slime molds follow the guidelines set for fungi in the Botanical Code. *Yeast* is not a taxonomic term, but is usually applied to fungi in the order Saccharomycetales. In a scientific paper, the word *yeast* can be used if the scientific name is used at first mention. Yeast used for cooking or brewing are not usually identified by genus or species names.

Table 12.16 Yeast Gene Conventions

Feature	Convention	Examples
Gene symbol	Three italic letters	ARG arg
Gene locus	Italicized number following the symbol	ARG_2
Dominant allele	Capitalized italic letter	ARG2
Recessive allele	Lowercase italic letters	Arg
Allele designation	Italicized number following the locus number and a hyphen	Arg2-14
Gene cluster	Italicize capital letter following the locus number	His4A his4B
Wild-type gene	Added plus symbol (sign)	ARG2+
Gene conferring resistance or susceptibility	Superscript R or S, not italicized	CUP^R1
phenotype	same characters as gene symbol, but not italicized; superscipt + and −	arg− arg+

iii. Lichens. A lichen is a made up of a fungus and an alga. Nomenclature for lichens reflects their fungal components. Lichens are commonly referred to by their vernacular names.

Flavoparmelia caperata belongs to the fungal genus *Flavoparmelia.*
Cladonia rangiferina is also known as Reindeer lichen.
Cetraria islandica is also known as Iceland moss.

iv. Algae. The word *algae* is not a taxonomic term. The same nomenclature style and format rules set forth for plants in the Botanical Code are also applied to algae. Typical taxonomic endings are shown below.

Table 12.17 Taxonomic Nomenclature for Algae

Taxon	Name		
Kingdom	Protista	Archaeplastida	Protista
Phylum	Heterokontophyta	Rhodophyta	Chlorophyta
Class	Phaeophyceae	Rhodophyceae	Ulvophyceae
Order	Fucales	Gigartinales	Ulvales
Family	Fucaceae	Gigartinaceae	Ulvaceae
Genus	*Fucus*	*Chondrus*	*Ulva*
Species	*Fucus serratus*	*Chondrus crispus*	*Ulva lactuca*

v. Sources for Specialized Terminologies

- Brodo IM, Sharnoff SD, Sharnoff S. Lichens of North America. New Haven (CT): Yale University Press; 2001.

- Greuter W, McNeill J, Barrie FR, Burdet HM, Demoulin V, Filgueiras TS, Nicholson DM, Silva PC, Skog JE, Trehance P, Turland NJ, Hawksworth DL. International code of botanical nomenclature. Konigstein (Germany): Koeltz Scientific Books; 2000.

- Trehane P, Brickell CD, Baum BR, Hetterscheid WLA, Leslie AC, McNeill J, Songberg SA, Vrugten F, editors. International code of nomenclature for cultivated plants. Regnum Vegetaile, v. 133. Windborne (UK): Quarterjack Publishing; 1995.

- Wehr JD, Sheath RG, editors. Freshwater algae of North America: ecology and classification. Boston (MA): Academic Press; 2003.

12.2.4.5 Human and Animal Life

i. Format for designating taxonomic categories. Nomenclature rules for taxonomic designation of animals are given by the *International Code of Zoological Nomenclature*. The International Commission on Zoological Nomenclature produces official lists of approved scientific names. *See above,* **Section 12.2.4.1: General Rules**.

ii. Format for author names. Inclusion of the author's name as part of the genus or species name is optional. When the author's name is included, it should also include the year in which it was named and be written in this format:

Enhydra lutris, Linnaeus 1758

A document's list of references should include the publication in which the taxonomic name was published.

iii. Vernacular names. Common names are not capitalized except in the case of proper names. Lists of scientific names with approved common names have been published for species in a number of phyla, including insects, reptiles, fish, birds, and mammals.

iv. Designation systems for laboratory animals. The International Index of Laboratory Animals provides information on inbred animal strains. A system for specifically designating inbred strains of mice has been established by the Committee on Standardized Genetic Nomenclature for Mice (CSGNM).

v. Sources for Specialized Terminologies

- American Ornithological Union. A.O.U. Check-list of North American birds [Internet]. 7th ed. McLean (VA): American Ornithologists' Union. Available from: http://www.aou.org/checklist/inde.php3

- Collins JT. Standard common and current scientific names for North American amphibians and reptiles. 3rd ed. Lawrence (KS): Society for the Study of Amphibians and Reptiles; 1990.

- Entomological Society of America, Committee on the Common Names of Insects. Common names of insects and related organisms [Internet]. Lanham (MD): ESA; c1995-2005. Available from: http://www.entsoc.org/ Pubs/Books/Common_ Names/index.htm

- FINS: the Fish Information Service. Fish index [Internet]. Cambridge (MA): Active Window Productions, Inc.; c1993-2000. Available from: http://fins.actwin.com/species/

- Hall ER. The mammals of North America. 2nd ed. New York (NY): John Wiley & Sons; 1981.

- Lyon MF. Rules for nomenclature of inbred strains. In: Lyon MF, Searle AG, editors. Genetic variants and strains of the laboratory mouse. 2nd ed. Oxford (UK): Oxford Univ. Press; 1989. pp 632-35.

- International Commission on Zoological Nomenclature. International code of zoological nomenclature. 4th ed. London (UK): The Natural History Museum, Intern'l Trust for Zoological Nomenclature; 1999.

- International Committee on Standardized Genetic Nomenclature for Mice; Rat Genome and Nomenclature Committee. Rules of nomenclature for mouse and rat strains [Internet]. [Bar Harbor (ME)]: The Jackson Laboratory. Available from: http://www.informatics.jax.org/mgihome/nomen/strains.shtml

- Melville RV, Smith JDD, editors. Official lists and indexes of names and works in zoology. London (UK): International Trust for Zoological Nomenclature; 1987.

- National Research Council (US), Comm. on Life Sciences, Inst. of Laboratory Animal Resources, Committee on Transgenic Nomenclature. Standardized nomenclature for transgenic animals. ILAR [Internet]. http://dels.nas.edu/ilar_n/ilarjournal/34_4/34_4Standardized Backup.shtml

- Systematic Biology. A quarterly of the Society of Systematic Biologists. Page R, ed. Philadelphia (PA): Taylor & Francis. 41:1, 1992– .

- Wilson DE. Cole FR. Common names of mammals of the world. Washington (DC): Smithsonian Institution Press; 2000.

12.2.5 Preferred Units in the Life Sciences

Preferred units have been established by the International System of Units (SI). SI units are recognized by international agreement, but they are not used exclusively in the United States, which may require dual reporting with metric units. For example, the Celsius scale (°C) is acceptable for reporting temperature rather than the SI unit, the kelvin (K). For SI derived units, multiplying prefixes, and conversion factors, *see* **Appendix B1**.

Table 12.18 Abbreviations for Common Measurement Units

Measurement	SI Unit	SI Abbreviation
Length	meter	m
Mass	kilogram	kg
Volume	liter	L
Quantity of substance	mole	mol
Molecular weight of substance	grams per mole	g/mol
Time	second	s
Temperature	kelvin	K
Electric current	ampere	A
Luminous intensity	candela	cd

12.2.5.1 Sources for Specialized Terminologies

- Alberts B, Johnson A, Lewis J, Raff M, Roberts K, Walter P. Molecular Biology of the Cell. 4th ed. New York, NY: Garland Science; 2002.

- ATCC: The Global Bioresource Center [Internet]. Manassas, VA: American Type Culture Collection; c2005. Available from: http://www.atcc.org

- Bennett RL, Steinhaus KA, Uhrich SB, O'Sullivan CK, Resta RG, Lochner-Doyle D, Markel DS, Vincent V, Hamanishi J. Recommendations for standardized human pedigree nomenclature. Pedigree Standardization Task Force of the National Society of Genetic Counselors. Am J Hum Genet. 1995;56(3):745-752.

- Council of Science Editors, Style Manual Committee. Scientific style and format: the CSE manual for authors, editors, and publishers. 7th ed. Reston, VA: The Council; 2006.

- Dodd JS, ed. The ACS Style Guide: A Manual for Authors and Editors. Washington, DC: American Chemical Society; 1997.

- Fishman AP, Alias JA, Fishman JA, Grippi MA, Kaiser LR, Senior RM, editor. Fishman's manual of pulmonary diseases and disorders. 3rd ed. New York (NY): McGraw-Hill; 2002.

- Interferon nomenclature. Arch Virol. 1983;77(2-4):283-285.

- International Anatomical Nomenclature Committee. Nomina anatomica: authorized by the 12th International Congress of Anatomists in London, 1985. 6th ed. Edinburgh (Scotland): Churchill Livingstone; 1989.

- *The International System of Units (SI).* Washington, DC: US Dept of Commerce, National Institue of Standards and Technology (NIST); 1991.

- Linnaeus C. Species plantarum: a facsimile of the first edition. London: The Ray Society. Vol 1,1957; Vol2, 1959.

- Kriz W, Bankir L. A standard nomenclature for structures of the kidney. The Renal Commission of the International Union of Physiological Sciences. Kidney Int. 1988;33(1):1-7.

- Medical Subject Headings [Internet]. Bethesda (MD): National Library of Medicine (US); 1999. Available from: http://www.nlm.nih.gov/mesh/

- Pappenheimer JR, Comroe JH, Cournand A, et al. Standardization of definitions and symbols in respiratory physiology. *Fed Proc.* 1950;9:602-605.

- Parfitt AM, Drezner MK, Glorieux FH, Kanis JA, Malluche H, Meunier PJ, Ott SM, Recker RR. Bonehistomorphometry: standardization of nomenclature, symbols, and units. Report of the ASBMR Histomorpho-metry Nomenclature Committee. J Bone Miner Res. 1987;2(6):595-610.

- Paul WE, Kashimoto T, Melchers F, Metcalf D, Mossman T, Oppen-heim J, Ruddle N, Van Snick J. Nomenclature for secreted regulatory proteins of the immune system (interleukins). WHO-IUIS Nomen-clature Subcommittee on Interleukin Designation. Bull World Health Organ. 1991;69(4):483-484.

- Proposed standard system of symbols for thermal physiology. J Appl Physiol. 1969:27(3):439-446.

- Pugh MB, ed. American Medical Association manual of style: a guide for authors and editors. 9th ed. Chicago, Illinois: Williams and Wilkins;1998.

- Unified Medical Language System [Internet]. Bethesda (MD): National Library of Medicine (US); 1999. Available from: http://www.nlm.nih.gov/research/umls/

12.3 Lists of Life Science Tables and Appended Material

12.3.1 Tables in This Chapter

Table 12.1 Abbreviations for Bases and Nucleosides

Table 12.2 Meaning of Mutation Symbols

Table 12.3 Nucleic Acid Abbreviation

Table 12.4 Amino Acid Abbreviations

Table 12.5 Abbreviations for Parent Aldoses

Table 12.6 Conventions for Sequence Type Abbreviation

Table 12.7 Abbreviations for Plasmids

Table 12.8 Field-Specific Abbreviations

Table 12.9 Symbols for Terms in Genetics

Table 12.10 Bacterial Nomenclature

Table 12.11 Standard Symbols for Taxonomy and Nomenclature for Bacteria

Table 12.12 Suffixes for Virus Taxonomy

Table 12.13 Abbreviations for Common Virus Species

Table 12.14 Suffixes for Taxonomic Ranks for Plants

Table 12.15 Suffixes for Taxonomic Ranks for Fungi

Table 12.16 Yeast Gene Conventions

Table 12.17 Taxonomic Nomenclature for Algae

Table 12.18 Abbreviations for Common Measurement Units

12.3.2 Contents of Appendix G in Part III

G1 Table of Chromosome Symbols and Abbreviations

G2 Table of Ranks In Plant Taxonomy

G3 Table of Common Viral Abbreviations

C4 Taxonomic Name Endings

C5 Life Science Glossary

C6 Life Science Journals and Their Abbreviations

Chapter 13. Style and Usage for Medical Science

Contents

Contents: Human Physiology and Anatomy Nomenclature, *continued*

Contents: Human Physiology and Anatomy Nomenclature, *continued*

Contents, *continued*

Chapter 13. Style and Usage for Medical Science

13.1 Preparing for Publication

As with all scientific writing, prior to preparing a manuscript, one should become familiar with the needs, standards, practices, and preferences of the intended journal of publication. This will not only increase the chances of publication, but it will also help with the writing process. For a listing of journals and their abbreviations, *see* **Part III Appendix H7**. For a general guide to preparing a manuscript, *see* **Part I, Chapter 2**.

It is also important for authors to keep their audience and their message in mind while writing. This will keep the focus and communication of the manuscript strong and clear.

There are seven types of articles which medical publications generally seek to publish:

- Reports of original data
- Review articles
- Descriptive articles
- Consensus statement and clinical practice guidelines
- Articles of opinion
- Correspondence
- Reviews of books, journals, and other media
- Other—personal reflections, news, obituaries, conference reports, etc.

13.1.1 Submission procedures

As with other scientific writing, peer review is an integral part of the publishing process in medical writing. Even for such pieces where peer review is not called for in other disciplines, the simple fact that human lives and welfare are materially dependent on the medical community demands that *all* types of submissions to medical journals be peer reviewed to assess the importance, quality, and validity of anything that appears in the journal.

Many journals now use an internet-based peer review system in order to speed up the process. Therefore, it is important to refer to the specific journal to see if they require any necessary steps for electronic manuscript submissions to facilitate peer reviewed.

Some articles may also require a number of accompanying documents before even considering to evaluate anything submitted for publication. These are summarized in **Chapter 2, Section 2.2.1.4**, above.

13.1.2 Preparing Text

Medical manuscripts should be submitted with an abstract summarizing the important points of the article, including the objectives of the piece, background information, study methods, primary results, and the conclusion of the piece (*see* **Chapter 2, Sections 2.2.1.4** on the parts of a paper, and **Section 2.2.1.5** on organizing the content). A structured abstract, running about 300 words, should be used for articles containing original information, and systematic and clinical reviews. Along with describing the context, objective, results, and conclusions of an article, structured abstracts may also contain sections describing the setting of the study, patients and other participants, interventions, outcome measurements, trial registration, data sources, selection of studies, guidelines for abstracting data, and evidence acquisitions. Unstructured abstracts, which run about 150 words, would be used for any other types of major article. Submissions such as letters, opinions, and special features do not require abstract because of their brevity.

13.1.3 Images, Illustrations, and Tables

Tables and figures are often used in a manuscript in order to convey large amounts of data in a more efficient way and in a smaller amount of space. When creating a table, keep in mind that tables and figures should be used to exemplify the text and should not be just a repetition. If it is uncertain whether to present information in text, a table, or a figure, it is preferred to use text for simple and concise qualitative information; figures should be used for complex relationships of qualitative information, or to present trends and patterns in quantitative information; and tables should be used for when displaying the exact values in quantitative information.

As with all submissions, refer to the specific journal guideline to which you are submitting for their preferred methods on images, figures, and tables. For example, some journals wish that authors submit all images digitally as an email attachment, while others might prefer hard copies. Some may ask that all images be uploaded onto a compact disc (a CD or DVD, while others will ask that they be submitted in multiple formats. (*See* **Chapter 2, Section 2.2.1.6** on preparation of tables and other graphic elements accompanying a scientific paper.)

Regarding the format of a table, many medical journals prefer that that they are submitted without lines or shading drawn in. Often, a journal will add their own horizontal lines and shading during the publication process. If footnotes are necessary in your figures, use a superscript numbering system rather than symbols (*. †, ‡, etc.) for greater clarity.

Illustrations and photographs are often used in medical writing to present images of clinical findings and procedures, or experimental results. They are used when a description may be difficult to convey in the text. Types of images may include radiographs, photomicrographs, and photographs of patients or biopsy specimens. Many journals require the author to submit the original gels along with their manuscript, in order to decrease the chances of authors submitting fraudulent or digitally manipulated images. Also, refer to the specific guidelines of the journal for their image resolution requirements. Images with low resolutions are often rejected due to print quality. Many journals also require that a signed form of consent be submitted along with any images of patients. If this is not possible, photos must be cropped to maintain the patients' anonymity.

13.2 Ethics and Validity

13.2.1 Overview

Editors and authors must ensure that studies involving human experimentation abide by ethical standards and protect an individual's right to privacy. Current rules and regulations for the protection of patients and research subjects have their basis in the Nuremberg Code, the World Medical Association's Declaration of Geneva, and the Declaration of Helsinki. The legal privacy doctrines of countries around the world, including the US Privacy Act in the United States, all protect an individual's right to privacy. (For and in-depth look on related ethics and legal concerns, refer to chapter 5 of the *AMA Manual of Style*.)

i. Institutional and legal guidelines for testing. In order to preserve the privacy and safety of human patients participating in experimental investigations, academic institutions and funding agencies require that subjects participate in the study voluntarily, and that they fully comprehend the nature of their participation, including all procedures and risks involved.

The experimental protocol for the study must be assessed and approved by a formal ethics committee. In addition, editors should request that authors explicitly inform readers in the "methods" section of their text that informed consent was obtained from all participating subjects, and that the research was approved by the appropriate ethics review board.

ii. Right to privacy and patient anonymity. To protect patients' rights to privacy and anonymity, superfluous identifying data (e.g. name, location, occupation) should be removed from published articles. In the case of studies of genetic pedigrees, information should be reported with patients' rights to privacy and anonymity under consideration. If obtaining written consent from individual members of a large pedigree is impossible, obtaining group consent may be considered.

iii. Relevance of identifying characteristics. Identifying characteristics (age, sex, race, etc.) that are not essential to the report should be omitted. Omitting identifying characteristics is preferable to altering them so as to avoid entering falsified data into medical literature.

13.2.2 Clinical Trials

In the United States, before drugs can be approved for human use, they are required to complete 1 to 5 years of preclinical studies followed by three phases of clinical trials. The International Committee of Medical Journal Editors requires the registration of clinical trials in an acceptable trial registry before patient enrollment begins.

i. Randomized double blind trials. A randomized double blind trial, also known as a randomized controlled trial (RCT), is optimal for studying and comparing two types of drugs or other forms of therapy. Randomization minimizes potential confounding factors through random distribution between the intervention group and the control group.

ii. Communicating protocol. Some journals require that authors complete the CONSORT checklist (Begg C, Cho M, Eastwood S, et al. "Improving the quality of reporting of randomized controlled trials: the CONSORT statement." *JAMA*, 1996; 276: 637-639). Completion of the checklist ensures the disclosure of information critical for interpreting the study.

All methods employed in a randomized double blind trial should be reported in detail so that the reader may judge the quality of the study and make comparisons to other studies. The CONSORT guidelines should be followed in order to ensure that the trial is reported completely.

13.2.3 Conflict of Interest

i. Authors. "Conflicts of interest" are competing interests that may result in biased decisions or reporting. Conflicts of interest can arise from personal or professional relationships, or the desire for professional or financial gain. Manuscript cover letters should fully disclose any potential conflicts of interest and their sources.

ii. Editors and reviewers. It is possible for conflicts of interest to arise during the review process, for example, if editors or reviewers have a positive or negative bias toward the author, or authors, of a manuscript. Editors should be aware of the potential for a conflict of interest for the reviewer, as objectivity could be compromised by friendship, collaboration, competition, or financial interest. Editors who cannot remain fully objective should step aside and allow associate or guest editors to assume editorial duties.

With proper justification, authors may request that editors exclude any candidate reviewers whom they feel may be biased toward the manuscript. Reviewers also have the responsibility to disclose any conflicts of interest influencing their review.

iii. Acknowledging sources of funding. A potentially serious conflict of interest is an author's financial expectations in regard to the results of his or her research. Many medical journals require that authors disclose their own financial interest in the manuscripts, as well as any financial support from grant or funding agencies. Once a manuscript has been accepted for publication, the editor (possibly after consulting the author) should decide what financial interests of the authors should be disclosed to the readers.

13.3 Usage

13.3.1 Human Physiology and Anatomy Nomenclature

13.3.1.1 Anatomic Descriptions

Official human anatomic terms are designated and approved by the International Anatomical Nomenclature Committee. Anatomic terms are treated as common nouns, and are neither capitalized nor italicized.

When referring to specific anatomic regions, do not state *right heart* or *left heart*; instead use descriptors like *left side of* or *part of* or *side of.* Also, do not use redundant terms like *upper arm* and *lower leg*, since the arm is part of the *upper extremities* and the leg is part of the *lower extremities*.

When possible, use specific anatomic descriptors.

proximal jejunum
femoral neck

13.3.1.2 Blood Group and Clotting Factor Designation Systems

i. Blood groups. Blood groups can be classified by a number of nomenclature systems, such as the ABO, Lewis, MNSs, and Rh systems.

The most commonly used is the ABO system, which indicates the red blood cell surface antigens present as either A, B, AB, or O.

The Rh system is another classification system often used in conjunction with the ABO system. The Rh system indicates the presence of the R antigen, with the two possible types being Rh-positive (Rh^+) or Rh-negative (Rh^-).

If an individual has red blood cells that contain both the B antigen and Rh antigen, their blood type can be written in one of the following ways:

B Rh-positive
B Rh^+
B^+

ii. Clotting factors. Human clotting factors are designated by the word *factor* followed by a roman numeral (I to XIII, but not VI), and also have descriptive names, as shown in **Table 13.1**. Preferred names are indicated by an asterisk.

Table 13.1 Clotting Factor Designations

Factor	Descriptive Name
factor I	fibrinogen*
factor II	prothrombin*
factor III	tissue factor
factor IV	calcium*
factor V*	Proaccelerin
factor VII*	Proconvertin
factor VIII*	antihemophilic factor (AHF)
factor IX*	plasma thromboplastin component (PTC)
factor X*	Stuart factor
factor XI*	plasma thromboplastin antecedent (PTA)
factor XII*	Hageman factor
factor XIII	fibrin stabilizing factor (FSF)

Activated clotting factors are indicated by a lowercase *a*, as in factor XIIIa.

13.3.1.3 Bone Abbreviations

A system for the standardization of nomenclature, symbols, and units in bone histomorphometry has been reported by the American Society for Bone and Mineral Research (ASBMR) Histomorphometry Nomenclature Committee.

13.3.1.4 Cardiology

i. Electrocardiograph nomenclature

The preferred abbreviation for "electrocardiogram" is ECG (not EKG). The main deflections in ECG tracings are designated P, Q, R, S, T, and U, and can be described in the following way:

Q wave	q wave
R wave	r wave
R′ wave	r′ wave

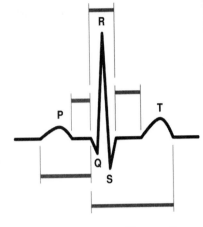

Fig. 13.1 Electrocardiograph Deflections

Segments between these deflection points can be designated as follows:

QS wave
ST segment
qRs complex

Deflection terms used as modifiers should be hyphenated.

P-wave irregularity
ST-segment duration

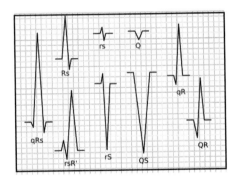

Fig. 13.2 Segments Between Electrocardiograph Deflections

ii. Echocardiography abbreviations. As with most abbreviations, terms relating to echocardiography should be written in their expanded form at first mention within the text, along with their abbreviation in parenthesis. Any following mentions may be written in the abbreviated form.

Table 13.2 Echocardiography Abbreviations

Term	Abbreviation
2-dimensional, 3-dimensional echocardiography	2DE, 3DE
aortic valve area	AVA
ejection fraction	EF
end diastole	d or ed
end systole	s or es
fractional shortening	FS
left, right ventricular internal dimension	LVID, RVID
mitral valve area	MVA
pressure half-time	PHT
posterior wall thickness	PW or PWT

iii. Heart disease classifications. Roman numerals are used to designate the different classes of heart diseases. Severity of the disease increases from lower to higher numbers and letters.

Forrester class I
Forrester class IV
Braunwald class IA
Braunwald class IIIB

iv. Cellular and molecular cardiology abbreviations. In most cases, cellular and molecular cardiology terms are written in full at their first mention and followed with their abbreviated form in parenthesis. However, some terms, such as NOS and P Cell, may written at first use without expansion. **Table 13.3** lists the cardiology terms that must be written out in their full form at their for use

Table 13.3 Cellular and Molecular Cardiology Abbreviations

Term	Abbreviation
apolipoprotein	apo
end diastole	d or ed
end systole	s or es
fractional shortening	FS
high-density lipoprotein	HDL
intermediate-density lipoprotein	IDL
low-density lipoprotein	LDL
very high-density lipoprotein	VHDL
high, intermediate, low, very high-density cholesterol	HDL-C, IDL-C, LDL-C, VHDL-C
high, intermediate, low, very high-density receptor	HDL-R, IDL-R, LDL-R, VHDL-R
troponin C, I, T	TnC, TnI, TnT
troponin C, I, T (cardiac form)	cTnC, cTnI, cTnT

13.3.1.5 Circulatory System Nomenclature

i. Main Circulation Symbols

Table 13.4 Main Circulation Symbols

Measurement	Symbol	SI Unit
area	A	mm^2
volume	V	L
flow rate	V/t	$L\ s^{-1}$
flow velocity	v	$cm\ s^{-1}$
flow acceleration	v/t	$cm\ s^{-2}$
pressure	P	kPa
resistance	R	$kPa\ L^{-1}$

ii. Modifying Circulation Symbols

Table 13.5 Modifying Circulation Symbols

Measurement	Symbol	Example
Arterial	*a*	P_a = Pressure, arterial
Venous	*ven*	P_{ven} = Pressure, venous
Capillary	*cap*	P_{cap} = Pressure, capillary
Systolic	*syst*	P_{syst} = Pressure, systolic
Diastolic	*diast*	P_{diast} = Pressure, diastolic

13.3.1.6 Genetics

See **Chapter 12 Life Science** for general conventions.

i. Pedigree symbols and diagrams. Symbols for pedigree diagrams are designated by the National Society of Genetic Counselors. Generations are numbered using roman numerals starting from the top of the diagram. Individuals in the same generation are assigned arabic numerals from left to right.

Males are indicated as squares, females as circles, and unknowns as diamonds. Parents are connected by a horizontal line. Offspring are connected to parents by a vertical line. In the case of multiple off-spring, this vertical line branches such that siblings are lined up horizontally in birth order from left to right.

Individuals expressing the trait in question (e.g. genetic disease) are indicated as solid symbols, while individuals without the trait are indicated as open symbols. A symbol with a slash through it indicates that the individual has died.

Symbols used for designating generations are listed in the table below. Subscripts indicate generation number.

Table 13.6 Symbols Used for Designating Generations

Gen. Symbol	Meaning	Example
$P_1, P_2, \ldots$	Parental generations	P_1 = Parents of F_1 P_2 = Grandparents of F_1
$F_1, F_2, \ldots$	Filial generations	F_1 = Children of P_1 F_2 = Children of F_1
$B_1, B_2, \ldots$	Backcross generations	B_1 = F_1 backcrossed with P_1 B_2 = B_1 backcrossed with P_1
$S_1, S_2, \ldots$	Self-fertilized generations (plants)	S_1 = Parental self-fertilization S_2 = S_1 self-fertilization

13.3.1.7 Hormone Abbreviations

Due to their long names, many hormones are referred to by their abbreviations.

gonadotropin-releasing hormone (GnRH)

A system for designating alternative names for hormones has been proposed by International Union of Pure and Applied Chemistry and the International Union of Biochemisty (IUPAC-IUB):

(1) Hormones from the adenohypophysis should end in *-tropin.*
(2) Hormones which are releasing factors from the hypothalamus should end in *-liberin.*
(3) Hormones which are release-inhibiting factors from the hypothalamus should end in *-statin.*

13.3.1.8 Immunologic Designations

i. Nomenclature for immunoglobulins. Immunoglobulin (Ig) molecules are divided into five classes or isotypes according to the type of heavy chain. The name of the heavy chain is the Greek letter corresponding to the name of the immunoglobulin.

Table 13.7 Nomenclature for Immunoglobulins

Immunoglobulin Name	Heavy Chain Name
IgG	γ
IgA	α
IgM	μ
IgD	δ
IgE	ε

Light chains of immunoglobulins are designated as κ or λ. Immunoglobulin heavy and light chains have variable and constant regions. Variable and constant regions are denoted by subscript letters and nonsubscript numbers, as shown in **Table 13.8.**

Table 13.8 Variables and Constant Regions in Immunglobulins

Immunoglobulin Region	Symbol	Examples			
Heavy Chain Variable Region	V_H	V_H1	V_H2	$V_\gamma 1$	$V_\varepsilon 2$
Heavy Chain Constant Region	C_H	C_H2	C_H3	$C_\gamma 2$	$C_\varepsilon 3$
Light Chain Variable Region	V_L	$V_\kappa 1$	$V_\kappa 2$	$V_\lambda 1$	$V_\lambda 2$
Light Chain Constant Region	C_L	$C_\kappa 2$	$C_\kappa 3$	$C_\lambda 2$	$C_\lambda 3$

ii. Nomenclature for lymphocytes. There are two main types of lymphocytes: T lymphocytes (which originate in the thymus) and B lymphocytes (which originate in the bone marrow). The terms *T lymphocyte, B lymphocyte, T cell,* and *B cell* can be used without expansion.

T lymphocytes are further designated:

helper T cells (T_H cells)
cytotoxic T cells (T_C cells)

Subtypes of helper T cells are further specified using non-subscript numerals.

T_H1 $\qquad$ T_H2 $\qquad$ T_H17

Clusters of differentiation (CD) cell markers define lymphocyte subsets. CD markers are specified either by a number, or by a number and a letter.

CD3 $\qquad$ CD38 $\qquad$ CD49b

The presence or absence of lymphocyte cell surface markers can be indicated by superscript plus and minus signs, respectively.

$CD5^+$ $\qquad$ (T cell expressing CD5)
$CD5^+CD6^-$ $\qquad$ (T cell expressing CD5 but not CD6)
$CD5^-CD6^-$ $\qquad$ (T cell expressing neither CD5 nor CD6)

iii. Nomenclature for cytokines. The largest group of cytokines, the interleukins (ILs), are numbered 1 through 16.

IL-2   IL-10

Interleukin receptors are indicated by the interleukin name followed by a capital R.

IL-2R   IL-10R

Interleukin receptor agonists are indicated by the interleukin receptor name followed by a capital A.

IL-6RA

Subunits of interleukin receptors are indicated by the interleukin receptor name followed by a Greek letter.

IL-2Rα
IL-10Rβ

Abbreviations for other cytokines are shown in the table below.

Table 13.9 Cytokine Abbreviations

Cytokine Name	Abbreviation
endothelial growth factor	EGF
interferon	IFN
leukemia inhibitory factor	LIF
lymphotoxin α	LTα
transforming growth factor β	TGFβ
tumor necrosis factor α	TNF-α
tumor necrosis factor β	TNF-β

13.3.1.9 Neurology

i. Cranial nerves. Cranial nerves may be referred to by name or by roman numeral. The following table lists the cranial nerve roman numeral designations and their corresponding English names.

Table 13.10 Cranial Nerve Designations

Cranial Nerve	English Name
I	olfactory
II	optic
III	oculomotor
IV	trochlear
V	trigeminal
VI	abducens
VII	facial
VIII	vestibulocochlear
IX	glossopharyngeal
X	vagus
XI	accessory
XII	hypoglossal

ii. Electroencephalographic terms. Guidelines for electroencephalographic (EEG) designations are established by the American Clinical Neurophysiology Society and the International Federation of Clinical Neurophysiology (IFCN). A comprehensive source of EEG terms is the IFCN.

The International 10-20 System is generally used for the placement of electrodes. Conventionally, letters refer to anatomic areas of the skull; odd numbers indicate electrodes placed on the left side of the skull; even numbers indicate those placed on the right side; and the letter z refers to midline electrodes. *See* **Table 13.11** for examples.

It should be noted that the extended and the neonatal electrode systems have differing electrode placements and designations.

Table 13.11 EEG Designations

Location	EEG Term
earlobe	A1, A2
central	Cz, C3, C4
lateral frontal	F7, F8
prefrontal	Fp1, Fp2
superior frontal	Fz, F3, F4
occipital	O1, O2
parietal	Pz, P3, P4
midtemporal	T3, T4
posterior temporal	T5, T6

iii. Molecular neuroscience abbreviations. As with most abbreviations, molecular neuroscience terms should be written in their expanded form at first mention within the text, along with their abbreviation in parenthesis. Any following mentions may be written in the abbreviated form.

Table 13.12 Molecular Neuroscience Abbreviations

Abbreviation	Molecular Neuroscience Term
α_{1A}, α_{1B}, α_{1D}, α_{2A}, α_{2B}, α_{2C}	α-adrenergic receptor subtypes
Ach	acetylcholine
AchE	acetylcholinesterase
AMPA	α-amino-3-hydroxy-5-methyl-4-isoxazole propionic acid class of glutamate receptor
β_1, β_2, β_3	β-adrenergic receptor subtypes
BDNF	brain-derived neurotrophic factor
CCR3, CCR5, CXCR4	chemokine receptors
CNTF	ciliary neurotrophic factor
D_1, D_2, D_3, D_4, D_5	dopamine receptors
DAT	dopamine transporters
GABA	γ-aminobutyric acid
$GABA_A$, $GABA_B$	GABA receptor classes
GAT-1, GAT-2, GAT-3, GAT-4	GABA family transporters
H_1, H_2, H_3	histamine receptors
IP_3	inostiol triphosphate
M_1, M_2, M_3, M_4, M_5	muscarinic receptors
MAO	monoamine oxidase
nAChR	nicotinic acetylcholine receptor

Table 13.12, *continued*

NGF	nerve growth factor
NMDA	*N*-methyl-D-aspartate class of glutamate receptor
NSF	*N*-ethylmaleamide sensitive factor
SERT	serotonin GABA family transporter
SNAP	soluble NSF attachment protein
SNARE	SNAP receptor

13.3.1.10 Obstetric designations

i. The GPA system of obstetric history. The GPA system is designed for designating obstetric history in shorthand. The letter *G* (for *gravida*) indicates the number of pregnancies, the letter *P* (for *para*) indicates the number of viable births, and the letter *A* (for *aborta*) indicates the number of spontaneous or induced abortions. For example, indicating four pregnancies, two of which were viable births, and two of which were abortions, could be written in either of the following ways:

> gravida 4, para 2, aborta 2

or

> G4, P2, A2

ii. The TPAL system of obstetric history. The TPAL system for designating obstetric history designates the number of term infants (T), premature infants (P), abortions (A), and living children (L). For example, indicating one term infant, one premature infant, one abortion, and two living children could be written as follows:

> TPAL: 1-1-1-2

or

> 1-1-1-2

iii. Apgar score for newborns. The Apgar score for assessing a newborn's well-being is the sum of five rated conditions, each rated on a scale of 0 to 2; the Apgar score is thus reported as a number from 0 to 10. Usually the score consists of two numbers representing assessment at different time intervals after birth. When reporting the Apgar score, these time intervals should be specified.

The newborn received an Apgar score of 8/10 at 1 and 10 minutes.

13.3.1.11 Ophthalmology Abbreviations

Table 13.13 Ophthalmology Abbreviations

Abbreviation	Opthalmology Term
D	diopter
Δ	prism diopter
DD	disc diameters
DA	disc areas
ERG	electroretinogram
a_1, a_2, b	waves of the ERG
PERG	pattern electroretinogram
a_{pt}, b_{pt}, c_{pt}	waves of the PERG
OD	oculus dexter (right eye)
OS	oculus sinister (left eye)
OU	oculus uterque (each eye)

13.2.1.12 Designations for Psychiatric Syndromes

Designations for psychiatric syndromes may be found in the *Diagnostic and Statistical Manual of Mental Disorders (DSM)*.

According to the DSM, psychiatric syndromes can be diagnosed on the following axes:

Axis I	Clinical disorders
	Other conditions that may be a focus of clinical attention
Axis II	Personality disorders
	Mental retardation
Axis III	General medical conditions
Axis IV	Psychosocial and environmental problems
Axis V	Global assessment of functioning

For information on referring to people suffering from psychiatric syndromes, *see* **Part I, Chapter 1, Section 1.8, iv**.

13.3.1.13 Pulmonary-Respiratory System Nomenclature

i. Common Respiratory Abbreviations

Table 13.14 Common Respiratory Abbreviations

Abbreviation	Respiratory Term
ERV	expiratory reserve volume
FIVC	forced inspiratory vital capacity
FVC	forced vital capacity
IRV	inspiratory reserve volume
IVC	inspiratory vital capacity
RV	residual volume
TLC	total lung capacity
VC	vital capacity

ii. Partial pressures. Symbols for the partial pressures of common gases may be used at first mention without being defined. The specific gas is designated in the subscript.

P_{CO} (partial pressure of carbon monoxide)

13.3.1.14 Radiology Terms

The AMA lists the following common radiology terms and their abbreviations:

Table 13.15 Radiology Terms

Abbreviation	Radiology Term
b value	gradient strength (s/mm^2)
MRI	magnetic resonance imaging
T1, T1ρ, T2, T2*	types of MRI relaxation time
TE	echo time
TR	repetition time

13.3.1.15 Thermal Regulation Symbols and Modifiers

Table 13.16 Thermal Regulation Symbols and Modifiers

Measurement	Symbol	SI Unit
absolute humidity	γ	$kg\ m^{-3}$
ambient temperature	T_a	°C
basal metabolic rate	BMR	W
Metabolic heat production	H	$W\ m^{-2}$
total body area	A_b	m^2
water vapor pressure	P_w	Pa

13.3.2 Diseases

13.3.2.1 Style Conventions for Diseases

Sources for standard nomenclature for animal and human diseases can be found in the Unified Medical Language System (UMLS) metathesaurus and the National Library of Medicine's Medical Subject Headings thesaurus.

13.3.2.2 Preferences for Eponymic Names for Diseases

In general, descriptive disease names are preferred over eponymic names. However, the author may use the eponymous term if he or she believes that readers will be more familiar with it. It is helpful to give both the eponymous and noneponymous terms at first mention, with one term in parenthesis following the other. The eponym should be capitalized, while the noun or adjective accompanying it should be lowercased.

Hürthle cells
Kveim test

When eponymic names are mentioned, the nonpossessive form should be used rather than the possessive form (in spite of common usage).

Correct: Hodgkin lymphoma
Incorrect: Hodgkin's lymphoma

13.3.2.3 Cancer Nomenclature Systems

i. Cancer stages. Cancer stages are designated using roman numerals (I-IV), with higher numbers indicating increasing severity. Cancer stages may be further subdivided using letter suffixes.

stage IA
stage IE
stage IVB

ii. The TNM staging system. The Tumor, Node, Metastasis (TNM) system is internationally recognized as the standardized system used for the classification of human tumors and the staging of cancer. The TNM classification system has been set forth by the Joint Committee on Cancer (AJCC) and the Union Internationale Contre le Cancer (UICC).

Table 13.17 TNM Staging Sequence

	Classification	Meaning
Tumor (T)	TX	Primary tumor cannot be assessed
	T0	No evidence of a primary tumor
	Tis	*In situ* carcinoma
	T1, T2, T3, T4	increasing size, penetrance of the tumor
Node (N)	NX	regional lymph nodes cannot be assessed
	N0	no regional lymph node metastasis
	N1, N2, N3	increasing lymph node metastasis
Metastasis (M)	MX	metastasis cannot be assessed
	M0	no metastasis
	M1	distant metastasis
Optional Descriptors	C1, C2, C3, C4, C5	certainty factor (C-factor)
	GX, G1, G2, G3, G4	histopathologic grading
	LX, L0. L1	lymphatic vessel invasion
	Rx, R0, R1, R2	residual tumor
	SX, S0, S1, S2	scleral invasion or serum markers
	VX, V0, V1, V2	venous invasion

13.3.2.4 Abbreviations for Protein Structure Abnormalities: Amyloidosis

Abbreviations for different systemic amyloidosis types consist of a two-letter code. The first letter *A* stands for *amyloid*, while the second letter stands for the protein that accumulates in that particular type of amyloidosis.

AL = primary amyloidosis
AA = secondary amyloidosis
ATTR = hereditary amyloidosis

13.3.2.5 Nomenclature for Bacterial, Viral, and Fungal Infections

Standardized nomenclature for animal parasitic diseases has been proposed by the World Association for the Advancement of Veterinary Parasitology. Nomenclature for fungal diseases has been proposed by the International Society for Human and Animal Mycology.

13.3.3 Drugs and Pharmaceutical Agents

13.3.3.1 Naming Drugs

i. Nonproprietary names. The nonproprietary name, also known as the generic name, is the official name of drug. The drug manufacturer is required by law to use the nonproprietary drug name on labels and advertising. In scientific literature, the nonproprietary name should be used whenever possible. When names other than the nonproprietary name are mentioned for any reason, the nonproprietary name should be listed first, followed by the alternative name in parenthesis.

Approval for nonproprietary names in the United States is obtained from the US Adopted Names (USAN) Council. The USAN follows standard guidelines for creating new nonproprietary names, and uses suffixes for new names that indicate a relationship to older drugs. For more drug suffixes, *see* **Table 13.18 Common Drug Suffixes**.

amikacin	miconazole
kanamycin	fenticonazole
neomycin	isoconazole
streptomycin	terconazole

The World Health Organization (WHO) International Nonproprietary Name (INN) Committee oversees individual drug nomenclature councils to maintain consistent nomenclature across different countries.

ii. Proprietary names. The proprietary name is the brand name of the drug selected by the drug's manufacturer. In scientific publications, the nonproprietary name is preferred over the proprietary name. If both the nonproprietary and proprietary names are mentioned, the nonproprietary name should be mentioned first, followed by the proprietary name, capitalized and in parenthesis. The proprietary and nonproprietary names of some common drugs are listed in **Table 13.19 Common Drugs and Pharmaceutical Agents**.

iii. Chemical names. The chemical name of a drug describes its chemical composition and structure. In scientific publications, the generic (nonproprietary) name is preferred over the chemical name.

N-(4-hydroxyphenyl)acetamide   acetaminophen

iv. Radiopharmaceuticals. Nonproprietary names for radiopharmaceutical drugs consist of the drug name followed by the element symbol and isotope number in the same type.

glycocholic acid C 14

Proprietary names for radiopharmaceutical drugs consist of the drug name (which may or may not include the element symbol), followed by the hyphenated isotope number.

Glofil-125

v. Drugs with inactive components. Nonproprietary names for drugs with an inactive component in addition to an active component have two parts indicating both components of the drug.

sildenafil citrate

vi. Drug names that include a percentage. Drug names may specify the percentage of active components in the drug by listing the percentage after the drug name.

hydrocortisone, 1%

13.3.3.2 Common Drug Suffixes

Table 13.18 Common Drug Suffixes

Suffix	Definition
-azine	antipsychotics, neuroleptics
-azole	antifungals
-azepam	antianxiety drugs
-ane	volatile general anesthetics
-barbital/-bital	barbiturate sedative hypnotics
-caine	local anesthetics
-cycline	antibiotic protein synthesis inhibitors
-gramostim	granulocyte macrophage colony-stimulating factors
-grastim	granulocyte colony-stimulating factors
-leukin	interleukin 2 type substances
-mab	monoclonal antibodies
-mostim	macrophage colony-stimulating factors
-mycin	aminoglycoside antibiotics
-navir	protease inhibitors
-olol	beta blockers
-operidol	neuroleptics
-oxin	inotropic agents
-phylline	methylxanthine
-pril	ACE inhibitors
-relin	hypothalamic peptide hormones that stimulate release of pituitary hormones
-relix	hypothalamic peptide hormones that inhibit release of pituitary hormones
-statin	HMG-CoA reductase inhibitors
-terol	beta2 agonists
-tidine	h2 antagonists
-tropin	pituitary hormones
-zosin	alpha1 blockers
-triptyline	tricyclic antidepressants
-ipramine	tricyclic antidepressants

13.3.3.3 Common Drugs and Pharmaceutical Agents

Table 13.19 Common Drugs and Pharmaceutical Agents

Nonproprietary/Generic Name	Proprietary/Brand Name
amphetamine	Adderall
fexofenadine hydrochloride	Allegra
zolpidem tartrate	Ambien
amoxicillin	Amoxil
acetylsalicylic acid	Aspirin
celecoxib	Celebrex
tadalafil	Cialis
loratadine	Claritin
duloxetine hydrochloride	Cymbalta
medroxyprogesterone acetate	Depo Provera
phenytoin sodium	Dilantin
metformin hydrochloride	Glucophage
human insulin	Humulin N
lamotrigine	Lamictal
escitalopram oxalate	Lexapro
atorvastatin calcium	Lipitor
eszopiclone	Lunesta
ibuprofen	Motrin
acetylcysteine	Mucomyst
naproxen	Naprosyn
amlodipine besylate	Norvasc
oxycodone HCl	Oxycontin
paroxetine hydrochloride	Paxil
famotide	Pepcid
progesterone	Prometrium
modafinil	Provigil
temazepam	Restoril
methylphenidate hydrochloride	Ritalin
methocarbamol	Robaxin
montelukast sodium	Singulair
acetaminophen	Tylenol
diazepam	Valium
valacyclovir hydrochloride	Valtrex
sildenafil citrate	Viagra
CIV alprazolam	Xanax
ranitidine hydrochloride	Zantac
azithromycin	Zithromax
sertaline hydrochloride	Zoloft
cetirizine hydrochloride	Zyrtec

Source: http://www.rxlist.com/script/main/hp.asp

Table 13.20 Common Vitamins

Familiar Term	Drug Name
vitamin A	beta carotene
vitamin B_1	thiamine hydrochloride
vitamin B_1 mononitrate	thiamine mononitrate
vitamin B_2	riboflavin
vitamin B_6	pyridoxine hydrochloride
vitamin B_8	adenosine phosphate
vitamin B_{12}	cyanobalamin
vitamin C	ascorbic acid
vitamin D	cholecalciferol
vitamin D_1	dihydrotachysterol
vitamin D_2	ergocalciferol
vitamin E	vitamin E
vitamin G	riboflavin
vitamin K_1	phytonadione
Folate	folic acid
Niacin	nicotinic acid
Niacinamide	nicotinamide

13.3.3.4 Pharmacokinetic abbreviations and modifiers

Table 13.21 Pharmacokinetic Abbreviations

Abbreviation	Definition
A^{SS}	amount of drug in the body at steady state
C	plasma drug concentration
CL	total plasma clearance of drug
C^{SS}	steady state plasma drug concentration during infusion at a constant rate
f_a	fraction of drug absorbed
f_u	fraction of drug unbound in plasma
R_0	constant infusion rate
t_{max}	time after administration of drug to reach peak concentration
$t_{1/2}$	elimination half-life
$t_{1/2a}$	absorption half-life

Table 13.22 Pharmacokinetic Modifiers

Modifiers (Sites of Measurement)	Abbreviation
blood	b
plasma	p
serum	s
tissue	t
urine	ur
Modifiers (Routes of Administration)	**Abbreviation**
excreted into urine	e
hepatic	H
metabolized	m
nonrenal	NR
renal	R
Modifiers (Routes of Elimination)	**Abbreviation**
intramuscular	im
intraperitoneal	ip
intravenous	iv
oral	o/po
rectal	pr
subcutaneous	sc
topical	top

13.3.4 Units of Measurement

13.3.4.1 SI Units Specific to Medicine

Most scientific journals prefer to use SI units as the primary method for reporting scientific measurements. Unfamiliarity with SI units in the United States has led to dual reporting of clinical laboratory values in SI and conventional units. In the case of dual reporting, the SI unit is mentioned first, followed by the conventional unit in parentheses.

Table 13.23 SI Units Specific to Medicine

Name	Abbreviation	Units	Definition
Molar	M	mol/L	one mole of solute in one liter of solution
Joule	J	J	energy content (i.e. food energy content)
cell count	none	cells $\times 10^9$/L	cell count per liter
Gray	Gy	J/kg ($m^2 \cdot s^{-2}$)	absorbed dose of radiation
radiation exposure	none	C/kg	coulomb per kilogram

13.3.4.2 Common Exceptions to SI Units

i. Temperature. The SI unit for measuring temperature is the Kelvin (K), but the more commonly used unit for measuring temperature is the Celsius scale (°C).

ii. Pressure. While the recommended SI unit for measuring pressure is the Pascal (P), physiologic pressure measurements (e.g. blood pressure) are more commonly taken in millimeters of mercury (mm Hg).

iii. Enzyme activity. The recommended SI unit for measuring enzyme activity is the katal, or the amount of enzyme generating 1 mole of product per second. However, enzymatic activity is more often reported as the amount of enzyme generating 1 micromole of product per minute.

iv. Drug concentration. The recommended SI units for measuring drug concentrations are millimoles per liter (mmol/L) or nanomoles per liter (nmol/L). However, in the United States, drug concentrations are commonly expressed as milligrams per liter (mg/L) or nanograms per milliliter (ng/mL).

13.4 Standard Medical Abbreviations

Table 13.24 Standard Medical Abbreviations

Name	Abbreviation
adverse drug reaction	ADR
autonomic nervous system	ANS
atrioventricular	AV
body mass index	BMI
basal metabolic rate	BMR
blood pressure	BP
beats per minute	BPM
Body weight	BW
biopsy	BX
cardiac index	CI
central nervous system	CNS
cardiac output	CO
chest X-ray	CXR
electrocardiogram	ECG
electromyogram	EMG
gastrointestinal	GI
heart rate	HR
intensive care unit	ICU
intramuscular	IM
intravenous	IV
magnetic resonance imaging	MRI
nasotracheal	NT
obstetrics	OB
operating room	OR
red blood cell	RBC
recommended daily allowance	RDA

Source: http://www.globalrph.com/abbrev.htm

13.5 Lists of Medicine Tables and Appended Material

13.5.1 Tables in This Chapter

13.5.2 Contents of Appendix H in Part III

H1 Academic Degrees, Certifications, and Honors

H2 Clinical, Technical, and Other Common Terms

H3 Selected Laboratory Tests, With Reference Ranges and Conversion Factors

H4 Glossary of Medical Terms

H5 Agencies and Organizations

H6 The National Library of Medicine's (NLM's) Table of Accepted Abbreviations

H7 Medical Journals and Their Abbreviations

PART III.

Appendices

Appendix A.
Tables and Conventions for Mathematics

Contents

Appendix A. Tables and Conventions for Mathematics

A1   Common Mathematical Functions and Their Abbreviations

Function	Abbreviation	Definition
cosine	cos	—
sine	sin	—
tangent	tan	—
secant	sec	—
cosecant	csc	—
cotangent	cot	—
inverse cosine	arccos	—
inverse sine	arcsin	—
inverse tangent	arctan	—
inverse secant	arcsec	—
inverse cosecant	arccsc	—
inverse cotangent	arccot	—
the exponential function	exp[argument] or $e^{argument}$	—
hyperbolic cosine	cosh	—
hyperbolic sine	sinh	—
hyperbolic tangent	tanh	—
natural logarithm	ln	$\ln x = \int_1^x \dfrac{du}{u}$
logarithm (base b)	Log$_b$	—

A1 Common Mathematical Functions and Their Abbreviations, continued

Function	Abbreviation	Definition
Legendre functions	first kind: $P_l(x)$ second kind: $Q_l(x)$	$$\frac{d^2 P_l}{dx^2} + \cot x \frac{dP_l}{dx} + l(l+1)P_l = 0$$ $$\frac{d^2 Q_l}{dx^2} + \cot x \frac{dQ_l}{dx} + l(l+1)Q_l = 0$$ If l is equal to an integer, then $P_l(x)$ becomes a Legendre polynomial.
associated Legendre functions	first kind: $P_l^m(x)$ second kind: $Q_l^m(x)$	$$\left(1-x^2\right)\frac{d^2 P_l^m}{dx^2} - 2x\frac{dP_l^m}{dx} + \left[l(l+1) - \frac{m^2}{1-x^2}\right]P_l^m$$ $$\left(1-x^2\right)\frac{d^2 Q_l^m}{dx^2} - 2x\frac{dQ_l^m}{dx} + \left[l(l+1) - \frac{m^2}{1-x^2}\right]Q_l^m$$ If l is equal to an integer, then $P_l^m(x)$ becomes an associated Legendre polynomial.
Bessel functions	first kind: $J_n(x)$ second kind: $Y_n(x)$	$$x^2\frac{d^2 J_n}{dx^2} + x\frac{dJ_n}{dx} + \left(x^2 - n^2\right)J_n = 0$$ $$x^2\frac{d^2 Y_n}{dx^2} + x\frac{dY_n}{dx} + \left(x^2 - n^2\right)Y_n = 0$$
modified Bessel functions	first kind: $I_h(x)$ second kind: $K_h(x)$	$$x^2\frac{d^2 I_h}{dx^2} + x\frac{dI_h}{dx} - \left(x^2 + h^2\right)I_h = 0$$ $$x^2\frac{d^2 K_h}{dx^2} + x\frac{dK_h}{dx} - \left(x^2 + h^2\right)K_h = 0$$
error function	$erf(x)$	$$erf(x) = \frac{2}{\sqrt{\pi}}\int_0^x e^{-u^2}\,du$$
hermite polynomial as used in physics	$H_n(x)$	$$\frac{d^2 H_n}{dx^2} - 2x\frac{dH_n}{dx} + 2nH_n = 0$$
inverse of a general function f	f^{-1}	

A2 Symbols Commonly Used in Mathematics

A2.1 Geometry and Trigonometry

Name	Symbol	Definition
two-dimensional Cartesian coordinates	(x,y)	the lengths of the projections of the position vector of a point along two mutually exclusive directions.
polar coordinates	(ρ,θ)	ρ is the distance from the origin and θ is the angle between the x axis and the line connecting the origin to the point.
two-dimensional elliptic coordinates	(μ,v)	$x = a \cosh \mu \cos v$ $y = a \sinh \mu \sin v$ lines of constant μ correspond to ellipses while lines of constant v correspond to hyperbole.
three-dimensional Cartesian coordinates	(x,y,z)	the lengths of the projections of the position vector of a point along three mutually exclusive directions.
cylindrical coordinates	(ρ,θ,z)	ρ is the distance of a point from the z axis while θ is the angle between the x axis and the projection of the position vector onto the xy plane.
spherical coordinates	(r, θ, φ)	r is the distance between the point and the origin, θ is the angle between the negative z axis and the position vector and φ is the angle between the x axis and the projection of the position vector onto the xy plane.
angle	$\angle ABC$	the angle created by the vertex formed by the line segments AB and BC.
pi	π	3.14...
unit vector|	$\hat{k}$	a unit vector is found by dividing a vector by its magnitude.
orthogonal or perpendicular	$\perp$	
parallel	$\parallel$	

A2 Symbols Commonly Used in Mathematics, *continued*

A2.2 Algebra and Number Theory

Name	Symbol
equal to	$=$
unequal to	$\neq$
identically equal to	$\equiv$
approximately equal to	$\cong$ or $\approx$
roughly equal to	$\sim$
greater than	$>$
less than	$<$
much greater than	$>>$
much less than	$<<$
greater than or equal to	$\geq$
less than or equal to	$\leq$
positive or negative	$\pm$
a function f having an argument of x	$f(x)$
the imaginary number i	$i \quad (= \sqrt{-1}\,)$
the real part of a complex number z = x + iy	$\mathrm{Re}(z)$
the imaginary part of a complex number z = x + iy	$\mathrm{Im}(z)$

A2.3 Logic and Set Theory

Name	Symbol
set delimiters	$\{\}$
union	$\cup$
intersection	$\cap$
subset	$\subseteq$
proper subset	$\subset$
null subset	ϕ
member of	$\in$
not a member of	$\notin$
and	$\wedge$
or	$\vee$
exclusive or	$\oplus$
not	$\neg$
there exists	$\exists$
for all	$\forall$

A3 Operators Commonly Used in Mathematics

A3.1 Calculus and Analysis

Name	Operator
limit of a function as its argument approaches the value b.	$\lim\limits_{x \to b} f(x)$
derivative of f(x) with respect to x	$\dfrac{df}{dx}$ or $f'(x)$
derivative of f(x) with respect to x evaluated at x = a.	$\dfrac{df}{dx}\bigg|_{a}$
partial derivative of a function with respect to x, keeping the set of variables $\{y_i\}$	$\left(\dfrac{\partial f}{\partial x}\right)_{\{y_i\}}$
mixed partial derivative of a function with respect to $\{x_i\}$	$\left(\dfrac{\partial^P f}{\partial^{P_1} x_1 \partial^{P_2} x_2 ... \partial^{P_n} x_n}\right)$ where $\sum\limits_{i=1}^{n} P_i$
indefinite integral	$\int f(x)dx$
definite integral from a to b	$\int_a^b f(x)dx$
definite multiple integral	$\int_h^k \int_c^g ... \int_a^b f(x)dx$
integral over an open manifold M	$\int_M fdM$
integral over a closed manifold M	$\oint_M fdM$
flux of the vector field **A** through the open surface S	$\int_S A \bullet dS$
flux of the vector field **A** through the closed surface S	$\oint_S A \bullet dS$
circulation of the vector field **A** around the closed curve **C**.	$\oint_C A \bullet dC$
gradient of the function f	∇f (The gradient of a function is a vector field)
curl of a vector field **A**	$\nabla x A$ (The curl of a vector field is a vector field)
divergence of a vector field **A**	$\nabla \bullet A$ (The divergence of a vector field is a function)

A3 Operators Commonly Used in Mathematics, *continued*

A3.2 Matrices and Vectors

Name	Operator
vector, or cross, product	$\mathbf{A} \times \mathbf{B}$
inner, or dot, product	$\mathbf{A} \cdot \mathbf{B}$
magnitude of the vector $\mathbf{A}$	$|\mathbf{A}|$
inverse of the matrix M	M^{-1}
transpose of the matrix M	M^T
adjoint of the matrix M	$M^\dagger$
determinant of the matrix M	det M or $|M|$

A3.3 Algebra and Number Theory

Algebra and Number Theory	
addition	$+$
subtraction	$-$
division	/ or $\div$
multiplication	$*$
factorial	!
absolute value of x	$|x|$
reciprocal of the number $x \neq 0$	$1/x$
addition or subtraction	$\pm$
the variable x raised to the nth power.	x^n
exponential function	$e^x \; \exp[x]$
logarithm of x to the base a	$\mathrm{Log}_a \, x$
logarithm of x to the base 10	Log x
natural logarithm	ln x

A4 Operators Using Special Bracket Notation

Operator	Notation	Comments
coordinates of a point in a two-dimensional space	(u_1,u_2)	The values of the coordinates depend upon the system of the coordinates used.
coordinates of a point in a three-dimensional space	(u_1,u_2,u_3)	The values of the coordinates depend upon the system of the coordinates used.
components of a two-dimensional vector	(V_1,V_2)	The values of the components depend upon the system of coordinates used.
components of a three-dimensional vector	(V_1,V_2,V_3)	The values of the components depend upon the system of coordinates used.
commutator of the operators M and N	$[M,N]$	$[M,N]$ = MN - NM
anti-commutator of the operators M and N	$\{M,N\}$	$\{M,N\}$ = MN + NM
Poisson bracket of the functions f and g	$\{f,g\}$	$$\{f,g\} = \sum_{i=1}^{N}\left[\frac{\partial f}{\partial q_i}\frac{\partial g}{\partial p_i} - \frac{\partial f}{\partial p_i}\frac{\partial g}{\partial q_i}\right]$$ where N is equal to the number of particles while $\{q_i,p_i\}$. Poisson brackets are commonly used in classical mechanics.
bra-ket notation	bra = $\langle\mu\lvert$ ket = $\lvert\varphi\rangle$ bra-ket = $\langle\mu\lvert\varphi\rangle$	Bra-ket notation is commonly used in quantum mechanics.
miller indices	(hlk)	The indices h, l and k constitutes a vector normal to a set of parallel planes. Often used in solid-state physics.
mean or expectation value of a statistical variable x	$\langle x\rangle$	The mean of a statistical variable x can also be written as $\bar{x}$.

A5 Greek Letters Used in Mathematics

Aα (Alpha)

α represents:
- the first angle in a triangle, opposite the side A
- one root of a quadratic equation, where β represents the other
- the statistical significance of a result
- the false positive rate in statistics

Bβ (Beta)

B represents the Beta function

β represents:
- the second angle in a triangle, opposite the side B
- one root of a quadratic equation, where α represents the other
- the false negative rate in statistics
- the beta coefficient, the non-diversifiable risk, of an asset in mathematical finance

Γγ (Gamma)

Γ represents:
- the gamma function, a generalization of the factorial
- the upper incomplete gamma function
- the modular group, the group of fractional linear transformations
- the gamma distribution, a continuous probability distribution defined using the gamma function

γ represents:
- the lower incomplete gamma function
- the third angle in a triangle, opposite the side C
- the Euler-Mascheroni constant in mathematics.
- second-order sensitivity to price in mathematical finance

Δδ (Delta)

Δ represents:
- a finite difference
- a difference operator
- the Laplace operator

δ represents:
- a variation in the calculus of variations
- the Kronecker delta function
- the Dirac delta function
- the maximum degree of any vertex in a given graph
- sensitivity to price in mathematical finance

Eε (Epsilon)

ε represents:
- a small positive quantity
- a random error in regression analysis

- in set theory, the limit ordinal of the sequence
$$\omega, \omega^{\omega}, \omega^{\omega^{\omega}}, \ldots$$
- in computer science, the empty string
- the Levi-Civita symbol
- set membership symbol $\in$ is based on ε

Ϝ ϝ *(Digamma)*

□ is sometimes used to represent the Digamma function, though the Latin letter F (which is nearly identical) is normally substituted.

Ζ ζ *(Zeta)*

ζ represents:
- the Riemann zeta function and other zeta functions

Η η *(Eta)*

η represents:
- the partial regression coefficient in statistics.
- the Minkowski metric tensor in relativity

Θ θ *(Theta)*

Θ represents:
- an asymptotically tight bound related to Big O notation.
- sensitivity to the passage of time in mathematical finance

θ represents:
- a plane angle in geometry

- the angle to the x axis in the xy-plane in spherical or cylindrical coordinates

Ι ι *(Iota)*

ι represents the index generator function in APL (in the form ⊠)

Κ κ *(Kappa)*

κ represents:
- the kappa curve
- the condition number of a matrix in numerical analysis
- the connectivity of a graph in graph theory
- curvature

Λ λ *(Lambda)*

Λ represents
- the set of logical axioms in the axiomatic method of logical deduction in first-order logic
- a diagonal matrix of eigenvalues in linear algebra

λ represents
- a unit of measure of volume equal to one microlitre (1 μL) or one cubic millimetre (1 mm³)
- function expressions in the lambda calculus
- a general eigenvalue in linear algebra
- the average life time or rate parameter in an exponential distribution
- the expected number of occurrences in a Poisson distribution in probability

- the lagrange multiplier in the mathematical optimization method, known as the shadow price in economics
- linear density

Mμ (Mu)

μ represents:
- the Möbius function in number theory
- the ring representation of a representation module
- the population mean or expected value in probability and statistics
- a measure in measure theory
- micro, an SI prefix denoting 10^{-6} (one millionth)
- the service rate in queuing theory

Nν (Nu)

Ξξ (Xi)

Ξ represents the grand canonical ensemble found in statistical mechanics

ξ represents a random variable

Oo (Omicron)

O represents big O notation (may be represented by an uppercase Latin O as well)

Ππ (Pi)

Π represents:
- the product operator in mathematics
- a plane
- Archimedes' constant, 3.14159..., the ratio of a circle's circumference to its diameter

π represents:
- the prime-counting function
- profit in game theory
- the state distribution of a Markov chain

Pρ (Rho)

ρ represents:
- the radius in a polar, cylindrical, or spherical coordinate system
- the correlation coefficient in statistics
- the sensitivity to interest rate in mathematical finance
- density (mass per unit volume)
- resistivity

Σσ (Sigma)

Σ represents:
- the summation operator
- the set of terminal symbols in a formal grammar

σ represents:
- the divisor function in number theory
- the sign of a permutation in the theory of finite groups

- the population standard deviation or spread in probability and statistics
- the selection operator in relational algebra

Tτ (Tau)

τ (lower-case) represents:
- an interval of time
- a mean lifetime
- a correlation coefficient
- the Golden ratio 1.618... (although φ (phi) is more common)
- Ramanujan's tau function in number theory
- the intertwining operator in representation theory
- the number of divisors of highly composite numbers

Yυ (Upsilon)

Φφ (Phi)

Φ represents the cumulative distribution function of the normal distribution in statistics

φ represents:
- the golden ratio 1.6180339887...
- Euler's totient function in number theory
- a holomorphic map on an analytic space
- the argument of a complex number
- the value of a plane angle
- the angle to the z axis in spherical coordinates
- the probability density function of the normal distribution in statistics

Xχ (Chi)

χ represents:
- the chi distribution in statistics (X^2 is the more frequently encountered chi-square distribution)
- the chromatic number of a graph in graph theory
- the Euler characteristic in algebraic topology
- a variable in algebraic equations

Ψψ (Psi)

Ωω (Omega)

Ω represents:
- the Omega constant
- an asymptotic lower bound related to Big O notation
- in probability theory and statistical mechanics, the set of possible distinct system states
- a solid angle
- the arithmetic function counting a number's prime factors

ω represents:
- the first infinite ordinal
- the set of natural numbers in set theory (although $\mathbb{N}$ or **N** is more common in other areas of mathematics)
- an asymptotically dominant quantity related to Big O notation
- in probability theory, a possible outcome of an experiment
- a differential form (especially on an analytic space

A6 Roman Letters Used in Mathematics

Aa

A represents:
- the first corner of a triangle
- the digit "10" in hexadecimal and other positional numeral systems with a radix of 11 or greater
- area

$\mathbb{A}$ represents the algebraic numbers or affine space in Algebraic Geometry

a represents:
- the first side of a triangle (opposite corner A)
- the acceleration in mechanics equations
- the x-intercept of a line
- the unit are for area (100 m^2)
- the unit prefix atto (10^{-18})
- the first term in a sequence or series (eg. $S_n = n(a+1)/2$)

Bb

B represents:
- the digit "11" in hexadecimal and other positional numeral systems with a radix of 12 or greater
- the second corner of a triangle
- a ball (also denoted by $\mathcal{B}$ or $\mathbb{B}$)
- a basis of a vector space or of a filter (both also denoted by $\mathcal{B}$)

B with various subscripts represents several variations of Brun's constant and Betti numbers

b represents:
- the second side of a triangle (opposite corner B)
- the y-intercept of a line
- a basis vector (usually with an index, sometimes with an arrow over it)

Cc

C represents:
- the third corner of a triangle
- the digit "12" in hexadecimal and other positional numeral systems with a radix of 13 or greater

C with indices denotes the number of combinations, a binomial coefficient

$\mathbb{C}$ represents the set of complex numbers

A vertically elongated C with an integer subscript n sometimes denotes the n-th coefficient of a formal power series.

c represents:
- the third side of a triangle (opposite corner C)
- the unit prefix centi (10^{-2})

c represents the speed of light in vacuum

$\mathfrak{c}$ denotes the cardinality of the set of real numbers (the "continuum"), or, equivalently, of the power set of natural numbers

Dd

D represents the digit "13" in hexadecimal and other positional numeral systems with a radix of 14 or greater

d represents
- the differential operator
- the unit day of time (86 400 s)
- the difference in an arithmetic sequence (eg. $S_n = n(2a+(n-1)d)/2$)

Ee

E represents:
- the digit "14" in hexadecimal and other positional numeral systems with a radix of 15 or greater
- an exponent in decimal numbers 1.2E3 is 1.2×10^3 or 1200
- the set of edges in a graph or matroid
- the unit prefix exa (10^{18})

e represents:
- Euler's number, a transcendental number equal to 2.71828182845... which is used as the base for natural logarithms
- a vector of unit length, especially in the direction of one of the coordinates axes

Ff

F represents
- the digit "15" in hexadecimal and other positional numeral systems with a radix of 16 or greater

- the probability distribution function in statistics
- the unit farad of electrical capacity
- f represents:
- the generic designation of a function
- the unit prefix femto (10^{-15})

Gg

G represents
- an arbitrary graph, as in: $G(V,E)$
- an arbitrary group
- the unit prefix giga (10^9)
- Newton's gravitational constant
- the Einstein tensor

g represents:
- the generic designation of a second function
- the acceleration due to gravity on Earth

Hh

H represents:
- a Hilbert space
- the homology and cohomology functors
- the (Shannon) entropy of information

h represents:
- a small increment in the argument of a function
- the unit hour for time (3600 s)
- the Planck constant (6.624 608 96(33) $\times 10^{-34}$ J·s)
- the unit prefix hecto (10^2)

ℍ represents the quaternions

$\mathcal{H}$ represents the Hamiltonian in Hamiltonian mechanics

Ii

I represents:
- the closed unit interval, which contains all real numbers from 0 to 1, inclusive
- the identity matrix

i represents:
- the imaginary unit, a complex number that is the square root of -1
- a subscript to denote the ith term (that is, a general term or index) in a sequence or list
- the index to the elements of a vector, written as a subscript after the vector name
- the index to the rows of a matrix, written as the first subscript after the matrix name
- an index of summation using the sigma notation

Jj

j represents:
- the index to the columns of a matrix, written as the second subscript after the matrix name
- in electrical engineering, the square root of -1, instead of i
- in electrical engineering, the principal cube root of 1:

$$-\frac{1}{2} + \frac{1}{2}i\sqrt{3}$$

Kk

K represents:
- the unit kelvin of temperature

- an unspecified (real) constant

k represents
- the unit prefix kilo- (10^3)
- the Boltzmann constant
- an integer, e.g. a dummy variable in summations, or an index of a matrix.
- an unspecified (real) constant

Ll

L represents:
- the unit litre of volume
- the space of all integrable real (or complex) functions
- the space of linear maps, as in $L(E,F)$ or $L(E) = \text{End}(E)$
- the Likelihood function
- a formal language

l represents:
- the length of a side of a rectangle or a rectangular prism (eg. $V = lwh$; $A = lw$)
- the last term of a sequence or series (eg. $S_n = n(a+l)/2$)

$\mathcal{L}$ (or sometimes just L) represents the Lagrangian

Mm

M represents:
- a manifold
- a metric space
- a matroid
- the unit prefix mega (10^6)

m represents:
- the number of rows in a matrix
- the slope in a linear regression or in any line
- the mass in mechanics equations

- the unit metre of length
- the unit prefix milli (10^{-3})

Nn

$\mathbb{N}$ represents the natural numbers

n represents
- the number of columns in a matrix
- the "number of" in algebraic equations.
- the unit prefix nano (10^{-9})
- the nth term of a sequence or series (eg. $t_n = a+(n-1)d$)

Oo

O represents
- the order of asymptotic behavior of a function (upper bound)
- $(0, 0, \ldots, 0)$ — the origin of the coordinate system in Cartesian coordinates

o represents
- the order of asymptotic behavior of a function (strict upper bound)
- the order of an element in a group

Pp

P represents:
- the pressure in physics equations
- the unit prefix peta (10^{15})

$\mathbb{P}$ represents
- the prime numbers
- projective space

p represents
- the unit prefix pico (10^{-12})
- a probability

Qq

$\mathbb{Q}$ represents the rational numbers

Rr

R represents the Ricci tensor

$\mathbb{R}$ represents the set of real numbers and various algebraic structures built upon the set of real numbers, such as $\mathbb{R}^n$

r represents:
- the radius of a circle or sphere
- the ratio of a geometric series (eg. ar^{n-1})

Ss

S represents
- a sum
- the unit siemens of electric conductance
- the unit sphere (with superscript denoting dimension)
- the scattering matrix

s represents:
- an arclength
- the unit second of time

S represents a system's action in physics

Tt

T represents:
- the top element of a lattice

- a tree (a special kind of graph)
- temperature in physics equations
- the unit tesla of magnetic flux density

- the unit prefix tera (10^{12})
- the stress-energy tensor

t represents:
- time in graphs, functions or equations
- a term in a sequence or series (e.g. $t_n = t_{n-1} + 5$)

Uu

U represents:
- a U-set which is a set of uniqueness
- a unitary operator

U(n) represents the unitary group of degree n

U represents the union operator

Vv

V represents:
- volume
- the unit volt of voltage
- the set of vertices in a graph

v represents the velocity in mechanics equations

Ww

W represents the unit watt of power

Xx

x represents
- an unknown variable, most often (but not always) from the set of real numbers
- the coordinate on the first or horizontal axis in a cartesian coordinate system

Yy

Y represents the unit prefix yotta (10^{24})

y represents:
- the unit prefix yocto (10^{-24})
- the coordinate on the second or vertical axis in a cartesian coordinate system

y represents a second unknown variable

Zz

Z represents:
- the unit prefix zetta (10^{21})
- a standarized normal random variable in Probability Theory and Statistics

$\mathbb{Z}$ represents the integers

z represents:
- the unit prefix zepto (10^{-21})
- the coordinate on the third or vertical axis in three dimensional space
- the argument of a complex function, or any other variable used to represent a complex value

A7 Style Guide to Mathematical Theorems and Properties

*Below is a list of theorems and properties in mathematics and related disciplines for spelling and style. (En-dashes separate names of individuals; hyphens are used within names. See **A.8 Glossary** for definitions of frequently used terms.)*

15 and 290 theorems
2π theorem

A

Abel's binomial theorem
Abel's limit theorem
Abelian and Tauberian theorems
Abelian group
Abel–Ruffini theorem
Abhanyankar's conjecture
Abouabdillah's theorem
AF+BG theorem
Alperin–Brauer–Gorenstein
    theorem
Analytic Fredholm theorem
Anderson's theorem
Angle bisector theorem
Ankeny–Artin–Chowla theorem
Apéry's theorem
Apollonius' theorem
Area theorem
Argand diagram
Aronszajn–Smith theorem
Arrow's impossibility theorem
Artin approximation theorem
Artin–Schreier theorem
Artin-Wedderburn theorem
Arzelà–Ascoli theorem
Atiyah–Bott fixed-point
    theorem
Atiyah–Singer index theorem
Atkinson's theorem
Attouch-Wets toplogy
Autonomous convergence
    theorem
Axiom of choice (*sixth of the
    Zermelo–Fraenkel axioms*)
Ax–Kochen theorem

B

Baire category theorem
Balian–Low theorem
Banach–Alaoglu theorem
Banach fixed point theorem
Banach space
Banach–Steinhaus theorem
Banach–Stone theorem
Banach–Tarski paradox
Bapat–Beg theorem
Barbier's theorem
Bass's theorem
Bayes' theorem
Beatty's theorem
Beck's monadicity theorem
Beck's theorem
Bell's theorem
Bendixson–Dulac theorem
Berger–Kazdan comparison
    theorem
Bernstein's theorem
Berry–Esséen theorem
Bertrand's ballot theorem
Bertrand's postulate
Bessel's inequality
Bézout's theorem
Bieberbach conjecture
Bing metrization theorem
Binomial theorem
Birkhoff's theorem
Blaschke selection theorem
Bloch's theorem
Bôcher's theorem
Bohr–Mollerup theorem
Bolyai–Gerwien theorem
Bolzano's theorem
Bolzano–Weierstrass theorem
Bombieri's theorem

A7, *continued*

Bombieri–Friedlander–Iwaniec
theorem
Bondy–Chvátal theorem
Bonnet theorem
Boolean algebra
Boolean field; –lattice; –ring
Boolean prime ideal theorem
Borel–Bott–Weil theorem
Borel field; –set
Borel fixed-point theorem
Borsuk–Ulam theorem
Bott periodicity theorem
Bounded inverse theorem
Bourbaki–Witt theorem
Brahmagupta theorem
Branching theorem
Brauer–Suzuki theorem
Brauer's three main theorems
Brouwer fixed point theorem
Browder–Minty theorem
Brown's representability
theorem
Bruck–Chowla–Ryser theorem
Brun's theorem
Brunn–Minkowski theorem
Buckingham π theorem
Burnside prolem
Busemann's theorem

C
Cantor's diagonal theorem
Cantor–Bernstein–Schroeder
theorem (*also known as
"Cantor–Bernstein" or
"Shroeder–Bernstein
theorem"*)
Cantor's theorem
Cantor set
Carathéodory–Jacobi–Lie
theorem

Carathéodory's theorem
(*conformal mapping*)
Carathéodory's theorem (*convex
hull*)
Carathéodory's theorem
(*measure theory*)
Carathéodory's extension
theorem
Caristi fixed point theorem
Carmichael's theorem
Carnot's theorem
Cartan–Dieudonné theorem
Cartan's theorem
Cartan's theorems A and B
Cartesian coordinates
Cartesian product
Casey's theorem
Castigliano's first and second
theorems
Cauchy condensation test
Cauchy–Hamilton theorem
Cauchy's integral formula
Cauchy's integral theorem
Cauchy principal value
Cauchy product; –remainder;
–sequence
Cauchy residue theorem
Cauchy–Riemann condition
Cauchy's root test
Cayley–Bacharach theorem
Cayley–Hamilton theorem
Cayley number; –octonion
Cayley's theorem
Central limit theorem
Ceva's theorem
Chain rule (for calculus)
Chebotarev's density theorem
Chen's theorem
Chern–Gauss–Bonnet theorem
Chern–Simons form

A7, *continued*

Chinese remainder theorem
Choi's map theorem
Choquet theory
Chowla–Mordell theorem
Church–Rosser theorem
Classification of finite simple
    groups
Clifford's theorem
Closed graph theorem
Cluster decomposition theorem
Coase theorem
Cochran's theorem
Cohn's irreducibility criterion
Compactness theorem
Conservativity theorem
Convolution theorem
Conway's game of life
Cook's theorem
Corona theorem
Cox's theorem
Cramer's rule
Cramer system
Cramér's theorem
Crystallographic restriction
    theorem
Cut-elimination theorem
Cybenko theorem

D
D'Alembert ratio test
Dandelin's theorem
Danskin's theorem
Darboux's theorem (*real
    analysis*)
Darboux's theorem (*symplectic
    topology*)
De Branges' theorem
De Finetti's theorem
De Gua's theorem
De Moivre's formula
De Morgans's laws
De Rham's theorem

Dedekind cut
Deduction theorem
Dehn filling; –surgery
Desargues' theorem
Descartes' theorem
diffeomorphism
Dilworth's theorem
Dimension theorem
    for vector spaces
Dini's theorem
Diophantine equation
Dirichlet function
Dirichlet test (*for convergence*)
Dirichlet's theorem on
    arithmetic progressions
Dirichlet's unit theorem
Disintegration theorem
Divergence theorem
Dominated convergence
    theorem
Donaldson's theorem
Donsker's theorem
Dugundji extension thorem
Dvoretzky's theorem
Dynkin's formula

E
Earnshaw's theorem
Ehresmann's theorem
Eilenberg–Zilber theorem
Eisenstein's theorem
Envelope theorem
Epsilon conjecture
Equidistribution theorem
Equipartition theorem
Erdős–Anning theorem
Erdős–Kac theorem
Erdős–Ko–Rado theorem
Erdős–Stone theorem
Euclidean algorithm
Euclidean metric; –plane; –ring;
    –space; –vectors

A7, *continued*

Euclid–Euler theorem
Euler's constant
Euler's formula
Euler's rotation theorem
Euler's theorem
Euler's theorem on
Extreme value theorem

F
Factor theorem
Faltings' theorem
Fáry's theorem
Fáry-Milnor theorem
Fatou's theorem
Fatou–Lebesgue theorem
Feit–Thompson theorem
Fejér's theorem
Fermat's last theorem
Fermat's little theorem
Fermat polygonal number
    theorem
Fields medal
Fisher separation theorem
Fitting's theorem
Five color theorem
Fixed point theorems
Fluctuation dissipation theorem
Fluctuation theorem
Four color theorem
Fourier coefficient
Fourier inversion theorem
Fourier series; –transform
Fourier theorem
Fredholm determinant
Freudenthal suspension theorem
Frobenius reciprocity theorem
Frobenius theorem
Fubini's theorem
Fuglede's theorem
Fundamental theorem
    of algebra

Fundamental theorem of
    arbitrage-free pricing
Fundamental theorem of
    arithmetic
Fundamental theorem of
    calculus
Fundamental theorem on
    homomorphisms

G
Galileo paradox
Galois group; –field
Gauss' theorem
Gauss's Theorema Egregium
Gauss–Bonnet theorem
Gauss–Lucas theorem
Gauss–Markov theorem
Gauss–Wantzel theorem
Gauss–Schmidt orthoganili-
    zation process
Gaussian field
Gelfand–Fuchs cohomology
Gelfand–Naimark theorem
Gelfond–Schneider theorem
Geometrization conjecture
Geometrization theorem
Gibbard–Satterthwaite
    theorem
Girsanov's theorem
Glaisher's theorem
Gleason's theorem
Glivenko's theorem
Goddard–Thorn theorem
Gödel's completeness
    theorem
Gödel's incompleteness
    theorem
Going-up and going-down
    theorems
Goldbach's conjecture
Goodstein's theorem

A7, *continued*

Great orthogonality
   theorem
Greenberg–Hastings model
Green–Tao theorem
Green's theorem
Gromov's compactness
   theorem
Gromov's theorem; –norm
Gromov–Ruh theorem

H
H-theorem
Haag's theorem
Haboush's theorem
Hadamard three-circle theorem
Hadwiger's theorem
Haefliger structure
Hahn embedding theorem
Hairy ball theorem
Hahn–Banach theorem
Hahn–Kolmogorov theorem
Haken manifold
Hales–Jewett theorem
Hamiltonian group
Hamiltonian (*mechanics*)
Ham sandwich theorem
HarishChandra's regularity
   theorem
Hartogs' theorem
Hasse diagram
Hasse–Minkowski theorem
Hausdorf space
Heegaard splitting
Heine–Borel theorem
Heine–Cantor theorem
Heine's theorem (*also known as
   "Lebesque–Borel
   theorem"*)
Hellinger–Toeplitz theorem
Helly's theorem
Herbrand-Ribet theorem

Herbrand's theorem
Hermite polynomials
Hermitian adjoint; –matrix;
   –transformation
Hermitian (inner) product
Higman's embedding theorem
Hilbert's basis theorem
Hilbert's Nullstellensatz
Hilbert–Schmidt theorem
Hilbert space
Hilbert–Speiser theorem
Hilbert's theorem
Hinge theorem
Hironaka theorem
Holland's schema theorem
Hopf algebra
Hopf–Rinow theorem
Hurewicz theorem
Hurwitz's theorem

I
Implicit function theorem
Increment theorem
Infinite monkey theorem
Intermediate value theorem
Intersection theorem
Inverse function theorem
Ising model
Isomorphism extension theorem
Isomorphism theorems: First
   isomorphism theorem;
   Second isomorphism
   theorem; Third isomorph-
   ism theorem (*also called
   "Zassenhaus' lemma"*)
Isoperimetric theorem

J
Jackson's theorem
Jacobian (of a function)
Jacobian matrix

A7, *continued*

Jacobi's theorem
Jacobson density theorem
Japanese theorem
Jones polynomial
Jordan arc; –content; –matrix;
    –measurable; –region
Jordan canonical form theorem
Jordan curve theorem
Jordan–Hölder theorem
Jordan–Schönflies theorem
Jung's theorem

K

Kantorovich theorem
Khinchin's theorem
Kirchhoff's theorem
Kirszbraun theorem
Kleene's recursion theorem
Klein four froup
Klein–Gordon equation
Knaster–Tarski theorem
Kneser theorem
Kodaira embedding theorem
Kolmogorov–Arnold–Moser
    theorem
Kolmogorov extension theorem
König's theorem
Korteweg–de Vries (KdV)
    equation
Kronecker delta
Kronecker's theorem
Kronecker–Weber theorem
Krull's principal ideal theorem
Krull–Schmidt theorem
Krylov–Bogolyubov theorem
Künneth theorem
Kurosh subgroup theorem

L

Ladner's theorem
Lagrange's identity
Lagrange's theorem (*groups*)
Lagrange's theorem
(*numbers*)Lagrange's four-square
    theorem
Lagrange inversion theorem
Lagrange remainder (of a
    Taylor series)
Lagrange reversion theorem
Lagrangian (*mechanics*)
Lambek–Moser theorem
Lami's theorem
Landau prime ideal theorem
Laplace's equation
Laplacian operator
Lasker–Noether theorem
Laurent expansion theorem
law of large numbers
Lax–Milgram theorem
Lax–Richtmyer theorem
Lebesgue covering dimension
Lebesgue's decomposition
    theorem
Lebesgue's density theorem
Lee–Hwa–Chung theorem
Lebesgue differentiation
    theorem
Lebesque integral
Lebesque measure
Lefschetz fixed point theorem
Legendre function
Legendre polynomial
Lehmann–Scheffé theorem
Leibniz convergence test
Leibniz n-th derivative formula
Levi's theorem
Lie algebra
Lie's third theorem
l'Hospital's rule
Lindemann–Weierstrass
    theorem
Lie–Kolchin theorem

A7, *continued*

Liénard's theorem
Linear congruence theorem
Linear speedup theorem
Linnik's theorem
Liouville's theorem (*complex analysis, conformal mapping*)
Lipschitz condition
Löb's theorem
Lochs' theorem
Löwenheim–Skolem theorem
Lumer–Phillips theorem
Lyapunov's central limit theorem

M

MacDonald identities
Mach wave reflection
Maclaurin series
Mahler's compactness theorem
Mahler's theorem
Mandelbrot set
Marcinkiewicz theorem
Marriage theorem
Master theorem
Maschke's theorem
Matiyasevich's theorem
Matukama's equation
max-flow-min-cut theorem
Maximum power theorem
Maxwell's demon
Maxwell's theorem
May's theorem
Mazur's torsion theorem
Mean value theorem (for functions; –for integrals)
Menelaus' theorem
Menger's theorem
Mercer's theorem
Mertens' theorems
Metrization theorems
Meusnier's theorem

Midy's theorem
Mihăilescu's theorem
Min-max theorem (*also known as "Variational theorem" and as "Courant-Fischer-Weyl min-max principle"*)
Minkowski inequality; –space
Minkowski's theorem
Minkowski–Hlawka theorem
Mitchell's embedding theorem
Mittag–Leffler's theorem
Modigliani–Miller theorem
Modularity theorem
Mohr–Mascheroni theorem
Monge's theorem
Monodromy theorem
Monotone convergence theorem
Mordell–Weil theorem
Moreau's theorem
Morera's theorem
Morley's categoricity theorem
Morley's trisector theorem
Mountain pass theorem
Multinomial theorem
Myers theorem
Myhill–Nerode theorem

N

Nagata–Smirnov metrization theorem
Nagell–Lutz theorem
Nash embedding theorem
Navier–Stokes equation
Newlander–Niremberg theorem
Nicomachus's theorem
Nielsen–Schreier theorem
No cloning theorem
No wandering domain theorem
Noether's theorem
No-ghost theorem

A7, *continued*

Norton's theorem
Nyquist–Shannon sampling
   theorem

O
Open mapping theorem
Ornstein theorem
Oseledec theorem
Ostrowski's theorem
Ostrowski–Hadamard gap
   theorem

P
Painlevé equation
Paley's theorem
Paley–Wiener theorem
Pappus's centroid theorem
Pappus's hexagon theorem
Paris–Harrington theorem
Parseval-Deschenés equation
Parseval's theorem
Pascal's theorem
Peano's axioms
Pentagonal number theorem
Perfect graph theorem
Peron–Frobenius theorem
Peter–Weyl theorem
Picard theorem
Picard–Lindelöf theorem
     Pigeon-hole principle (*also
     known as "Kitchen drawer
     principle"*)
Pick's theorem
Pitman-Koopman-Darmois
   theorem
Plancherel theorem
Poincaré conjecture
Poincaré–Bendixson theorem
Poincaré–Birkhoff-Witt theorem
Poincaré duality theorem
Pompeiu's theorem

Poncelet–Steiner theorem
Post's theorem
Preimage theorem
prime number theorem
primitive element theorem
principal axis theorem
Proth's theorem
Prym variety
Ptolemaios' theorem
Pythagorean metric
Pythagorean theorem

Q
Quillen–Suslin theorem
Quadratic reciprocity theorem

R
Raabe's test
Radon's theorem
Radon–Nikodym theorem
Ramanujan–Skolem's
   theorem
Ramsey's theorem
Rank–nullity theorem
Rao–Blackwell theorem
Rational root theorem
Rédei's theorem
Reeh–Schlieder theorem
residue theorem
Reuleaux polytope; –triangle
Reynolds transport theorem
Rice's theorem
Riemann integral
Riemann–Stieltjes integral
Riemann mapping theorem
Riemann–Roch theorem
Riemann–Schottky problem
Riemann space; –matrix;
   –metric
Reisz–Fischer theorem
Riesz representation theorem

A7, *continued*

Riesz–Thorin theorem
Robertson–Seymour theorem
Robinson's joint consistency
    theorem
Rokhlin's theorem
Rolle's condition
Rolle's theorem
Rosser's theorem
Roth's theorem
Rouché's theorem
Routh–Hurwitz theorem
Runge's theorem
Russell's paradox

S
Sahlqvist correspondence
    theorem
Sarkovskii's theorem
Savitch's theorem
Schauder fixed point theorem
Schreier refinement theorem
Schur's lemma
Schur's theorem
Schwartz inequality
Seiberg–Witten theory
Seifert–van Kampen theorem
Separating axis theorem
Shannon's expansion theorem
Shannon's theorem
Sheffer stroke
Shift theorem
Simplicial approximation
    theorem
Simpson's rule
sine–Gordon equation
Skoda–El Mir theorem
Skolem–Noether theorem
Sokhatsky–Weierstrass theorem
Soundness theorem
Space hierarchy theorem
Spectral theorem

Speedup theorem
Sperner's theorem
Spin–statistics theorem
Sprague–Grundy theorem
Squeeze theorem
Stallings–Zeeman theorem
Stanley's reciprocity theorem
Stark–Heegner theorem
Stewart's theorem
Stirling's formula
Stokes' theorem
Stolper–Samuelson theorem
Stone's representation theorem
Stone's theorem on one-
    parameter unitary groups
Stone–Tukey theorem
Stone–von Neumann theorem
Stone–Weierstrass theorem
Strassman's theorem
Structured program theorem
Sturm's theorem
Sturm–Picone comparison
    theorem
Subspace theorem
Supporting hyperplane theorem
Swan's theorem
Sylow subgroup
Sylow's theorem
Sylvester's Law
Sylvester's theorem
Sylvester–Gallai theorem
Szőkefalvi-Nagy's dilation
    theorem
Szemerédi's theorem
Szemerédi–Trotter theorem

T
Takagi existence theorem
Tarski's indefinability theorem
Taylor expansion (*also called*
    "Taylor series")

A7, *continued*

Taylor's theorem
Thales' theorem
Thébault's theorem
Thevenin's theorem
Thue's theorem
Thue–Siegel–Roth theorem
Tietze extension theorem
Tijdeman's theorem
Tikhonov fixed point theorem
Time hierarchy theorem
Titchmarsh convolution
    theorem
Tonelli's theorem
Tsen's theorem
Tunnell's theorem
Tutte theorem
Turán's theorem
Turing machine
Tychonoff's theorem

U
Ugly duckling theorem
Uniformization theorem
Unique factorization theorem
Universal coefficient theorem
Unmixedness theorem

V
Van der Waerden's theorem
Vantieghems theorem
Venn diagram
Verma module
Vinogradov's theorem
Virasoro algebra
Virial theorem
Vitali convergence theorem
Vitali theorem
Vitali–Hahn-Saks theorem
Viviani's theorem
Von Neumann's theorem

W
Wallis formula
Wedderburn theorem
Weierstrass–Casorati theorem
Weierstrass M-test
Weierstrass preparation theorem
Well-ordering theorem
Whitehead theorem
Whitney embedding theorem
Wiener's tauberian theorem
Wiener-Ikehara theorem
Wigner–Eckart theorem
Wilson's theorem
WKB approximation
Wolstenholme's theorem

Y
Young–Baxter equation

Z
Z* theorem
Zariski's main theorem
Zermelo–Fraenkel Axioms (*the
    sixth of which is known as
    the "Axiom of choice"; the
    seventh of which is known
    as the "Axiom of infinity"*)
Zermelo's well-ordering
    theorem
ZJ theorem
Zeckendorf's theorem
Zorn's lemma

A8 Glossary of Mathematical Terms. *(Terms used in mathematical literature at many levels of technical sophistication.)*

A

abscissa. The horizontal axis in Cartesian coordinates—*see x-axis.*

absolute value. The distance of a number from zero; the value of a number with positive or negative sign removed. (Indicated by vertical lines on either side of the number—e.g., $|3|$; $|-4| = 4$.)

acute angle. A positive angle measuring less than 90 degrees.

acute triangle. A triangle each angle of which measures less than 90 degrees.

additive identity. The number zero is called the additive identity because the sum of zero and any number is that number.

additive inverse. The additive inverse of any number x is the number that gives zero when added to x. For example, the additive inverse of 5 is -5.

adjacent angles. Two angles that share both a side and a vertex.

algebra. A seminal branch of mathematics, dealing with relations, structure, and quantity.

analysis. A seminal branch of mathematics, dealing with functions, change, and limits. **Calculus** is a subdiscipline within the general study of analysis.

analytic function. A function that is everywhere differentiable.

angle. The union of two rays with a common endpoint, called the vertex.

arc. A portion of the circumference of a circle.

area. The number of square units that covers a shape or figure.

Argand diagram. A representation of complex numbers and functions in which the real component is represented by the x-axis, and the imaginary component is represented by the y-axis.

associative. The property of a sequence of relations in an equation that can be performed in any order.

associative property of addition. $(a + b) + c = a + (b + c)$

associative property of multiplication. $(a \times b) \times c = a \times (b \times c)$

asymptote. A curve that a function approaches as a variable of the function of that curve goes to infinity. A **"linear asymptote"** is a straight line that a curve approaches ever closer (but never reaches) as a variable approaches infinity.

average. A number that represents the characteristics of a data set. In arithmetic, the sum of a set of numbers divided by the number of numbers in the set is the set's average.

axiom. A statement known (or assumed) to be true without proof.

axis of symmetry. A line that passes through a figure so that the section of the figure on one side of the line is a mirror reflection of the section on the other side of the line.

B

base. In geometry, the bottom of a plane figure or three-dimensional figure. In algebra, a term or number raised to a power by an exponent. In number theory, a positional number system is said to be of a base of a certain number if a new position is required to accommodate that number (or multiples of it). The decimal system commonly used is "of base 10"; computer programming uses a system "of base 2."

basis. The minimum number of vectors that "span" (or "cover") a vector space.

bisect. To divide into two congruent parts.

box and whisker plot. A type of graphic plot of data that displays the range of each plotted point of the data set.

C

calculus of variation. A calculus-like mathematical analysis of paths and processes that do not lend themselves to Cartesian expression.

Cartesian coordinates. A system in which points on a plane are identified by an ordered pair of numbers, representing the distances to two or three perpendicular axes. (So named after its inventor, René Descartes.)

category theory. A branch of algebra that takes as primary the abstract character of collections rather than the items that comprise those collections.

catenary. The curve described by a flexible chain or cord, held fixed at both ends and allowed to hang down (as, for example, the cables of a suspension bridge). Named by Thomas Jefferson; mathematically described by Jakob Bernoulli; first applied architecturally by Robert Hooke.

cellular automaton (plural: **automata**). A mathematical model in which an aggregate of cells all have states described by parameters and rules of interaction with neighboring cells. The model has been useful in both computer science and in life science.

central angle. An angle the vertex of which is at the center of a circle.

chord. A line segment that connects two points on a curve.

circle. The set of points in a plane that are a fixed distance from a given point, called the **center**.

circumference. The distance around a circle.

coefficient. In a mathematical expression, a constant that multiplies a variable.

collinear. Points are collinear if they lie on the same line.

combination. A selection in which order is not important.

combinatorics. A branch of mathematics concerned with the counting, arrangement, enumeration, and analysis of discrete, usually finite numbers, of objects. **Combinatorial geometry** applies this discipline to shapes and geometrical structures and their features.

common factor. A factor shared by two or more numbers.

common multiple. A multiple of two or more numbers.

commutative property of addition: $a + b = b + a$.

commutative property of multiplication: $a*b = b*a$.

complementary angles. Two angles whose sum is 90 degrees.

complex; complex analysis. A mathematical term that consists of a real part and an "imaginary" part (see below). The study of functions of complex numbers and variable is "complex analysis."

composite number. A natural number that is not prime.

conditional. An "if X then Y" statement, in which the "X" is the *antecedent* (or the *hypothesis*) and the "Y" is the *consequent* (or *conclusion*)

cone. A three-dimensional figure with one vertex and a circular base.

congruent. Figures or angles that have the same size and shape.

connective. A sign between mathematical objects indicating a relation among or operation between those objects.

constant. A term—in an equation or in the world—the value of which does not change.

convex. In geometry, a space that contains all points on a line (or curve) connecting any two points in that space is said to be a "convex" space. A "convex body" in any space is a (non-zero) closed, bounded set that contains all points on all straight lines described by any two points contained in the body.

coordinate plane. The plane determined by a horizontal number line, called the x-axis, and a vertical number line, called the y-axis, intersecting at a point called the origin. Each point in the coordinate plane can be specified by an ordered pair of numbers, called its "coordinates."

coplanar. Points that lie within the same plane.

counting numbers. The natural numbers, or the numbers used to count.

counting principle. If a first event has **n** outcomes and a second event has **m** outcomes, then the first event followed by the second event has **n** times **m** outcomes.

cross product. A product found by multiplying the numerator of one fraction by the denominator of another fraction and the denominator of the first fraction by the numerator of the second.

cube. A solid figure with six square faces. Also, a number (or expression) on which an exponent of 3 operates. A number or expression that, when multiplied twice in succession yields the number is its "cube root."

curvature. In geometry, the rate at which a line or surface deviates from a straight line. More generally, it is used to describe the "intrinsic structure" of a space and the "natural" behavior of objects in it.

curve. Any continuous collection of points in space (real or mathematical).

cylinder. A three-dimensional figure having two parallel bases that are congruent circles.

D

data. Information that is gathered in an experiment or in observation of natural phenomena.

decimal number. The numbers in the base–10 number system, having one or more places to the right of a decimal point.

decision theory. Study of algorithms in computational programs. The "decision problem" is the general endeavor to determine the successful completion—how and if accomplished—of a task by a programmed computer in a finite number of steps.

degree. A unit of measure of an angle.

denominator. The bottom part of a fraction.

dependent events. Two events in which the outcome of the second is influenced by the outcome of the first.

diagonal. The line segment connecting two nonadjacent vertices in a polygon.

diameter. The line segment joining two points on a circle and passing through the center of the circle.

difference. The result of subtracting two numbers.

differential equation. A mathematical equation for an unknown function of one or several variables that relates the values of the function itself and of its derivatives of various orders. Differential equations are generally of two kinds: **"ordinary"** (ODE), where the unknown function is a function of a *single* independent variable; and **"partial"** (PDE), where the unknown function is a function of *multiple* independent variables and their partial derivatives.

differential geometry. The branch of mathematics that applies the tools and methods of calculus to the study of curves and space.

digit. The ten symbols, 0, 1, 2, 3, 4, 5, 6, 7, 8, and 9. The number 215 has three digits: 2, 1, and 5.

dimension. In geometry, the minimum number of base or unit vectors required to describe any point in a space is the *dimension* of that space.

Diophantine equation. A polynomial equation for which variables may only be intergers.

distributive property of addition: $a(b + c) = ab + ac$

dividend. In $a / b = c$, **a** is the dividend.

divisor. In $a / b = c$, **b** is the divisor.

E

ellipse. The set of all points in a plane such that the sum of the distances to two fixed points is a constant.

ensemble. A mathematical assemblage of copies of a system used to determine the state or development of the system from one state to another.

equation. A mathematical statement that asserts that two expressions have the same value; any number sentence with an equal sign (=).

equilateral triangle. A triangle that has three equal sides.

equivalent. Identical. To emphasize the strength of the identity, a three-stroke equal sign ($\equiv$) may be used.

equivalent equations. Two equations whose solutions are the same.

equivalent fractions. Fractions that reduce to the same number.

Euclidean space. A finite-dimensional (vector) space that satisfies the Pythagorean theorem for measuring distances (i.e., has a Pythagorean metric).

evaluate. To substitute number values into an expression and perform the operations in it.

even number. A natural number that is divisible by 2.

event. In probability, a set of outcomes.

exponent. A number that indicates the operation of repeated multiplication.

F

factor. A number which, when multiplied a specified number of times, yields a second number, is said to be a factor of that second number. "Factoring" a number is finding what are the various numbers that are factors of that number.

factorial. The product of all positive integers less than or equal to a non-negative number N is called "N factorial" and is symbolized as "N!".

face. A flat surface of a three-dimensional figure.

field. In algebra, a set of elements combined with the operations of addition, multiplication, and non-zero division that satisfy the rules of ordinary arithmetic. (Technically, a field is a "commutative division ring.")

formula. An equation that states a rule, definition, or fact.

fractal. An irregular curve or shape for which a suitably chosen part is similar in shape to a given larger or smaller part when magnified or reduced to the same size (so that the scale of the shape gives no indication of actual length).

fraction. A number used to name a part of a group or a whole. The number below the bar is the *denominator*, and the number above the bar is the *numerator*.

frequency. The number of times a particular item appears in a data set.

frequency table. A data listing which also lists the frequencies of the data.

G

geodesic. A generalized form of the concept of the shortest distance between two points being a straight line. A geodesic is the shortest distance between two points within a space and in accordance to the metric rules of that space. On a sphere, for example, a geodesic is a "great circle," or a circumference connecting two points.

Gödel's Proof. Popular characterization of the finding of 20th-century logician Kurt Gödel that any mathematical system sufficiently complex to include arithmetic must have propositions (assertions) that cannot be proven to be true or false.

Goldbach's conjecture. All even numbers may be expressed as the sum of two prime numbers.

graph. A type of drawing used to represent data.

greatest common factor (GCF). The largest number that divides two or more numbers evenly.

H

homology/cohomology. Homology in mathematics is measure of the "connectedness" of a space or algebraic structure. Cohomology is more specific and indicates ways in which structures have "holes" or areas within the general space in which they are defined that are not consistent with the definition or characterization of the object. In topology, cohomology defines the number of holes that characterize a topological surface.

horizontal. A line with zero slope.

hypotenuse. The side opposite the right angle in a right triangle.

I

identity property of addition: The sum of any number and 0 is that number.

identity property of multiplication: The product of 1 and any number is that number.

improper fraction. A fraction in which the numerator is greater than the denominator.

independent events. Two events in which the outcome of the second is not affected by the outcome of the first. When two variables in an equation may take on values independent of one another, they are called "independent variables."

inequality. A mathematical expression which asserts that two quantities are not equal.

infinity. A limitless quantity.

inscribed angle. An angle placed inside a circle with its vertex on the circle and the sides of which are chords of the circle.

inscribed polygon. A polygon placed inside a circle so that each vertex of the polygon touches the circle.

integers. The set of the positive and negative natural numbers, and zero.

integrate. The process of summing over infinitesimals. The sum is called an "integral."

intercept. The x-intercept of a line or curve is the point where it crosses the x-axis, and the y-intercept of a line or curve is the point where it crosses the y-axis.

intercepted arc. The arc of a circle within an inscribed angle.

interpolation. Methods for estimating values that lie between two known values.

intersecting lines. Lines that have one and only one point in common.

inverse. -5 is the additive inverse of 5, because their sum is zero. 1/3 is the multiplicative inverse of 3, because their product is 1.

inverse operations. Two operations that have the opposite effect, such as addition and subtraction.

irrational number. A number that cannot be expressed as the ratio of two integers.

isosceles triangle. A triangle with at least two equal sides.

K

kernel. In algebra, the element of a group (or another algebraic structure) that allows or performs the identity operation (i.e., that when operated on by any element of the system leaves it unchanged). In computer science, a kernel is the basic element of the operating system that is common to all applications.

Kronecker delta function. A matrix function defined by $\delta_{ij} = 0$ if $i \neq j$, and $\delta_{ij} = 1$ if $i = j$.

L

least common denominator (LCD). The smallest multiple of the denominators of two or more fractions.

least common multiple (LCM). The smallest nonzero number that is a multiple of two or more numbers.

like fractions. Fractions that have the same denominator.

line. A straight line is a mathematical idealization (and may be considered postulated to exist without definition), but can be described as a set of points for which the extension of any two points will include only members of that set, and any member of that set can be found through such an extension.

line of symmetry. Line that divides a geometric figure into two congruent portions.

line segment. Two points on a line, and all the points between those two points.

locus. A path of points describing a line, curve, or shape.

logic. The discipline of constructing arguments that display sound reasoning.

lowest terms. (Also, "simplest form"), when the GCF of the numerator and the denominator of a fraction is 1, or when all operations that can be performed within a mathematical expression have been performed.

M

Mandelbrot set. A fractal which, when plotted using computer graphics, resembles heart-shaped discs to which smaller, similarly-shaped discs are attached, to which other, smaller, similarly-shaped discs are attached, in a (potentially) infinite series.

manifold. A generalized space in which every point in the vicinity of any selected point has the same dimensional properties of the space as a whole.

matrix. A mathematical array of numbers or expressions that is arranged in rows and columns with set rules on the manner in which these elements are to be manipulated to yield numerical or functional results.

mean. In a data set, the sum of all the data points, divided by the number of data points; average.

median. The middle number in a data set when the data are put in order; a type of average.

metric. A mathematical formula that provides a real nonnegative number (a "distance") with each pair of elements in a set such that the number is zero if and only if the two elements are identical; the number is independent of the order in which the elements are applied in the formula; and the number associated with any pair of elements obeys the associative rules of arithmetic. In a vector space, the metric is expressed in terms of a **norm**.

midpoint. A point on a line segment that divides the segment into two congruent segments.

mixed number. A number written as a whole number and a fraction.

mode. A type of average; the number (or numbers) that occurs most frequently in a set of data.

modus ponens. the logical principle that if A implies B, and A is true, then B is true.

multiple. A multiple of a number is the product of that number and any other whole number. Zero is a multiple of every number.

multiplicative identity. The number 1 is the multiplicative identity because multiplying 1 times any number gives that number.

multiplicative inverse. The reciprocal of a number.

mutually exclusive events. Two or more events that cannot occur at the same time.

N

natural numbers. The counting numbers.

negative number. A real number that is less than zero.

norm. *See* **metric.**

normal. Perpendicular.

number line. A line on which every point represents a real number.

number theory. Branch of mathematics dealing with numbers, specifically integers, relations between integers (as with Diophantine equations) and the appearance, characterization, and distribution of prime numbers.

numerator. The top part of a fraction.

O

obtuse angle. An angle whose measure is greater than 90 degrees.

obtuse triangle. A triangle with an obtuse angle.

octagon. A polygon with 8 sides.

odd number. A whole number that is not divisible by 2.

operation. Addition, subtraction, multiplication, and division are the basic arithmetic operations.

opposites. Two numbers that lie the same distance from 0 on the number line but in opposite directions.

ordered pair. Set of two numbers in which the order has an agreed-upon meaning, such as the Cartesian coordinates (x, y), where the first coordinate represents the horizontal position, and the second coordinate represents the vertical position.

origin. The point (0, 0) on a coordinate plane, where the x-axis and the y-axis intersect.

outcome. In probability, a possible result of an experiment.

P

parallel. Two lines are parallel if they are in the same plane and never intersect.

parallelogram. A quadrilateral with opposite sides parallel.

pentagon. A five-sided polygon.

percent. A fraction, or ratio, in which the denominator is assumed to be 100. The symbol % is used for percent.

perimeter. The sum of the lengths of the sides of a polygon.

permutation. A way to arrange things in which order is important.

perpendicular. Two lines are perpendicular if the angle between them is 90 degrees.

pi. The ratio of the circumference of a circle to its diameter, symbolized by the Greek letter, π (an irrational number, the first ten digits of which are 3.141592653, a mnemonic for which is, "How I need a drink, alcoholic of course, after the heavy lectures involving quantum mechanics").

plane. A flat surface that stretches into infinity.

point. A location in a plane or in space, having no dimensions.

polygon. A closed plane figure made up of several line segments that are joined together.

polyhedron. A three-dimensional solid that is bounded by plane polygons.

positive number. A real number greater than zero.

power. A number that indicates the operation of repeated multiplication.

prime number. A number whose only factors are itself and 1.

probability. For an experiment, the total number of successful events divided by the total number of possible events.

product. The result of two numbers being multiplied together.

proper fraction. A fraction whose numerator is less than its denominator.

proportion. An equation of fractions in the form: $a/b = c/d$

protractor. A device for measuring angles.

pyramid. A three-dimensional figure that has a polygon for its base and whose faces are triangles having a common vertex.

Pythagorean Theorem. The theorem that relates the three sides of a right triangle: $a^2 + b^2 = c^2$

Q

quadrant. One of the quarters of the plane of the Cartesian coordinate system.

quadrilateral. A polygon with 4 sides.

quotient. The answer to a division problem.

R

radical sign. The symbol for the root of a number: $\sqrt{\ }$. " $\sqrt[3]{x}$ " signifies the cube root of x; with no number on the radical sign, the assumption is that a square root is indicated.

radius. The distance from the center to a point on a circle; the line segment from the center to a point on a circle.

range. In statistics, the difference between the largest and the smallest numbers in a data set.

rate. A ratio that compares different kinds of units.

ratio. A pair of numbers that compares different types of units.

rational number. A number that can be expressed as the ratio of two integers.

ray. Part of a line, with one endpoint, and extending to infinity in one direction.

real numbers. The combined set of rational numbers and irrational numbers.

reciprocal. The number which, when multiplied times a particular fraction, gives a result of 1.

rectangle. A quadrilateral with four 90-degree angles.

reflection. A transformation resulting from a flip.

regular polygon. A polygon in which all the angles are equal and all of the sides are equal.

repeating decimal. A decimal in which the digits endlessly repeat a pattern. Such a repeating decimal, though infinite, is rational.

rhombus. A parallelogram with four equal sides.

right angle. An angle whose measure is 90 degrees.

right triangle. A triangle that contains a right angle.

ring. In algebra, a set of elements with the operations of addition and multiplication. When a ring also obeys rules of arithmetic and has a non-zero operation of division, it becomes a "field."

root. The root of an equation is the same as the solution to the equation. The root of a number is that number which when multiplied a specified number of times yields the number.

rotation. A transformation in which a figure is rotated through a given angle, about a point.

S

sample space. For an experiment, the sample space is the space that includes all possible outcomes.

scale drawing. A drawing that is a reduction or enlargement of the original.

scalene triangle. A triangle with three unequal sides.

scattergram. A graph with points plotted on a coordinate plane.

scientific notation. A method for writing extremely large or small numbers compactly in which the number is shown as the product of two factors, one of which is usually a power of 10.

sequence. An ordered collection of quantities described as a progression in which some quality of the components explains membership in the progression.

When the sequence of terms are added in order of their appearance in the sequence, the result is a **series.**

The simple collection of objects, related or unrelated, with no operation performed, is a **set.**

similar. Two polygons are similar if their corresponding sides are proportional.

simplifying. Reducing to lowest terms.

skew lines. Lines that are not in the same plane and that do not intersect.

slope. The steepness of a line expressed as a ratio, using any two points on the line.

solution. The value of a variable that makes an equation true.

sphere. A three-dimensional figure with all points in space a fixed distance from a given point, called the center.

square. A quadrilateral with four equal sides and four 90 degree angles.

square root. The square root of x is the number that, when multiplied by itself, gives the number, x.

statistics. The science of collecting, organizing, and analyzing data.

stem and leaf plot. A technique for organizing data for comparison.

straight angle. An angle that measures 180 degrees.

supplementary angles. Two angles are supplementary if their sum is 180 degrees.

surface area. For a three-dimensional figure, the sum of the areas of all the faces.

T

terminating decimal. A fraction whose decimal representation contains a finite number of digits.

translation. A transformation, or change in position, resulting from a slide with no turn.

transformation. A change in the position, shape, or size of a geometric figure.

transversal. A line that intersects two other lines.

trapezoid. A quadrilateral that has exactly two sides parallel.

tree diagram. A diagram that shows outcomes of an experiment.

triangle. A three-sided polygon.

trigonometry. The branch of mathematics that studies the relationships of the sides of triangles and the functional relationships between the angles and the sides of triangles and other polygons.

U

unit price. Price per unit of measure. "Unit" is used in this manner in connection with other variables and quantities, such as, areas, commodities, etc.

V

variable. A letter used to represent a number value in an expression or an equation. A "dependent variable" is one the value of which is determined by another variable in the expression; an "independent variable" is not determined by values or changes in any other variable.

vertex. The point on an angle where the two sides intersect.

vertical angles. A pair of opposite angles that is formed by intersecting lines.

volume. A measurement of space, or capacity.

W

whole numbers. The set of numbers that includes zero and all of the natural numbers.

X

x-axis. The horizontal axis in a Cartesian coordinate plane.

x-intercept. The value of x at the point where a line or curve crosses the x-axis.

Y

y-axis. The vertical axis in a Cartesian coordinate system.

y-intercept. The value of y at the point where a curve crosses the y-axis.

Z

zero. The additive identity; the number that, when added to another number **n**, gives **n**.

zero property of multiplication. The product of zero and any number is zero.

A9 Mathematics Journals and Their Abbreviations

Full Title	Abbreviation
Acta Mathematica Academiae Paedagogicae Nyí regyháziensis.	Acta Math. Acad. Paedagog. Nyházi.
Acta Mathematica Universitatis Comenianae.	Acta Math. Univ. Comenianae.
Acta Mathematica	Acta Math.
Acta Mathematica	Acta Math.
Acta Numerica	Acta Numerica
Acta Scientiarum Mathematicarum. (Szegediensis)	Acta Sci. Math. (Szeged)
Advances in Applied Mathematics.	Adv. Appl. Math.
Advances in Difference Equations.	Adv. Difference Equ.
Advances in Econometrics.	Adv. Econom.
Advances in Geometry	Adv. Geom.
Advances in Geometry	Adv. Geom.
Advances in Mathematics.	Adv. Math.
Advances in Theoretical and Mathematical Physics.	Adv. Theor. Math. Phys.
Algebraic and Geometric Topology	AGT
Algebra and Logic	Algebra Logic
Algebra & Number Theory	Algebra Number Theory
Algebra Colloquium	Algebra Colloq
American Journal of Mathematics	Am. J. Math.
American Mathematical Monthly	Am. Math. Mon.
Analysis & PED	Anal. PED
Analytical Methods in Special Functions	Anal. Methods Spec. Funct.
Annals of Mathematics	Ann. of Math.
Applied Mathematics and Computation	Appl. Math. Comput.
Applied Sciences	Appl. Sci.
Atlantis Studies in Mathematics for Engineering and Science	Atlantis Stud. Math. Eng. Sci.
Archive for Rational Mechanics and Analysis	Arch. Ration. Mech. Anal.
Asian Journal of Mathematics, The.	Asian J. Math.
Aspects of Mathematics.	Aspects Math
Balkan Journal of Geometry and its Applications	Balkan J. Geom. Appl.
Banach Center Publications	Banach Cent. Publ.
Banach Journal of Mathematical Analysis	Banach J. Math. Anal.

A9, *continued*

BIT	BIT
Boletin Asociacio Matematical Vanezolana	Bol. Asoc. Mat. Venez.
Boundary Value Problems	Bound. Value Prob.
Bulletin of the London Mathematical Society	Bull. Lond. Math. Soc.
Bulletin of Statistics & Economics	Bull. Stat. Econ.
Calculus of Variations and Partial Differential Equations	Calc. Var. Partial Differential Equations
Cambridge History of Science, The	Camb. Hist. Sci.
Cambridge Monographs on Applied and Computational Mathematics	Camb. Monogr. Appl. Comput. Math.
Cambridge Monographs on Particle Physics, Nuclear Physics, and Cosmology	Camb. Monogr. Part. Phys. Nucl. Phys. Cosmol.
Cambridge Studies in Mathematical Biology	Camb. Stud. Math. Biol.
Canadian Journal of Mathematics	Can. J. Math.
Canadian Mathematical Bulletin	Can. Math. Bull.
Chinese Journal of Contemporary Mathematics	Chinese J. Contemp. Math.
College Mathematics Journal, The	College Math. J.
Combinatorics, Probability and Computing	Comb. Probab. Comput.
Commentarii Mathematici Helvetici	Comment. Math. Helv.
Communications in Algebra.	Commun. Algebra
Communications in Computational Physics	Commun. Comput. Phys.
Communications in Mathematical Analysis	Commun. Math. Anal.
Communications in Mathematical Physics	Commun. Math. Phys.
Communications in Mathematical Sciences	Commun. Math. Sci.
Communications of the Korean Mathematical Society	Commun. Korean Math. Soc.
Communications on Pure and Applied Analysis	Commun. Pure Appl. Anal.
Communications on Pure and Applied Mathematics	Comm. Pure Appl. Math.
Communications on Stochastic Analysis	Commun. Stoch. Anal.
Compel	Compel
Complex Analysis and Operator Theory	Complex Anal. Oper. Theory
Compositio Mathematica	Compos. Math.

A9, *continued*

Computational and Applied Mathematics	Comput. Appl. Math.
Computers and Artificial Intelligence	Comput. Artif. Intell.
Computer Assisted Mechanics and Engineering Sciences	Comput. Assist. Mech. Eng. Sci.
Conformal Geometry and Dynamics	Conform. Geom. Dyn.
Creation in Mathematical Sciences	Creation Math. Sci.
Creation in Mathematics	Creation Math.
Creative Mathematics and Informatics	Creat. Math. Inform.
CRM Monograph Series	CRM Monogr. Ser.
Cryptologia	Cryptologia
Culture and History of Mathematics	Cult. Hist. Math.
Current Development in Theory and Applications of Wavelets	Curr. Dev. Theory Appl. Wavelets
Current Issues in Electronic Modeling	Curr. Issues Electron. Model.
Cybernetics and Systems	Cybern. Syst.
Cybernetics and Systems Analysis	Cybern. Syst. Anal.
Czechoslovak Journal of Physics	Czech. J. Phys.
Data Handling in Science and Technology	Data Handl. Sci. Technol.
Data & Knowledge Engineering	Data Knowl. Eng.
Decisions in Economics and Finance	Decis. Econ. Finance
Demonstratio Mathematica	Demonstr. Math.
Developments in Boundary Element Methods	Dev. Boundary Elem. Methods
Developments in Civil and Foundation Engineering	Dev. Civ. Found. Eng.
Developments in Mathematics	Dev. Math.
Differential and Integral Equations	Differ. Integral Equ.
Differential Equations	Differ. Equ.
Differential Equations & Nonlinear Mechanics	Differ. Equ. Nonlinear Mech.
Differential Geometry and its Applications	Differ. Geom. Appl.
Differential Geometry - Dynamical Systems	Differ. Geom. Dyn. Syst.
Discrete Applied Mathematics	Discrete Appl. Math.
Discrete Dynamics in Nature and Society	Discrete Dyn. Nat. Soc.
Discrete Mathematics	Discrete Math.
Discrete Mathematics and Applications	Discrete Math. Appl.
Discrete Mathematics and Theoretical Computer Science	Discrete Math. Theor. Comput. Sci.
Discussiones Mathematicae	Discuss. Math.

A9, *continued*

Divulgaciones Matemáticas	Divulg. Mat.
Documentation Mathématique	Doc. Math.
Documents Mathématiques (Paris).	Doc. Math., Paris
Doklady Mathematics	Dokl. Math.
Duke Mathematical Journal	Duke Math. J.
Dynamics and Control	Dyn. Control
Dynamics and Stability of Systems	Dyn. Stab. Syst.
Dynamics Reported	Dyn. Rep.
Econometric Reviews	Econ. Rev.
Econometrics Journal, The	Econom. J.
Econometric Theory	Econom. Theory
Electrical Engineering and Electronics	Electr. Eng. Electron.
Electronic Journal of Combinatorics, The	Electron. J. Comb.
Electronic Journal of Differential Equations	Electron. J. Differ. Equ.
Electronic Journal of Linear Algebra	Electron. J. Linear Algebra
Electronic Journal of Mathematical and Physical Sciences (EJMAPS)	Electron. J. Math. Phys. Sci. (EJMAPS)
Electronic Journal of Probability	Electron. J. Probab.
Electronic Journal of Qualitative Theory of Differential Equations	Electron. J. Qual. Theory Differ. Equ.
Electronic Research Announcements in Mathematical Sciences	Electron. Res. Announc. Math. Sci.
Electronic Research Announcements of the American Mathematical Society	Electron. Res. Announc. Am. Math. Soc.
Electronic Transactions on Numerical Analysis	Electron. Trans. Numer. Anal.
EMS Tracts in Mathematics	EMS Tracts Math.
Encyclopedia of Mathematics and Its Applications	Encycl. Math. Appl.
Engineering Analysis with Boundary Elements	Eng. Anal. Bound. Elem.
Engineering Application of Fracture Mechanics	Eng. Appl. Fract. Mech.
Ergodic Theory and Dynamical Systems	Ergodic Theory Dyn. Syst.
European Series in Applied and Industrial Mathematics (ESAIM): Control, Optimization and Calculus of Variations.	ESAIM, Control Optim. Calc. Var.
European Series in Applied and Industrial Mathematics (ESAIM): Mathematical Modelling and Numerical Analysis	ESAIM, Math. Model. Numer. Anal.

A9, *continued*

European Series in Applied and Industrial Mathematics (ESAIM): Probability and Statistics	ESAIM, Probab. Stat.
European Series in Applied and Industrial Mathematics (ESAIM): Proceedings	ESAIM, Proc.
Euclides	Euclides
European Consortium for Mathematics in Industry	Eur. Consortium Math. Ind.
European Finance Review	Eur. Finance Rev.
European Journal of Applied Mathematics	Eur. J. Appl. Math.
European Journal of Economic and Social Systems	Eur. J. Econ. Soc. Syst.
European Journal of Physics	Eur. J. Phys.
European Journal of Pure and Applied Mathematics	Eur. J. Pure Appl. Math.
Euromath Bulletin	Euromath Bull.
Experimental Economics	Exp. Econ.
Experimental Mathematics	Experiment. Math.
Far East Journal of Applied Mathematics	Far East J. Appl. Math.
Far East Journal of Dynamical Systems	Far East J. Dyn. Syst.
Far East Journal of Electronics and Communications	Far East J. Electron. Commun.
Far East Journal of Mathematical Sciences	Far East J. Math. Sci.
Far East Journal of Theoretical Statistics	Far East J. Theor. Stat.
Fibonacci Quarterly, The	Fibonacci Q.
Fields Institute Communications.	Fields Inst. Commun.
Fields Institute Monographs	Fields Inst. Monogr.
Filomat	Filomat
Finance and Stochastics	Finance Stoch.
Finite Elements in Analysis and Design	Finite Elem. Anal. Des.
Finite Elements in Fluids	Finite Elem. Fluids
Finite Fields and their Applications	Finite Fields Appl.
Fixed Point Theory	Fixed Point Theory
Fixed Point Theory and Applications	Fixed Point Theory Appl.
Fluid Dynamics	Fluid Dyn.
Fluid Dynamics Research	Fluid Dyn. Res.
Focus on Learning Problems in Mathematics	Focus Learn. Probl. Math.
Forum Geometricorum	Forum Geom.
Foundations of Artificial Intelligence	Found. Artif. Intell.

A9, *continued*

Foundations of Physics	Found. Phys.
Fractional Calculus & Applied Analysis	Fract. Calc. Appl. Anal.
Frontiers in Applied Mathematics	Front. Appl. Math.
Frontiers in Artificial Intelligence and Applications	Front. Artif. Intell. Appl.
Functional Analysis and its Applications	Funct. Anal. Appl.
Fundamenta Mathematicae	Fundam. Math.
Games and Economic Behavior	Games Econ. Behav.
Gazette des Mathématiciens. Société Mathématique de France	Gaz. Math., Soc. Math. Fr.
General Mathematics	Gen. Math.
General Relativity and Gravitation	Gen. Relativ. Gravitation
Genetic Algorithms and Evolutionary Computation	Genet. Algorithms Evol. Comput.
Genetic Programming and Evolvable Machine	Genet. Program. Evol. Mach.
GeoInformatica	GeoInformatica
Geometry & Topology	Geom. Topol.
Geometry and Topology Monographs	Geom. Topol. Monogr.
Geophysical and Astrophysical Fluid Dynamics	Geophys. Astrophys. Fluid Dyn.
Geophysical Journal	Geophys. J.
Geophysical Journal International	Geophys. J. Int.
Glasgow Mathematical Journal	Glasg. Math. J.
Glasnik Matematički	Glas. Mat
Graduate Series in Analysis	Grad. Ser. Anal.
Graduate Studies in Mathematics	Grad. Stud. Math.
Grammars	Grammars
Granular Matter	Granul. Matter
Graphical Models and Image Processing	Graph. Models Image Process.
Grazer Mathematische Berichte	Grazer Math. Ber.
Groups, Geometry, and Dynamics	Groups Geom. Dyn.
Hacettepe Journal of Mathematics and Statistics	Hacet. J. Math. Stat.
Hacettepe Bulletin of Natural Sciences and Engineering	Hacettepe Bull. Nat. Sci. Eng.
Hadronic Journal	Hadronic J.
Hadronic Press Monographs in Mathematics	Hadronic Press Monogr. Math.
Hadronic Press Monographs in Theoretical Physics	Hadronic Press Monogr. Theor. Phys.
Handbuch der Informatik	Handb. Inform.
Handbook of Logic in Computer Science	Handb. Log. Comput. Sci.

A9, *continued*

Handbook of Numerical Analysis	Handb. Numer. Anal.
Handbooks in Operations Research and Management Science	Handb. Oper. Res. Manage. Sci.
Heat and Technology	Heat Technol.
Hiroshima Mathematical Journal	Hiroshima Math. J.
History of Logic	Hist. Logic
Historia Mathematica	Hist. Math.
Homology, Homotopy and Applications	Homology, Homotopy Appl.
Hunan Annals of Mathematics	Hunan Ann. Math.
IAPQR Transactions	IAPQR Trans.
IBM Journal of Research and Development	IBM J. Res. Dev.
IBM Systems Journal	IBM Syst. J.
IEE Computing Series	IEE Comput Ser.
IEE Digital Electronics and Computing Series	IEE Digital Electron. Comput. Ser.
IEE Proceedings. Circuits Devices and Systems.	IEE Proc., Circuits Devices Syst.
IEE Proceedings. Computers and Digital Techniques	IEE Proc., Comput. Digit. Tech.
IEEE Annals of the History of Computing	IEEE Ann. Hist. Comput.
IEEE Transactions on Automatic Control	IEEE Trans. Autom. Control
Illinois Journal of Mathematics	Ill. J. Math.
IMA Journal of Applied Mathematics	IMA J. Appl. Math.
International Journal of Algebra and Computation	Internat. J. Algebra Comput.
IMA Journal of Mathematics Applied in Medicine and Biology	IMA J. Math. Appl. Med. Biol.
IMA Journal of Numerical Analysis	IMA J. Numer. Anal.
Industrial Mathematics	Ind. Math.
Indagationes Mathematicae	Indag. Math.
Indiana University Mathematics Journal	Indiana Univ. Math. J.
Information and Computation	Inf. Comput.
Information and Control	Inf. Control
Insurance Mathematics & Economics	Insur. Math. Econ.
International Game Theory Review	Int. Game Theory Rev.
International Journal for Mathematics Teaching and Learning	Int. J. Math. Teach. Learn.
International Journal of Applied Mathematical Sciences	Int. J. Appl. Math. Sci.
International Journal of Applied Mathematics	Int. J. Appl. Math.

A9, *continued*

International Journal of Applied Mathematics and Computer Science	Int. J. Appl. Math. Comput. Sci.
International Journal of Computing and Applications	Int. J. Comput. Appl.
International Journal of Computational Fluid Dynamics	Int. J. Comput. Fluid Dyn.
International Journal of Ecological Economics & Statistics	Int. J. Ecol. Econ. Stat.
International Journal of Information Management	Int. J. Inf. Manag.
International Journal of Mathematics	Int. J. Math.
International Journal of Mathematics and Analysis	Int. J. Math. Anal.
International Journal of Mathematics and Computer Science	Int. J. Math. Comput. Sci.
International Journal of Mathematics and Statistics	Int. J. Math. Stat.
International Journal of Nonlinear Modelling in Science and Engineering	Int. J. Nonlinear Model. Sci. Eng.
International Journal of Science & Engineering	Int. J. Sci. Eng.
International Journal of Tomography & Statistics	Int. J. Tomogr. Stat.
Inventiones Mathematicae	Invent. Math.
Journal of Algebra	J. Algebra
Journal of Applied Mathematics	J. Appl. Math.
Journal of Applied Mathematics and Mechanics	J. Appl. Math. Mech.
Journal of the Australian Mathematical Society	J. Aust. Math. Soc.
Journal of Classification	J. Classif.
Journal of Computational Mathematics	J. Comput. Math.
Journal of Differential Equations	J. Differ. Equations
Journal of Dynamical Systems and Geometric Theories	J. Dyn. Syst. Geom. Theor.
Journal of Economic Growth	J. Econ. Growth
Journal of Fluid Mechanics	J. Fluid Mech.
Journal of Functional Analysis	J. Funct. Anal.
Journal of Geometry	J. Geom.
Journal of Graph Algorithms and Applications	J. Graph Algorithms Appl.
Journal of Graph Theory	J. Graph Theory
Journal of Inequalities and Applications	J. Inequal. Appl.
Journal of Integer Sequences	J. Integer Seq.

A9, *continued*

Journal of the Japan Statistical Society	J. Jap. Stat. Soc.
Journal of the Korean Mathematical Society	J. Korean Math. Soc.
Journal of Knot Theory and its Ramifications	J. Knot Theory Ramifications
Journal of Lie Theory	J. Lie Theory
Journal of Mathematical Physics	J. Math. Phys.
Journal of Mathematics and Music	J. Math. Music
Journal of Number Theory	J. Number Theory
Journal of Operator Theory	J. Oper. Theory
Journal of the American Mathematical Society	J. Amer. Math. Soc.
Karachi Journal of Mathematics	Karachi J. Math.
Knowledge Engineering Review, The	Knowl. Eng. Rev.
Korean Journal of Computational & Applied Mathematics, The	Korean J. Comput. Appl. Math.
Kyungpook Mathematical Journal	Kyungpook Math. J.
Kyushu Journal of Mathematics	Kyushu J. Math.
Large Scale Systems	Large Scale Syst.
Lectures in Applied Mathematics	Lect. Appl. Math.
Lectures on Mathematics in the Life Sciences	Lect. Math. Life Sci.
Lecture Notes in Computer Science	Lect. Notes Comput. Sci.
Lecture Notes in Mathematics	Lect. Notes Math.
Lobachevskii Journal of Mathematics	Lobachevskii J. Math.
Manuscripta Geodaetica	Manuscr. Geod.
Manuscripta Mathematica	Manuscr. Math.
Matematichki Vesnik	Mat. Vesn.
Mathematical Inequalities & Applications	Math. Inequal. Appl.
Mathematical Intelligencer, The	Math. Intell.
Mathematical Journal of Okayama University	Math. J. Okayama Univ.
Mathematical Methods of Statistics	Math. Methods Stat.
Mathematical Modelling and Scientific Computing	Math. Model. Sci. Comput.
Mathematical Notes	Math. Notes
Mathematical Physics Electronic Journal	Math. Phys. Electron. J.
Mathematical Problems in Engineering	Math. Probl. Eng.
Mathematical Proceedings of the Cambridge Philosophical Society	Math. Proc. Camb. Philos. Soc.
Mathematical Sciences Research Journal	Math. Sci. Res. J.

A9, *continued*

Mathematical Structures in Computer Science	Math. Struct. Comput. Sci.
Mathematicians of Our Time	Math. Time
Mathematics Magazine	Math. Mag.
Mathematics of Computation	Math. Comp.
Methods of Operations Research	Methods Oper. Res.
Missouri Journal of Mathematical Sciences	Missouri J. Math. Sci.
Nagoya Mathematical Journal	Nagoya Math. J.
Natural Computing	Nat. Comput.
Neural, Parallel & Scientific Computations	Neural Parallel Sci. Comput.
New Mathematical Library	New Math. Libr.
New York Journal of Mathematics	New York J. Math.
News Bulletin of Calcutta Mathematical Society	News Bull. Calcutta Math. Soc.
Nexus Network Journal	Nexus Netw. J.
Nonlinear Analysis Forum	Nonlinear Anal. Forum
Notices of the American Mathematical Society	Notices Am. Math. Soc.
Ohio State University Mathematical Research Institute Publications.	Ohio State Univ. Math. Res. Inst. Publ.
Open Economies Review	Open Econ. Rev.
Oxford Applied and Engineering Mathematics	Oxf. Appl. Eng. Math.
Oxford Engineering Science Series	Oxf. Eng. Sci. Ser.
Oxford Graduate Texts in Mathematics	Oxf. Grad. Texts Math.
Pacific Journal of Mathematics	Pac. J. Math.
Panamerican Mathematical Journal	Panam. Math. J.
Periodical of Ocean University of China	Period. Ocean Univ. China
Proceedings of Symposia in Applied Mathematics	Proc. Symp. Appl. Math.
Proceedings of Symposia in Pure Mathematics	Proc. Symp. Pure Math.
Proceedings of the American Mathematical Society	Proc. Amer. Math. Soc.
Proceedings of the Mathematical and Physical Society of Egypt	Proc. Math. Phys. Soc. Egypt
Proceedings of the Royal Society of Edinburgh. Section A. Mathematics	Proc. R. Soc. Edinb., Sect. A, Math.
Progress in Mathematics	Prog. Math.
Proyecciones	Proyecciones
Publications Mathématiques de l'IHÉS	Publ. Math.
Pure and Applied Mathematics	Pure Appl. Math.

A9, *continued*

Quaestiones Mathematicae	Quaest. Math.
Questions and Answers in General Topology	Quest. Answers Gen. Topology
Random & Computational Dynamics	Random Comput. Dyn.
Ratio Mathematica	Ratio Math.
Representation Theory	Represent. Theory
Research Notes in Mathematics	Res. Notes Math.
Review of Economic Dynamics	Rev. Econ. Dyn.
Reviews in Mathematical Physics	Rev. Math. Phys.
Ricerche Economiche	Ric. Econ.
Rocky Mountain Journal of Mathematics	Rocky Mt. J. Math.
Romanian Journal of Physics	Rom. J. Phys.
Russian Academy of Sciences. Doklady. Mathematics	Russ. Acad. Sci., Dokl., Math.
Russian Journal of Mathematical Physics	Russ. J. Math. Phys.
Russian Mathematics	Russ. Math.
Scientific Annals of Computer Science	Sci. Ann. Comput. Sci.
SIAM Journal on Algebraic and Discrete Methods	SIAM J. Algebraic Discrete Methods
SIAM Journal on Applied Dynamical Systems	SIAM J. Appl. Dyn. Syst.
SIAM Journal on Computing	SIAM J. Comput.
SIAM Journal on Control and Optimization	SIAM J. Control Optim.
SIAM Journal on Mathematical Analysis	SIAM J. Math. Anal.
SIAM Journal on Matrix Analysis and Applications	SIAM J. Matrix Anal. Appl.
Siberian Advances in Mathematics	Sib. Adv. Math.
Siberian Journal of Differential Equations	Sib. J. Differ. Equ.
Southwest Journal of Pure and Applied Mathematics	Southwest J. Pure Appl. Math.
St. Petersburg Mathematical Journal	St. Petersburg Math. J.
Studies in Applied Mathematics	Stud. Appl. Math.
Studies in the History of Mathematics and Physical Sciences	Stud. Hist. Math. Phys. Sci.
Systems Engineering and Electronics	Syst. Eng. Electron.
Systems Science and Mathematical Sciences	Syst. Sci. Math. Sci.
Taiwanese Journal of Mathematics	Taiwanese J. Math.
Texts in Applied Mathematics	Texts Appl. Math.

A9, *continued*

Thai Journal of Mathematics	Thai J. Math.
Theory and Applications of Categories	Theory Appl. Categ.
Theoretical and Computational Fluid Dynamics	Theor. Comput. Fluid Dyn.
Theory of Probability and its Applications	Theory Probab. Appl.
Theory of Probability and Mathematical Statistics	Theor. Probability and Math. Statist.
Tokyo Journal of Mathematics	Tokyo J. Math.
Topics in Applied Physics	Top. Appl. Phys.
Topology	Topology
Topology and its Applications	Topology Appl.
Transactions of the American Mathematical Society	Trans. Amer. Math. Soc.
TRU Mathematics	TRU Math.
Turkish Journal of Mathematics	Turk. J. Math.
Ukrainian Mathematical Journal	Ukr. Math. J.
Ultra Scientist of Physical Sciences	Ultra Sci. Phys. Sci.
Uniform Distribution Theory	Unif. Distrib. Theory
University of Arkansas Lecture Notes in the Mathematical Sciences, The	Univ. Arkansas Lect. Notes Math. Sci.
University Lecture Series	Univ. Lect. Ser.
Vietnam Journal of Mathematics	Vietnam J. Math.
Virtual Reality	Virtual Real.
Visual Mathematics	Vis. Math.
WIT Transactions on Modelling and Simulation	WIT Trans. Modul. Simul.
World Scientific Advanced Series in Dynamical Systems	World Sci. Adv. Ser. Dyn. Syst.
World Scientific Series in Applicable Analysis	World Sci. Ser. Appl. Anal.
World Scientific Series in Robotics and Intelligent Systems	World Sci. Ser. Robot. Intell. Syst.
Yokohama Mathematical Journal	Yokohama Math. J.

Appendix B. Tables and Conventions for Physics

Contents

B1 International System of Units (SI) Tables

B1.1 SI Base Units

Quantity	Name	Symbol
length	meter	m
mass	kilogram	kg
time	second	s
electric current	ampere	A
thermodynamic temperature	kelvin	K
amount of substance	mole	mol
luminous intensity	candela	cd

B1.2 Standard SI Prefixes

Factor	Name	Symbol
$10^{24} = (10^3)^8$	yotta	Y
$10^{21} = (10^3)^7$	zetta	Z
$10^{18} = (10^3)^6$	exa	E
$10^{15} = (10^3)^5$	peta	P
$10^{12} = (10^3)^4$	tera	T
$10^9 = (10^3)^3$	giga	G
$10^6 = (10^3)^2$	mega	M
$10^3 = (10^3)^1$	kilo	k
10^2	hector	h
10^1	deca	da
10^0	-	-
10^{-1}	deci	d
10^{-2}	centi	c
$10^{-3} = (10^3)^{-1}$	milli	m
$10^{-6} = (10^3)^{-2}$	micro	μ
$10^{-9} = (10^3)^{-3}$	nano	n
$10^{-12} = (10^3)^{-4}$	pico	p
$10^{-15} = (10^3)^{-5}$	femto	f
$10^{-18} = (10^3)^{-6}$	atto	a
$10^{-21} = (10^3)^{-7}$	zepto	z
$10^{-24} = (10^3)^{-8}$	yocto	y

B1.3 SI-Derived Units with Special Names

Quantity	Name	Sym -bol	Expression in terms of base units	Expression in terms of other units
area	square meter	–	m^2	–
volume	cubic meter	–	m^3	–
speed, velocity	meter per second	–	m/s	–
acceleration	meter per second squared	–	m/s^2	–
wave number	reciprocal meter	–	m^{-1}	–
mass density	kilogram per cubic meter	–	kg/m^3	–
specific volume	cubic meter per kilogram	–	m^3/kg	–
current density	ampere per square meter	–	A/m^2	–
magnetic field strength	ampere per meter	–	A/m	–
amount of substance concentration	mole per cubic meter	–	mol/m^3	–
luminance	candela per square meter	–	cd/m^2	–
plane angle	radian	rad	$m \cdot m^{-1} = 1$	–
solid angle	steradian	sr	$m^2 \cdot m^{-2} = 1$	–
frequency	hertz	Hz	s^{-1}	–
force	newton	N	$m \cdot kg \cdot s^{-2}$	–
pressure, stress	pascal	Pa	$m^{-1} \cdot kg \cdot s^{-2}$	N/m^2

B1.3, *continued*

Quantity	Name	Sym-bol	Expression in terms of base units	Expression in terms of other units
energy, work, quantity of heat	joule	J	$m^2 \cdot kg \cdot s^{-2}$	N·m
power, radiant flux	watt	W	$m^2 \cdot kg \cdot s^{-3}$	J/s
electric charge, quantity of electricity	coulomb	C	s·A	–
electric potential, potential difference, electromotive force	volt	V	$m^2 \cdot kg \cdot s^{-3} \cdot A^{-1}$	W/A
capacitance	farad	F	$m^{-2} \cdot kg^{-1} \cdot s^4 \cdot A^2$	C/V
electric resistance	ohm	Ω	$m^2 \cdot kg \cdot s^{-3} \cdot A^{-2}$	V/A
electric conductance	siemens	S	$m^{-2} \cdot kg^{-1} \cdot s^3 \cdot A^2$	A/V
magnetic flux	weber	Wb	$m^2 \cdot kg \cdot s^{-2} \cdot A^{-1}$	V·s
magnetic flux density	tesla	T	$kg \cdot s^{-2} \cdot A^{-1}$	Wb/m^2
inductance	henry	H	$m^2 \cdot kg \cdot s^{-2} \cdot A^{-2}$	Wb/A
Celsius temperature	degree Celsius	°C	K	–
luminous flux	lumen	lm	–	cd·sr
illuminance	lux	lx	$m^{-2} \cdot cd \cdot sr$	lm/m^2
activity (of a radionuclide)	becquerel	bq	s^{-1}	–
absorbed dose, specific energy, kerma	gray	Gy	$m^2 \cdot s^{-2}$	J/kg
dose equivalent	sievert	Sv	$m^2 \cdot s^{-2}$	J/kg
catalytic activity	katal	kat	$s^{-1} \cdot mol$	

B1.4 Units in Use with SI

Name	Symbol	Value in SI Unit
minute	min	1 min = 60 s
hour	h	1 h = 60 min = 3 600 s
day	d	1 d = 24 h = 86 400 s
degree	°	$1\ ° = (\pi/180)$ rad
minute	'	$1\ ' = (1/60)° = (\pi/10\ 800)$ rad
second	"	$1\ " = (1/60)\ ' = (\pi/648\ 000)$ rad
liter	l, L	$1\ L = 1\ dm^3 = 10^{-3}\ m^3$
metric ton	t	$1\ t = 10^3\ kg$

B1.5 Units Approved for Use with SI

Name	Symbol	Value in SI Unit
nautical mile		1 nautical mile = 1 852 m
knot	kn	1 knot = 1 nautical mile per hour = (1852/3600) m/s
angstrom	Å	$1\ Å = 0.1\ nm = 10^{-10}\ m$
are	a	$1\ a = 1\ dam^2 = 10^2\ m^2$
hectare	ha	$1\ ha = 1\ hm^2 = 10^4\ m^2$
barn	b	$1\ b = 100\ fm^2 = 10^{-28}\ m^2$
bar	bar	$1\ bar = 0.1\ MPa = 10^5\ Pa$
galileo	Gal	$1\ Gal = 1\ cm/s^2 = 10^{-2}\ m/s^2$
curie	Ci	$1\ Ci = 3.7 \times 10^{10}\ Bq$
roentgen	R	$1\ R = 2.58 \times 10^{-4}\ Gy$
rad	rad	$1\ rad = 1\ cGy = 10^{-2}\ Gy$
rem	rem	$1\ rem = 1\ cSv = 10^{-2}\ Sv$

B1.6 SI to Non-SI Conversion Factors

To convert Column 1 into Column 2, multiply by	Column 1 SI Unit	Column 2 non-SI Unit	To convert Column 2 into Column 1, multiply by
Length			
0.621	kilometer, km $(10^{-3}$ m$)$	mile, mi	1.609
1.094	meter, m	yard, yd	0.914
3.28	meter, m	foot, ft	0.304
1.0	micrometer, μm $(10^{-6}$ m$)$	micron, μ	1.0
3.94×10^{-2}	millimeter, mm $(10^{-3}$ m$)$	inch, in	25.4
10	nanometer, nm $(10^{-9}$ m$)$	angstrom, Å	0.1
Area			
2.47	hectare, ha	acre	0.405
247	square kilometer, km^2 $(10^3$ m$)^2$	acre	4.05×10^{-3}
0.386	square kilometer, km^2 $(10^3$ m$)^2$	square mile, mi^2	2.590
2.47×10^{-4}	square meter, m^2	acre	4.05×10^3
10.76	square meter, m^2	square foot, ft^2	9.29×10^{-2}
1.55×10^{-3}	square millimeter, mm^2 $(10^{-3}$ m$)^2$	square inch, in^2	645
Volume			
9.73×10^{-3}	cubic meter, m^3	acre-inch	102.8
35.3	cubic meter, m^3	cubic foot, ft^3	2.83×10^{-2}
6.10×10^4	cubic meter, m^3	cubic inch, in^3	1.64×10^{-5}
2.84×10^{-2}	liter, L $(10^{-3}$ m$^3)$	bushel, bu	35.24
1.057	liter, L $(10^{-3}$ m$^3)$	quart (liquid), qt	0.946
3.53×10^{-2}	liter, L $(10^{-3}$ m$^3)$	cubic foot, ft^3	28.3
0.265	liter, L $(10^{-3}$ m$^3)$	gallon	3.78
33.78	liter, L $(10^{-3}$ m$^3)$	ounce (fluid), oz	2.96×10^{-2}
2.11	liter, L $(10^{-3}$ m$^3)$	pint (fluid), pt	0.473

B1.6, *continued*

Mass			
2.20×10^{-3}	gram, g (10^{-3} kg)	pound, lb	454
3.52×10^{-2}	gram, g (10^{-3} kg)	ounce (avdp), oz	28.4
2.205	kilogram, kg	pound, lb	0.454
0.01	kilogram, kg	quintal (metric), q	100
1.10×10^{-3}	kilogram, kg	ton (2000 lb), ton	907
1.102	megagram, Mg (tonne)	ton (U.S.), ton	0.907
1.102	tonne, t	ton (U.S.), ton	0.907
Yield and Rate			
0.893	kilogram per hectare, kg ha^{-1}	pound per acre, lb acre^{-1}	1.12
7.77×10^{-2}	kilogram per cubic meter, kg m^{-3}	pound per bushel, lb bu^{-1}	12.87
1.49×10^{-2}	kilogram per hectare, kg ha^{-1}	bushel per acre, 60 lb	67.19
1.59×10^{-2}	kilogram per hectare, kg ha^{-1}	bushel per acre. 56 lb	62.71
1.86×10^{-2}	kilogram per hectare, kg ha^{-1}	bushel per acre, 48 lb	53.75
0.107	liter per hectare, kg ha^{-1}	gallon per acre	9.35
893	tonne per hectare, Mg ha^{-1}	pound per acre, lb acre^{-1}	1.12×10^{-3}
893	megagram per hectare, Mg ha^{-1}	pound per acre, lb acre^{-1}	1.12×10^{-3}
0.446	megagram per hectare, Mg ha^{-1}	ton (2000 lb) per acre, ton acre^{-1}	2.24
2.24	meter per second, m s^{-1}	mile per hour	0.447
Specific Surface			
10	square meter per kilogram, m^2 kg^{-1}	square centimeter per gram, cm^2 g^{-1}	0.1
1000	square meter per kilogram, m^2 kg^{-1}	square millimeter per gram, mm^2 g^{-1}	0.001

B1.6, *continued*

Pressure			
9.90	megapascal, MPa (10^6 Pa)	atmosphere	0.101
10	megapascal, MPa (10^6 Pa)	bar	0.1
1.00	megagram, per cubic meter, Mg m^{-3}	gram per cubic centimeter, g cm^{-3}	1.00
2.09×10^{-2}	pascal, Pa	pound per square foot, lb ft^{-2}	47.9
1.45×10^{-4}	pascal, Pa	pound per square inch, lb in^{-2}	6.90×10^3
Temperature			
1.00 (K − 273)	kelvin, K	Celsius, °C	1.00 (°C + 273)
(9/5 °C) +32	Celsius, °C	Fahrenheit, °F	5/9 (°F − 32)
Energy, Work, Quantity of Heat			
9.52×10^{-4}	joule, J	British thermal unit, Btu	1.05×10^3
0.239	joule, J	calorie, cal	4.19
10^7	joule, J	erg	10^{-7}
0.735	joule, J	foot-pound	1.36
2.387×10^{-5}	joule per square meter, W m^{-2}	calorie per square centimeter (langley)	4.19×10^4
10^5	newton, N	dyne	10^{-5}
1.43×10^{-3}	watt, per square meter, W m^{-2}	calorie per square centimeter minute (irradiance), cal cm^{-2} min^{-1}	698
Transpiration and Photosynthesis			
3.60×10^{-2}	milligram per square mater second, mg m^{-2} s^{-1}	gram per square decimeter hour, g dm^{-2} h^{-1}	27.8

B1.6, *continued*

5.56×10^{-3}	milligram (H_2O) per square mater second, mg m^{-2} s-^{1}	micromole (H_2O) per square centimeter second, μ mol cm^{-2} s^{-1}	180
10^{-4}	milligram per square mater second, mg m^{-2} s-^{1}	Milligram per square centimeter second, mg cm^{-2} s^{-1}	10^4
35.97	milligram per square mater second, mg m^{-2} s-^{1}	milligram per square decimeter hour, mg dm^{-2} h^{-1}	2.78×10^{-2}
Plane Angle			
57.3	radian, rad	degrees (angle), °	1.75×10^{-2}
Electrical Conductivity, Electricity, and Magnetism			
10	siemen per meter, S m^{-1}	millimho per centimeter, mimho cm^{-1}	0.1
10^4	tesla, T	gauss, G	10^{-4}
Water Measurement			
9.73×10^{-3}	cubic meter, m^3	acre-inch, acre-in	102.8
9.81×10^{-3}	cubic meter per hour, m^3 h^{-1}	cubic foot per second, ft^3 s^{-1}	101.9
4.40	cubic meter per hour, m^3 h^{-1}	U.S. gallon per minute, gal min^{-1}	0.227
8.11	hectare meter, ha m	acre-foot, acre-ft	0.123
97.28	hectare meter, ha m	acre-inch, acre-in	1.03×10^{-2}
8.1×10^{-2}	hectare centimeter, ha cm	acre-foot, acre-ft	12.33

B1.6, *continued*

Concentrations			
1	centimole per kilogram, cmol kg^{-1}	milliequivalent per 100 grams, meq 100 g^{-1}	1
0.1	gram per kilogram, g kg^{-1}	percent, %	10
1	milligram per kilogram, mg kg^{-1}	parts per million, ppm	1
Radioactivity			
2.7×10^{-11}	becquerel, Bq	curie, Ci	3.7×10^{10}
2.7×10^{-2}	becquerel per kilogram, Bq kg^{-1}	picocurie per gram, pCi g^{-1}	37
100	gray, Gy (absorbed dose)	rad, rd	0.01
100	sievert, Sv (equivalent dose)	rem (roentgen equivalent man)	0.01

B2 Electromagnetic Spectrum

A remarkable achievement of 19th-century physics was the discovery that various radiation phenomena could be seen and explained as the product of the interaction and promulgation of electric and magnetic fields. This includes the radiation that makes up the visible radiation, but extends far in doth directions on the wavelength-frequency scale (*see* the illustration on the next page for a rough depiction of the spectrum).

The speed of light, symbolized by c, is approximately 3×10^8 meters per second (m/s), and the equation $c = \lambda \upsilon$, where λ is the wavelength (measured in meters—for some disciplines and for shorter wavelength radiation, lengths are measured in units of a micrometer, 10^{-6} m, symbolized by "μm"; or the nanometer, 10^{-9} m, symbolized by nm. In spectroscopy, wavelengths are often expressed in terms of Angstrom units (Å), where 10 Å = 1 nm. Infrared spectroscopists use the "wave number" ($\overline{\nu}$), defined as the number of waves in one centimeter (i.e., $\overline{\nu} = 1/\lambda$), and is therefore measured in reciprocal centimeters (cm^{-1}).

In astronomy, the relationship between the wavelength of radiation and its energy (a finding of modern physics) is used to express radiation in terms of its energy content, E, given by the equation $E=h\upsilon$, (where h is Planck's constant—*see* **Appendix B).** The energy in electron volts (eV) is given by the equation $E = 1.234 / \lambda$, where λ is given in micrometers. The electron volt is a non-SI unit that converts to SI by the conversion factor, 1 eV = $1.60217646 \times 10^{-19}$ joules (or 0.1602…aJ).

These measurements have taken on a new importance with the advent of space-based astronomical observation, which circumvents the restrictions imposed on the inability of radiation from space other than visible and radio frequencies to penetrate the earth's atmosphere.

Electromagnetic radiation is also observed to have two components of polarization, which (when light is polarized) are respectively characterized as "circular " or "linear," or "perpendicular" or "parallel."

For the various names of radiation, the following rules apply:

• The proper form is "X-ray," and not "x-ray," "X ray," or the many variations possible, though the rule is widely ignored, even in technical papers. Other designations are not capitalized or hyphenated ("gamma rays," "cosmic rays") except when used adjectively ("gamma-ray burst").

• The abbreviations UV and IR are used for ultraviolet and infrared radiation, respectively; these segments of the spectrum are divided into segments that are labeled A, B, and C, and which appear with hyphens ("UV-A," "IR-B"). Designations "near UV" and "far UV," or biological designations of "actinic UV" and "vacuum UV" are not hyphenated.

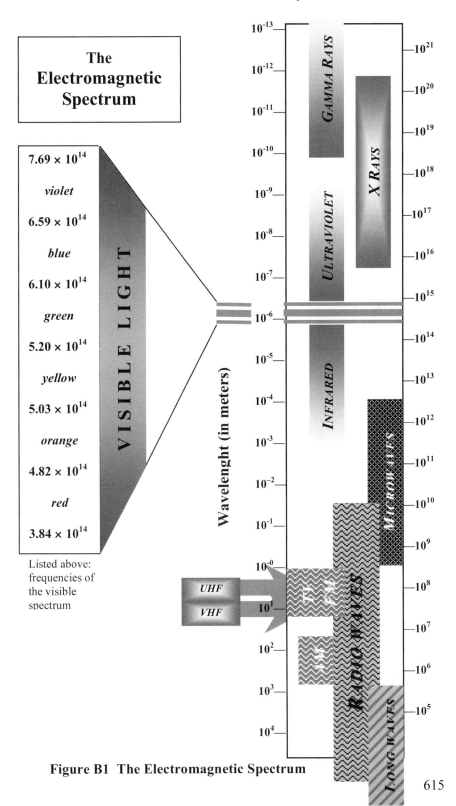

Figure B1 The Electromagnetic Spectrum

B3 Style Guide to Physical Effects and Phenomena

AC Kerr effect
Acoustoelectric effect
Alexandrite effect
Alignment effects
Analogue sates
Anisotropy effect
Anisotropy effects
Arago's experiment
Archimede's principle
Auger effect
Avalanche effect
Azbel-Kaner effect
Babinet's principle
Back-Goudsmit effect
Barkhausen effect
Barkhausen effect, anomalous
Barkhausen effect in thin films
Barnett effect
Baratropic phenomenon
Barrier-layer photoelectric effect
Bauschinger effect
Beck effect
Becquerel effect
Becquerel effect
Bénard effect
Berremann effect
Binaural effect
Birefringence
Blears effect
Bobeck effect
Boersch effect
Bohm-Aharonov effect
Böning effect
Borrmann effect
Bragg effect
Branley-Lenard effect
Breaking effects
Brillouin scattering

Bulk effect
Burstein effect
Cage effect
Casimir effect
Channelling effect
Cherenkov effect
Coanda effect
Cold-conductor effect
Colligative effects
Comovement effect
Compton effect
Compton effect, inverse
Cooling effect
Coriolis force
Corona effect
Cotton-Mouton effect
Cottrell effect
Coupling effects
Coupling effect, quantum-mechanical
Cross effects
Crystal effects
Crystal photoeffect
Current conduction effects
Debye effects
Debye-Falkenhagen effect
Debye-Sears effect
Déchêne effect
Deck effect
Degradation effect
ΔE effect
Dember effect
Destriau effect
Diffraction
Diffraction filter-beam focusing effect
Diffusion
Diffusion thermo-effects
Dipole effects

B3, *continued*

Dispersion effects
Dissociation-voltage effect
Doppler effect, acoustic
Dorn effect
Drag effect
Dynamoelectric principle
Early effect
Edison effect
Effects in electrolytes
Einstein effects
Einstein-de Haas effect
Electrocaloric effect
Electrocapillarity
Electrokinetic effects
Electrooptic effects
Electroosmosis
Electrophoresis
Electrostriction
Electro-tunnelling effect
Electroviscous effect
Emitter-dip effect
Eötvös effect
Eötvös experiment
Exchange effects
Fano effect
Faraday effect
Ferroelectric Barkhausen effect
Field effect
Field effect
Fizeau's experiment
Flow current
Foldy effect
Foucault's experiment
Fractional quantum Hall effect
Franck-Condon principle
Franck-Hertz experiment
Franz-Keldysh effect
Funkel effect
Galvanomagnetic effects
Corbino effect (1911)

Dissymmetry effect
Ettingshausen effect
Hall effect
Hall effect, anomalous
Magnetoresistance effect
Nernst effect
Thomson effect
Galvanomagnetic effects in
  semiconductors
Gay-Lussac experiment
Goos-Hähnchen effect
Gorsky effect
Gravitational frequency shift
Ground effect
Gudden-Pohl effect
Gunn effect
Gyromagnetic effects
Haas effect
De Haas-van Alphen effect
Hanle effect
Heal effect
Hertz effect
Hertz's experiments
Holography
Hopkinson effect Hopping
  effect
Hubble effect
Hughes effect
Hyper Raman effect
Immersion effect
Interference
Ioffe effect
Isotope effect
Jaccarino-Peter effect
Jahn-Teller effect
Johnsen-Rahbeck effect
Josephson effects
Joule effect
Joule-Thomson effect
Kelvin-Helmholtz effect

B3, *continued*

Kerr effect
Kikuchi effect
Kirk effect
Kirkendall effect
Kirlian effect
von-Klitzing effect
Knight shift
Knudsen effect
Kohn effect
Kondo effect
Kossel effect
Kundt effect
Lamb shift
Langmuir effect
Lasers
Laue effect
Leidenfrost phenomenon
Lenard effect
Lens effect
Light deflection
Light-hydraulic effect
Liquid crystal effects
Bistability effect
Electrohydrodynamic effects
Field effects
Guest-host effect
Memory effect
OPD effect
Schadt-Helfrich effect
Texture conversion effect
Thermooptic effect
Lossew effect
Luminescence effects
Macaluso-Corbino effect
Magnetic aftereffects
Magnetic contraction effect
Magnetoacoustic effect
Magnetocaloric effect
Magnetoelectric effect
Magnetohydrodynamic effect

Magnetomechanical effects
Magnetooptic effects
Magnetooptic Kerr effect
Magnetoplasma effects
Magnetostriction
Magnus effect
Majorana effect
Malus's experiment
Marangoni effect
Masking effect
Matteucci effect
Maxwell effect
Meissner effect
Memory effect
Merrington effect
Michelson's experiment
Mie effect
Modal noise effect
Mössbauer
Neugebauer effect
Nikischov effect
Nonlinear-optical effects
Nuclear Hanle effect
Nuclear magnetic
  (electrodynamic) resonance
  effects
Nuclear photoeffect
Nuclear spin effects
Nucleation effect
Onnes effects
Optical Doppler effect
Optical Kerr effect
Optoacoustic effect
Optogalvanic effect
Overhauser effect
Ovshinksky effects
Packing effect
Pair-production effect
Particle scattering
Bhabha scattering

B3, *continued*

Born scattering
Delbrück scattering
Mott scattering
Rutherford scattering
Paschen-Back effect
Peak effect
Penning effect
Perihelion rotation, secular
Photocapacitative effects
Photodielectric effect
Photoelectric effects
External photoelectric effect
Hallwachs effect
Hertz effect
Internal photoelectric effect
Selective photoelectric effect
Photoelectric effect at high light
  intensities
Photo-Hall effect
Photorefractive effects
Piezocaloric effect
Piezoelectric effect
Piezomagnetic effect
Piezoresistance effect
Pinch effect
Plasma effects
Pockels effect
Poisson effect
Polarization effect in mass
  spectroscopy
Pole effect
Pomeranchuk effect
Poole-Frenkel effect
Portevin-Le Chatelier effect
Poynting effect
Primakoff effect
Printout effect
Proximity effect
Pyroelectric effect
Radiometer effect

Rainbow effect
Raman-induced Kerr effect
Raman-Smekal effect
Ramsauer effect
Ranque effect
Rayleigh scattering
Rayleigh-Taylor effect
Rectifier effect
Red Shift
Rehbinder effect
Relaxation effects
Renninger effect
Resistance-pressure-tension
  effect
Richardson effect
Riehl effect
Röntgen-Eichenwald
  experiment
Rotation effect, magnetic
Rowland effect
Runaway effect
Sagnac effect
Salting-out effect
Scattering effects
Schoch effect
Schottky effect
Screening effect
Self-focusing
Senftleben effect
Shapiro's experiment
Shot effect
Shpol'skii effect
Shubnikov-de Haas effect
Size effects
Skin effect
Skin effect, anomalous
Smith-Purcell-Salisbury effect
Smoluchowski effect
Snoek effect
Sondheimer effect

B3, *continued*

Staebler-Wronski effect
Stark effect
Stavermann effect
Stern-Gerlach experiment
Stroboscope effect
Substituent effects
Suhl effect
Superconductivity
Synergetic phenomena
Szilard-Chalmers effect
Taylor effect
Telescope effect
Thermistor effect
Thermoelectric effects
    Bridgman effect
    Peltier effect
    Phonon-drag effect
    Seebeck effect
Thermoelectric homogeneous
  effect
Thomson effect
Thomson effect, inverse
Thermomagnetic effects
    Ettingshausen-Nernst
      effect
      Maggi-Righi-effect
      OEN effect
      Righi-Leduc effect

Thirring-Lense effect
Time effects
Touschek effects
Track adaptation effect
Transistor effect
Transit-time effects
Transport effects
Trouten-Noble experiment
Tunnelling effect
Tyndall effect
Übler effect
Ueling effect
Undercooling effect
Valency effect
Varistor effect
Villari effect
Void effect
Voigt effect
Volta effect
Weissenberg effect
Wiedemann effect
Wiegand effect
Wien effect
Wigner effect
Zeeman effect
Zeeman effect, inverse
Zener effect

B4 Coordinate Systems

The choice of the cordinate system in which to formulate a theory or solve a problem can often be decisive. In physics, three systems dominate most formulations. The most straightforward is the **cartesian system**, based on the 2-dimensional system of analytic geometry (**Figure B2**), formulated (independently) by Descartes and Fermat in the seventeenth century. The synthesis of algebra and geometry was critical for the develpment of calculus and Newtonian physics.

In the cartesian system, a point in space is identified by coordinates that are defined by axiis that stretch to infinity. Unit vectors are defined as vectors **i, j,** and **k,** lying along the x-, y-, and z-axis, respectively. A distance vector from the origin of the system to point (x_i, y_i, z_i) would thus be defined as:

$$\mathbf{r}(0,i) = x_i\mathbf{i} + y_i\mathbf{j} + z_i\mathbf{k}$$

to which one may apply the equations of classical mechanics.

Another method of identifying a point in 3-dimensional space is through the use of **spherical coordinates** (**Figure B4**—derived from the 2-dimensional polar coordinate system), in which a point P is described by three numbers (r, θ, φ): a radial length, r, from the origin to the point; the "zenith" angle, θ, made by the radial line r and the positive z-axis; and the "azimuth" angle φ, made by the orthogonal (vertical) projection of the point to the x-y plane and the x-axis.

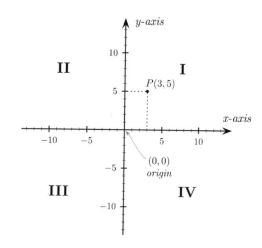

Figure B2. The 2-dimensional cartesian system, with the four quadrants marked.

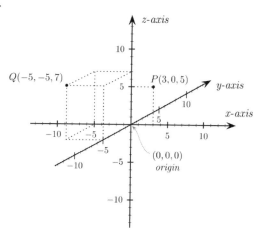

Figure B3. The 3-dimensional cartesian system, with each point uniquely identified by three numbers, x_i, y_i, and z_i.

B4 , *continued*

Lengths and thus vectors may seem a bit more complicated in the spherical system, but several formulations and equations take a simpler form in spherical coordinates. The position vector is given by:
$(r,\ \theta,\ \varphi)$:

$$\mathbf{A} = A_r\,\mathbf{e}_r + A_\theta\,\mathbf{e}_\theta + A_\varphi\,\mathbf{e}_\varphi$$

Spherical coordinates are particularly useful in field-theoretic calculations, a standard part of electromagnetic theory. For example, the Laplacian in spherical coordinates, $\Delta\psi$, for some $\psi\,(r,\ \theta,\ \varphi)$ is:

$$\frac{1}{r}\frac{\partial^2 (r\psi)}{\partial r^2} + \frac{1}{r^2 \sin\theta}\frac{\partial}{\partial\theta}\left(\sin\theta\,\frac{\partial\,\psi}{\partial\theta}\right)$$

$$+ \frac{1}{r^2 \sin^2\theta}\frac{\partial}{\partial\theta}\left(\frac{\partial^2\,\psi}{\partial\varphi^2}\right)$$

which is separable.

A third coordinate system that is often used, particularly in quantum theory, is the **cyclindrical coordinate system (Figure B5)**, named for the cylinder described by the three principal axiis. Its La-placian is even simpler:

$$\Delta\psi = \frac{1}{\rho}\frac{\partial}{\partial\rho}\left(\rho\,\frac{\partial\,\psi}{\partial\rho}\right)$$

$$+ \frac{1}{\rho^2}\frac{\partial^2\,\psi}{\partial\varphi^2} + \frac{\partial^2\,\psi}{\partial z^2}$$

and more useful (in the right configuration).

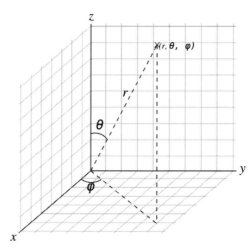

Figure B4. The 3-dimensional spherical coordinate system, each point uniquely identified by three numbers $(r,\ \theta,\ \varphi)$, two of which are measured in radians.

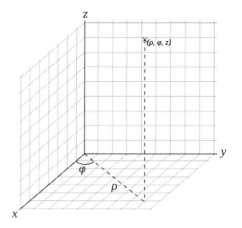

Figure B5. The 3-dimensional cylindrical system, with each point uniquely identified by three numbers (ρ, φ_i, *and z*), only one of which are measured in radians.

622

B5 Abbreviations Commonly Used in Physics

4WM	Four Wave Mixing	AIGER	American Industry Government Emissions Research Consortium
AAMI	Association for the Advancement of Medical Instrumentation	AIP	American Institute of Physics
AAPM	American Association of Physicists in Medicine	ALG	Advanced Lithography Group
ACR (ACRii)	Absolute Cryogenic Radiometer (second upgrade)	AM	amplitude modulation
ACS	American Chemical Society	AMMAC	Mexican Metrology Association
ACTS	Automated Computer Time Service	AMO	Atomic, Molecular and Optical
ADCL	Accredited Dosimetry Calibration Laboratory	AMS	Accelerator-Mass Spectrometry
ADMIT	Analytical Detection Methods for the Irradiation Treatment of foods	ANS	American Nuclear Society
AECL	Atomic Energy Canada Limited	ANSI	American National Standards Institute
AEDC	Arnold Engineering Development Center	ANSOM	Apertureless Near-Field Scanning Optical Microscopy
AFB	Air Force Base	ANVIS	Aviator Night Vision Imaging System
AFGL	Air Force Geophysics Laboratory	APAS	Astrophysical, Planetary and Atmospheric Sciences
AFM	Atomic Force Microscope	APOMA	American Precision Optics Manufacturers Association
AFOSR	Air Force Office of Scientific Research	APRF	Army Pulse Radiation Facility
AFPL	Air Force Phillips Laboratory	APS	American Physical Society *or* Advanced Photon Source
AFRRI	Armed Forces Radiobiology Research Institute	APT	Annular Proton Telescope
AI	Associative Ionization	ARO	Army Research Office
AIAA	American Institute for Aeronautics and Astronautics	ARPES	Angle Resolved Photoelectron Spectroscopy
AIGER	American Industry Government Emissions Research Consortium	AIP	American Institute of Physics

B5, *continued*

ALG	Advanced Lithography Group	ASSI	Airglow Solar Spectrometer Instrument
AM	amplitude modulation	ASTER	Advance Spaceborne Thermal Emission and Reflectance Radiometer
AM1	First Launch in the morning series of EOS platforms	ASET	Association of Super-advanced Electronics Technologies
AMMAC	Mexican Metrology Association	ASTM	American Society for Testing and Materials
AMO	Atomic, Molecular and Optical	ATD	Above-Threshold Dissociation
AMS	Accelerator-Mass Spectrometry	ATI	Above-Threshold Ionization
ANS	American Nuclear Society	ATLAS	Atmospheric Laboratory for Applications and Science
ANSI	American National Standards Institute	ATP	Advanced Technology Program
ANSOM	Apertureless Near-Field Scanning Optical Microscopy	ATW	Accelerator Transmutation of Waste
ANVIS	Aviator Night Vision Imaging System	AURA	Association of Universities for Research in Astronomy
APAS	Astrophysical, Planetary and Atmospheric Sciences	AXAF	Advanced X-ray Astrophysical Facility
APHIS	Animal and Plant Health Inspection Service	AXAF-I	Imaging Advanced X-ray Astrophysical Facility
APOMA	American Precision Optics Manufacturers Association	AXAF-S	Spectroscopy Advanced X-ray Astrophysical Facility
APRF	Army Pulse Radiation Facility	BARC	Bhabha Atomic Research Centre
APS	American Physical Society (*or* Advanced Photon Source)	BB	Blackbody
APT	Annular Proton Telescope	BB	Broad-bandwidth
ARO	Army Research Office	BBIR	Broad Band Infrared
ARPES	Angle Resolved Photoelectron Spectroscopy	BBO	Beta-Barium Borate
ART	Algebraic Reconstruction Technique	BBXRT	Broad-Band X-Ray Telescope
ASCA	Advanced Satellite for Cosmology & Astrophysics	BCC	Broadband Calibration Chamber
ASCA	Japan-NASA X-ray Satellite	BCS	Bardeen-Cooper-Schrieffer theory of superconductivity
ASME	American Society for Mechanical Engineers	BCS	Bragg Crystal Spectrometer

B5, *continued*

BEB	Binary-encounter-Bethe	CAD	Computer Aided Design
BEC	Bose-Einstein Condensation	CAM	Computer Aided Machining
BED	Binary-encounter-dipole	CAMOS	Committee on Atomic, Molecular and Optical Sciences
BEEM	Ballistic Electron Emission Spectroscopy	CARB	Center for Advanced Research in Biotechnology
BEV	Bundesamt für Eich- und Vermessungswesen, Vienna, Austria	CARS	Coherent Anti-Stokes Raman Spectroscopy
BFRL	Building & Fire Research Laboratory	CASS	Calibration Accuracy Support System
BIB	Blocked Impurity Band	CAST	Council of Agricultural Science and Technology
BIFL	Burst Integrated Fluorescence	CBNM	Central Bureau for Nuclear Measurements
BIPM	Bureau International des Poids et Mesures	CCD	Charged Coupled Device
BL	Beam Line at SURF-III	CCDM	Consultative Committee for the Definition of the Meter
BMDO	Ballistic Missile Defense Organization	CCDS	Consultative Committee for the Definition of the Second
BNL	Brookhaven National Laboratory	CCE	Consultative Committee on Electricity
BNM	National Bureau of Metrology, France	CCEMRI	Consultative Committee for Ionizing Radiations, CIPM
BOMAB	Bottle Manikin Absorber	CCG	Calibration Coordination Group
BRAN	Boulder Research and Administrative Network	CCP6	Collaborative Computational Project 6
BRDF	Bidirectional Reflectance Distribution Function	CCPR	Consultative Committee on Photometry and Radiometry
BSDF	Bidirectional Scattering Distribution Function	CCRI	Comité Consultatif des Rayonnements Ionisants
BSS	Beta Secondary System	CCTF	Consultative Committee for Time and Frequency
BSS.2	Beta-Particle Secondary Standards System	CDRH	Center for Devices and Radiological Health
BTI	Bubble Technology, Inc.	CEL	Correlated Emission Laser
BWO	Backward Wave Oscillator	CEMRC	Carlsbad Environmental Monitoring Research Center
BXR	BMDO Transfer Radiometer	CERES	Clouds and Earth's Radiant Energy System

B5, *continued*

CFS	Constant-Final-State Spectroscopy	CRCPD	Conference of Radiation Control Program Directors
CHEM	Chemistry	CRDS	Cavity-Ring-Down Spectroscopy
CIAQ	Committee on Indoor Air Quality	CRI	Cambridge Research Instrumentation
CIE	Commission Internationale De L'Éclairage	CRP	Coordinated Research Program
CIPM	International Committee of Weights and Measures, France	CRT	Cathode Ray Tube
CIRMS	Council on Ionizing Radiation Measurements and Standards	CRYRING	Electron Accelerator and Storage Ring Facility (Stockholm, Sweden)
CIRRPC	Committee on Interagency Radiation Research and Policy Coordination	CSDA	Continuous-Slowing-Down Approximation
CIS	Constant-Initial-State Spectroscopy	CSEWG	Cross Section Evaluation Working Group
CMC	Calibration Measurement Capabilities	CSI	Compton Scatter Imaging
CMR	Colossal Magnetoresistance	CSIC	Consejo Superior de Investigaciones Científicas
CNIF	Californium Neutron Irradiation Facility	CSIR	Council of Scientific and Industrial Research
CNRF	Cold Neutron Research Facility	CSTL	Chemical Science and Technology Laboratory
CODATA	Committee on Data for Science and Technology	CT	Computed Tomographic
CORM	Council for Optical Radiation Measurements	CTI	Critical Technologies Institute
COSPAR	Committee on Space Research	CVD	Chemical Vapor Deposition
CPIC	International Physics Center, Elba	cw (CW)	continuous wave
CPT	conjugation, parity, and time	CXRO	Center for X-Ray Optics at Berkeley, CA
CPU	Central Processing Unit	CY	Calendar Year
CR	Cascaded Rectifier Accelerator	DARPA	Defense Advanced Research Project Agency
CRADA	Cooperative Research and Development Agreement	DEC	Digital Electronics Corporation

B5, *continued*

DMA	Defense Mapping Agency	DNA	Deoxyribose Nucleic Acid (*see next entry*)
CRDS	Cavity-Ring-Down Spectroscopy	DNA	Defense Nuclear Agency (*see previous entry*)
CRI	Cambridge Research Instrumentation	DOC (DoC)	Department of Commerce
CRP	Coordinated Research Program	DOD (DoD)	Department of Defense
CRT	Cathode Ray Tube	DOE (DoE)	Department of Energy
CRYRING	Electron Accelerator and Storage Ring Facility (Stockholm, Sweden)	DOELAP	Department of Energy Laboratory Accreditation Program
CSDA	Continuous-Slowing-Down Approximation	DORT	Discrete Ordinates Code
CSEWG	Cross Section Evaluation Working Group	DPPC	dipalmitoylphosphatidylcholine
CSI	Compton Scatter Imaging	DRAM	Dynamic Random Access Memory
CSIC	Consejo Superior de Investigaciones Científicas	DRIP	Detector Response Intercomparison Program
CSIR	Council of Scientific and Industrial Research	DR-SFG	Doubly Resonant Sum Frequency Generation
CSTL	Chemical Science and Technology Laboratory	DSA	Digital Subtraction Angiography
CT	Computed Tomographic	DSC	differential scattering cross section
CTI	Critical Technologies Institute	DVM	Digital Voltmeter
CU	University of Colorado	DUV	Deep Ultraviolet
CVD	Chemical Vapor Deposition	EBIS	Electron Beam Ion Source
cw (CW)	continuous wave	EBIT	Electron Beam Ion Trap
CXRO	Center for X-Ray Optics at Berkeley, CA	ec	electron-capture
CY	Calendar Year	ECP	Effective Core Potential
DARPA	Defense Advanced Research Project Agency	ECPR	Electrically-Calibrated Pyroelectric Radiometer
DEC	Digital Electronics Corporation	ECR	Electron Cyclotron Resonance
DMA	Defense Mapping Agency	ECRIS	Electron-Cyclotron-Resonance Ion Source

B5, *continued*

ECS	Energy-Corrected-Sudden	ESA	European Space Agency
ECSED	Electronic Commerce in Scientific and Engineering Data	ESB	Electrical Substitution Bolometer
EDX	Energy-Dispersive X-ray Analysis	ESDIAD	Electron-Stimulated Desorption Ion Angular Distributions
EEEL	Electronics & Electrical Engineering Laboratory	ESO	European Southern Observatory
EELS	Electron Energy Loss Spectroscopy	ESR	Experimental Storage Ring *or* Electrical Substitution Radiometer
EEO	Equal Employment Opportunity	ESR	Electron Spin Resonance (EPR now preferred)
EEP	Einstein Equivalence Principle	ETI	Environmental Technology Initiative
EM	Environmental Management	ETRAN	Monte Carlo computer code for Electrons and Photons through Extended Media
ENEA	Ente per le Nuove Tecnologie, L'Energia E L'Ambiente	EURADOS	European Radiation Dosimetry Group
ENDF	Evaluated Nuclear Data File	EURATOM	European Nuclear Energy Organization
ENDL	Evaluated Nuclear Data Library	EUROMET	A European collaboration in measurement standards
ENSDF	Evaluated Nuclear Structure Data File	EUV	Extreme Ultraviolet
EOM	Electro-Optic Modulator	EUVE	Extreme Ultraviolet Explorer
EOS	Earth Observing System	EUVL-LLC	EUV Lithography - Limited Liability Corporation
EPA	Environmental Protection Agency	EXAFS	Edge X-ray Absorption Fine Structure
EPIC	Earth Polychromatic Imaging Camera	FAA	Federal Aviation Administration
ERATO	Exploratory Research for Advanced Technology Office	FACSS	Federation of Analytical Chemistry and Spectroscopy Societies
EROS	Electric Resonance Optothermal Spectrometer	FAD	FASCAL Absolute Detector

B5, *continued*

FARCAL	Facility for Advanced Radiometric Calibrations	FT	Fourier Transform
FASCAL	Facility for Automatic Spectroradiometric Calibrations	FT-IRAS	Fourier Transform-Infrared Reflection Absorption Spectroscopy
FCCSET	Federal Coordinating Council Science, Engineering and Technology	FTIR	Fourier Transform Infrared
FCDC	Fundamental Constants Data Center	FTMS	Fourier Transform Microwave Spectroscopy
FCPM	Fundamental Constants and Precision Measurements	FTMW	Fourier Transform Microwave
FDA	U.S. Food and Drug Administration	FTS	Fourier Transform Spectroscopy
FEA	Field-Emitter Array	FUSE	Far Ultraviolet Spectroscopic Explorer
FED	Field-Emitter Display	FUV	Far Ultraviolet
FEL	Free Electron Laser	FWHM	Full Width at Half Maximum
FEMA	Federal Emergency Management Agency	GAMS 4 (GAMS4)	NIST High Resolution Spectrometer
FET	Field Effect Transistor	GDRIMS	Glow-Discharge Resonance Ionization Mass Spectrometry
FIMS	Fissionable Isotope Mass Standards	GEC	Gaseous Electronics Conference
FIR	Far Infrared	GFP	Green Fluorescent Protein
FLIR	Forward Looking Infrared Radiometer	GHRS	Goddard High Resolution Spectrograph
FM	Frequency Modulation	GIM	Grazing-Incidence Monochromator
FNR	Ford Nuclear Reactor	GINGA	Japanese X-ray Satellite
FOS	Faint Object Spectrograph	GMR	Giant Magnetoresistance
FOV	Field of View	GOES	Geostationary Operational Environmental Satellite
FPC	Fullerene Production Chamber	GPIB	General Purpose Instrumentation Bus
FR	Filtered Radiometer	GPS	Global Positioning System
FRET	Fluorescence Resonant Energy Transfer	GRI	Gas Research Institute

B5, *continued*

GRT	Germanium Resistance Thermometer	HST	Hubble Space Telescope
GSFC	Goddard Space Flight Center	HTBB	High-Temperature Black Body
GSI	Gessellschaft für Schwerionenforschung	HTD	Heat Transfer Division
GVHD	Graft-Versus Host Disease	HTS	High-Temperature Superconductivity
HACR	High Accuracy Cryogenic Radiometer	HUT	Hopkins Ultraviolet Telescope
HALO	Hypersonic Aircraft Launch Option	HVL	Half-Value Layer
HBM	Hybrid Bilayer Membrane	IAEA	International Atomic Energy Agency
HCCD	Mono-deuterated Acetylene	IAG	International Association of Gravity
HDR	High Dose Rate	IAU	International Astronomical Union
HECT	High-Energy Computed Tomography	IC	Integrated Circuit
HFIR	High Flux Isotope Reactor	ICAM-DATA	International Conference on Atomic and Molecular Data
HID	High Intensity Discharge	ICP	Inductively Coupled Plasma
HIRDLS	High Resolution Dynamics Limb Sounder	ICPEAC	International Conference on the Physics of Electronic and Atomic Collisions
HPGe	High Purity Germanium	ICRM	International Committee for Radionuclide Metrology
HPLC	High Pressure Liquid Chromatographic	ICRP	International Commission on Radiological Protection
HPS	Health Physics Society	ICRU	International Commission on Radiation Units and Measurements
HPSSC	Health Physics Society Standards Committee	ID (id or i.d.)	inside diameter
HR3DCT	High Resolution 3-D Computed Tomography	IDMS	Isotope Dilution Mass Spectrometry
HRTS	High Resolution Telescope Spectrograph	IEC	International Electrotechnical Commission
HSST	Heavy Section Steel Technology	IEEE	Institute of Electrical and Electronics Engineers

B5, *continued*

IEN	Istituto Elettrotecnico Nazionale (Italy)	IQI	Image Quality Indicators
IES	Illumination Engineering Society	IR	Infrared
IESNA	Illumination Engineering Society of North America	IRAS	Infrared Astronomical Satellite (*see next entry*)
IGC	International Gravity Commission	IRAS	Infrared Reflection Absorption Spectroscopy
IHPRPTP	Integrated High Payoff Rocket Propulsion Technology Program	IRB	Institutional Review Board
ILL	Institut Laue Langevin	IRDCF	Infrared Detector Calibration Facility
ILS	International Laser Spectroscopy	IRMM	Institute of Reference Materials and Measurements
IMECE	International Mechanical Engineering Congress and Exposition	ISCC	Inter-Society Color Council
IMGC	Istituto di Metrologia "G. Colonnetti" (Italy)	ISO	International Organization for Standardization
IMS	Institute for Molecular Science	ISP	International Specialty Products
INDC	International Nuclear Data Committee	ISCC	Information System to Support Calibrations
INISO-TTC	Experimental Radiochromic Film	ISS	International Space Station
INM	Institute National de Metrologie	ISSI	International Space Science Institute (Bern, Switzerland)
INMM	Institute for Nuclear Materials Management	ITAMP	International Meeting of Theory of Atomic and Molecular Physics
INMRI-ENEA	Istituto Nazionale di Metrologia delle Radiazioni Ionizzanti - Ente per le Nuove Tecnologie	ITL	Information Technology Laboratory
IPNS	Intense Pulsed Neutron Source	ITEP	Institute for Theoretical and Experimental Physics
IPSN	L'Institut de Protection et de Sûreté Nucléaire	ITER	International Thermonuclear Experimental Reactor
IPTS	International Practical Temperature Scale	ITS	International Temperature Scale
IQEC	International Quantum Electronics Conference	ITU	International Telecommunication Union

B5, *continued*

IUCr	International Union of Crystallography	LANL	Los Alamos National Laboratory
IUE	International Ultraviolet Explorer	LANSCE	Los Alamos Neutron Scattering Center
IVBT	Intravascular Brachytherapy	LASP	Laboratory for Atmospheric and Space Physics, University of Colorado
IVR	Intramolecular Vibrational Relaxation	LBIR	Low Background Infrared Radiometry
IVR	Intramolecular Vibrational Redistribution	LBL	Lawrence Berkeley Laboratory
IWG	Investigators Working Group	LBRS	Low Background Reference System
KCDB	Key Comparison and Calibration Database	LCD	Liquid Crystal Display
JAERI	Japan Atomic Energy Research Institute	LCIF	Laser-Collision-Induced Fluorescence
JANNAF	Joint Army-Navy-NASA-Air Force	LED	Light Emitting Diode
JCCRER	U.S.-Russia Joint Coordinating Committee on Radiation Effects Research	LEED	Low Energy Electron Diffraction
JCGM	Joint Committee for Guides on Metrology	LEI	Laser-Enhanced Ionization
JCMT	James Clerk Maxwell Telescope	LET	Linear Energy Transfer
JET	Joint European Torus	LHC	Large Hadron Collider
JILA	formerly Joint Institute Laboratory for Astrophysics	LIF	Laser Induced Fluorescence
JEOL	JEOL USA, Inc.	LIGO	Laser Interferometric Gravitational-Wave Observatory
JPL	Jet Propulsion Laboratory	LISA	Laser Interferometer Space Antenna
LAGOS	Laser Gravitational-Wave Observatory in Space	LLNL	Lawrence Livermore National Laboratory
LAMPF	Los Alamos Meson Physics Facility	LMR	Laser Magnetic Resonance
LAMSCAL	Large Area Monochromatic Source for Calibrations	LMRI	Laboratoire de Mesure des Rayonnements Ionisants (France)

B5, *continued*

LO	Laser Optics	MCT	Mercury-cadmium-telluride
LORAN-C	A Radio Navigation System Operated by the U.S. Coast Guard	MCU	Mobile Calibration Unit
LPM	Light-Particle Monitor	MDRF	Materials Dosimetry Reference Facility
LPRI	Laboratoire Primaire des Rayonnements Ionisants, Gif-sur-Yvette, France	MEA	Materials Engineering Associates
LPRT	Light-Pipe Radiation Thermometers	MEA	membrane electrode assembly
LPTF	Laboratoire Primaire du Temps et des Fréquencies	MEDEA	Microelectronics Development for European Applications
LS	Liquid Scintillation	MEIBEL	Merged Electron-Ion Beam Energy Loss
LSC	Liquid Scintillation Counting	MEL	Manufacturing Engineering Laboratory
LTE	Local Thermodynamic Equilibrium	MEMS	Micro Electro Mechanical Systems
LT	Low-Temperature	MET	Medium Energy Telescope
LTEC	Lamp Testing Engineers Conference	MEVVA	Metal Vapor Vacuum Arc
LTG	Low-Temperature Growth	MIDAS	Modular Interactive Data Acquisition System
LVIS	Low Velocity Intense Source	MIL	Military
LWIR	Long Wave Infrared	MIM	Metal-Insulator-Metal (Diode)
MARLAP	Multi-Agency Radiological Laboratory Procedures	MIRD	Medical Internal Radiation Dose (committee)
MARS	Multiple-Angle Reference System	MIRF	Medical and Industrial Radiation Facility
MBE	Molecular Beam Epitaxy	MISR	Multi-angle Imaging Spectroradiometer
MBIR	Medium Background Infrared Facility	MIT	Massachusetts Institute of Technology
MBOS	Molecular-Beam Optothermal Spectrometer	MJD	Modified Julian Date
MCNP	Monte Carlo Neutron Photon (computer code)	MMI	Mallinckrodt Medical, Inc.
MCQDT	Multi Channel Quantum Defect Theory	MOBY	Marine Optical Buoy

B5, *continued*

MOCVD	Metal Organic Chemical Vapor Deposition	MSEL	Materials Science and Engineering Laboratory
MODIL	Manufacturing Operations Development & Integration Laboratory	MSX	Midcourse Space Experiment
MODIS	Moderate Resolution Imaging Spectrometer	MTG	Methanol To Gasoline
MOEMS	Micro Optp Electro Mechanical Systems	MURR	University of Missouri Research Reactor
MOKE	Magneto-Optical Kerr Effect	MW	Microwave
MOPITT	Measurement of Pollution In The Troposphere	NAMP	National Analytical Management Program
MOS	Marine Optical System *or* Marine Optical Spectrograph	NAMT	National Advanced Manufacturing Testbed
MOSFET	Metal Oxide Semiconductor Field Effect Transistor	NAPM	National Association of Photographic Manufacturers
MOT	Magneto Optical Trap	NAS	National Academy of Sciences
MQDT	Multichannel Quantum Defect Theory	NAS/NRC	National Academy of Sciences/National Research Council
MPD	Multiphoton Detector	NASA	National Aeronautics and Space Administration
MPI	Multiphoton Ionization	NATO	North Atlantic Treaty Organization
MPP	Multi-Pinned Phasing	NBS	National Bureau of Standards, (*now* NIST)
MQA	Measurement Quality Assurance	NBS-4	Older Primary Frequency Standard (*retains NBS name*)
MQDT	Multichannel Quantum Defect Theory	NBS-6	Previous Primary Frequency Standard (*retains NBS name*)
MQSA	Mammography Quality Standards Act	NBSR	National Bureau of Standards' Reactor (*retains NBS name*)
MRA	Mutual Recognition Arrangement	NCAR	National Center for Atmospheric Research
MRI	Magnetic Resonance Imaging	NCI	National Cancer Institute
MRT	Minimal Resolvable Temperature	NCNR	NIST Center for Neutron Research

B5, *continued*

NCRP	National Council on Radiation Protection and Measurements	NGS	National Geological Society
NCSCANS	National Steering Committee for the Advanced Neutron Source	NICE-OHMS	Noise-Immune Cavity Enhanced Optical Heterodyne Molecular Spectroscopy
NCSL	National Conference of Standards Laboratories	NI&D	Nuclear Interactions and Dosimetry
ND	Neutron Density	NIDR	National Institute of Dental Research
NDT	Nondestructive Testing	NIF	National Ignition Facility
Nd:YAG	Neodymium: Yttrium-Aluminum-Garnet (YAG doped with Nd)	NIH	National Institutes of Health
NEANDC	Nuclear Energy Agency Nuclear Data Committee	NIM	Normal-Incidence Monochromator *or* National Instrumentation Methods
NEANSC	Nuclear Energy Agency Nuclear Science Committee	NIOF	Neutron Interferometry and Optics Facility
NEC	Nippon Electric Corporation	NIPDE	National Initiative for Product Data Exchange
NED	Nuclear Effects Directorate	NIR	Near Infrared
NEI	Nuclear Energy Institute	NIST	National Institute of Standards and Technology
NELAC	National Environmental Laboratory Accreditation Conference	NIST-7	Former NIST Primary Frequency Standard
NEOS	Newport Electro-Optic Systems	NIST-F1	Current NIST Primary Frequency Standard
NERI	Nuclear Energy Research Initiative	NISTAR	NIST Advanced Radiometer
NESDIS	National Environmental Satellite Data and Information Service	NMI	National Metrology Institute *or* National Measurement Institute
NEWRAD	New Radiometry	NML	National Measurement Laboratory (Japan)
NG6, NG-6	Monochromatic Beam Line near the end of NG6	NMR	Nuclear Magnetic Resonance
NG6M	Neutron Guide No. 6	NMS	Natural Matrix Standard
NG7	Neutron Guide No. 7	NOAA	National Oceanic and Atmospheric Administration

B5, *continued*

NOAO	National Optical Astronomy Observatory	NSCANS	National Steering Committee for the Advanced Neutron Source
NOBCChE	National Organization for the Professional Advancement of Black Chemists and Chemical Engineers	NSF	National Science Foundation
NORA	Non-Overlapping Redundant Array	NSLS	National Synchrotron Light Source, Brookhaven National Laboratory
NORAMET	A North American regional collaboration in national measurement standards and services	NSOM	Near-Field Scanning Optical Microscopy
NPL	National Physical Laboratory (U.K.)	NTSB	National Transportation Safety Board
NPOESS	National Polar-orbiting Operational Environmental Satellite	NVIS	Night Vision Imaging System
NPSS	Nuclear and Plasma Sciences Society	NVLAP	National Voluntary Laboratory Accreditation Program
NRC	National Research Council	OAI	Optical Associates, Inc.
NRC	Nuclear Regulatory Commission	OCLI	Optical Cooling Laboratory Incorporated
NRCC	National Research Council Canada	OD	Optical Density
NREL	National Renewable Energy Laboratory	OE	Optical Engineering
NRIP	NIST Radiochemistry Intercomparison Program	OFS	Österreichisches Forschungszentrum
NRL	Naval Research Laboratory	OIML	International Organization of Legal Metrology
NRLM	National Research Laboratory of Metrology (Japan)	OMEGA	24-Beam Laser Facility at Rochester
NRRS	Near Resonance Rayleigh Scattering	OMH	National Office of Measures (Hungary)
NSA	National Security Agency	ONR	Office of Naval Research
NSBP	National Society of Black Physicists	OPA	Optical Parametric Amplifier

B5, *continued*

OPO	Optical Parametric Oscillator	PEM	polymer-electrolyte-membrane
OPTCON	*Joint project of:* Optical Society of America; Society of Photooptical Instrumentation Engineers; and Institute of Electrical and Electronics Engineers	PES	Photoelectron Spectroscopy
ORM	Office of Radiation Measurement	PET	Positron Emission Tomography
ORELA	Oak Ridge Electron Linear Accelerator	PFID	Perturbed Free Induction Decay
ORNL	Oak Ridge National Laboratory	PID	Proportional, Integral, and Derivative
OSA	Optical Society of America	PIXE	Particle Induced X-ray Emission
OSL	Optical Stimulated Luminescence	PL	Physics Laboratory
OSRD	Office of Standard Reference Data	PM	phase modulation
OSTP	Office of Science and Technology Policy	PMG	Precision Measurement Grant
OTD	Optical Technology Division	PMMA	polymethylmethacrylate
PA	Proton Affinity	PMS	Particle Measurement System
PADE	Parallel Applications Development Environment	PMT	Photomultiplier Tube
PARCS	Primary Atomic Reference Clock in SPACE	PNL	Pacific Northwest Laboratory
PC	Personal Computer	PNNL	Pacific Northwest National Laboratory
PCB	polychlorinated biphenyl	PNR	Polarized Neutron Reflectivity
PDC	Parametric down-conversion	POC	Physical Optics Corporation
PDE	Product Data Exchange	POP	Plasma Oscillation Probe
PDML	Photovoltaic Device Measurement Laboratory	POPA	Panel on Public Affairs of American Physical Society
PE	Performance Evaluation	PREP	Professional Research Experience Program
PECVD	Plasma-Enhanced Chemical Vapor Deposition	PRF	Petroleum Research Fund

B5, *continued*

PRL	Physical Review Letters	RBS	Rutherford Backscattering
PRM	Precision Radiation Measurement	RDP	Rubidium Di-Hydrogen Phosphate
PS	Polystyrene	REDA	Resonant-Excitation-Double-Autoionization
PSD	Photon-Stimulated Desorption	REI	Rad Elec, Inc.
PSL	Polystyrene-latex	RESL	Radiological and Environmental Science Laboratory
PT	Performance Testing	RF (rf)	Radio Frequency
PTB	Physikalisch-Technische Bundesanstalt (Germany)	RHEED	Reflection High Energy Electron Diffraction
PTCA	Percutaneous Transluminal Coronary Angioplasty	RIMS	Resonance Ionization Mass Spectrometry
PTFE	Polytetrafluoroethylene	RKR	Rydberg-Klein-Rees
PTI	Proxima Therapeutics, Inc.	RNA	Ribonucleic Acid
PUD	Paired Uranium Detector	ROSAT	Roentgensatellit
PWR	Pressurized-Water Reactor	ROSPEC	Rotating Spectrometer for Neutrons
PWS	Primary Working Standard	RS-232	An IEEE Standard Bus
PZT	Piezoelectric Transducer	RTC	Radiochromic Film Task Group
QA/QC	Quality Assurance/Quality Control	RTP	Rapid Thermal Processing
QCD	Quantum Chromodynamics	SACR	Space-based Active Cavity Radiometer
QED	Quantum Electrodynamics	SAM	Self-Assembled Monolayer
QELS	Quantum Electronics and Laser Science	SANS	Small-Angle Neutron Scattering
QFT	Quantum Field Theory	SBIR	Small Business Innovation Research
QMD	Quantum Metrology Division	SCAMPY	Scanning Micro Pyrometer
QNLC	quenched narrow-line laser cooling	SCATMECH	Consortium of 14 U.S. Semiconductor Manufacturers
QPD	Quantum Physics Division	SCC	Standards Coordinating Committee *or* Spectral Calibration Chamber
RAC	Research Advisory Committee	SCF	Spectral Comparator Facility
R&D	Research & Development	SCLIR	Secondary Calibration Laboratories for Ionizing Radiation

B5, *continued*

SDI	Strategic Defense Initiative		SME	Solar Mesosphere Explorer
SDIO	Strategic Defense Initiative Organization		SM	Single Molecule
SDL	Space Dynamics Laboratory		SNOM	Scanning Near Field Optical Microscope
SEAWIFS (SeaWiFS)	Sea-Viewing of Wide Field Sensor		SOG	Spin-on Glass
SEBA	Standards' Employees Benefit Association		SOLSPEC	Solar Spectrometer
SEM	Scanning Electron Microscope		SOLSTICE	Solar Stellar Irradiance Comparison Experiment
SEMATECH	Consortium of 14 U.S. Semiconductor Manufacturers		SPIE	Society of Photo-optical Instrumentation Engineers
SEMPA	Scanning Electron Microscopy with Polarization Analysis		SPP	Storage Photostimulable Phosphor
SFA	Sachs Freeman and Associates		SQL	Structured Query Language
SFCP	Special Foreign Currency Program		SRAM	Static Random Access Memory
SFG	Sum Frequency Generation		SRDP	Standard Reference Data Program
SI	Système International d' Unités *or* International System of Units		SRM	Standard Reference Material
SIA	Semiconductor Industry Association		SSBUV	Shuttle Solar Backscatter Ultraviolet
SID	Society for Information Display		SSC	Superconductor Super Collider
SIMS	Secondary Ion Mass Spectrometry		SSDL	Secondary Standard Dosimetry
SIRCUS	Spectral Irradiance and Radiance Calibration with Uniform Sources		SSPM	Solid State Photomultipliers
SIRREX-3	SeaWiFS Intercalibration Round-Robin Experiment		SSRCR	State Suggested Regulations for Controlling Ionizing Radiations
SKACR	Superconducting Kinetic-inductance Absolute Cryogenic Radiometer		SSTR	Solid State Track Recorder
SLM	Synthetic Layer Microstructure		SSUV	Shuttle Solar Ultraviolet

B5, *continued*

STARR	Spectral Tri-function Automated Reference Reflectometer	TDCR	Triple-to-Double Coincidence Ratio
STD	Standard	TEPC	Tissue Equivalent Proportional Counter
STM	Scanning Tunneling Microscope	TEXT	Texas Experimental Tokamak
STP	Standard Temperature Pressure	TFTC	Thin-Film Thermocouple
STScI	Space Telescope Science Institute	TFWM	transient four-wave-mixing
SUNY	State University of New York	TGM	Toroidal-Grating Monochromator
SURF	Summer Undergraduate Research Fellowship program	TIMED	Thermosphere Ionosphere Mesosphere Energetics and Dynamics
SURF-III (SURF II)	The NIST Synchrotron Ultraviolet Radiation Facility Electron Storage Ring	TLC	Thin Layer Chromatography
SUSIM	Solar Ultraviolet Spectral Irradiance Monitor	TLD	Thermoluminescent Detector
SVGL	Silicon Valley Group Lithography	TMA	Tri-methyl-aluminum
SWIXR	Short Wave Infrared Transfer Radiometer	TOF	Time-of-Flight Spectrometer
SWIR	Short Wave Infrared	TOMS	Total Ozone Mapping Spectrometer
SXR	SeaWiFS Transfer Radiometer	TOP	Time-Orbiting-Potential Trap
TAG	Technical Advisory Group	TPB	Tetraphenyl Butadiene
TAI	International Atomic Time	TPD	Temperature Programmed Desorption
TAMOC	Theoretical Atomic, Molecular, and Optical Physics Community	TQM	Total Quality Management
TASSII	Total and Spectral Solar Irradiance Investigation	TRIGA	Training, Research and Isotope Reactor, General Atomics
TC	Technical Committee	TRU	Transuranic
TCAP	Time-Correlated Associated Particle	TuFIR	Tunable Far Infrared (Radiation)

B5, *continued*

TXR	Thermal-infrared Transfer Radiometer	VCSEL	Vertical-Cavity Surface-Emitting Laser
UA	University of Arizona	VDG	Van de Graaff
UARS	Upper Atmosphere Research Satellite	VEEL	Vibrational and Electronic Energy Levels
UCN	Ultra Cold Neutron	VET	Vibration Energy Transfer
UDC	University of the District of Columbia	VIS	Visible
UHV	Ultrahigh Vacuum	VLA	Very Large Array
UPS	Ultraviolet Photoelectron Spectroscopy	VLBI	Very Long Baseline Interferometer (or Interferometry)
URL	Uniform Resource Locator	VNIIFTRI	National Scientific and Russian Research Institute for Physical, Technical and Radiotechnical Measurements
USAIDR	U.S. Army Institute of Dental Research		
USCEA	U.S. Council for Energy Awareness	VNIIM	Mendeleyev Institute of Metrology
USDA	U.S. Department of Agriculture	VNIIOF	All-Union Research Institute for Optical and Physical Measurements
USFDA	U.S. Food and Drug Administration	VR	Vibrationally Resonant
USGCRP	United States Global Change Research Program	VRML	Virtual Reality Markup Language
USNA	U.S. Naval Academy	VR-SFG	Vibrationally Resonant Sum-frequency Generation
USNC	United States National Committee	VUV	Vacuum Ultraviolet
USNO	U.S. Naval Observatory	VXR	Visible Transfer Radiometer
USSR	Union of Soviet Socialist Republics	WAFAC	Wide-Angle Free-Air Chamber
UTC	Coordinated Universal Time	WERB	Washington Editorial Review Board
UV	Ultraviolet	WG	Working Group
UV-B (UVB)	Ultraviolet-B	WHC	Washington Hospital Center
VCO	Voltage-Controlled Oscillator	WHO	World Health Organization

B5, *continued*

WIPP	Waste Isolation Pilot Plant	XPS	X-ray Photoelectron Spectroscopy
WISE	Women in Science and Engineering	XROI	X-Ray Optical Interferometer
WKB	Wentzel-Kramers-Brillouin	XSW	X-ray Standing Wave
4WM	Four Wave Mixing	XTE/PCA	X-ray Timing Explorer/Proportional Counter Array
WMO	World Metrological Organization		
WPMA	Working Party on Measurement Activities	XUV	Extreme Ultraviolet
WSTC	Westinghouse Science and Technology Center	YAEL	Yankee Atomic Environmental Laboratory
WWW (www)	World Wide Web	YAG	Yttrium-Aluminum-Garnet
WYSIWYG	What You See Is What You Get	Yb:YAG	Ytterbium: Yttrium-Aluminum-Garnet (YAG doped with Yb)
XANES	X-ray Absorption Near-Edge Structure	YBCO	Yttrium-Barium-Cuprate

Appendix C. Tables and Conventions for Astronomy

Contents

C1 Astronomy Units and Symbols

Symbol	Meaning
α	Right ascension *or* the brightest star in a constellation
AU	Astronomical unit (mean Earth–Sun distance); UA in French
B	Besselian
β	Second brightest star in a constellation
c	Speed of light
ΔT	Increment to be added to universal time to give terrestrial dynamical time
δ	Declination *or* fourth brightest star in a constellation
E	Color excess
ET	Ephemeris time (a measure of time for which a constant rate was defined); used from 1958 to 1983
G	Gravitation constant
γ	Third brightest star in a constellation
H_0	Hubble constant
H II region	Volume of hydrogen photoionized (into protons and electrons) by the ultraviolet radiation from a central, hot object

Symbol	Meaning
HA	Hour angle
k	Curvature index of space *or* Gaussian gravitational constant
kpc	kiloparsec
L	Luminosity, stellar
L $\odot$	Luminosity, solar
λ	Celestial longitude
M	Absolute magnitude
$M_\oplus$	Mass of Earth
$M_\odot$	Mass, solar
mJy	milli-Jansky [unit of luminous flux]
Mpc	megaparsec, meaning a million parsecs or 10^6 pc
μg	microgauss
pc	parsec
Q	Aphelion (the point in solar orbit farthest from the Sun)
R	Cosmic scale factor (a measure of the size of the Universe as a function of time
RA	Right ascension
$R_\oplus$	Radius of Earth
rv	Radial velocity
t	Hour angle
UT	Universal time
v	Visual magnitude
z	Red-shift parameter

C2 Astronomy Abbreviations

A

ACS	Advanced Camera for Surveys [See HST]
ACIS	Advanced CCD Imaging Spectrometer (on Chandra)
ACT	Atacama Cosmology Telescope [See CMB]
ACRS	Astrographic Catalog Reference Stars
AD	Accretion Disks
ADONIS	Adaptive Optics Near Infrared System
AEGIS	All-wavelength Extended Groth Strip International Survey
AGB	Asymptotic Giant Branch
AGN	Active Galactic Nuclei
ALFALFA	Arecibo Legacy Fast ALFA Survey
ALMA	Atacama Large Millimeter Array
AMIGA	Analysis of the Interstellar Medium of Isolated Galaxies
AMR	Age-Metallicity Relationship
ANNz	Artificial Neural Networks
AOT	Astronomical Observation Template [See ISOPHOT]
ARCADE	Absolute Radiometer for Cosmology, Astrophysics and Diffuse Emission [See CMB]
ASAS	All Sky Automated Survey
ASCA	Advanced Satellite for Cosmology and Astrophysics
ASM	All-Sky Monitor [See RXTE]
ATCA	Australia Telescope Compact Array
AXP	Anomalous X-ray Pulsar

B

BAL	Broad Absorption Line [See QSOs]
BATSE	Burst And Transient Source Experiment
BBN	Big Bang Nucleosynthesis
BCD	Blue Compact Dwarf (galaxies)
BDs	Brown Dwarfs
BDMC	Boulby Dark Matter Collaboration [See WIMP]
BE	Baldwin Effect (decrease of line equivalent width with luminosity)
BEAST	Background Emission Anisotropy Scanning Telescope
BFS	Blank Field Sources (XR sources without optical counterpart)
BH	Black Hole
BLCCs	Blue Luminous Compact Clusters
BLR	Broad Line Region
BROs	Baryon-Rich Outflows [See GRB]
BWP	"Black Widow" Pulsar

C2, *continued*

C

CAIRNS	Cluster And Infall Region Nearby Survey
CASE	Cluster AgeS Experiment
CAST	CERN Axion Solar Telescope
CBI	Cosmic Background Imager
CCAT	Cornell Caltech Atacama Telescope
CCF	Cross-Correlation Function
CCO	Compact Central Object [See SNR]
CCSNe	Core-Collapse SuperNovae
CDF	Chandra Deep Field (North or South)
CDM	Cold Dark Matter
CFL	Color-Flavor Locked (state)
CFRS	Canada-France-Redshift Survey
CGB	Cosmic Gamma-ray Background
CGP	Close-in Giant Planets
CGPS	Canadian Galactic Plane Survey
CHIPS	Cosmic Hot Interstellar Plasma Spectrometer
CHVC	Compact High Velocity Clouds
CIG	Catalog of Isolated Galaxies (of Karachentseva)
CIRs	Corotating Interaction Regions
CIRB	Cosmic Infrared Background (sometimes: CIB)
CIZA	Clusters in the Zone of Avoidance
CLASS	Cosmic Lens All Sky Survey
CMB	Cosmic Microwave Background
CMBR	Cosmic microwave background radiation
CMD	Colour-Magnitude Diagram
CME	Coronal Mass Ejections
CoMRS	Center-of-Mass Reference System
COSMOS	Cosmic Evolution Survey
CN	Classical Novae
CND	CircumNuclear Disk
CORALS	Complete Optical and Radio Absorption Line System
CR	Cosmic Ray
CRG	Cosmic-Ray Gas
CSE	CircumStellar Envelope
CSM	Circum-Stellar Material
CSOs	Compact Symmetric Objects
CSS	Compact Steep Spectrum (radio sources)
CUDSS	Canada-UK Deep Submillimeter Survey
CVs	Cataclysmic Variables
CWDBs	Close White Dwarf Binaries [See LISA]
CXO	Chandra X-ray Observatory
CYDER	Calan-Yale Deep Extragalactic Research
CZT	Cadmium-Zinc-Telluride (detectors of hard X-rays)

C2, *continued*

D

DACs	Discrete Absorption Components
DASI	Degree Angular Scale Interferometer
DEBRA	Diffuse Extragalactic BackGround Radiation
DEEP	Deep Extragalactic Evolutionary Probe (spectroscopic survey)
DEEP	Deep Extragalactic Exploratory Probe (photometric survey)
DIG	Diffused Ionized Gas
DIOS	Diffuse Intergalactic Oxygen Surveyor [See WHIM]
DIRB	Diffuse InfraRed Background
DLA	Damped Lyman-Alpha (systems, absorbers)
DM	Dispersion Measure
DMH	Dark Matter Haloes
DNB	Diffusive Nuclear Burning (in neutron stars)
DNO	Dwarf Nova Oscillations
DNS	Double Neutron Stars
DPSG	Dusty Post-Starburst Galaxies
dSph	dwarf Spheroidal

E

EAS	Extensive Air Showers
EDR	Early Data Release [See SDSS]
EECC	Envelope Expansion with Core Collapse
EGP	Extrasolar Giant Planets
EGRB	Extragalactic Gamma-Ray Background
EHB	Extreme Horizontal Branch
EHECR	Extremely High Energy Cosmic Rays
EIS	ESO Imaging Survey
ELAIS	European Large Area ISO Survey
ELGs	Emission-Line Galaxies
ELTs	Extremely Large Telescopes
EMBH	ElectroMagnetic Black Hole (model)
EMD	Emission Measure Distribution
EMF	ElectroMotive Force
EOS	Equation Of State
EPIC	European Photon Imaging Camera (on XMM Newton)
ERG	Extremely Red Galaxies
ERO	Extremely Red Object
ESP	ExtraSolar Planets
ESO	European Southern Observatory (Chile)
EXIST	Energetic X-ray Imaging Survey Telescope

C2, *continued*

F

FCOS	Fornax Compact Object Survey [See UCD]
FIR	Far InfraRed (spectral range)
FISICA	Florida Image Slicer for Infrared Cosmology and Astrophysics
FIRAS	Far InfraRed Absolute Spectrophotometer
FIRST	Faint Images of the Radio Sky at Twenty centimeters
FLS	(Spitzer) First Look Survey
FOBOS	Fan Observatory Bench Optical Spectrograph
FOS	Faint Object Spectrograph [See HST]
FP	Fundamental Plane (of the galaxies)
FUSE	Far Ultraviolet Spectroscopic Explorer (satellite)
FWHM	Full Width at Half Maximum

G

GaBoDS	Garching-Bonn Deep Survey
GALEX	Galaxy Evolution Explorer
GBHC	Galactic Black Hole Candidates
GCs	Globular Clusters
GCVFs	Galaxy Circular Velocity Functions
GDDS	Gemini Deep Deep Survey (of galaxies at $0.8 < z < 2.0$)
GEMS	The Group Evolution Multiwavelength Study
GLAST	Gamma-ray Large Area Space Telescope
GLE	Ground Level Event
GLIMPSE	Galactic Legacy Infrared Mid-Plane Survey Extraordinaire
GMASS	Galaxy Mass Assembly ultra-deep Spectroscopic Survey
GMRT	Giant MeterWave Radiotelescope
GOODS	Great Observatories Origins Deep Survey
GR	General Relativity
GraS	Gravitational Suppression
GRB	Gamma Ray Burst
GRBM	Gamma-Ray Burst Monitor (See on board BeppoSAX)
GSSS	Guide Star Selection System [See HST]
GTP	Gravitational Thomas Precession (gravitomagnetic effect)
GUP	Generalized Uncertainty Principle
GWB	Gravitational Wave Background

C2, *continued*

H

HARPS	High Accuracy Radial velocity Planet Searcher
HAT	Hungarian-made Automated Telescope
HDS	High Dispersion Spectrograph (on Subaru)
HE	Heating Efficiency [See SNe]
HEG	High Energy Grating (on Chandra)
HERO	Hyper Extremely Red Object
HESS	High Energy Stereoscopic System
HETE 2	High Energy Transient Explorer 2 (launched on Oct 9 2000)
HFGW	High-Frequency Gravitational Waves
HFQPOs	High Frequency Quasi-Periodic Oscillations
HH	Herbig-Haro (outflows, objects)
HID	Hardness-Intensity Diagram [See XRB]
HIPASS	HI Parkes All-Sky Survey
HIRES	HIgh REsolution Spectrograph (on Keck I telescope)
HLIRGs	HyperLuminous Infrared Galaxies
HMSCs	High-Mass Starless Cores
HOD	Halo Occupation Distribution
HSB	High Surface Brightness (galactic disks)
HST	Hubble Space Telescope
HUDF	HST Ultra Deep Field
HVC	High-Velocity Clouds
HXR	Hard X-Rays
HzRGs	High-redshift Radio Galaxies

I

ICM	IntraCluster Medium
IDV	IntraDay Variability [See AGN]
IFMR	Initial - Final Mass Relationship
IGM	InterGalactic Medium
IGWs	Inflationary Gravitational Waves
IMBH	Intermediate-Mass Black Hole
IOTA	Infrared-Optical Telescope Array
IRAC	InfraRed Array Camera [See SST]
IRAS	InfraRes Astronomical Satellite
IRFM	InfraRed Flux Method (for derivation of T_eff)
IRCS	Infrared Camera and Spectrograph (on Subaru)
IRTS	InfraRed Telescope in Space
ISCO	Innermost Stable Circular Orbit [See BH]
ISM	InterStellar Medium
ISO	Infrared Space Observatory
ISW	Integrated Sachs-Wolfe (effect)

C2, *continued*

J

JDEM	Joint Dark Energy Mission
JEDI	Joint Efficient Dark-energy Investigation
JWST	James Webb Space Telescope

K

KBO	Kuiper Belt Objects
KELT	Kilodegree Extremely Little Telescope
KISS	KPNO International Spectroscopic Survey [See KPNO]
KPNO	Kitt Peak National Observatory

L

LAST	Local apparent sidereal time
LBG	Lyman Break Galaxies
LCBGs	Luminous Compact Blue Galaxies
LCDM	Lambda Cold Dark Matter (cosmology)
LDP	Lithium Depletion Boundary
LFI	Low Frequency Instrument
LHAF	Luminous Hot Accretion Flows
LIC	Local Interstellar Cloud (in Sun's vicinity)
LIGO	Laser Interferometer Gravitational-Wave Observatory
LINER	Low Ionization Nuclear Emission-line Regions
LIRGs	Luminous InfraRed Galaxies
LIS	Low-Ionization Structures [See PNe]
LISA	Laser Interferometer Space Antenna
LISM	Local InterStellar Medium
LLAGN	Low Luminosity AGN [See AGN]
LMC	Large Magellanic Cloud
LMST	Local mean sidereal
LMXB	Low Mass X-ray Binary
LOFAR	LOw Frequency ARray (radio telescope)
LOSVD	Line-Of-Sight Velocity Distribution
LSB	Low Surface Brightness (galactic disks)
LSC	Local SuperCluster (of galaxies)
LSDS	Lenses Structure & Dynamics Survey
LSP	Long Secondary Period
LSS	Large Scale Structure
LST	Local sidereal time
LTE	Local Thermodynamical Equilibrium
LWS	Long Wavelength Spectrometer [See ISO; FIR]

C2, *continued*

M

MACHO	MAssive Compact Halo Objects
MAGIC	Major Atmospheric Gamma ray Imaging Cherenkov Telescope
MASIV	Micro-Arcsecond Scintillation-Induced Variability (survey of the northern sky)
MASTER	Mobile Astronomical System of TElescope-Robots
MAXIMA	Millimeter wave Anisotropy eXperiment IMaging Array
mCV	magnetic Cataclysmic Variables
MDO	Massive Dark Object [See SMBH]
MECO	Magnetospheric, Eternally Collapsing Objects [See GR]
MEGA	Microlensing Exploration of the Galaxy and Andromeda
MERLIN	Multi Element Radio-Linked Interferometer
MGC	Millennium Galaxy Catalogue
MHD	MagnetoHydroDynamics
MIRI	Mid-InfraRed Instrument
MK	Morgan-Kinnon (spectral classification)
MLT	Mixing-Length Theory
MODEST	MOdeling DEnse STellar systems (organization)
MOND	Modified Newtonian Dynamics
MOS	Multi-Object Spectroscopy
MOST	Microvariability and Oscillations of STars (ultra-precise photometric space satellite)
MPFS	MultiPupil Fiber Spectrograph
MRI	MagnetoRotational Instability
MSID	Magnetized Singular Isothermal Disk
MSP	MilliSecond Pulsar
MTB	Multi-Temperature Blackbody (model)
MUNICS	Munich Near--Infrared Cluster Survey
MYC	Massive Young Clusters

N

NEOs	Near Earth Objects
NFGS	Nearby Field Galaxies Survey
NGST	Next Generation Space Telescope
NIRS	Near-InfraRed Spectrometer [See IRTS]
NLR	Narrow Line Region
NOG	Nearby Optical Galaxy (catalog)
NOAO	National Optical Astronomy Observatory (USA)
NRAO	National Radio Astronomy Observatory (USA)
NSVS	Northern Sky Variability Survey
NTF	Nonthermal Radio Filaments
NVSS	NRAO VLA Sky Survey [See NRAO]

C2, *continued*

O

OGLE	Optical Gravitational Lensing Experiment
OSER	Optical Scintillation by Extraterrestrial Refractors
OVRO	Owens Valley Radio Observatory
OWL	Orbiting Wide-field Light-collectors

P

PAH	Particles of Aromatic Hydrogencarbonates
PAST	PrimevAl Structure Telescope
PAO	Pierre Auger Observatory
PBH	Primordial Black Holes
PBL	Polarized Broad Lines
PCA	Proportional Counter Array [See RXTE]
PCA	Principal Component Analysis
PDR	PhotoDissociation Region
PDS	Phoenix Deep Survey (GHz frequences)[See SFGs]
PEBS	Positron Electron Balloon Spectrometer
PLZ	Period-Luminosity-Metallicity (relation)
PMAS	Potsdam Multi-Aperture Spectrophotometer
PMS	Pre-Main-Sequence (stars)
PNe	Planetary Nebulae
PPA	Polarization Position Angle
PPNs	ProtoPlanetary Nebulae
PRGs	Polar-Ring Galaxies
PSF	Point Spread Function
PWN	Pulsar Wind Nebulae

Q

QED	Quantum ElectroDynamics
QPO	Quasi-Periodic Oscillations
QSFs	Quasi-Stationary Flocculi [See SNe]
QSO	Quasi-Stellar Objects (Quasars)
QSSs	Quasi-Soft (X-ray) Sources

R

RAPTOR	RAPid Telescopes for Optical Response
RATS	RApid Temporal Survey
RBSC	ROSAT (all sky survey) Bright Source Catalog [See ROSAT]
RBW	Relativistic Blast Wave
REMP	Relativistic ElectroMagnetic Particle

C2, *continued*

REXBs	Radio Emitting X-ray Binaries
RGB	Red Giant Branch
RLQs	Radio-Loud Quasars
RLT	Receiver Lab Telescope
RM	Rotation Measure
ROSAT	ROentgen SATellite
RPM	Reduced Proper-Motion (diagram)
RQQs	Radio-Quiet Quasars
RTG	Relativistic Theory of Gravity
RXTE	Rossi X-ray Timing Explorer

S

SAFIR	Single Aperture Far-InfraRed (observatory)
SALT	Southern African Large Telescope
SACY	Search for Associations Containing Young-stars
SAURON	Spectrographic Areal Unit for Research on Optical Nebulae [See WHT]
SBF	Surface Brightness Fluctuations (method for distance determination)
SCDM	Standard Cold Dark Matter (model)
SCUBA	Submillimeter Common-User Bolometer Array
SDF	Subaru Deep Field [See SCUBA]
SDIS	Space Telescope Imaging Spectrograph
SDSS	Sloan Digital Sky Survey
SED	Spectral Energy Distribution
SFGs	Star Forming Galaxies
SFH	Star Formation History
SGB	SubGiant Branch
SGPS	Southern Galactic Plane Survey [See ATCA]
SGRs	Soft Gamma Repeaters [See AXP]
SHADES	Scuba HAlf Degree Extragalactic Survey
SIM	Space Interferometry Mission
SINGG	Survey for Ionization in Neutral Gas Galaxies [See NOAO]
SINGS	Spitzer Infrared Nearby Galaxies Survey
SIRs	Slipping Interaction Regions
SIRTF	Space Infrared Telescope Facility
SKA	Square-Kilometre-Array
SLACS	Sloan Lens ACS (survey)
SLRs	Short-Lived Radiounclides
SLUGS	SCUBA Local Universe Survey
SMBH	SuperMassive Black Hole
SMC	Small Magelanic Cloud
SMGs	SubMillimeter-selected Galaxies

C2, *continued*

SMS	Super-Massive Stars
SNAP	SuperNova/Acceleration Probe (satellite)
SNe	SuperNovae
SNLS	Supernova Legacy Survey
SNR	SuperNova Remnant
SPH	Smoothed Particle Hydrodynamics
SRN	Supernova Relic Neutrino
SRS	Southern reference system
SRVs	Semi-Regular Variables
SSC	Synchrotron-Self Compton (radiation)
SSSs	SuperSoft (X-ray) Sources
SSSGs	Small Scale System of Galaxies
SST	Spitzer Space Telescope
STIS	Space Telescope Imaging Spectrograph [See HST]
SUSY	SUperSYmmetric (particles)
SWIRE	Spitzer Wide-area InfraRed Extragalactic [legacy survey]
SXRB	Soft X-Ray Background
SZE	Sunyaev-Zeldovich Effect

T

TALON	Telescope ALert Operations Network
TASS	The Amateur Sky Survey
TAI	International atomic time
TDB	Barycentric dynamical time
TDF	Temperature Distribution Function
TDGs	Tidal Dwarf Galaxies
TDT	Terrestrial dynamical time
TFR	Tully-Fisher Relation (luminosity-rotation speed)
THINGS	The HI Nearby Galaxy Survey [See VLA]
TNOs	Trans-Neptunian objects
T-ReCS	Thermal-Region Camera and Spectrograph [on Gemini South telescope]
TTD	Transit-Time Damping [See MHD]

U

UCD	Ultra Compact Dwarf (galaxies)
UHECR	UltraHigh-Energy Cosmic Rays
UIB	Unidentified Infrared Bands
UIT	Ultraviolet Imaging Telescope

C2, *continued*

UKIDSS	The UKIRT Infrared Deep Sky Survey
UKIRT	United Kingdom Infra-Red Telescope (Hawaii)
ULIGs	Ultraluminous Infrared Galaxies
ULX	UltraLuminous X-ray sources
USS	Ultra-Steep Spectrum (Radio sources)

V

VAMP	VAriable-Mass Particles [See CDM]
VERITAS	Very Energetic Radiation Imaging Telescope Array System
VGPS	VLA Galactic Plane Survey
VHE	Very High Energy (gamma rays)
VLA	Very Large Array [See NRAO]
VLBA	Very Large Baseline Array [See NRAO]
VLBL	Very-long-baseline radio interferometry
VLA	Very Large Array [See NRAO]
VLBI	Very Long Baseline Interferometry
VLT	Very Large Telescope
VMS	Very Massive Stars (M>140 M_sol)
VOBOZ	VOronoi BOund Zones
VPF	Void Probability Function
VSL	Varying Speed of Light (theories)
VSNET	Variable Star NETwork
VSNG	Variable Stars in Nearby Galaxies
VSOP	VLBI Space Observatory Programme [See VLBI]
VSSs	Very Soft (X-ray) Sources
VST	VLT Survey Telescope
VVDS	VIMOS VLT Deep Survey

W

WDS	Wind-Driven Shell
WFI	Wide Field Imager (camera at the ESO/MPG 2.2-m telescope, La Silla)
WFMOS	Wide-Field Multi-Object Spectrograph
WFPC	Wide Field Planetary Camera [See HST]
WHAM	Wisconsin H-Alpha Mapper
WHAT	Wise observatory Hungarian-made Automated Telescope
WHIM	Warm-Hot Intergalactic Medium
WHT	William Herschel Telescope (IAC)

C2, *continued*

WIM	Warm Ionized Medium
WIMP	Weakly Interacting Massive Particles
WIRE	Widefield InfraRed Explorer
WLRG	Weak-Line Radio Galaxies [See LINER]
WMAP	Wilkinson Microwave Anisotropy Probe
WNM	Warm Neutral Medium
WSRT	Westerbork Synthesis Radio Telescope
WXM	Wide-field X-ray Monitor [See HETE 2]

X

XBONGs	X-ray Bright, Optically Normal Galaxies
XDCS	X-ray Dark Cluster SurveyXDR
XDRs	X-rays Dominated Regions [See CND]
XEUS	X-ray Evolving Universe Spectroscopy (mission)
XRBs	X-Ray Binaries
XRFs	X-Ray Flashes
XRS	X-Ray Spectrometer
XSC	eXtended Source Catalog (of 2MASS)

Y

YSO	Young Stellar Objects

Z

ZAHB	Zero Age Horizontal Branch
ZAMS	Zero Age Main Sequence
ZHR	Zenithal hourly rate
ZOA	Zone of Avoidance

C3 Hertzsprung–Russell diagram

The Hertzsprung-Russell diagram (abbreviated "H-R diagram" or "HR diagram") is arguably the most famous diagram in the history of astronomy. It provides a picture of the distribution of stars in the universe and an indication of the evolutionary stages of stars over the course of the many millions of years of their lifetimes. First published by Danish astronomer Ejnar Hertzsprung in 1911, and developed independently by Princeton astronomer Henry Norris Russell in 1913, the diagram has been developed to provide information regarding stellar distances, the composition of star cluster s, and the possible future of stars, especially the Sun. The diagram is also referred to as the "color-magnitude diagram" because of the coordinates used in its construction.

Structure. The H-R diagram is basically a graph of the absolute magnitude of a stellar object plotted against its temperature. The absolute magnitude of a star is its apparent magnitude measured at a distance of 10 parsecs (= 3.26156 light years, or 30.857×10^{15} m). This is plotted on the ordinate, which can also be laid out in terms of the luminosity of the object, measured in terms of the energy output of the Sun (i.e., the Sun has a luminosity of 1, symbolized by $L_\odot$). On the abscissa (x-axis) is the temperature, which is also coordinated with the spectral type. (Color version of the H-R diagram have the background shading going through the visible spectrum from blue on the lft to red on the right. The Sun is in the middle of the yellow region, as observed.)

When the stars of any group or populations are plotted on this graph, a pattern emerges that displays the kind of stars that exist in the universe and indicates their likely development. Just as a graph of the height vs the weight of a human population would provide a snapshot of the likely developmental scheme that humans experience (growing to a point and then maintaining a height for a period of time, if not also a weight), the graph shows that most stars lie on a diagonal line in this graph known as the "Main Sequence," and it is here that we can expect a star to spend most of its life. These stars are known as "dwarfs' because they are comparatively smaller than the "giants that populat the upper right quadrant of the graph.

As the fuel in a star burns, the star becomes larger (as the byproducts of the fusion that powers the star accumulates) and the star becomes both brighter (because of its size), and both cooler and redder (also because of its size and the mixture of matter that is not generating energy. It is then that the star becomes either a red giant or a red supergiant (depending on where it started on the Main Sequence).

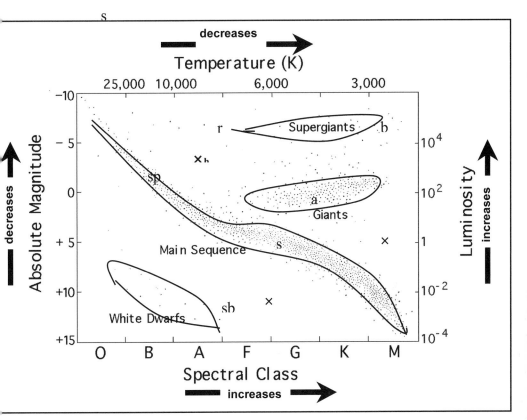

Figure C1 The H-R Diagram, with positions marked for the Sun (s) and several other stars: Betelgeuse (b); Rigel (r); Arcturus (a) Sirius B (sb); and Spics (sp). Explaining the gaps in the graph (the ×'s—×ₕ being the "Hertzsrprung gap") is a concern of modern theory of stellar evolution.

At some point toward the end of its life, when its hydrogen fuel is spent, the gravitational pull of the star's matter will force an implosion that will end in an explosion—either a nova or a supernova—and the star will become a white dwarf—a cinder remnant of the original star. Meanwhile, the matter produced by the nova will spread newly created elements of all kinds through the universe to seed new stars and to provide raw material for the creation of planets and new stars.

The H-R diagram above is based on observations in the vicinity of the Sun, but an H-R diagram of a star cluster (such as the one at right) will display a different distribution and can thus provide clues regarding the age, distance, and makeup of the cluster.

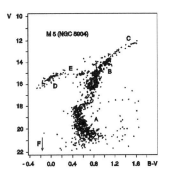

C4 Glossary of Astronomy and Physics Terms

absolute magnitude – A measure of the intrinsic (or absolute) brightness of a star. Defined as equal to the apparent magnitude of a star viewed from a standard distance of 10 parsecs. From the difference between the observed apparent magnitude and the intrinsic absolute magnitude one can deduce the distance to the star. The symbol used for Absolute magnitude is a capital M.

absolute space – Abstract space postulated by Newton to provide a background frame of reference for all motion.

absorption lines – Dark lines in the midst of a bright continuous spectrum, (created when a cooler gas absorbs photons.)

absorption spectrum – A unique series of dark lines that appear in the field of a object due to the type and amount of light being absorbs.

acceleration – The rate of change of velocity with respect to time.

Accretion Disk – A disk of gas which forms around a white dwarf, a neutron star, or a black hole, becoming hot as it spirals in, and emits light and X-rays.

aeon – A unit of time used to measure one thousand million years.

alternating current – An electric current that regularly changes in direction.

ampère's law – Relates a circulating magnetic field to an electric current passing through a loop.

angström – A length unit used for measuring the wavelength of electromagnetic radiation. One angström (Å) is a hundred millionth part of a centimeter.

angular momentum – Angular momentum is a measure of the rotational motion of an object. Defined in terms of the motion of a body with respect to some point in space and the angle between the direction of the motion and the direction toward that defining point. The "conservation of angular momentum" means that the angular momentum of a closed system remains constant as long as no external torque acts.

C4, *continued*

apparent magnitude – The brightness of a star as it *appears* to the eye or to the telescope—measured in units of magnitude. The symbol for apparent magnitude is a lower case m.

astrophysics – A subject that deals with the physical properties of astronomical objects.

asymptotic giant branch – The part of the HR diagram (the upper right hand corner) where stars move to after Helium burning ceases in their cores. The carbon core of the star shrinks, the outer layers expand, and the star becomes a large red giant.

atom – Smallest unit of a chemical element, the limit of classical physics on the small length scales.

atomic clock – A highly precise timekeeping device that measures the movement of electrons in atoms to tell the passage of time.

Big Bang – The state of extremely high (classically, infinite) density and temperature from which the universe began expanding. The beginning point of time and space for the universe.

blackbody – A theoretical object that is a perfect absorber of light (hence the name since it would appear completely black if it were cold), and also a perfect emitter of light. Light is emitted by solid objects because those objects are composed of atoms and molecules which can emit and absorb light. They emit light because they are wiggling around due to their heat content (thermal energy). So a blackbody emits a certain spectrum of light that depends only on its temperature. The higher the temperature, the more light energy is emitted and the higher the frequency (shorter the wavelength) of the peak of the spectrum.

black hole – A celestial object predicted to exist by the properties of the theory of general relativity, which is gravitationally collapsing with such force that not even light can escape from it.

BL Lac Object (also Blazar) – A type of active galaxy characterized by very rapid (day to day) variability by large percentages in total luminosity, no emission lines, strong nonthermal radiation, and starlike appearance.

C4, *continued*

bolometric magnitude – The magnitude that a star would have if all of its energy (at all wavelengths) were included in the measurement.

boltzmann constant – A physical constant relating temperature to energy.

brightness – Energy output per unit time per unit area (measured in ergs per second per square centimeter). The apparent brightness of a star is called the apparent magnitude, which is what is measured with a telescope.

bulge, galactic – A thick spheroidal region at the center of the Galaxy, containing a great deal of warm gas and metal-rich older stars.

causality – The principal in which causes follow effects.

centrifugal force – A force that causes a circular moving body to move away from its rotating center.

cepheid variable stars – A type of luminous giant star whose luminosity varies in a periodic way. Cepheids exhibit a rapid rise in luminosity followed by a slow decline. The period of the cycle is related to the luminosity of the cepheid by the Period-Luminosity relationship. The more luminous the cepheid, the longer the period, a property that makes cepheids useful for obtaining distances. Cepheids come in two types, Type I, which are metal rich and the more luminous; and Type II, which are metal poor.

Chandrasekhar mass – The maximum mass (calculated to be approximately 1.4 solar masses) above which a star's mass cannot support itself against collapse by electron degeneracy pressure. This establishes the maximum mass of a white dwarf.

chromosphere – The layer of the Sun's atmosphere that lies above the photosphere; its width is on the order of about 2000 km. Its gas density is lower than the photospheres, but its temperature is higher. (The temperature continues to rise with altitude, into the corona lying above the chromosphere.)

classical physics – A branch of physics that describes natural phenomena without dealing with quantum theory.

C4, *continued*

closed universe – A model of the universe with a spherical geometry (hence finite in space—and since it will eventually stop expanding and recollapse, is finite in time as well).

CNO cycle – A series of nuclear reactions that convert 4 hydrogen into 1 helium nucleus. The process is so named because it involves Carbon turning into Nitrogen and then into Oxygen. The process ends with the ejection of an alpha particle (helium nucleus) and the return to Carbon. Thus, carbon is neither destroyed nor created but merely acts as a catalyst in the H to He process. The CNO cycle is important only in stars more massive than the Sun.

color filter – A filter that measures the wavelength of specific light by partially absorbing selective light patterns.

color index – The difference in a star's brightness (magnitude) is measured in two different wavelength bands. Paradoxically, the larger the magnitude the fainter the brightness.

compact radio source – A celestial object emitting radio wavelength emissions, such as the core of a radio galaxy.

conduction – The process of heat transport through the physical collisions of the particles making up a substance. The thermal analog of electrical conductivity. Substances with large heat conductivity can transfer heat rapidly (e.g. a hot metal plate). Some substances have low conductivity; they are insulators (e.g., an insulating blanket of foam). Conductivity is an important heat transfer mechanism within white dwarf stars, but not in stars such as the Sun.

contact binary – Two stars in a binary system that are so close to one another that they share a common gas envelope. There gravitational fields also overlap, creating a "peanut" shaped object—a double-cored, "figure 8" stellar object.

continuous spectrum – A smooth spectrum of emitted radiation with all wavelengths present over a broad range. Blackbodies give off continuous spectra, as does the sub when sunlight passes through a prism. By contrast, if some discrete lines are missing from the spectrum that is an "absorption" spectrum. If only discrete lines are observed, that is an "emission" spectrum.

C4, *continued*

continuum emission – Emission of light over a continuous range of wave-lengths. This contrasts with spectral lines which represent sharp peaks or troughs at discrete set of wavelengths.

convection – Heat transfer caused by hot material physically moving from a hot region to a cooler region (causing cool material to move into the hot region). Convection is important for heat transport in some types of stars and in certain regions of stars. In the Sun convection is important in the top layers.

core – The center region of a star where the temperature, pressure and density are greatest. In main sequence stars, the core is where nuclear reactions occur. White dwarf stars are what is left when cores of stars have ejected their outer layers.

core collapse – Catastrophic gravitational collapse of a star when it can no longer maintain sufficient pressure for hydrostatic equilibrium.

corona – The "atmosphere" of the Sun consisting of hot, very thin gas, and extending out away from the Sun for a substantial distance. This gas emits light which cannot be seen against the direct glare of the Sun except during a total eclipse, when direct light is blocked by the moon and the white glow of the corona becomes visible.

Cosmic Background Radiation – The blackbody radiation, mostly in the microwave band, made up of relic photons left over from the hot, early moments of the Big Bang.

cosmic rays – Very high energy atomic nuclei (mostly protons) traveling through space at speeds close to the speed of light. When they hit atoms in the Earth's upper atmosphere, they generate short-lived exotic atomic particles.

Cosmological Principle – The working principle of modern cosmology, namely that there is no center to the universe, that the universe is the same in all directions and the same everywhere, when considered on the largest scale. This means that what we observe of the universe from Earth is what the universe is really like from every vantage point.

cosmology – An area of study devoted to the origin, function, and evolution of the Universe.

C4, *continued*

Coulomb barrier – According to Coulomb's law, like electric charges repel with a force inversely proportional to the square of their distance apart. This means that two such charges are prevented from getting very close to each other. This repulsive effect is expressed by saying that there exists a Coulomb barrier between the like charges.

critical density – The mass density of the universe which will (theoretically) eventually stop the expansion of the universe. The critical density is the boundary value between models of the universe that expands forever (open models) and those that ultimately collapse (closed models.

crystal – A regular ordering of atoms, molecules, or ions.

dark matter – Astronomical mass that does not produce significant light and thus is not observable. Examples of dark matter include planets, black holes, and white dwarfs, plus more exotic mass like weakly interacting particles (WIMPs).

density – The mass of an object divided by its volume. The Sun is composed of compressible (hot) gases and is much denser at its center than at its surface.

detached binary – A binary star system where the two stars are well separated from each other. Each star evolves on its own like an isolated star.

disk, galactic – The flattened, rotating mass of the Galaxy, centered on the galactic nucleus and consisting of mainly dust and gas, and some newly formed stars. The spiral arms of galactic disks are a prominent feature of spiral galaxies.

"distance ladder" – The techniques used by astronomers to deduce distances to progressively more distant astronomical objects.

distance modulus – The formula for calculating the distance to a star using the difference between the absolute and apparent magnitude of a star, $(m - M)$. The formula is $(m - M) = 5 \log(d/10)$, where d is the distance, measured in parsecs. A star is distance modulus of m-M = 0 when it is exactly 10 pc away, since the apparent and absolute magnitudes are equal.

C4, *continued*

Doppler shift – The observed change in frequency of a wave (light, sound, etc.) due to the motion of a source relative to the receiver. Wave sources moving toward the receiver have shorter wavelengths. Wave sources moving away are observed to be emitting lengthened wavelengths.

dust – Near microscopic grains of stuff (e.g., carbon grains—soot—and silicate grains—sand—that are about 0.1-1.0 micron in size) that are a major component of the interstellar medium. Dust blocks visible light and scatters incident starlight, particularly in the blue wavelengths area of the spectrum, causing interstellar reddening.

eccentricity – A mathematical parameter, often denoted by the letter e, specifying the nature of a planetary orbit. Thus $e = 0$ corresponds to a circular orbit. As e increases from 0 to 1 the orbit becomes more and more elliptical. $e = 1$ represents a parabolic orbit while higher values of e describe hyperbolic orbits. Planets are known to have only elliptic orbits.

eclipsing binary – A binary star system where one star passes in front of the other during their orbits, as observed from earth. This causes the total light from the system to vary—dimmer during eclipse and brighter otherwise. The way that the light changes as a function of time (the graph of which is known as a "light curve") provides direct information about the size of the stars and the orientation and size of their orbits.

elastic collision – A collision, during which no kinetic is lost.

electrical resistance – A measure of the degree to which an object opposes the passage of electric current.

electric current – The flow of electric charge through an object.

electrodynamics – Subject dealing with motion of electrical charges under the influence of electric and magnetic fields.

electromotive force – The amount of energy gained per unit charge that passes through a device in the opposite direction to the electric field existing across that device.

C4, *continued*

electron – An elementary particle (of the class of particles lepton) with a negative charge. One of the components of atoms, the electrons orbit around the nucleus, and the distribution and number of electrons determine the chemical properties of an element.

electronvolt (eV) – Work needed to be done to move an electron against an electrical potential barrier of one volt. In terms of the conventional energy unit, the joule, 1 eV equals approximately 1.602×10^{-19} joules.

electron degeneracy pressure – Quantum mechanics restricts the number of electrons in an atom that can have low energy. Basically, each electron must occupy its own energy state. When electrons are packed together in a white dwarf, many electrons are forced into high energy states, which increases the pressure.

elementary particles – Particles believed to be the primary constituents of matter. Experiments over the last few decades have revealed the existence of more than a hundred such particles.

elliptical galaxy – A galaxy classification that gets its names from its overall shape. Ellipticals are subclassified by their degree of ellipticity as they appear to an observer. E1 through E7 have increasing degrees of ellipticity: E0 types are completely spherical; E7 types are very elliptical (elongated). Ellipticals are smooth and structureless, and contain mainly old Pop. II type stars; they range in size from the rare Giants, which contain a trillion stars and can be as big as a Megaparsec across, to the more common dwarf ellipticals, which can contain a million stars and can be as small as a kiloparsec across.

emission lines – The bright lines created when a hot gas emits photons characteristic of the elements of which it is composed.

Emission Nebula – A glowing cloud of hot, mainly hydrogen, interstellar gas, energized by nearby or embedded young hot stars.

emission spectrum – The series of bright lines in the spectrum of a luminous object. The lines represent packets of energy emitted by excited (i.e. energetic) atoms or molecules in the object at characteristic wavelengths.

C4, *continued*

energy – Usually defined as "the capacity to do work" Energy is "conserved" within a closed system, which means that it is neither created nor destroyed, but simply moved from one place to another (possibly changing from one form of energy to another).

entropy – A measure of disorder in a physical system. The law of increase of entropy means a physical system goes from order to disorder.

envelope – The outer, thinner portion of a star, surrounding the hot, dense core. The outer 3/4 (approximately) of the radius of the Sun is considered to be the envelope.

epoch – A typical moment of time in the history of the Universe.

equilibrium – A balance in the rates of opposing forces or processes—so that emission balances absorption of photons; creation balances destruction of matter; etc. As a result, there is no net change.

escape velocity – The outward velocity a body must have to leave the surface of a body mass M and radius R and not fall back. The formula for the escape velocity is $(2GM/R)^{1/2}$ for all bodies.

Euclidean geometry – Geometry based on Euclid's postulates.

event horizon – A boundary dividing space into regions that cannot observe one another. In the case of a black hole, it is the surface part which light cannot escape. No signal or information from within a black hole's event horizon can reach the outside universe.

evolutionary track – The evolutionary track of a star on the HR diagram.

expansion factor – The amount by which the universe has "scaled up" in size due to the expansion of space—equal to $(1 + z)$ where z is the cosmological redshift.

fission, nuclear – The release of nuclear energy from the breaking apart of large, heavy elements (e.g. Uranium) into two or more smaller atoms. Nuclear fission is the basis for nuclear power reactors, but also of nuclear weapons.

C4, *continued*

flux – The rate at which something passes through a surface. In astronomy, flux to expresses the amount of energy radiated per second passing across an area of a square centimeter.

force – An external cause for acceleration in a physical system.

frequency – A property of a wave—the number of wave crests that pass a given point per second. Frequency is measured in units of inverse time (e.g., "cycles per second"). A cycle per second— known as a "Hertz"— is the unit of frequency. Since light moves at a constant speed, the frequency of a light wave is related to the wavelength. The greater the wavelength, the smaller the frequencies, and vice versa.

Friedmann Universe models – Mathematical models of the Universe, first worked out in 1922 by physicist A. Friedmann by using Einstein's general theory of relativity. These models require the Universe to originate in a big bang.

fusion, nuclear – The release of nuclear energy by the fusing or combining of two light elements to form one heavier element. The Sun generates its power from the fusing of four hydrogen atoms into one helium atom. Nuclear fusion is the source of energy in H-bombs.

giant molecular cloud – A region of dense interstellar medium that is cold enough for molecules to form. There clouds are very cold (10-20K), with relatively high densities (trillion of particles per cubic meter), and very large. These regions are believed to be where new stars form.

globular cluster – A dense, rich, spherical cluster of stars, held together by gravity, and containing up to hundreds of thousands of stars within a sphere of 100 pc. Globular clusters are generally found in the halo of the Galaxy, and contain old Population II type stars.

gravitational lens – A massive object—a star or galaxy—that causes light passing near it to bend.

gravitational radiation – According to the theory of general relativity, masses which generates a gravitational field, can propagate waves of gravity electric and magnetic fields (i.e., light) to in electromagnetism, with gravitational radiation having energy and traveling at the speed of light.

C4, *continued*

gravity – An attractive force between particles with mass.

gravity, surface – A spherical object of Mass M and radius R exerts a downward gravitational force on a unit mass at its surface (the surface gravity) equal to GM/R^2. Increasing the mass or decreasing the radius increases the surface gravity. The surface gravity of the Earth (called one "g") generates an acceleration of 9.8 meters per second squared.

ground state (of an electron) – The ground state of a physical system is the state in which it has the lowest possible energy.

hadron – A class of particles involved in the strong interaction (the force that binds atomic nuclei together). Hadrons consist of those particles (baryons, mesons) which are, in turn, made up of quarks.

halo, galactic – The extended region surrounding a galaxy. The halo contains globular clusters and other old stars, and has considerable mass but low luminosity, suggesting that it must contain a lot of dark matter.

Hamiltonian mechanics – A reformulation of Lagrangian mechanics.

helium flash – When certain low-mass stars become red giants, their cores are maintained by electron-degeneracy pressure. When helium begins to burn in the triple-alpha reaction, the temperature in the core rises, but the pressure does not because electron degeneracy pressure is insensitive to pressure. The nuclear reaction rate increases as the temperature rises until an explosion occurs, during which the core to expands. This lowers the temperature, until the core is supported by ordinary pressure, and kept hot by slower, more stable triple-alpha reactions.

helium shell flash – Helium burns in the shell of an triple alpha reaction until it blows those layers away from the star to create a planetary nebula.

hertz – Unit of frequency named after the nineteenth-century scientist Hertz, who first produced and detected electromagnetic waves in the laboratory.

HII ("H-Two") Region – Highly ionized hydrogen—so highly, it radiates an emission nebula.

670

C4, *continued*

homogeneous – Having the same physical properties at all points of space.

horizontal branch – Stars that are burning Helium in their core lie along a nearly horizontal line in the HR diagram referred to as the Horizontal Branch. Compare to the main sequence, which is the line of stars that are burning hydrogen in their cores.

H-R Diagram – A graph that uses two stellar properties (such as luminosity versus surface temperature) to characterize where a star is in its evolution. *See* **Appendix C3**.

Hubble constant – The constant of proportionality (designated H) of the recession velocity of a star or galaxy and its distance in the Hubble law.

Hubble Law – An equation that describes the distance to a galaxy (R) as a function of the velocity with which that galaxy is receding from us (v) due to the overall expansion of the universe. The equation is $v = H_o R$ where H_o is a constant of proportionality known as Hubble's constant. The present "best" estimate of the Hubble constant is about 70 kilometers per second per Megaparsec.

Hubble time – The inverse of the Hubble constant. The Hubble time, also called the Hubble age or the Hubble period, provides an estimate for the age of the universe by presuming that the universe has always expanded at the same rate as it is expanding today.

Hydrostatic Equilibrium – This refers to the balancing of forces in a fluid (which describes a star's interior). Stars are in hydrostatic equilibrium because their enormous self-gravitational forces are balanced by the force pushing up, preventing the star's collapse.

ideal gas – A gas consisting of identical particles of negligible volume, with no intermolecular forces.

inertia – A historical concept used for describing massive, moving objects.

inertial frame of reference – Frame of reference in which Newton's first law of motion holds good, i.e. a frame with respect to which a body under no forces remains at rest or has uniform motion in a straight line.

C4, *continued*

initial mass function (IMF) – The distribution of masses created by the process of star formation.

interference, wave – The phenomenon of waves combining to produce a combined wave—greater where the waves are moving in the same direction; less when they are moving in opposite directions; then two combined ("interfering") light sources produce waves that in some places produce large amplitudes (i.e., brightness) and other places produce zero (i.e., darkness). When these sources of light are projected onto a screen, this wave interference effect produces alternating light and dark spots or lines.

interstellar extinction – As light from a star travels through interstellar space, it typically encounters a great deal of dust, which causes the intensity of the light to diminish. This can be a critical effect in measuring the apparent magnitude of stars. Much of the dark portions of the milky way in the sky are due to this effect.

interstellar reddening – Light from stars encounters interstellar space dust, in which dust scatters mainly the short wavelength (blue) components, allowing the longer wavelength (red) end of the spectrum through. Starlight is thus "reddened" as it travels through space.

interstellar medium – The material that floats in space between the stars. It is mostly hydrogen gas and dust. Even at its densest, the interstellar medium is a better vacuum than human technology can create in the laboratory. Yet, because the universe is so vast, the interstellar medium still accounts for a huge portion of the universe.

ionization – Splitting of neutral atoms or molecules into components carrying equal and opposite charges.

irregular galaxy – A galaxy type from the Hubble classification scheme. These galaxies tend to be smaller than others, containing ("only") 100 million to 10 billion stars, configured in irregular shapes.

irregular galaxy cluster – Clusters of galaxies containing all types of galaxies, both regular and irregular. Our local group is an example of an irregular cluster of galaxies.

C4, *continued*

isotopes – Atomic nuclei with the same number of protons but different numbers of neutrons. Thus deuterium is an isotope of hydrogen.

isotropic – Having the same property in all directions.

jets, radio – Narrow beams of plasma ejected from the cores of galaxies that produce radio (synchrotron) radiation. These jets can extend outward distances larger than the size of the galaxy producing them. These jets are believed to be generated by accretion disks surrounding supermassive black holes at the center of galaxies.

Joule's law – Equation for the heat generated by a current flowing in a conductor.

Kelvin scale – The temperature scale which uses the same degree scale as the Celsius or Centigrade system, but begins at absolute zero, the coldest temperature possible corresponding to the lowest possible energy state of a system. Temperature in degrees Kelvin provides measure of a system's average energy.

kinetic energy – The energy of mass in macroscopic motion. In classical physics, the kinetic energy of mass m traveling at velocity v is equal to one half the mass times the velocity squared, i.e., $1/2\ mv^2$.

Kirchhoff's Laws – A set of rules that describes when radiating matter emits a continuous spectrum of light, an emission spectrum, or an absorption spectrum. Solid bodies and dense gases or liquids radiate continuous spectra. Cool gas absorbs certain wavelengths of a continuous spectrum of light and thus produces an absorption spectrum. Warm gas emits an emission spectrum. The specific wavelengths emitted or absorbed by a substance are uniquely determined by its chemical makeup.

lagrangian – A function describing the equations of motion for a system.

lagrangian mechanics – An abstract reformulation of classical mechanics.

lepton – The class of particles that are not involved in the strong interaction (the force that binds atomic nuclei together). The best-known lepton is the electron; another is the neutrino.

C4, *continued*

light curve – A graph of the amount of light detected from an object (i.e. its apparent magnitude) as a function of time. Light curves provide evidence of eclipsing binaries, variable stars, and possibly exoplanets, and track the progress of nova and supernova explosions.

light-cone – Region in space-time diagram accessible to light emitted from a point. Signals traveling faster than light lie outside this cone while signals traveling slower than light lie within it.

light-year – The distance that light travels within the timeframe of one year. This is approximately 9×10^{12} km, i.e. nine million million kilometers.

light-second – The distance light travels within the timeframe of one second, i.e. approximately 300 000 km.

"lookback" time – The time required for light to travel from an emitting object to the receiver. Thus, when we look at a distant object we are "looking back" in time.

luminosity – Total amount of energy radiated by a star (or any object) per second. It has units of energy (e.g., ergs) per second. One solar luminosity is 4×10^{33} ergs per second. For comparison, the luminosity of a 400-Watt light bulb is 10^{-24} solar luminosities.

luminosity class – Stars are classified by how luminous they are, corresponding to regions on the H-R diagram. The luminosity classes are: I – Supergiants; II - Bright giant; III – Giant; IV – Subgiant; V - Main Sequence.

luminosity function – The relative number of astronomical objects that have a given luminosity. In the case of stars, low luminosity stars are the most abundant, and the number declines rapidly with increasing luminosity.

MACHO (Massive Compact Halo Object) – Objects such as white dwarfs, neutron stars, or black holes that could account for much of the dark matter in galaxy halos.

C4, *continued*

magnitude – An astronomical unit of brightness. Originally corresponding to the eye's response to starlight, the magnitude system is logarithmic, with 5 magnitudes corresponding to a factor of 100 in brightness.

main sequence – The area or line in the H-R diagram along which lie stars that are burning hydrogen in their cores. The majority of a star's lifetime is spent as a main sequence star. Stars in the main sequence are luminosity class V.

main sequence turnoff – When stars run out of hydrogen in their core(s) they begin to change and they move off the main sequence toward the red giant branch of the H-R diagram. When stars form(s), they are all on the main sequence. The point where stars leave (for any of a variety of reasons) the main sequence is the main sequence turnoff point.

mass – The measure of how much "stuff" something has. Mass determines the "inertia" of an object (its resistance to being accelerated by a force) and how much gravitational force it exerts on other objects. In classical physics, mass was "conserved"—neither created nor destroyed. Einstein posited that mass can be converted into energy (and vice versa). But the conservation of mass is still a good approximation since mass-energy conversions generally involve very small amounts of mass. The mass of astronomical objects is often measured in terms of the Sun's mass, which is 2×10^{33} grams.

mass luminosity relation – A main sequence star's luminosity is roughly proportional to its mass to the 3.5 power; i.e., $L \sim M^{3.5}$.

Maxwell's equations – Four equations that describe electric and magnetic fields, and their interaction with matter.

metals – Astronomers refer to all elements other than hydrogen and helium as "metals" (even though these elements are not all metals as defined in chemistry).

momentum – The product of mass and velocity.

nucleus, galactic – The central region of a galaxy, usually containing high densities of stars, and often a supermassive black hole.

C4, *continued*

Neutrino – Any of three species of very weakly-interacting lepton with an extremely small, possibly zero, mass. Electron neutrinos are generated in the interior of the Sun (and other stars). Such neutrinos generally do not interact with matter, but a few have been detectors coming from the Sun here on Earth.

neutron – A charge-neutral particle of the hadron type, which, with the proton, make up the nucleus of the atom. Neutrons are unstable outside a nucleus, but stable within it. The number of protons in the nucleus determines what element that nucleus is; different numbers of neutrons in the nucleus create different isotopes of a given element.

neutron degeneracy pressure – Because each neutron (in a atom or in a plasma) must occupy its own energy state, when neutrons are packed together, as they are in a neutron star, the number of available low energy states is small and many neutrons are forced into high energy states. These high energy neutrons create pressure that supports the neutron star. But, because the neutron is much more massive than the electron, neutron degeneracy pressure is much larger than electron degeneracy pressure and can support stars more massive than the Chandrasekhar mass limit.

neutron star – The core remnant left over after a supernova explosion.

non-Euclidean geometry – A geometry based on postulates differing from Euclid's in one or more respects.

nova – An explosion that occurs when hydrogen is transferred from one star in a semidetached binary system to the other. As hydrogen builds up on the surface of the recipient star the temperature rises, until the star explodes.

nucleosynthesis – The process by which nuclear reactions produce the elements of the periodic table.

nucleus – Central dense region. The nucleus of an atom contains protons and neutrons. The nucleus of a galaxy may contain densely packed stars, gas, and dust.

ohm's law – Relationship between the current flowing in a conductor and the voltage difference between its ends.

C4, *continued*

Omega – The ratio of the actual density of the universe to its critical density; currently a matter of intense research. A value greater than 1 means the universe is denser than the critical value and is thus a closed universe; a value less than one means ours is an open universe.

opacity – The property of a substance that determines how hard it is for radiation to get through that substance (hence how "opaque" that substance is). The atmosphere has low opacity to light. By contrast, fog has a much higher opacity. The opacity of a substance determines how well it can transport heat by radiative transport.

open cluster – A loose, irregular grouping of several hundred stars, in a volume around 10 pc across and generally found in the disk of a galaxy. Open clusters consist of Population I stars formed relatively recently. Examples include the Pleiades and the Hyades.

open universe – A model of the universe which expands forever and is infinite in space and time.

parallax – The apparent shift in the direction to an object as seen from two different locations. This shift can be used to determine distances (through "triangulation"). Stellar parallax occurs as the Earth orbits the Sun and our line of sight to nearby stars varies. Stellar distances are so great, that parallax shifts are less than an arc second, which is unobservable to the unaided eye.

parsec – A unit of distance used to describe the vast scales of the universe, the parsec is equal to about 3.262 light years, or 3.09×10^{16} meters. A star that is one parsec away produces a parallax angle of one second of arc.

peculiar velocity – Any velocity of a galaxy with respect to Earth that does not obey the Hubble law velocity due to the expansion of space. Peculiar velocities are usually due to the gravitational influences of nearby galaxies.

perfect cosmological principle (PCP) – The principle according to which the Universe in the large is unchanging in time. The Universe is also expected to have the symmetries imposed by the ordinary cosmological principle.

C4, *continued*

period-luminosity relation – A relationship between the pulsation period of a variable star (e.g., a Cepheid) and its luminosity (or absolute magnitude). Generally, the more luminous the star the longer the pulsation period. The relationship permits distances to be measured. *See* **distance modulus**.

photodisintegration – The process by which atomic nuclei are broken apart into their constituent protons and neutrons by high energy gamma rays (photons). Photodisintegration takes place during the core collapse phase of a Type II supernova explosion.

photoelectric effect – The ability of light incident upon certain metals to cause currents to flow; (this is the basis of photocells). Einstein explained that light causes electrons to be knocked loose from the surface, which is consistent with the area of light as a particle. Not as a wave. (For this, Einstein was awarded the Nobel prize in 1921.)

photometry – The measurement of light. Different types of photometry are defined by the wavelength at which measurements are made. For example "UBV Photometry" measures the light within three standard regions defined by filters. These are Ultraviolet, Blue and Visual (hence UBV). There are many different photometry systems and standards.

photon – Experiments indicate that light of a given energy (i.e., frequency) cannot be broken up indefinitely. Instead, light exists in discrete bundles of energy that are multiples of hf, where h is Planck's constant and f is the frequency. These discrete bundles of light are what are called photons.

photosphere – The surface layer of the sun, which is what is visible to us on Earth. The Sun does not have a "surface" in the sense we usually think of, since it is a ball of gas. But the photosphere looks like a surface because it is the point where light from the hot gas of the Sun escapes into space without further scattering.

planetary nebula – At the end of the life of a lower mass star, it is a huge red giant on the asymptotic giant branch of the H-R diagram. The core is composed of carbon and oxygen; helium burning has ceased. Helium Shell Flashes take place near the core, which causes the outer layers to be ejected. This material is the matter out of which planets can form.

C4, *continued*

plasma – Gas containing ionized atoms or molecules and free electrons.

Population I Stars – Relatively young stars, containing a larger amount of metals, found mainly in the disk of a galaxy.

Population II Stars – Relatively old stars, containing a smaller amount of metals, found mainly in the halo of a galaxy and in globular clusters.

positron – Dirac's investigations established the existence of "anti-matter" which has the property of annihilating matter and giving rise to radiation. The positron is the opposite of the electron in this respect. It is positively charged and has the same mass as the electron, but is made of antimatter.

potential energy – Energy stored when doing work against a force.

principle of least action – A way of deducing new laws of physics. The "action" is a mathematical entity defined in terms of physical quantities. The principle states that, in Nature, these quantities are related in such a way that, if they are changed slightly, the resulting change in the action is zero. This property is usually sufficient to deduce the relationship between the physical quantities.

proper motion – The motion that an object has in the plane of the sky. The direction is in the plane perpendicular to the radial line (*see* **radial velocity**).

proton – A particle of the hadron family, which, with the neutron, makes up atomic nucleus. The proton has a positive electrical charge.

proton-proton chain – The series of steps of nuclear fusion, by which the sun converts four hydrogen nuclei into one helium nucleus and thereby generates energy in its core.

protostar – A nascent star, prior to settling down to the main sequence and burning hydrogen in its core.

pulsar – A rotating magnetic neutron star that produces regular pulses of radiation. The pulses are produced every time the rotation brings the magnetic pole region of the neutron star into view. Pulsars are the universe's "light houses," sweeping beams of radiation through space.

pulsar glitch – A sudden change in the period of a pulsar due to a sudden shift in the crust of the neutron star (a "starquake").

pulsar, millisecond – Pulsars with periods measured in milliseconds (thousandths of a second). The shortest have periods of about one and two milliseconds. Millisecond pulsar periods are very constant; the most accurate timepieces known. Most millisecond pulsars are found in binary systems.

QSO (quasi-stellar object) – Compact extragalactic object which presents a star-like appearance in spite of being much more (million times or more) than a star. A quasar is a QSO which also emits radio waves.

quantum physics – Physics based on quantum theory. This theory imposes certain fundamental limitations on measurements of physical quantities and leads in many cases to a discrete behavior of matter where classical (pre-quantum) physics predicted a continuous behavior.

Quasar – Short for "quasi-stellar object," a star-like (i.e., unresolved) object that has a very large luminosity and located at very large distances from us (as indicated by their high cosmological redshifts).

radial velocity – The velocity of an object directly toward or away from the observer. Radial velocity is determined using the doppler shift.

Radian – A unit of angle equal to about 57 degrees. The length along the arc of a circle covered by one radian is equal to the radius of the circle. The complete angle around the circle (360 degrees) is equal to 2 pi radians.

radiative transport – The direct transport of energy via by electromagnetic radiation.

radio galaxy – A galaxy that is emitting most of its energy in the form of radio waves rather than light in or near the visible bands of the spectrum which stars emit most of their radiation. The mechanism for this is still under investigation.

C4, *continued*

radius – The radius of a star or planet is the distance from the center of the star or planet out to its surface. The radius of a sphere is equal to half the diameter. Star sizes are often compared to the solar radius, which is 7×10^{10} cm.

red giant – A star with low surface temperature (which makes it red), and large size (giant). These stars are located on the upper-right hand corner of the H-R diagram. The red giant phase in a star's life occurs after it has left the main sequence. The Sun is destined to become a red giant in about 5 billion years.

redshift, cosmological – A shift of the spectral lines observed from a celestial object towards the red end of the spectrum, redshift caused by the expansion of space.

reflection nebula – A nebula composed of dust particles that scatter and reflect light incident from other sources, rather than glowing from their own intrinsic emission. Dust scatters short wavelengths more easily, so reflection nebulae have a characteristic blue appearance.

regular galaxy cluster – Great assemblages of galaxies into huge spherical distributions that have large numbers of galaxies concentrated in their centers. They tend to contain thousands of galaxies, many of which are bright, elliptical, and S0 type galaxies.

relativity, general – Einstein's theory of relativity incorporating the force of gravity into the special theory of relativity. This theory incorporates gravity into the nature of space and time. Among other things, it predicts the existence of gravitational radiation and black holes.

relativity, special – A set of rules relating observations in one inertial frame of reference to the observations of the same phenomenon in another inertial frame of reference. This theory postulates that the speed of light is the same for all observers, and shows the equivalence of matter and energy through the equation, $E = mc^2$.

Roche lobe – The region surrounding a star in a binary system inside of which the star's material matter is gravitationally bound to the star. When a star exceeds its Roche lobe, it may become a semidetached binary.

rotation curve – A graph of the orbital velocity of the disk of a galaxy versus the radius of the galaxy. This curve can be used to obtain the mass of the galaxy (by using Kepler's laws for orbital dynamics). Many rotation curves already plotted suggest that galaxies have much more matter than is accounted for by visible stars (*see* **dark matter**).

r-process – The huge numbers of neutrons given off during a supernova explosion allow leads to a rapid (hence "r") absorption of neutrons by light elements, transforming them into elements higher up in the periodic table—an important step in the process of nucleosynthesis.

RR lyrae variables – A star that has a regularly varying luminosity. Since these stars all have about the same luminosity, they are useful making for obtaining distances, though not as useful as Cepheid variables, because they are not as luminous.

Schwarzschild radius – The radius of the event horizon surrounding a nonrotating black hole. Its size is given by the equation $R_s = 2GM / c^2$. For a star of one solar mass, this is about 3 kilometers.

semidetached binary – A binary system where one star, much larger then its companion, transfers some of its outer layer to over to its binary companion star. Semidetached binary stars can form accretion disks.

Seyfert galaxy – Seyferts are spiral galaxies that have bright cores. Seyferts have strong emission lines, and the emission lines are very broad, implying velocities from 500 to 4000 km/sec. Seyferts are classified into two types based on the width of their emission lines. Seyferts with very broad hydrogen emission lines are Type I; and Seyferts with more narrow hydrogen emission lines are Type II. Many Seyferts have strong radio sources at their core.

shell burning – In the later stages in the life of a star, life regions of the envelope become hot enough to begin nuclear burning, burning in shells throughout the star. There can be more than one region of shell burning, each shell with its own nuclear reactions.

singularity – In general relativity (or in classical physics), a location at which some physical quantities such as density become infinite. A singularity lies at the center of a black hole.

C4, *continued*

sink and source – Concepts in thermodynamics. The source is where energy originates, and the sink is where it is deposited. The subject of thermodynamics deals with the modes of transfer of energy from source to sink, and with the properties of the source and the sink themselves.

space-time – The three dimensions of space and one of time became closely linked when Einstein formulated the theory of relativity (*see* **relativity.**) "Space-time" is used to represent the combination of space and time.

spectral type – A system of star classification based on the strength of various types of emission lines in their spectrum. The spectral type is a measure of the surface temperature of the star, since the temperature determines which emission lines will be present and how strong they will be. From hottest to coolest, stars are grouped into categories O, B, A, F, G, K, and M. Each letter is subdivided into 10 numbers, from hotter to cooler 0, 1, 2, 3, 4, etc.

spectrum, electromagnetic – The full range of possible frequencies and wavelengths of light and radiation. A "spectral line" refers to emission or absorption at a particular wavelength of light.

spectrum, nonthermal – A continuous spectrum produced by processes other then ordinary thermal radiation, as with dense, hot matter (e.g., blackbody radiation). An example is Synchrotron radiation.

spectroscopy – The study of the features of a star's spectrum, performed by measuring the intensity of the star's light at different wavelengths. The resulting spectrum of light provides the location of emission and absorption lines, which is used to determine the composition of the star, its doppler shift, its spectral type, and its luminosity class.

spiral galaxy – A galaxy consisting of a flattened rotating disk of stars, a central bulge and a surrounding halo. The disk is prominent due to the presence of young, hot stars which are often arranged in spiral patterns.

s-process – The absorption of neutrons by elements in massive stars, causing them to transform into isotopes, and, through subsequent nuclear decay, into other elements. The flux of neutrons is small enough that the process happens slowly (hence "s" process), but is an important part of the nucleosynthesis of the elements.

C4, *continued*

"standard candle" – Any astronomical object of known luminosity that can be used to calculate distances. Cepheid variables, Main sequence stars, and type I supernovae have all be used as standard candles.

statistical mechanics – This is an attempt to understand the irreversible behavior of macroscopic systems in terms of the statistical description of the large number of microscopic constituents making up the systems. In the statistical description details of the individual members are lost and only averages of ensembles are left.

Stefan-Boltzmann law – This is a law of blackbody radiation, according to which, the amount of energy given off by a blackbody per second per unit area is proportional to the fourth power of the blackbody temperature.

supermassive black hole – A black hole that has a million—or sometimes a billion—solar masses. Huge black holes of this kind lie at the center of many galaxies.

supernova – The explosion of a star. Supernovae come in two varieties: Type I, caused by sudden nuclear burning in a white dwarf star; Type II, caused by the collapse of the core of a supermassive star at the end of its nuclear-burning life. In either case, the star is destroyed and the light given off in the explosion is briefly comparable to the total light given off by an entire galaxy.

supernova remnant – What remains after a supernova explosion, seen as a great glowing cloud expanding into space.

Synchrotron Radiation – Radiation emitted by electrons moving close to the speed of light in the presence of magnetic fields. The magnetic force field causes the electron (which has a negative electrical charge) to spiral around the magnetic field. The electron produces electromagnetic radiation, observed mainly as radio waves.

temperature – The average kinetic energy of molecules.

theoretical physics – Part of physics which seeks to explain the outcome of scientific experiments or the observations of natural phenomena in terms of mathematical models based on certain fundamental laws of nature.

C4, *continued*

thermal equilibrium – (1) The concept that energy radiated from an object (e.g., a star) is replaced by other energy, so that the temperatures remain constant. (2) A state in which energy is equally distributed among all particles, which gives rise to the concept that the statistical properties of the system can be described by a single parameter, the temperature.

thermal pulse – A sudden increase in a star's temperature caused by a dramatic increase in the nuclear burning rate. Though not quite an explosion, it causes a readjustment in the star.

thermodynamics – A branch of science that deals with the exchange of heat energy and its relation to work and the mechanical behavior of physical systems.

thermodynamic equilibrium – The ultimate state of disorder of a physical system. In the equilibrium state the emission and absorption of heat reach a balance and the "entropy" of the system attains its maximum value.

thermonuclear reactions – Reactions which take place when different atomic nuclei are brought together at high temperatures. The reactions result in changes of structure (break-up or fusion) of participating nuclei accompanied by a release or absorption of heat.

triple alpha reaction – The process by which helium (known as an alpha particle) is converted into carbon. When temperatures are high and the density of helium is large, three helium atoms can combine to form one Carbon atom, which is why it is called: 3 helium reaction.

Tully-Fisher relation – An empirical relationship between the width of the 21-cm line of hydrogen emissions from spiral galaxies, and the mass of the galaxy.

21 Centimeter Emission – The radio wavelength emission that emanates from a neutral hydrogen atom. Thus, cold neutral hydrogen in space emits this radiation, which can be detected using a radio telescope.

Visual Binary – A binary star system that can be seen as separate stars in direct observations. In some binary systems, the two stars are so close that they cannot be distinguished (resolved) by observation.

C4, *continued*

waves – Periodic change in a physical quantity with respect to spatial displacement and passage of time is called a wave. In a plane wave the change in the physical quantity takes place only in one direction of space. In a spherical wave the disturbance has a centre at a point in space; at all points in space at any given moment of time.

Wavelength – In a wave, the distance from one peak to the next. It is measured in units of distance. The wavelengths of visible light is measured in hundreds of nanometers (billionths of a meter). For light, the shorter the wavelength, the higher the energy of the light wave.

white dwarf – The remnant of a star, at the end of its life. White dwarfs consist of a carbon and oxygen core supported by electron degeneracy pressure. The surface has a very high temperature and radiates mainly in the ultraviolet (hence it is "white hot"), but it is only about the size of the Earth (which makes it a dwarf). The maximum mass that can be supported by electron degeneracy pressure—and thus the maximum possible mass of a white dwarf—is known as the Chandrasekhar mass and is equal to 1.4 solar masses.

Wien's Law – The law that says that for blackbody radiation, the higher the temperature, the higher the frequency and the shorter the wavelength of the light it emits.

WIMP – An acronym for a Weakly-Interacting Massive Particle. A particle with a nonzero mass which exists only in the weak nuclear interaction. Such particles (currently hypothetical) could fill space and provide gravitational force without any luminosity. They are thus candidates for **dark matter**.

X-ray burster – A semidetached binary system where matter is accreting onto a neutron star. As hydrogen flows to a neutron star, the hydrogen is promptly burned into helium. The helium accumulates until the temperature is high enough for a helium-burning explosion.

Young's modulus – A measure of the stiffness of a body or material.

zero age main sequence (ZAMS) – The theoretical line of the main sequence which corresponds to the becoming of hydrogen burning, and thus the beginning (zero age) of a star's life.

C5 Astronomical Catalogues and Their Abbreviations

Full Title	Abbreviation
Abell catalogue	Abell
Aitken Double Star Catalogue	ADS
Astrographic Catalogue	AC
Astronomische Gesellschaft Katalog	AG, AGK, AGKR
Astronomische Nachrichten	AN
Bonner Durchmusterung	BD
Bordeaux Astrographic Catalog	BAC
Boss general catalogue of 33342 stars	GC (Boss)
Bright Star Catalogue (Harvard Revised Catalogue)	HR
Burnham Double Star Catalogue	BDS
Caldwell catalogue	C
Cape Photographic Catalogue	CPC
Cape Photographic Durchmusterung	CPD
Catalog of Components of Double and Multiple Stars	CCDM
Catalog of Stellar Identifications	CSI
Collinder catalog	Col
Cordoba Durchmusterung	CD / CoD
CoRoT Catalogue	CoRoT
CoRoT Catalogue	CoRoT-Exo
Dearborn Observatory	DO
Dominion Observatory List A	DA
E. E. Barnard's List of Dark Nebulae	B
Fourth Fundamental Catalogue	FK4
Fifth Fundamental Catalogue	FK5
General Catalog of Variable Stars	GCVS
General Catalogue of Nebulae and Clusters	GC
General Catalogue of Stellar Radial Velocities	GCRV
General Catalogue of Trigonometric Parallaxes	GCTP
General Catalogue of Trigonometric Stellar Parallaxes and Supplement	PLX
Gliese-Jahreiß catalogue or Gliese-Jahreiss catalogue	Gl / GJ
Guide Star Catalog	GSC
Guide Star Catalog II	GSC2 / GSC II
Hamburg/ESO Survey	HE
Henize Catalogues of Hα-Emission Stars and Nebulae in the Magellanic Clouds	Hen
Henry Draper Catalogue	HD

C5 *continued*

Full Title	Abbreviation
Henry Draper Extension	HDE
HI Parkes All-Sky Survey	HIPASS
Hipparcos Catalogue	HIP
Index Catalog	C
Index Catalogue of Visual Double Stars	IDS
Infrared Astronomical Satellite	IRAS
Luyten Five-Tenths catalogue	LFT
Luyten Half-Second catalogue	LHS
Luyten Proper-Motion Catalogue	LPM
Luyten Two-Tenths catalogue	LTT
Massive astrophysical compact halo object	MACHO
Minor Planet Circulars	MPC
Nearby Stars Database	NStars
New Catalogue of Suspected Variable Stars	NSV
New General Catalogue	NGC
New Luyten Two-Tenths Catalogue	NLTT
Northern HIPASS Catalog	NHICAT
Optical Gravitational Lensing Experiment	OGLE
Positions and Proper Motions Star Catalogues	PPM
Research Consortium on Nearby Stars	RECONS
Revised New General Catalogue	RNGC
ROSAT observations	RX
Sloan Digital Sky Survey	SDSS
Smithsonian Astrophysical Observatory Star Catalog	SAO
Spitzer Space Telescope c2d Legacy Source	SSTc2d
Struve the Father double star	STF
Trans-Atlantic Exoplanet Survey	TrES
Two Micron All Sky Survey	2MASS
Tycho Catalogue	TYC
Tycho-2 Catalog	TYC2
Uppsala General Catalogue	UGC
Uranometria (Bayer designation)	BAY
US Naval Observatory	USNO
Van Biesbroeck catalog	vB
Washington Double Star Catalog	WDS
Woolley Nearby Star Catalogue	Wo
XEST, Optical/UV Monitor	XEST-OM
XMM-Newton Bright Serendipitous Survey	XBSS
XMM-Newton, Bright Source	XBS
XMM-Newton Extended Survey of the Taurus Molecular	XEST
X-ray Timing Explorer	XTE

C6 Astronomy and Physics Journals and Their Abbreviations

Full Title	Abbreviation
Accounts of Chemical Research	Acc. Chem. Res.
Acta Astronomica	Acta Astronom.
Acta Chemica Scandinavica	Acta Chem. Scand.
Acta Crystallographica	Acta Crystallogr.
Acta Crystallographica, Section A: Crystal Physics, Diffraction, Theoretical and General Crystallography	Acta Crystallogr. Sec. A
Acta Crystallographica, Section B: Structural Crystallography and Crystal Chemistry	Acta Crystallogr. Sec. B
Acta Mathematica Academiae Scientiarum Hungaricae	Acta Math. Acad. Sci. Hung.
Acta Metallurgica	Acta Metall.
Acta Physica	Acta Phys.
Acta Physica Austriaca	Acta Phys. Austriaca
Acta Physica Polonica	Acta Phys. Pol.
Acustica	Acustica
Advances in Applied Mechanics	Adv. Appl. Mech.
Advances in Atomic and Molecular Physics	Adv. At. Mol. Phys.
Advances in Chemical Physics	Adv. Chem. Phys.
Advances in Magnetic Resonance	Adv. Magn. Reson.
Advances in Physics	Adv. Phys.
Advances in Space Research	Adv. Space. Res.
Advances in Quantum Chemistry	Adv. Quantum Chem.
AIAA Journal	AIAA J.
AIChE Journal	AIChE J.
AIP Conference Proceedings	AIP Conf. Proc.
Akusticheskii Zhurnal	Akust. Zh.
American Journal of Physics	Am. J. Phys.
Anales de Física	An. Fis.
Analytical Chemistry	Anal. Chem.
Annalen der Physik	Ann. Phys.
Annales de Chimie et de Physique	Ann. Chim. Phys.
Annales de Geophysique	Ann. Geophys.
Annales de l'Institut Henri Poincare	Ann. Inst. Henri Poincare
Annales de l'Institut Henri Poincare, Section A: Physique Theorique	Ann. Inst. Henri Poincare, A
Annales de l'Institut Henri Poincare, Section B: Calcul des Probabilite's et Statistique	Ann. Inst. Henri Poincare, B
Annales de Physique Paris	Ann. Phys. Paris
Annals of Fluid Dynamics	Ann. Fluid Dyn.
Annals of Mathematics	Ann. Math.
Annals of Physics New York	Ann. Phys. N.Y.

C6, *continued*

Annual Review of Astronomy and Astrophisics	Annu. Rev. Astron. Astrophys.
Annual Review of Atomic and Molecular Physics	Annu. Rev. At. Mol. Phys.
Annual Review of Earth and Planetary Sciences	Annu. Rev. Earth Planet. Sci.
Annual Review of Fluid Mechanics	Annu. Rev. Fluid Mech.
Annual Review of Nuclear Science	Annu. Rev. Nucl. Sci.
Apeiron	Apeiron
Applied Optics	Appl. Opt.
Applied Physics A: Materials Science & Processing	Appl. Phys. A, Mater. Sci. Process.
Applied Physics B: Lasers and Optics	Appl. Phys. B, Laser Optic.
Applied Physics Letters	Appl. Phys. Lett.
Applied Spectroscopy	Appl. Spectrosc.
Arkiv foer Fysik	Ark. Fys.
Astronomical Journal	Astron. J.
Astronomicheskii Zhurnal	Astron. Zh.
Astronomische Nachrichten	Astron. Nachr.
Astronomy and Astrophysics	Astron. Astrophys.
Astrophysical Journal	Astrophys. J.
Astrophysical Journal, Letters to the Editor	Astrophys. J. Lett.
Astrophysical Journal, Supplement Series	Astrophys. J. Suppl. Ser.
Astrophysical Letters	Astrophys. Lett.
Astrophysics and Space Science	Astrophys. Space Sci.
Atomic Data and Nuclear Data Tables	At. Data. Nucl. Data Tables
Atomnaya Energiya	At. Energ.
Australian Journal of Physics	Aust. J. Phys.
Baltic Astronomy	Baltic Astronomy
Bell System Technical Journal	Bell Syst. Tech. J.
Berichte der Bunsengesellschaft für Physikalische Chemie	Ber. Bunsenges. Phys. Chem.
Biophysical Journal	Biophys. J.
Biophysics	Biophysics
British Journal of Applied Physics	Br. J. Appl. Phys.
Bulletin of the Academy of Sciences of the USSR, Physical Series	Bull. Acad. Sci. USSR, Phys. Ser.
Bulletin of the American Astronomical Society	Bull. Am. Astron. Soc.
Bulletin of The American Physical Society	Bull. Am. Phys. Soc.
Bulletin of the Astronomical Institutes of the Netherlands	Bull. Astron. Inst. Neth.
Bulletin of the Astronomical Society of India	Bull. Astron. Soc. India.
Bulletin of the Chemical Society of Japan	Bull. Chem. Soc. Jpn.
Bulletin of the Seismological Society of America	Bull. Seismol. Soc. Am.
Canadian Journal of Chemistry	Can. J. Chem.
Canadian Journal of Physics	Can. J. Phys.
Canadian Journal of Research	Can. J. Res.
Celestial Mechanics and Dynamical Astronomy	Celestial Mech. Dyn. Astron.
Central European Journal of Physics	Cent. Eur. J. Phys.
Chaos	Chaos

C6, *continued*

Chemical Physics	Chem. Phys.
Chemical Physics Letters	Chem. Phys. Lett.
Chemical Reviews	Chem. Rev.
Chinese Astronomy	Chin. Astron.
Chinese Journal of Physics (translation of Wuli Xuebao)	Chin. J. Phys.
Classical and Quantum Gravity	Classical Quant. Grav.
Comments on Astrophysics and Space Physics	Comments Astrophys. Space Phys.
Comments on Atomic and Molecular Physics	Comments At. Mol. Phys.
Comments on Nuclear and Particle Physics	Comments Nucl. Part. Phys.
Comments on Plasma Physics and Controlled Fusion	Comments Plasma Phys. Controlled Fusion
Comments on Solid State Physics	Comments Solid State Phys.
Communications in Mathematical Physics	Commun. Math. Phys.
Communications on Pure and Applied Mathematics	Commun. Pure Appl. Math.
Complex Systems	Complex Syst.
Comptes Rendus Hebdomadaires des Se´ances de l'Acade´mie des Sciences	C. R. Acad. Sci.
Comptes Rendus Hebdomadaires des Se´ances de l'Acade´mie des Sciences, Serie A: Sciences Mathe´matiques	C. R. Acad. Sci. Ser. A
Comptes Rendus Hebdomadaires des Se´ances de l'Acade´mie des Sciences, Serie B: Sciences Physiques	C. R. Acad. Sci. Ser. B
Computer Physics Communications	Comput. Phys. Commun.
Cryogenics	Cryogenics
Czechoslovak Journal of Physics	Czech. J. Phys.
Discussions of the Faraday Society	Discuss. Faraday Soc.
Doklady Akademii Nauk SSSR	Dokl. Akad. Nauk SSSR
Earth and Planetary Science Letters	Earth. Planet. Sci. Lett.
Electronics Letters	Electron. Lett.
European Journal of Physics	Eur. J. Phys.
European Physical Journal-Applied Physics, The	Eur. Phys. J. Appl. Phys.
Europhyics Letters	Europhys. Lett.
Few-Body Systems	Few-Body Syst.
Fields and Quanta	Fields Quanta
Fizika Elementarnykh Chastits i Atomnogo Yadra Journal of Particles and Nuclei	Fiz. Elem. Chastits At. Yadra J. Part. Nucl.
Fizika i Tekhnika Poluprovodnikov	Fiz. Tekh. Poluprovodn.
Fizika Metallov i Metallovedenie	Fiz. Met. Metalloved.
Fizika Nizkikh Temperatur	Fiz. Nisk. Temp.
Fizika Plazmy	Fiz. Plazmy
Fizika Tverdogo Tela (Leningrad)	Fiz. Tverd. Tela (Leningrad)
Fortschritte der Physik	Fortschr. Phys.
Foundations of Physics	Found. Phys.
General Relativity and Gravitation	Gen. Relativ. Gravit.
Geochimica et Cosmochimica Acta	Geochim. Cosmochim. Acta
Helvetica Chimica Acta Helv.	Chim. Acta

C6, *continued*

Helvetica Physica Acta	Helv. Phys. Acta
High Temperature	High Temp.
Hyperfine Interactions	Hyperfine Interact.
IBM Journal of Research and Development	IBM J. Res. Dev.
Icarus	Icarus
IEEE Journal of Quantum Electronics	IEEE J. Quantum Electron.
IEEE Transactions on Electron Devices	IEEE Trans. Electron Devices
IEEE Transactions on Information Theory	IEEE Trans. Inf. Theory
IEEE Transactions on Instrumentation and Measurement	IEEE Trans. Instrum. Meas.
IEEE Transactions on Magnetics	IEEE Trans. Magn.
IEEE Transactions on Microwave Theory and Techniques	IEEE Trans. Microwave Theory Tech.
IEEE Transactions on Nuclear Science	IEEE Trans. Nucl. Sci.
IEEE Transactions on Plasma Science	IEEE Trans. Plasma Sci.
IEEE Transactions on Sonics and Ultrasonics	IEEE Trans. Sonics Ultrason.
Industrial and Engineering Chemistry	Ind. Eng. Chem.
Infrared Physics	Infrared Phys.
Inorganic Chemistry	Inorg. Chem.
Inorganic Materials	Inorg. Mater.
Instruments and Experimental Techniques	Instrum. Exp. Tech.
International Journal of Energy Research	Int. J. Energy Res.
International Journal of Magnetism	Int. J. Magn.
International Journal of Quantum Chemistry	Int. J. Quantum Chem.
International Journal of Quantum Chemistry, Part 1	Int. J. Quantum Chem. 1
International Journal of Quantum Chemistry, Part 2	Int. J. Quantum Chem. 2
International Journal of Theoretical Physics	Int. J. Theor. Phys.
Atmospheric and Oceanic Physics	Atmos. Oceanic Phys.
Physics of the Solid Earth	Phys. Solid Earth
Inorganic Materials	Inorg. Mater.
Japanese Journal of Applied Physics	Jpn. J. Appl. Phys.
Japanese Journal of Physics	Jpn. J. Phys.
JETP Letters	JETP Lett.
Journal de Chimie Physique	J. Chim. Phys.
Journal de Physique (Paris)	J. Phys. (Paris)
Journal de Physique et le Radium	J. Phys. Radium
Journal of Applied Crystallography	J. Appl. Crystallogr.
Journal of Applied Physics	J. Appl. Phys.
Journal of Applied Spectroscopy	J. Appl. Spectrosc.
Journal of Astrophysics & Astronomy	J. Astrophys. Astron.
Journal of Atmospheric and Terrestrial Physics	J. Atmos. Terr. Phys.
Journal of Atmospheric Sciences	J. Atmos. Sci.
Journal of Chemical Physics	J. Chem. Phys.
Journal of Computational Physics	J. Comput. Phys.
Journal of Cosmology and Astroparticle Physics	J. Cosmo. Astropart. Phys.
Journal of Crystal Growth	J. Cryst. Growth
Journal of Electron Spectroscopy and Related Phenomenon	J. Electron. Spectrosc. Relat. Phenom.

C6, *continued*

Journal of Experimental and Theoretical Physics	J. Exp. Theor. Phys.
Journal of Fluid Mechanics	J. Fluid Mech.
Journal of Geophysical Research	J. Geophys. Res.
Journal of Inorganic and Nuclear Chemistry	J. Inorg. Nucl. Chem.
Journal of Low Temperature Physics	J. Low Temp. Phys.
Journal of Luminescence	J. Lumin.
Journal of Macromolecular Science, (Part B) Physics	J. Macromol. Sci. Phys.
Journal of Magnetism and Magnetic Materials	J. Magn. Magn. Mater.
Journal of Mathematical Physics (New York)	J. Math. Phys. (N.Y.)
Journal of Molecular Spectroscopy	J. Mol. Spectrosc.
Journal of Non-Crystalline Solids	J. Non-Cryst. Solids
Journal of Nonlinear Science	J. Nonlinear Sci.
Journal of Nuclear Energy	J. Nucl. Energy
Journal of Nuclear Energy, Part C: Plasma Physics, Accelerators, Thermonuclear Research	J. Nucl. Energy, Part C
Journal of Nuclear Materials	J. Nucl. Mater.
Journal of Optics A: Pure and Applied Optics	J. Optic. A, Pure Appl. Optic.
Journal of Physical and Chemical Reference Data	J. Phys. Chem. Ref. Data
Journal of Physical Chemistry	J. Phys. Chem.
Journal of Physics A: Mathematical and General	J. Phys. A
Journal of Physics and Chemistry of Solids	J. Phys. Chem. Solids
Journal of Physics B: Atomic, Molecular and Optical	J. Phys. B
Journal of Physics C: Solid State Physics	J. Phys. C
Journal of Physics D: Applied Physics	J. Phys. D
Journal of Physics E: Scientific Instruments	J. Phys. E
Journal of Physics F: Metal Physics	J. Phys. F
Journal of Physics G: Nuclear and Particle Physics	J. Phys. G.
Journal of Physics: Condensed Matter	J. Phys. Condens. Matter
Journal of Physics (Moscow)	J. Phys. (Moscow)
Journal of Plasma Physics	J. Plasma Phys.
Journal of Polymer Science	J. Polym. Sci.
Journal of Polymer Science, Polymer Letters Edition	J. Polym. Sci. Polym. Lett. Ed.
Journal of Polymer Science, Polymer Physics Edition	J. Polym. Sci. Polym. Phys. Ed.
Journal of Quantitative Spectroscopy & Radiative Transfer	J. Quant. Spectrosc. Radiat. Transfer
Journal of Research of the National Bureau of Standards	J. Res. Natl. Bur. Stand.
Journal of Research of the National Bureau of Standards, Section A: Physics and Chemistry	J. Res. Natl. Bur. Stand. Sec. A
Journal of Research of the National Bureau of Standards, Section B: Mathematical Sciences	J. Res. Natl. Bur. Stand. Sec. B
Journal of Research of the National Bureau of Standards, Section C: Engineering and Instrumentation	J. Res. Natl. Bur. Stand. Sec. C

C6, *continued*

Journal of Scientific Instruments	J. Sci. Instrum.
Journal of Sound and Vibration	J. Sound Vib.
Journal of Statistical Physics	J. Stat. Phys.
Journal of the Acoustical Society of America	J. Acoust. Soc. Am.
Journal of the American Ceramic Society	J. Am. Ceram. Soc.
Journal of the American Chemical Society	J. Am. Chem. Soc.
Journal of the American Institute of Electrical Engineers	J. Am. Inst. Electr. Eng.
Journal of the Audio Engineering Society	J. Audio Eng. Soc.
Journal of the Chemical Society	J. Chem. Soc.
Journal of the Electrochemical Society	J. Electrochem. Soc.
Journal of the Mechanics and Physics of Solids	J. Mech. Phys. Solids
Journal of the Optical Society of America A: Optics, Image Science, and Vision	J. Opt. Soc. Am. A, Opt. Image Sci. Vis.
Journal of the Optical Society of America B: Optical Physics	J. Opt. Soc. Am. B.
Journal of the Physical Society of Japan	J. Phys. Soc. Jpn.
Journal of Vacuum Science and Technology	J. Vac. Sci. Technol.
Kolloid Zeitschrift & Zeitschrift für Polymere	Kolloid Z. Z. Polym.
Kongelige Danske Videnskabernes Selskab, Matematisk-Fysiske Meddelelser	K. Dan. Vidensk. Selsk. Mat. Fys. Medd.
Kristallografiya	Kristallografiya
Kvantovaya Elektronika	Kvant. Elektron.
Laser Physics	Laser Phys.
Laser Physics Letters	Laser Phys. Lett.
Laser Physics Review	Laser Phys. Rev.
Lettere al Nuovo Cimento	Lett. Nuovo Cimento
Lick Observatory Bulletin	Lick Obs. Bull.
Magnetic Resonance in Medicine	Magn. Reson. Med.
Materials Research Bulletin	Mater. Res. Bull.
Mathematical Physics and Applied Mathematics	Math. Phys. Appl. Math.
Medical Physics	Med. Phys.
Memoirs of the Royal Astronomical Society	Mem. R. Astron. Soc.
Molecular Crystals and Liquid Crystals	Mol. Cryst. Liq. Cryst.
Molecular Physics	Mol. Phys.
Monthly Notices of the Royal Astronomical Society	Mon. Not. R. Astron Soc.
National Bureau of Standards (U.S.), Circular	Natl. Bur. Stand. (U.S.), Circ.
National Bureau of Standards (U.S.), Miscellaneous Publication	Natl. Bur. Stand. (U.S.), Misc. Publ.
National Bureau of Standards (U.S.), Special Publication	Natl. Bur. Stand. (U.S.), Spec. Publ.
Nature (London)	Nature (London)
Nature Materials	Nat. Mater.
Nature Photonics	Nat. Photonics.
Nature Physics	Nat. Phys.
Naturwissenschaften	Naturwissenschaften
New Astronomy	New Astron.
New Astronomy Review	New Astron. Rev.

C6, *continued*

Nuclear Data, Section A	Nucl. Data, Sec. A
Nuclear Data, Section B	Nucl. Data, Sec. B
Nuclear Fusion	Nucl. Fusion
Nuclear Instruments	Nucl. Instrum.
Nuclear Instruments & Methods	Nucl. Instrum. Methods
Nuclear Physics	Nucl. Phys.
Nuclear Physics A	Nucl. Phys. A
Nuclear Physics B	Nucl. Phys. B
Nuclear Science and Engineering	Nucl. Sci. Eng.
Nukleonika	Nukleonika
Nuovo Cimento	Nuovo Cimento
Nuovo Cimento A	Nuovo Cimento A
Nuovo Cimento B	Nuovo Cimento B
Optica Acta	Opt. Acta
Optics and Spectroscopy	Opt. Spectrosc.
Optics Communications	Opt. Commun.
Optics Express	Optic. Express.
Optics Letters	Opt. Lett.
Optics News	Opt. News
Optik (Stuttgart)	Optik (Stuttgart)
Optika i Spectroskopiya	Opt. Spectrosk.
Optiko-Mekhanicheskaya Promyshlennost	Opt.-Mekh. Prom.
Philips Research Reports	Philips Res. Rep.
Philosophical Magazine	Philos. Mag.
Philosophical Magazine Letters	Philos. Mag. Lett.
Philosophical Transactions of the Royal Society of London	Philos. Trans. R. Soc. London
Philosophical Transactions of the Royal Society of London, Series A: Mathematical and Physical Sciences	Philos. Trans. R. Soc. London, Ser. A
Physica A: Statistical Mechanics and its Applications	Phys. A, Stat. Mech. Appl.
Physica B: Condensed Matter	Phys. B, Condens. Matter.
Physical Chemistry Chemical Physics	Phys. Chem. Chem. Phys.
Physical Review	Phys. Rev.
Physical Review A: Atomic, Molecular, and Optical	Physics Phys. Rev. A
Physical Review B: Condensed Matter	Phys. Rev. B
Physical Review C: Nuclear Physics	Phys. Rev. C
Physical Review D: Particles and Fields	Phys. Rev. D
Physical Review E: Statistical Physics, Plasmas, Fluids, and Related Interdisciplinary Topics	Phys. Rev. E
Physical Review Letters	Phys. Rev. Lett.
Physical Scripta	Phys. Scripta
Physica Status Solidi	Phys. Status Solidi
Physica Status Solidi A: Applied Research	Phys. Status Solidi A
Physica Status Solidi B: Basic Research	Phys. Status Solidi B
Physica (Utrecht)	Physica (Utrecht)
Physics and Chemistry of Solids	Phys. Chem. Solids

C6, *continued*

Physics in Medicine and Biology	Phys. Med. Biol.
Physics Letters	Phys. Lett.
Physics Letters A	Phys. Lett. A
Physics Letters B	Phys. Lett. B
Physics (New York)	Physics (N.Y.)
Physics of Fluids	Phys. Fluids
Physics of Metals and Metallography	Phys. Met. Metallogr.
Physics Reports	Phys. Rep.
Physics Teacher	Phys. Teach.
Physics Today	Phys. Today
Physikalische Zeitschrift	Phys. Z.
Physikalische Zeitschrift der Sowjetunion	Phys. Z. Sowjetunion
Physik der Kondensierten Materie	Phys. Kondens. Mater.
Pis'ma v Astronomicheskii Zhurnal	Pisma Astron. Zh.
Pis'ma v Zhurnal Eksperimental'noi i Teoreticheskoi Fiziki	Pis'ma Zh. E ´ ksp. Teor. Fiz.
Pis'ma v Zhurnal Tekhnicheskoi Fiziki	Pis'ma Zh. Tekh. Fiz.
Planetary and Space Science	Planet. Space Sci.
Plasma Physics	Plasma Phys.
Plasma Physics and Controlled Fusion	Plasma Phys. Control. Fusion
Plasma Sources Science and Technology	Plasma Sourc Sci. Tech.
Pribory i Tekhnika Eksperimenta	Prib. Tekh. Eksp.
Proceedings of the Cambridge Philosophical Society	Proc. Cambridge Philos. Soc.
Proceedings of the IEEE	Proc. IEEE
Proceedings of the IRE	Proc. IRE
Proceedings of the National Academy of Sciences of the United States of America	Proc. Natl. Acad. Sci. USA
Proceedings of the Physical Society, London	Proc. Phys. Soc. London
Proceedings of the Physical Society, London, Section A	Proc. Phys. Soc. London, Sec. A
Proceedings of the Physical Society, London, Section B	Proc. Phys. Soc. London, Sec. B
Proceedings of the Royal Society of London	Proc. R. Soc. London
Proceedings of the Royal Society of London, Series A: Mathematical and Physical Sciences	Proc. R. Soc. London, Ser. A
Progress of Theoretical Physics	Prog. Theor. Phys.
Publications of the Astronomical Society of Japan	Publ. Astron. Soc. Jpn.
Publications of the Astronomical Society of the Pacific	Publ. Astron. Soc. Pac.
Quantum Electronics	Quantum Electron.
Quantum Optics	Quantum Opt.
Radiation Effects	Radiat. Eff.
Radio Engineering and Electronic Physics	Radio Eng. Electron. Phys.
Radio Engineering and Electronics	Radio Eng. Electron.
Radiotekhnika i Elektronika	Radiotekh. Elektron.
Radiotekhnika i Elektronika	Radiotekh. Elektron.
RCA Review	RCA Rev.
Reports on Progress in Physics	Rep. Prog. Phys.
Review of Scientific Instruments	Rev. Sci. Instrum.

C6, *continued*

Reviews of Modern Physics	Rev. Mod. Phys.
Revista Mexicana de Astronomía y Astrofísica	Rev. Mex. Astron. Astrofis.
Revue d'Optique, Theorique et Instrumentale	Rev. Opt. Theor. Instrum.
Russian Journal of Physical Chemistry	Russ. J. Phys. Chem.
Science	Science
Scientific American	Sci. Am.
Solar Physics	Sol. Phys.
Solid State Communications	Solid State Commun.
Solid-State Electronics	Solid-State Electron.
Soviet Astronomy	Sov. Astron.
Soviet Astronomy Letters	Sov. Astron. Lett.
Soviet Journal of Atomic Energy	Sov. J. At. Energy
Soviet Journal of Low Temperature	Sov. J. Low Temp. Phys.
Soviet Journal of Nuclear Physics	Sov. J. Nucl. Phys.
Soviet Journal of Optical Technology	Sov. J. Opt. Technol.
Soviet Journal of Particles and Nuclei	Sov. J. Part. Nucl.
Soviet Journal of Plasma Physics	Sov. J. Plasma Phys.
Soviet Journal of Quantum Electronics	Sov. J. Quantum Electron.
Soviet Physics—Acoustics	Sov. Phys. Acoust.
Soviet Physics—Crystallography	Sov. Phys. Crystallogr.
Soviet Physics—Doklady	Sov. Phys. Dokl.
Soviet Physics—JETP	Sov. Phys. JETP
Soviet Physics Journal	Sov. Phys. J.
Soviet Physics—Semiconductors	Sov. Phys. Semicond.
Soviet Physics—Solid State	Sov. Phys. Solid State
Soviet Physics—Technical Physics	Sov. Phys. Tech. Phys.
Soviet Physics—Uspekhi	Sov. Phys. Usp.
Soviet Radiophysics	Sov. Radiophys.
Soviet Technical Physics Letters	Sov. Tech. Phys. Lett.
Space Journal	Space. J.
Spectrochimica Acta	Spectrochim. Acta
Spectrochimica Acta, Part A: Molecular Spectroscopy	Acta, Part A Spectrochim. Acta, Part A
Spectrochimica Acta, Part B: Atomic Spectroscopy	Acta, Part B Spectrochim. Acta, Part B
Surface Science	Surf. Sci.
Teplofizika Vysokikh Temperatur	Teplofiz. Vys. Temp.
Theoretica Chimica Acta	Theor. Chim. Acta
Theoretical and Mathematical Physics	Theor. Math. Phys.
Thin Solid Films	Thin Solid Films
Transactions of the American Crystallographic Association	Trans. Am. Crystallogr. Assoc.
Transactions of the American Geophysical Union	Trans. Am. Geophys. Union
Transactions of the American Institute of Mining, Metallurgical and Petroleum Engineers	Trans. Am. Inst. Min. Metall. Pet. Eng.
Transactions of the American Nuclear Society	Trans. Am. Nucl. Soc.
Transactions of the American Society for Metals	Trans. Am. Soc. Met.
Transactions of the American Society of Mechanical Engineers	Trans. Am. Soc. Mech. Eng.
Transactions of the British Ceramic Society	Trans. Br. Ceram. Soc.

C6, *continued*

Transactions of the Faraday Society	Trans. Faraday Soc.
Transactions of the Metallurgical Society of AIME	Trans. Metall. Soc. AIME
Transactions of the Society of Rheology	Trans. Soc. Rheol.
Ukrainian Physics Journal	Ukr. Phys. J.
Ultrasonics	Ultrasonics
Vistas in Astronomy	Vistas Astron.
Wuli Xuebao	Wuli Xuebao
Yadernaya Fizika	Yad. Fiz.
Zeitschrift für Analytische Chemie	Z. Anal. Chem.
Zeitschrift für Angewandte Physik	Z. Angew. Phys.
Zeitschrift für Anorganische und Allgemeine Chemie	Z. Anorg. Allg. Chem.
Zeitschrift für Astrophysik	Z. Astrophys.
Zeitschrift für Elektrochemie	Z. Elektrochem.
Zeitschrift für Kristallographie, Kristallgeometrie, Kristallphysik,Kristallchemie	Z. Kristallogr. Kristallgeom. Kristallphys. Kristallchem.
Zeitschrift für Metallkunde	Z. Metallkd.
Zeitschrift für Naturforschung	Z. Naturforsch.
Zeitschrift für Naturforschung, Teil A: Physik, Physikalische Chemie, Kosmophysik	Z. Naturforsch. Teil A
Zeitschrift für Physik	Z. Phys.
Zeitschrift für Physik A: Atoms and Nuclei	Z. Phys. A
Zeitschrift für Physik B: Condensed Matter and Quanta	Z. Phys. B
Zeitschrift für Physik C: Particles and Fields	Z. Phys. C
Zeitschrift für Physikalisch-Chemische Materialforschung	Z. Phys. Chem. Materialforsch.
Zeitschrift für Physikalische Chemie, Abteilung A: Chemische Thermodynamik, Kinetik, Elektrochemie, Eigenschaftslehre	Z. Phys. Chem. Abt. A
Zeitschrift für Physikalische Chemie, Abteilung B: Chemie der Elementarprozesse, Aufbau der Materie	Z. Phys. Chem. Abt. B
Zeitschrift für Physikalische Chemie (Frankfurt am Main)	Z. Phys. Chem. (Frankurt am Main)
Zeitschrift für Physikalische Chemie (Leipzig)	Z. Phys. Chem. (Leipzig)
Russian Journal of Physical Chemistry	Russ. J. Phys. Chem.
Zhurnal Prikladnoi Spektroskopii	Zh. Prikl. Spektrosk.
Zhurnal Tekhnicheskoi Fiziki	Zh. Tekh. Fiz.

Appendix D. Tables and Conventions for Chemistry

Contents

Note: For a glossary of chemistry terms, see **Appendix E2**, **Glossary of Chemistry Terms**.

D1 Chemistry Symbols and Abbreviations

D1.1 General Chemistry

Name	Symbol	SI Unit
amount of substance	n	mol
atomic weight	A_r	[dimensionless]
concentration	c	$mol \cdot m^{-3}$
degree of dissociation	α	[dimensionless]
extent of reaction	ζ	mol
mass fraction	w	[dimensionless]
molality	m, b	$mol \cdot kg^{-1}$
molar mass	M	$kg \cdot mol^{-1}$
molar volume	V_m	$m^3 \cdot mol^{-1}$
molarity	M	$mol \cdot L^{-1}$
mole fraction	x	[dimensionless]
molecular weight	M_r	[dimensionless]
number concentration	C, n	m^{-3}
number density of entities	C, n	m^{-3}
number of entities	N	[dimensionless]
partial pressure of substance B	p_B	Pa
relative atomic mass	A_r	[dimensionless]
relative molecular mass	M_r	[dimensionless]
stoichiometric coefficient	ν	[dimensionless]
surface concentration	Γ	$mol \cdot m^{-2}$
volume fraction	Φ	[dimensionless]

D1.2 Melting and Boiling Point Abbreviations

Term	Abbreviation
melting point	mp
boiling point	bp
literature value	lit.
decomposition	dec

D1.3 Chemical Kinetics

Name	Symbol	SI Unit
absolute temperature	T	K
Arrhenius or activation energy	E_a, E_A	$J{\cdot}mol^{-1}$
Boltzmann constant	k, k_B	$J{\cdot}K^{-1}$
energy of activation	E	$J{\cdot}mol^{-1}$
half-life	$t_{1/2}$	s
photochemical yield, quantum yield	Φ	[dimensionless]
rate constant (first order)	k	s^{-1}
rate constant (second order)	k	$mol^{-1}{\cdot}sec^{-1}$
rate of concentration change of substance X	r_X	$mol{\cdot}m^{-3}{\cdot}s^{-1}$
rate of conversion	$\dot{\xi}$	$mol{\cdot}sec^{-1}$
rate of reaction	v	$mol{\cdot}m^{-3}{\cdot}s^{-1}$
relaxation time	τ	s
standard enthalpy of activation	$\Delta^{\ddagger}H^{\circ}$, $\Delta H^{\ddagger}$	$J{\cdot}mol^{-1}$
standard entropy of activation	$\Delta^{\ddagger}S^{\circ}$, $\Delta S^{\ddagger}$	$J{\cdot}mol^{-1}{\cdot}K^{-1}$
standard Gibbs energy of activation	$\Delta^{\ddagger}G^{\circ}$, $\Delta G^{\ddagger}$	$J{\cdot}mol^{-1}$
standard internal energy of activation	$\Delta^{\ddagger}U^{\circ}$, $\Delta U^{\ddagger}$	$J{\cdot}mol^{-1}$
volume of activation	$\Delta^{\ddagger}V$, $\Delta V^{\ddagger}$	$J{\cdot}mol^{-1}$

D1.4 Polymer Chemistry

Name	Symbol	SI Unit
bulk modulus	K	Pa
complex permittivity	ε^*	$F{\cdot}m^{-1}$
crack-tip radius	ρ_c	m
electrophoretic mobility	μ	$m^2{\cdot}V^{-1}{\cdot}s^{-1}$
fracture strain	γ_f, ε_f	[dimensionless]
modulus of elasticity	E	Pa
tensile strength	σ	Pa
viscosity	v	$Pa{\cdot}s$
yield stress	σ_y	Pa
Young's modulus	E	Pa

D1.5 Electrochemistry

Name	Symbol	SI Unit
charge number of an ion	z	dimensionless
conductivity	κ	$S \cdot m^{-1}$
diffusion rate constant	k_d	$m \cdot s^{-1}$
electric current	I	A
electric current density	j	$A \cdot m^{-2}$
electric mobility	u	$m^2 \cdot V^{-1} \cdot s^{-1}$
electrode potential	E	V
electrolytic conductivity	κ	$S \cdot m^{-1}$
elementary charge	e	C
Faraday constant	F	$C \cdot mol^{-1}$
ionic strength (concentration basis)	I_c	$mol \cdot m^{-3}$
ionic strength (molality basis)	I_m	$mol \cdot kg^{-1}$
mass-transfer coefficient	k_d	$m \cdot s^{-1}$
molar conductivity	Λ	$S \cdot m^2 \cdot mol^{-1}$
reduction potential	E°	V
standard electrode potential	E°	V
standard electromotive force (emf)	E°	V
surface charge density	σ	$C \cdot m^{-2}$
transport number	t	dimensionless

D1.6 Subatomic Particles

Subatomic particle	Symbol	Subatomic particle	Symbol
alpha particle	α	neutrino	$\nu, \nu_e, \nu_\mu, \nu_\tau$
beta particle	β	neutron	N
deuteron	d	photon	Γ
electron	e	pion	Π
hellion	h	proton	P
kaon	K	tau	T
muon	μ	triton	T

A more complete table of subatomic particles may be found in **Chapter 7, Section 7.2.2.**

D1.7 Chirality

Term	Symbol
anticlockwise (counterclockwise)	*A*
clockwise	*C*
cube	*CU*
dodecahedron	*DD*
octahedron	*OC*
trigonal phase	*TP*
trigonal prism	*TPR*
trigonal pyramid	*TPY*

D1.8 Spectroscopy

D1.8.1 NMR Spectroscopy

Abbreviation	Definition
δ	chemical shift downfield from the standard (in parts per million or ppm)
J	coupling constant (in hertz or Hz)
s	singlet
d	doublet
t	triplet
q	quartet
br	broadened

D1.8.2 IR Spectroscopy

Abbreviation	Definition
v_{max}(with a ~ on top)	wavenumber of maximum absorption peaks (in cm^{-1})
w	weak
m	medium
s	strong
vw	very weak
vs	very strong
br	broad

D1.8.3 Mass Spectroscopy

Abbreviation	Definition
m/z	mass-to-charge ratio
M	molecular weight of the molecule
M^+	molecular ion
HRMS	high-resolution mass spectrometry
FAB	fast atom bombardment
EIMS	electron-impact mass spectrometry

D1.8.4 UV-Visible Spectroscopy

Abbreviation	Definition
λ_{max}	wavelength of maximum absorption in nanometers
ε	extinction coefficient or molar absorptivity
sh	shoulder
ν (with a ~ on top)	wavenumber (in μm^{-1})

D2 Word Division of Chemical Names

To divide the name of a chemical between lines, divide between any already hyphenated elements, between an affix and a root, or within a component according to the following list. *For general guidelines for word division, see* **Chapter 3, Section 3.3**.

ace-naph-tho	an-thra-cene	car-bo-ni-um
ace-tal	an-thra-ce-no	car-bon-yl
acet-al-de-hyde	an-thryl	car-box-ami-do
acet-amide	ar-se-nate	car-boxy
acet-ami-do	ar-si-no	car-box-yl
acet-amin-o-phen	aryl	car-byl-a-mi-no
acet-an-i-lide	az-i-do	chlo-ride
ace-tate	azi-no	chlo-rite
ac-et-azol-amide	azo	chlo-ro
ace-tic	benz-ami-do	chlo-ro-syl
ace-to	ben-zene	chlo-ryl
ace-to-ace-tic	benz-hy-dryl	cu-mene
ace-tone	ben-zo-yl	cy-a-nate
ace-to-ni-trile	ben-zyl	cy-a-nide
ace-tyl	ben-zyl-i-dene	cy-a-na-to
acet-y-late	bi-cy-clo	cy-a-no
acet-y-lene	bo-ric	cy-clo
acro-le-in	bo-ryl	cy-clo-hex-ane
ac-ryl-am-ide	bro-mide	cy-clo-hex-yl
ac-ry-late	bro-mo	deca
acryl-ic	bu-tane	di-azo
ac-ry-lo	bu-ten-yl	di-bo-ran-yl
ad-i-po-yl	bu-tyl	di-car-bon-yl
al-a-nine	bu-tyl-ene	di-im-ino
al-kyl	bu-tyl-i-dene	di-oxy
al-lyl	car-ba-mate	di-oyl
ami-di-no	car-bam-ic	diyl
amide	car-ba-mide	do-de-cyl
ami-do	carb-an-ion	ep-oxy
amine	car-ba-ryl	eth-ane
ami-no	car-ba-zole	eth-a-no
am-mine	car-bi-nol	eth-a-nol
aam-mo-nio	car-bol-ic	eth-a-no-yl
am-mo-ni-um	car-bon-ate	eth-en-yl
an-thra	car-bon-ic	eth-yl

D2, *continued*

eth-yl-ene
eth-yl-i-dene
eth-yn-yl
fluo-res-cence
fluo-ride
fluo-ro
form-al-de-hyde
form-ami-do
for-mic
form-imi-do-yl
for-myl
fu-ran
ger-myl
gua-ni-di-no
gua-nyl
halo
hep-ta
hep-tane
hep-tyl
hexa
hex-ane
hex-yl
hy-dra-zide
hy-dra-zine
hy-dra-zi-no
hy-dra-zo
hy-dra-zo-ic
hy-dric
hy-dride
hy-dri-od-ic
hy-dro
hy-dro-chlo-ric
hy-dro-chlo-ride
hy-dro-chlo-ro
hy-drox-ide
hy-droxy
hy-drox-yl
ico-sa-he-dral
imi-da-zole
imide
imi-do

imi-do-yl
imi-no
in-da-mine
in-da-zole
in-dene
in-de-no
in-dole
io-date
io-dide
iodo
io-do-syl
io-dyl
iso-cy-a-na-to
iso-cy-a-nate
iso-cy-a-nide
iso-pro-pen-yl
iso-pro-pyl
mer-cap-to
mer-cu-ric
meth-an-ami-do
meth-ane
meth-ano
meth-yl
meth-yl-ate
meth-yl-ene
meth-yl-i-dene
mono
mono-ac-id
mono-amine
naph-tha-lene
naph-tho
naph-thyl
neo-pen-tyl
ni-trate
ni-tric
ni-trile
ni-trilo
ni-trite
ni-tro
ni-troso
no-na

oc-ta
oc-tane
oc-tyl
ox-idase
ox-ide
ox-ido
ox-ime
oxo
ox-o-nio
oxy
palm-i-toyl
pen-ta
pen-tane
pen-tyl
pen-tyl-i-dene
per-chlo-rate
per-chlo-ride
per-chlo-ryl
per-man-ga-nate
per-ox-idase
per-ox-ide
per-oxy
phen-ac-e-tin
phen-an-threne
phen-an-thro
phen-an-thryl
phen-a-zine
phe-no
phe-nol
phe-none
phen-ox-ide
phen-oxy
phen-yl
phen-yl-ene
phos-phate
phos-phide
phos-phine
phos-phi-no
phos-phin-yl
phos-phite
phos-pho

D2, *continued*

phos-pho-nio
phos-pho-no
phos-phor-anyl
phos-pho-li-pase
phos-pho-lip-id
phos-pho-ni-um
phos-pho-ric
phos-pho-rus
phos-pho-ryl
plum-byl
pro-pane
pro-pen-yl
pro-pen-yl-ene
pro-pyl
pro-pyl-ene
pro-pyl-i-dene
pu-rine
py-ran
pyr-a-zine
pyr-a-zole
pyr-i-dine
pyr-id-a-zine
pyr-role
quin-o-line
qui-none
sel-e-nate
se-le-nic
sel-e-nide
sel-e-nite
se-le-no
si-lane
sil-anyl
sil-ox-anyl
sil-ox-yl
si-lyl
spi-ro
stan-nic
stan-nite
stan-nous
stan-nyl
stib-ino

sty-rene
sty-ryl
sul-fa-mo-yl
sul-fate
sul-fe-no
sul-fe-nyl
sul-fide
sul-fi-do
sul-fi-no
sul-fi-nyl
sul-fite
sul-fo
sul-fon-ami-do
sul-fo-nate
sul-fone
sul-fon-ic
sul-fo-nio
sul-fo-nyl
sulf-ox-ide
sul-fu-ric
sul-fu-rous
su-fu-ryl
tet-ra
thio
thio-nyl
thio-phene
thi-oxo
thi-oyl
tol-u-ene
tol-u-ide
tol-yl
tri-a-zine
tri-a-zole
tri-yl
urea
ure-ide
ure-ido
uric
vi-nyl
vi-nyl-i-dene
xan-thene

xan-tho
xy-lene
xy-li-dine
xy-lyl
xy-li-din-yl
yl-i-dene

D3 Atomic Weights of the Elements

See also **Section 9.2.3.1 Periodic Table of the Elements**

Element Name	Atomic Number	Chemical Symbol	Atomic Weight	Notes
Hydrogen	1	H	1.00794(7)	1, 2, 3
Helium	2	He	4.002602(2)	1, 2
Lithium	3	Li	[6.941(2)]	1, 2, 3, 4
Beryllium	4	Be	9.012182(3)	
Boron	5	B	10.811(7)	1, 2, 3
Carbon	6	C	12.0107(8)	1, 2
Nitrogen	7	N	14.0067(2)	1, 2
Oxygen	8	O	15.9994(3)	1, 2
Fluorine	9	F	18.9984032(5)	
Neon	10	Ne	20.1797(6)	1, 3
Sodium	11	Na	22.98976928(2)	
Magnesium	12	Mg	24.3050(6)	
Aluminum	13	Al	26.9815386(8)	
Silicon	14	Si	28.0855(3)	2
Phosphorus	15	P	30.973762(2)	
Sulfur	16	S	32.065(5)	1, 2
Chlorine	17	Cl	35.453(2)	3
Argon	18	Ar	39.948(1)	1, 2
Potassium	19	K	39.0983(1)	1
Calcium	20	Ca	40.078(4)	1
Scandium	21	Sc	44.955912(6)	
Titanium	22	Ti	47.867(1)	
Vanadium	23	V	50.9415(1)	
Chromium	24	Cr	51.9961(6)	
Manganese	25	Mn	54.938045(5)	
Iron	26	Fe	55.845(2)	
Cobalt	27	Co	58.933195(5)	
Nickel	28	Ni	58.6934(4)	
Copper	29	Cu	63.546(3)	2
Zinc	30	Zn	65.38(2)	
Gallium	31	Ga	69.723(1)	
Germanium	32	Ge	72.64(1)	
Arsenic	33	As	74.92160(2)	
Selenium	34	Se	78.96(3)	
Bromine	35	Br	79.904(1)	

D3, *continued*

Element Name	Atomic Nunber	Chemical Symbol	Atomic Weight	Notes
Krypton	36	Kr	83.798(2)	1, 3
Rubidium	37	Rb	85.4678(3)	1
Strontium	38	Sr	87.62(1)	1, 2
Yttrium	39	Y	88.90585(2)	
Zirconium	40	Zr	91.224(2)	1
Niobium	41	Nb	92.90638(2)	
Molybdenum	42	Mo	95.96(2)	1
Technetium	43	Tc	[98]	5
Ruthenium	44	Ru	101.07(2)	1
Rhodium	45	Rh	102.90550(2)	
Palladium	46	Pd	106.42(1)	1
Silver	47	Ag	107.8682(2)	1
Cadmium	48	Cd	112.411(8)	1
Indium	49	In	114.818(3)	
Tin	50	Sn	118.710(7)	1
Antimony	51	Sb	121.760(1)	1
Tellurium	52	Te	127.60(3)	1
Iodine	53	I	126.90447(3)	
Xenon	54	Xe	131.293(6)	1, 3
Caesium	55	Cs	132.9054519(2)	
Barium	56	Ba	137.327(7)	
Lanthanum	57	La	138.90547(7)	1
Cerium	58	Ce	140.116(1)	1
Praseodymium	59	Pr	140.90765(2)	
Neodymium	60	Nd	144.242(3)	1
Promethium	61	Pm	[145]	5
Samarium	62	Sm	150.36(2)	1
Europium	63	Eu	151.964(1)	1
Gadolinium	64	Gd	157.25(3)	1
Terbium	65	Tb	158.92535(2)	
Dysprosium	66	Dy	162.500(1)	1
Holmium	67	Ho	164.93032(2)	
Erbium	68	Er	167.259(3)	1
Thulium	69	Tm	168.93421(2)	
Ytterbium	70	Yb	173.054(5)	1
Lutetium	71	Lu	174.9668(1)	1
Hafnium	72	Hf	178.49(2)	

D3, *continued*

Element Name	Atomic Nunber	Chemical Symbol	Atomic Weight	Notes
Tantalum	73	Ta	180.94788(2)	
Tungsten	74	W	183.84(1)	
Rhenium	75	Re	186.207(1)	
Osmium	76	Os	190.23(3)	1
Iridium	77	Ir	192.217(3)	
Platinum	78	Pt	195.084(9)	
Gold	79	Au	196.966569(4)	
Mercury	80	Hg	200.59(2)	
Thallium	81	Tl	204.3833(2)	
Lead	82	Pb	207.2(1)	1, 2
Bismuth	83	Bi	208.98040(1)	
Polonium	84	Po	[209]	5
Astatine	85	At	[210]	5
Radon	86	Rn	[222]	5
Francium	87	Fr	[223]	5
Radium	88	Ra	[226]	5
Actinium	89	Ac	[227]	5
Thorium	90	Th	232.03806(2)	1, 5
Protactinium	91	Pa	231.03588(2)	5
Uranium	92	U	238.02891(3)	1, 3, 5
Neptunium	93	Np	[237]	5
Plutonium	94	Pu	[244]	5
Americium	95	Am	[243]	5
Curium	96	Cm	[247]	5
Berkelium	97	Bk	[247]	5
Californium	98	Cf	[251]	5
Einsteinium	99	Es	[252]	5
Fermium	100	Fm	[257]	5
Mendelevium	101	Md	[258]	5
Nobelium	102	No	[259]	5
Lawrencium	103	Lr	[262]	5
Rutherfordium	104	Rf	[267]	5
Dubnium	105	Db	[268]	5
Seaborgium	106	Sg	[271]	5
Bohrium	107	Bh	[272]	5
Hassium	108	Hs	[270]	5
Meitnerium	109	Mt	[276]	5

D3, *continued*

Element Name	Atomic Nunber	Chemical Symbol	Atomic Weight	Notes
Darmstadtium	110	Ds	[281]	5
Roentgenium	111	Rg	[280]	5
Ununbium	112	Uub	[285]	5, 6
Ununtrium	113	Uut	[284]	5, 6
Ununquadium	114	Uuq	[289]	5, 6
Ununpentium	115	Uup	[288]	5, 6
Ununhexium	116	Uuh	[293]	5, 6
Ununoctium	118	Uuo	[294]	5, 6

1. Geological specimens are known in which the element has an isotopic composition outside the limits for normal material. The difference between the atomic weight of the element in such specimens and that given in the Table may exceed the stated uncertainty.
2. Range in isotopic composition of normal terrestrial material prevents a more precise value being given; the tabulated value should be applicable to any normal material.
3. Modified isotopic compositions may be found in commercially available material because it has been subject to an undisclosed or inadvertant isotopic fractionation. Substantial deviations in atomic weight of the element from that given in the Table can occur.
4. Commercially available Li materials have atomic weights that range between 6.939 and 6.996; if a more accurate value is required, it must be determined for the specific material [range quoted for 1995 table 6.94 and 6.99].
5. Element has no stable nuclides. The value enclosed in brackets, e.g. [209], indicates the mass number of the longest-lived isotope of the element. However three such elements (Th, Pa, and U) do have a characteristic terrestrial isotopic composition, and for these an atomic weight is tabulated.
6. The names and symbols for elements 112-118 are under review. The temporary system recommended by J Chatt, *Pure Appl. Chem.*, 51, 381-384 (1979) is used above.

D4 Chemistry Journals and Their Abbreviations

Full Title	Abbreviation
Accounts of Chemical Research	Acc. Chem. Res.
ACS Symposium Series	ACS Symp. Ser.
Acta Chemica Scandinavica	Acta Chem. Scand.
Acta Crystallographica, Section C: Crystal Structure Communications	Acta Crystallogr., Sect. C: Cryst. Struct. Commun.
Acta Crystallographica, Section D: Biological Crystallography	Acta Crystallogr., Sect. D: Biol. Crystallogr.
Acta Crystallographica, Section E: Structure Reports Online	Acta Crystallogr., Sect. E: Struct. Rep. Online
Acta Materialia	Acta Mater.
Acta Pharmacologica Sinica	Acta Pharmacol. Sin.
Acta Physica Polonica, B	Acta Phys. Pol., B
Advanced Functional Materials	Adv. Funct. Mater.
Advanced Materials (Weinheim, Germany)	Adv. Mater. (Weinheim, Ger.)
Advanced Synthesis and Catalysis	Adv. Synth. Catal.
Advances in Catalysis	Adv. Catal
Advances in Organometallic Chemistry	Adv. Organomet. Chem.
Advances in Physical Organic Chemistry	Adv. Phys. Org. Chem.
Advances in Space Research	Adv. Space Res.
AIChE Journal	AIChE J.
AIP Conference Proceedings	AIP Conf. Proc.
Aldrichimica Acta	Aldrichimica Acta
American Journal of Clinical Nutrition	Am. J. Clin. Nutr.
American Journal of Human Genetics	Am. J. Hum. Genet.
American Journal of Obstetrics and Gynecology	Am. J. Obstet. Gynecol.
American Journal of Pathology	Am. J. Pathol.
American Journal of Physiology	Am. J. Physiol.
American Journal of Respiratory Cell and Molecular Biology	Am. J. Respir. Cell Mol. Biol.
American Journal of Veterinary Research	Am. J. Vet. Res.
American Mineralogist	Am. Mineral.
Analyst (Cambridge, United Kingdom)	Analyst (Cambridge, U. K.)
Analytical and Bioanalytical Chemistry	Anal. Bioanal.Chem.
Analytical Biochemistry	Anal. Biochem.
Analytical Chemistry	Anal. Chem.
Analytical Letters	Anal. Lett.
Analytical Sciences	Anal. Sci.
Anesthesiology	Anesthesiology
Angewandte Chemie, International Edition	Angew. Chem., Int. Ed.

D4, *continued*

Annals of the New York Academy of Sciences	Ann. N. Y. Acad. Sci.
Annual Reports Section A of the Royal Society of Chemistry	Ann. Rep. Prog. Chem. Sect. A
Annual Reports Section B of the Royal Society of Chemistry	Ann. Rep. Prog. Chem. Sect. B
Annual Reports Section C of the Royal Society of Chemistry	Ann. Rep. Prog. Chem. Sect. C
Annual Review of Physical Chemistry	Annu. Rev. Phys. Chem.
Anti-Cancer Drugs	Anti-Cancer Drugs
Anticancer Research	Anticancer Res.
Antimicrobial Agents and Chemotherapy	Antimicrob. Agents Chemother.
Antioxidants & Redox Signaling	Antioxid. Redox Signaling
Applied Catalysis, A: General	Appl. Catal., A
Applied Catalysis, B: Environmental	Appl. Catal., B
Applied Geochemistry	Appl. Geochem.
Applied Microbiology and Biotechnology	Appl. Microbiol. Biotechnol.
Applied Optics	Appl. Opt.
Applied Organometallic Chemistry	Appl. Organomet. Chem.
Applied Physics A: Materials Science & Processing	Appl. Phys. A: Mater. Sci. Process.
Applied Physics B: Lasers and Optics	Appl. Phys. B: Lasers Opt.
Applied Physics Letters	Appl. Phys. Lett.
Applied Radiation and Isotopes	Appl. Radiat. Isot.
Applied Spectroscopy	Appl. Spectrosc.
Aquaculture	Aquaculture
Aquatic Toxicology	Aquat. Toxicol.
Archives of Biochemistry and Biophysics	Arch. Biochem. Biophys.
Archives of Environmental Contamination and Toxicology	Arch. Environ. Contam. Toxicol.
Archives of Pharmacal Research	Arch. Pharmacal Res.
Archives of Virology	Arch Virol.
ARKIVOC (Gainsville, FL, United States)	ARKIVOC (Gainsville, FL, U. S.)
Arteriosclerosis, Thrombosis, and Vascular Biology	Arterioscler., Thromb., Vasc. Biol.
Asian Journal of Chemistry	Asian J. Chem.
Astronomical Journal	Astron. J.
Astrophysical Journal	Astrophys. J.
Atherosclerosis (Amsterdam, Netherlands)	Atherosclerosis (Amsterdam, Neth.)
Atmospheric Chemistry and Physics	Atmos. Chem. Phys.
Atmospheric Environment	Atmos. Environ.
Australian Journal of Chemistry	Aust. J. Chem.

D4, *continued*

Australian Journal of Education in Chemistry	Aust. J. Edu. Chem.
Azerbaidzhanskii Kmimicheskii Zhurnal	Azerb. Khim. Zh.
Bandaoti Xuebao	Bandaoti Xuebao
Beilstein Journal of Organic Chemistry	Beilstein J. Org. Chem.
Biochemical and Biophysical Research Communications	Biochem. Biophys. Res. Commun.
Biochemical Journal	Biochem. J
Biochemical Pharmacology	Biochem. Pharmacol.
Biochemical Society Transactions	Biochem. Soc. Trans.
Biochemical Systems and Ecology	Biochem. Syst. Ecol.
Biochemistry (Moscow, Russian Federation)	Biochemistry (Moscow, Russ. Fed.)
Biochimica et Biophysica Acta	Biochim. Biophys. Acta
Bioconjugate Chemistry	Bioconjugate Chem.
Bioelectrochemistry	Bioelectrochem.
Bioinformatics	Bioinformatics
Biological & Pharmaceutical Bulletin	Biol. Pharm. Bull.
Biological Chemistry	Biol. Chem.
Biological Trace Element Research	Biol. Trace Elem. Res.
Biology of Reproduction	Biol. Reprod.
Biomacromolecules	Biomacromolecules
Biomaterials	Biomaterials
Bioorganic & Medicinal Chemistry	Bioorg. Med. Chem.
Bioorganic & Medicinal Chemistry Letters	Bioorg. Med. Chem. Lett.
Biophysical Chemistry	Biophys. Chem.
Biophysical Journal	Biophys. J.
Biopolymers	Biopolymers
Bioresource Technology	Bioresour. Technol.
Bioscience, Biotechnology, and Biochemistry	Biosci., Biotechnol., Biochem.
Biosensors & Bioelectronics	Biosens. Bioelectron.
BioTechniques	BioTechniques
Biotechnology and Bioengineering	Biotechnol. Bioeng.
Biotechnology Letters	Biotechnol. Lett.
Biotechnology Progress	Biotechnol. Prog.
Blood	Blood
Brain Research	Brain Res.
British Journal of Cancer	Br. J. Cancer
British Journal of Clinical Pharmacology	Br. J. Clin. Pharmacol.
British Journal of Nutrition	Br. J. Nutr.
British Journal of Pharmacology	Br. J. Pharmacol.

D4, *continued*

Bulletin of Environmental Contamination and Toxicology	Bull. Environ. Contam. Toxicol.
Bulletin of the Chemical Society of Japan	Bull. Chem. Soc. Jpn.
Bulletin of the Korean Chemical Society	Bull. Korean Chem. Soc.
Bunseki Kagaku	Bunseki Kagaku
Canadian Journal of Chemistry	Can. J. Chem.
Cancer Cell	Cancer Cell
Cancer Letters (Amsterdam, Netherlands)	Cancer Lett. (Amsterdam, Neth.)
Cancer Research	Cancer Res.
Cancer Science	Cancer Sci.
Carbohydrate Polymers	Carbohydr. Polym.
Carbohydrate Research	Carbohydr. Res.
Carcinogenesis	Carcinogenesis
Cardiovascular Research	Cardiovasc. Res.
Catalysis Communications	Catal. Commun.
Catalysis Letters	Catal. Lett.
Catalysis Reviews	Cat. Rev.
Catalysis Today	Catal. Today
Catalysts and Catalysed Reactions	CCR
Cell (Cambridge, MA, United States)	Cell (Cambridge, MA, U. S.)
Cell Cycle	Cell Cycle
Cellular and Molecular Life Sciences	Cell. Mol. Life Sci.
Cement & Concrete Composites	Cem. Concr. Compos.
Cement and Concrete Research	Cem. Concr. Res.
Ceramic Engineering and Science Proceedings	Ceram. Eng. Sci. Proc.
Ceramic Transactions	Ceram. Trans.
Ceramics International	Ceram. Int.
Ceramics-Silikaty	Ceram. Silik.
Cereal Chemistry	Cereal Chem.
ChemBioChem	ChemBioChem
Chemical & Engineering News	Chem. Eng. News
Chemical & Pharmaceutical Bulletin	Chem. Pharm. Bull.
Chemical Communications (Cambridge, United Kingdom)	Chem. Commun. (Cambridge, U.K.)
Chemical Engineering and Processing	Chem. Eng. Process.
Chemical Engineering Journal (Amsterdam, Netherlands)	Chem. Eng. J. (Amsterdam, Neth.)
Chemical Engineering Research and Design	Chem. Eng. Res. Des.
Chemical Engineering Science	Chem. Eng. Sci.
Chemical Geology	Chem. Geol.

D4, *continued*

Chemical Physics	Chem. Phys.
Chemical Physics Letters	Chem. Phys. Lett.
Chemical Research in Chinese Universities	Chem. Res. Chin. Univ.
Chemical Research in Toxicology	Chem. Res. Toxicol.
Chemical Reviews (Washington, DC, United States)	Chem. Rev. (Washington, DC, U. S.)
Chemical Society Reviews	Chem. Soc. Rev.
Chemie Ingenieur Technik	Chem. Ing. Tech.
Chemische Berichte	Ber.
Chemistry & Biology	Chem. Biol.
Chemistry Education Research and Practice	CERP
Chemistry Letters	Chem. Lett.
Chemistry of Heterocyclic Compounds (New York, NY, United States)	Chem. Heterocycl. Compd. (N. Y., NY, U. S.)
Chemistry of Materials	Chem. Mater.
Chemistry--A European Journal	Chem.--Eur. J.
ChemMedChem	ChemMedChem
Chemosphere	Chemosphere
ChemPhysChem	ChemPhysChem
Chinese Chemical Letters	Chin. Chem. Lett.
Chinese Journal of Chemistry	Chin. J. Chem.
Chinese Science Bulletin	Chin. Sci. Bull.
Chromatographia	Chromatographia
Circulation Research	Circ. Res.
Clinica Chimica Acta	Clin. Chim. Acta
Clinical and Experimental Immunology	Clin. Exp. Immunol.
Clinical and Experimental Pharmacology and Physiology	Clin. Exp. Pharmacol. Physiol.
Clinical Biochemistry	Clin. Biochem.
Clinical Chemistry (Washington, DC, United States)	Clin. Chem. (Washington, DC, U. S.)
Clinical Chemistry and Laboratory Medicine	Clin. Chem. Lab. Med.
Clinical Immunology (San Diego, CA, United States)	Clin. Immunol. (San Diego, CA, U. S.)
Clinical Science	Clin. Sci.
Collection of Czechoslovak Chemical Communications	Collect. Czech. Chem. Commun.
Colloid and Polymer Science	Colloid Polym. Sci.
Colloids and Surfaces, A: Physicochemical and Engineering Aspects	Colloids Surf., A
Colloids and Surfaces, B: Biointerfaces	Colloids Surf., B

D4, *continued*

Combustion and Flame	Combust. Flame
Communications in Soil Science and Plant Analysis	Commun. Soil Sci. Plant Anal.
Comparative Biochemistry and Physiology, Part A: Molecular & Integrative Physiology	Comp. Biochem. Physiol., Part A: Mol. Integr. Physiol.
Comparative Biochemistry and Physiology, Part B: Biochemistry & Molecular Biology	Comp. Biochem. Physiol., Part B: Biochem. Mol. Biol.
Comptes Rendus Chimie	C. R. Chim.
Computers & Chemical Engineering	Comput. Chem. Eng.
Computers and Chemistry	Comput. Chem. (Oxford)
Coordination Chemistry Reviews	Coord. Chem. Rev.
Corrosion Science	Corros. Sci.
Crystal Growth & Design	Cryst. Growth Des.
CrystEngComm	CrystEngComm
Cuihua Xuebao	Cuihua Xuebao
Current Biology	Curr. Biol.
Current Medicinal Chemistry	Curr. Med. Chem.
Current Microbiology	Curr. Microbiol.
Current Pharmaceutical Design	Curr. Pharm. Des.
Current Science	Curr. Sci.
Cytokine+	Cytokine+
Czechoslovak Journal of Physics	Czech J. Phys.
Dalton Transactions	Dalton Trans.
Desalination	Desalination
Developmental Biology (San Diego, CA, United States)	Dev. Biol. (San Diego, CA, U S)
Developmental Brain Research	Dev. Brain Res.
Developmental Cell	Dev. Cell
Diabetes	Diabetes
Diabetologia	Diabetologia
Diamond and Related Materials	Diamond and Related Materials
Diffusion and Defect Data--Solid State Data, Pt. B: Solid State Phenomena	Diffus. Defect Data, Pt. B
Digestive Diseases and Sciences	Dig. Dis. Sci.
DNA Repair	DNA Repair
Dokladi na Bulgarskata Akademiya na Naukite	Dokl. Bulg. Akad. Nauk
Doklady Earth Sciences	Dokl. Earth Sci.
Dopovidi Natsional'noi Akademii Nauk Ukraini	Dopov. Nats. Akad. Nauk Ukr.
Drug Metabolism and Disposition	Drug Metab. Dispos.
Dyes and Pigments	Dyes Pigm.
Earth and Planetary Science Letters	Earth Planet. Sci. Lett.

D4, *continued*

Ecotoxicology and Environmental Safety	Ecotoxicol. Environ. Saf.
Education in Chemistry	Educ. Chem.
Electroanalysis	Electroanalysis
Electrochemical and Solid-State Letters	Electrochem. Solid-State Lett.
Electrochemistry (Tokyo, Japan)	Electrochemistry (Tokyo, Jpn.)
Electrochemistry Communications	Electrochem. Commun.
Electrochimica Acta	Electrochim. Acta
Electronics Letters	Electron. Lett.
Electrophoresis	Electrophoresis
EMBO Journal	EMBO J.
EMBO Reports	EMBO Rep.
Endocrinology	Endocrinology
Energy & Fuels	Energy & Fuels
Environmental Chemistry	Environ. Chem.
Environmental Health Perspectives	Environ. Health Perspect.
Environmental Science and Technology	Environ. Sci. Technol.
Environmental Technology	Environ. Technol.
Environmental Toxicology and Chemistry	Environ. Toxicol. Chem.
Enzyme and Microbial Technology	Enzyme Microb. Technol.
Eukaryotic Cell	Eukaryotic Cell
European Food Research and Technology	Eur. Food Res. Technol.
European Journal of Biochemistry	Eur. J. Biochem.
European Journal of Endocrinology	Eur. J. Endocrinol.
European Journal of Immunology	Eur. J. Immunol.
European Journal of Inorganic Chemistry	Eur. J. Inorg. Chem.
European Journal of Medicinal Chemistry	Eur. J. Med. Chem.
European Journal of Organic Chemistry	Eur. J. Org. Chem.
European Journal of Pharmaceutical Sciences	Eur. J. Pharm. Sci.
European Journal of Pharmacology	Eur. J. Pharmacol.
European Physical Journal A: Hadrons and Nuclei	Eur. Phys. J. A
European Physical Journal B: Condensed Matter and Complex Systems	Eur. Phys. J. B
European Physical Journal C: Particles and Fields	Eur. Phys. J. C
European Physical Journal D: Atomic, Molecular and Optical Physics	Eur. Phys. J. D
European Polymer Journal	Eur. Polym. J.
Europhysics Letters	Europhys. Lett.
Experimental Cell Research	Exp. Cell Res.

D4, *continued*

Experimental Eye Research	Exp. Eye Res.
Experimental Neurology	Exp. Neurol.
Expert Opinion on Investigational Drugs	Expert Opin. Invest. Drugs
Faraday Discussions	Faraday Discuss.
Faraday Transactions	J. Chem. Soc., Faraday Trans.
Farmaco	Farmaco
FASEB Journal	FASEB J.
FEBS Letters	FEBS Lett.
Federal Register	Fed. Regist.
FEMS Immunology and Medical Microbiology	FEMS Immunol. Med. Microbiol.
FEMS Microbiology Letters	FEMS Microbiol. Lett.
Fenxi Huaxue	Fenxi Huaxue
Fenxi Shiyanshi	Fenxi Shiyanshi
Ferroelectrics	Ferroelectrics
Fish Physiology and Biochemistry	Fish Physiol. Biochem.
Fluid Phase Equilibria	Fluid Phase Equilib.
Food Additives & Contaminants	Food Addit. Contam.
Food and Chemical Toxicology	Food Chem. Toxicol.
Food Chemistry	Food Chem.
Food Hydrocolloids	Food Hydroco
Free Radical Biology & Medicine	Free Radical Biol. Med.
Free Radical Research	Free Radical Res.
Fresenius Environmental Bulletin	Fresenius Environ. Bull.
Fuel	Fuel
Fuel Processing Technology	Fuel Process. Technol.
Fusion Engineering and Design	Fusion Eng. Des.
Gangtie	Gangtie
Gaodeng Xuexiao Huaxue Xuebao	Gaodeng Xuexiao Huaxue Xuebao
Gaofenzi Cailiao Kexue Yu Gongcheng	Gaofenzi Cailiao Kexue Yu Gongcheng
Gaofenzi Xuebao	Gaofenzi Xuebao
Gaoneng Wuli Yu Hewuli	Gaoneng Wuli Yu Hewuli
Gaoxiao Huaxue Gongcheng Xuebao	Gaoxiao Huaxue Gongcheng Xuebao
Gastroenterology	Gastroenterology
Gene	Gene
Gene Expression Patterns	Gene Expression Patterns
Gene Therapy	Gene Ther.
General and Comparative Endocrinology	Gen. Comp. Endocrinol.
Genes & Development	Genes Dev.
Genetics	Genetics
Genome Research	Genome Res.
Genomics	Genomics

D4, *continued*

Geochemical Transactions	Geochem. Trans.
Geochimica et Cosmochimica Acta	Geochim. Cosmochim. Acta
Geophysical Research Letters	Geophys. Res. Lett.
Green Chemistry	Green Chem.
Guangpuxue Yu Guangpu Fenxi	Guangpuxue Yu Guangpu Fenxi
Guangxue Xuebao	Guangxue Xuebao
Guisuanyan Xuebao	Guisuanyan Xuebao
Hecheng Xiangjiao Gongye	Hecheng Xiangjiao Gongye
Helvetica Chimica Acta	Helv Chim Acta
Heterocycles	Heterocycles
Hormone and Metabolic Research	Horm. Metab. Res.
Huagong Xuebao (Chinese Edition)	Huagong Xuebao (Chin. Ed.)
Huanjing Kexue Xuebao	Huanjing Kexue Xuebao
Huaxue Tongbao	Huaxue Tongba
Huaxue Xuebao	Huaxue Xuebao
Human Molecular Genetics	Hum. Mol. Genet.
Human Reproduction	Hum. Reprod.
Hydrobiologia	Hydrobiologia
Hyomen Gijutsu	Hyomen Gijutsu
Hyperfine Interactions	Hyperfine Interact
Hypertension	Hypertension
IEEE Electron Device	IEEE Electron Device Lett.
IEEE Journal of Quantum Electronics	IEEE J. Quantum Electron.
IEEE Transactions on Electron Devices	IEEE Trans. Electron Devices
IEEE Transactions on Magnetics	IEEE Trans. Magn.
IEEE Transactions on Nuclear Science	IEEE Trans. Nucl. Sci.
Igaku no Ayumi	Igaku no Ayumi
Immunology	Immunology
Immunology Letters	Immunol. Lett.
Indian Journal of Chemistry, Section B: Organic Chemistry Including Medicinal Chemistry	Indian J. Chem., Sect. B: Org. Chem. Incl. Med. Chem.
Indian Journal of Environmental Protection	Indian J. Environ. Prot.
Infection and Immunity	Infect. Immun.
Inflammation Research	Inflammation Res.
Inorganic Chemistry	Inorg. Chem.
Inorganic Chemistry Communications	Inorg. Chem. Commun.
Inorganic Materials	Inorg. Mater.
Inorganica Chimica Acta	Inorg. Chim. Acta
Insect Biochemistry and Molecular Biology	Insect Biochem. Mol. Biol.
Institute of Physics Conference Series	Inst. Phys. Conf. Ser.
Intermetallics	Intermetallics

D4, *continued*

International DATA Series, Selected Data on Mixtures, Series A: Thermodynamic Properties of Non-Reacting Binary Systems of Organic Substances	Int. DATA Ser., Sel. Data Mixtures, Ser. A
International Immunology	Int. Immunol.
International Immunopharmacology	Int. Immunopharmacol.
International Journal of Antimicrobial Agents	Int. J. Antimicrob. Agents
International Journal of Biochemistry & Cell Biology	Int. J. Biochem. Cell Biol.
International Journal of Cancer	Int. J. Cancer
International Journal of Heat and Mass Transfer	Int. J. Heat Mass Transfer
International Journal of Hydrogen Energy	Int. J. Hydrogen Energy
International Journal of Mass Spectrometry	Int. J. Mass Spectrom.
International Journal of Molecular Sciences	Int. J. Mol. Sci.
International Journal of Nanoscience	Int. J. Nanosci.
International Journal of Pharmaceutics	Int. J. Pharm.
International Journal of Quantum Chemistry	Int. J. Quantum Chem.
International Reviews in Physical Chemistry	Int. Rev. Phys. Chem.
Ion Exchange Letters	*Ion Exch. Lett.*
ISIJ International	ISIJ Int.
Izvestiya Akademii Nauk, Seriya Fizicheskaya	Izv. Akad. Nauk, Ser. Fiz.
Japanese Journal of Applied Physics, Part 1: Regular Papers, Short Notes & Review Papers	Jpn. J. Appl. Phys., Part 1
Japanese Journal of Applied Physics, Part 2: Letters & Express Letters	Jpn. J. Appl. Phys., Part 2
JETP Letters	JETP Lett.
Jiegou Huaxue	Jiegou Huaxue
Jinshu Xuebao	Jinshu Xuebao
Jisuanji Yu Yingyong Huaxue	Jisuanji Yu Yingyong Huaxue
Journal of Agricultural and Food Chemistry	J. Agric. Food Chem.
Journal of Alloys and Compounds	J. Alloys Compd.
Journal of Analytical and Applied Pyrolysis	J. Anal. Appl. Pyrolysis

D4, *continued*

Journal of Analytical Atomic Spectrometry	J. Anal. At. Spectrom
Journal of Animal Science (Savoy, IL, United States)	J. Anim. Sci. (Savoy, IL, U. S.)
Journal of Antimicrobial Chemotherapy	J. Antimicrob. Chemother
Journal of AOAC International	J. AOAC Int.
Journal of Appled Electrochemistry	J. Appl. Electrochem.
Journal of Applied Crystallography	J. Appl. Crystallogr.
Journal of Applied Electrochemistry	J. Appl. Electrochem.
Journal of Applied Physics	J. Appl. Phys.
Journal of Applied Physiology	J. Appl. Physiol.
Journal of Applied Polymer Science	J. Appl. Polym. Sci.
Journal of Applied Spectroscopy	J. Appl. Spectrosc.
Journal of Bacteriology	J. Bacteriol.
Journal of Biochemistry (Tokyo, Japan)	J. Biochem. (Tokyo, Jpn.)
Journal of Biological Chemistry	J. Biol. Chem.
Journal of Biological Inorganic Chemistry	J. Biol. Inorg. Chem.
Journal of Biomedical Materials Research, Part A	J. Biomed. Mater. Res., Part A
Journal of Biomedical Materials Research, Part B: Applied Biomaterials	J. Biomed. Mater. Res., Part B
Journal of Biomolecular NMR	J. Biomol. NMR
Journal of Bioscience and Bioengineering	J. Biosci. Bioeng.
Journal of Biotechnology	J. Biotechnol.
Journal of Bone and Mineral Research	J. Bone Miner. Res.
Journal of Cardiovascular Pharmacology	J. Cardiovasc. Pharmacol.
Journal of Catalysis	J. Catal.
Journal of Cell Biology	J. Cell Biol.
Journal of Cell Science	J. Cell Sci.
Journal of Cellular Biochemistry	J. Cell. Biochem.
Journal of Cellular Physiology	J. Cell. Physiol.
Journal of Chemical and Engineering Data	J. Chem. Eng. Data
Journal of Chemical Ecology	J. Chem. Ecol.
Journal of Chemical Education	J. Chem. Educ.
Journal of Chemical Engineering of Japan	J. Chem. Eng. Jpn.
Journal of Chemical Information and Computer Sciences	J. Chem. Inf. Comput. Sci
Journal of Chemical Information and Modeling	J. Chem. Inf. Model.
Journal of Chemical Physics	J. Chem. Phys.
Journal of Chemical Research	J. Chem. Res., Synop .

D4, *continued*

Journal of Chemical Research, Synopses	J. Chem. Res., Synop.
Journal of Chemical Technology and Biotechnology	J. Chem. Technol. Biotechnol.
Journal of Chemometrics	J. Chemom.
Journal of Chromatography, A	J. Chromatogr., A
Journal of Chromatography, B: Analytical Technologies in the Biomedical and Life Sciences	J. Chromatogr., B: Anal. Technol. Biomed. Life Sci.
Journal of Clinical Endocrinology and Metabolism	J. Clin. Endocrinol. Metab.
Journal of Clinical Investigation	J. Clin. Invest.
Journal of Clinical Microbiology	J. Clin. Microbiol.
Journal of Cluster Science	J. Cluster Sci.
Journal of Combinatorial Chemistry	J. Comb. Chem.
Journal of Comparative Neurology	J. Comp. Neurol.
Journal of Computational Chemistry	J. Comput. Chem.
Journal of Controlled Release	J. Controlled Release
Journal of Coordination Chemistry	J. Coord. Chem.
Journal of Crystal Growth	J. Cryst. Growth
Journal of Dairy Science	J. Dairy Sci.
Journal of Electroanalytical Chemistry	J. Electroanal. Chem.
Journal of Electronic Materials	J. Electron. Mater.
Journal of Endocrinology	J. Endocrinol.
Journal of Environmental Engineering (Reston, VA, United States)	J. Environ. Eng. (Reston, VA, U. S.)
Journal of Environmental Monitoring	J. Environ. Monit.
Journal of Environmental Quality	J. Environ. Qual.
Journal of Environmental Radioactivity	J. Environ. Radioact.
Journal of Environmental Science and Health, Part A: Environmental Science and Engineering	J. Environ. Sci. Health, Part A
Journal of Experimental and Theoretical Physics	J. Exp. Theor. Phys.
Journal of Experimental Biology	J. Exp. Biol.
Journal of Experimental Botany	J. Exp. Bot.
Journal of Experimental Medicine	J. Exp. Med
Journal of Fluorine Chemistry	J. Fluorine Chem.
Journal of Food Protection	J. Food Prot.
Journal of Food Science	J. Food Sci.
Journal of General Virology	J. Gen. Virol.
Journal of Geophysical Research, [Atmospheres]	J. Geophys. Res., [Atmos.]
Journal of Hazardous Materials	J. Hazard. Mater.
Journal of Heterocyclic Chemistry	J. Heterocycl. Chem.

D4, *continued*

Journal of Histochemistry and Cytochemistry	J. Histochem. Cytochem.
Journal of Hypertension	J. Hypertens.
Journal of Immunological Methods	J. Immunol. Methods
Journal of Immunology	J. Immunol.
Journal of Infectious Diseases	J. Infect. Dis.
Journal of Inorganic and Nuclear Chemistry	J. Inorg. Nucl. Chem.
Journal of Inorganic Biochemistry	J. Inorg. Biochem.
Journal of Investigative Dermatology	J. Invest. Dermatol.
Journal of Leukocyte Biology	J. Leukocyte Biol.
Journal of Lipid Research	J. Lipid Res.
Journal of Liquid Chromatography & Related Technologies	J. Liq. Chromatogr. Relat. Technol.
Journal of Low Temperature Physics	J. Low Temp. Phys.
Journal of Luminescence	J. Lumin.
Journal of Macromolecular Science, Part A Pure and Applied Chemistry	J. Macromol. Sci. Part A Pure Appl. Chem.
Journal of Magnetic Resonance	J. Magn. Reson.
Journal of Magnetism and Magnetic Materials	J. Magn. Magn. Mater.
Journal of Mass Spectrometry	J. Mass Spectrom.
Journal of Materials Chemistry	J. Mater. Chem.
Journal of Materials Research	J. Mater. Res.
Journal of Materials Science	J. Mater. Sci.
Journal of Materials Science Letters	J. Mater. Sci. Lett.
Journal of Materials Science: Materials in Electronics	J. Mater. Sci.: Mater. Electron.
Journal of Materials Science: Materials in Medicine	J. Mater. Sci.: Mater. Med.
Journal of Mathematical Chemistry	J. Math. Chem.
Journal of Medicinal Chemistry	J. Med. Chem.
Journal of Membrane Science	J. Membr. Sci.
Journal of Molecular Biology	J. Mol. Biol.
Journal of Molecular Catalysis A: Chemical	J. Mol. Catal. A: Chem.
Journal of Molecular Catalysis B: Enzymatic	J. Mol. Catal. B: Enzym.
Journal of Molecular Evolution	J. Mol. Evol.
Journal of Molecular Liquids	J. Mol. Liq.
Journal of Molecular Spectroscopy	J. Mol. Spectrosc.
Journal of Molecular Structure	J. Mol. Struct.
Journal of Natural Products	J. Nat. Prod.
Journal of Neurochemistry	J. Neurochem.
Journal of Neuroimmunology	J. Neuroimmunol.

D4, *continued*

Journal of Neurophysiology	J. Neurophysiol.
Journal of Neuroscience	J. Neurosci.
Journal of Neuroscience Research	J. Neurosci. Res.
Journal of Non-Crystalline Solids	J. Non-Cryst. Solids
Journal of Nuclear Materials	J. Nucl. Mater.
Journal of Nutrition	J. Nutr.
Journal of Organic Chemistry	J. Org. Chem.
Journal of Organometallic Chemistry	J. Organomet. Chem.
Journal of Pharmaceutical and Biomedical Analysis	J. Pharm. Biomed. Anal.
Journal of Pharmacological Sciences (Tokyo, Japan)	J. Pharmacol. Sci. (Tokyo, Jpn.)
Journal of Photochemistry and Photobiology, A: Chemistry	J. Photochem. Photobiol., A
Journal of Physical Chemistry A	J. Phys. Chem. A
Journal of Physical Chemistry B	J. Phys. Chem. B
Journal of Physical Chemistry C	J. Phys. Chem. C
Journal of Physical Organic Chemistry	J. Phys. Org. Chem.
Journal of Physics A: Mathematical and General	J. Phys. A: Math. Gen.
Journal of Physics and Chemistry of Solids	J. Phys. Chem. Solids
Journal of Physics B: Atomic, Molecular and Optical Physics	J. Phys. B: At., Mol. Opt. Phys.
Journal of Physics D: Applied Physics	J. Phys. D: Appl. Phys.
Journal of Physics G: Nuclear and Particle Physics	J. Phys. G: Nucl. Part. Phys.
Journal of Physics: Condensed Matter	J. Phys.: Condens. Matter
Journal of Physiology (Oxford, United Kingdom)	J. Physiol. (Oxford, U. K.)
Journal of Plant Physiology	J. Plant Physiol.
Journal of Polymer Science, Part A: Polymer Chemistry	J. Polym. Sci., Part A: Polym. Chem.
Journal of Polymer Science, Part B: Polymer Physics	J. Polym. Sci., Part B: Polym. Phys.
Journal of Power Sources	J. Power Sources
Journal of Quantitative Spectroscopy & Radiative Transfer	J. Quant. Spectrosc. Radiat. Transfer
Journal of Radioanalytical and Nuclear Chemistry	J. Radioanal. Nucl. Chem.
Journal of Raman Spectroscopy	J. Raman Spectrosc.
Journal of Separation Science	J. Sep. Sci.
Journal of Solid State Chemistry	J. Solid State Chem.
Journal of Steroid Biochemistry and Molecular Biology	J. Steroid Biochem. Mol. Biol.

D4, *continued*

Journal of Surgical Research	J. Surg. Res.
Journal of the American Chemical Society	J. Am. Chem. Soc.
Journal of the Brazilian Chemical Society	J. Braz. Chem. Soc.
Journal of the Chemical Society	J. Chem. Soc.
Journal of the Chemical Society, Perkin Transactions 1	J. Chem. Soc., Perkin Trans. 1
Journal of the Electrochemical Society	J. Electrochem. Soc.
Journal of the National Cancer Institute (1988)	J. Natl. Cancer Inst.
Journal of the Optical Society of America B: Optical Physics	J. Opt. Soc. Am. B
Journal of the Physical Society of Japan	J. Phys. Soc. Jpn.
Journal of the Science of Food and Agriculture	J. Sci. Food Agric
Journal of Thermal Analysis and Calorimetry	J. Therm. Anal. Calorim.
Journal of Vacuum Science & Technology, A: Vacuum, Surfaces, and Films	J. Vac. Sci. Technol., A
Journal of Vacuum Science & Technology, B: Microelectronics and Nanometer Structures--Processing, Measurement, and Phenomena	J. Vac. Sci. Technol., B: Microelectron. Nanometer Struct.--Process., Meas., Phenom.
Journal of Virological Methods	J. Virol. Methods
Journal of Virology	J. Virol.
Kagaku to Seibutsu	Kagaku to Seibutsu
KEK Proceedings	KEK Proc.
Key Engineering Materials	Key Eng. Mater.
Kidney International	Kidney Int.
Kogyo Zairyo	Kogyo Zairyo
Korean Journal of Chemical Engineering	Korean J. Chem. Eng.
Laboratory Investigation	Lab. Invest.
Langmuir	Langmuir
Lecture Notes in Physics	Lect. Notes Phys.
Liebigs Annalen der Chemie	Liebigs Ann. Chem.
Life Sciences	Life Sci.
Liquid Crystals	Liq. Cryst.
Macromolecular Chemistry and Physics	Macromol. Chem. Phys.
Macromolecular Rapid Communications	Macromol. Rapid Commun.
Macromolecular Symposia	Macromol. Symp.
Macromolecules	Macromolecules
Magnetic Resonance in Chemistry	Magn. Reson. Chem.
Marine Pollution Bulletin	Mar. Pollut. Bull.

D4, *continued*

Materials Chemistry and Physics	Mater. Chem. Phys.
Materials Letters	Mater. Lett.
Materials Research Bulletin	Mater. Res. Bull.
Materials Research Society Symposium Proceedings	Mater. Res. Soc. Symp. Proc.
Materials Science & Engineering, A: Structural Materials: Properties, Microstructure and Processing	Mater. Sci. Eng., A
Materials Science & Engineering, B: Solid-State Materials for Advanced Technology	Mater. Sci. Eng., B
Materials Science & Engineering, C: Biomimetic and Supramolecular Systems	Mater. Sci. Eng., C
Materials Science and Technology	Mater. Sci. Technol.
Materials Science Forum	Mater. Sci. Forum
Materials Transactions	Mater. Trans.
Measurement Science & Technology	Meas. Sci. Technol.
Mechanisms of Development	Mech. Dev.
Mendeleev Communications	Mendeleev Commun.
Metabolism, Clinical and Experimental	Metab., Clin. Exp.
Metallurgical and Materials Transactions A: Physical Metallurgy and Materials Science	Metall. Mater. Trans. A
Meteoritics & Planetary Science	Meteorit. Planet. Sci.
Methods in Enzymology	Methods Enzymol.
Microbiology (Reading, United Kingdom)	Microbiology (Reading, U. K.)
Microchimica Acta	Microchim. Acta
Microelectronic Engineering	Microelectron. Eng.
Microporous and Mesoporous Materials	Microporous Mesoporous Mater.
Minerals Engineering	Miner. Eng.
Modern Physics Letters A	Mod. Phys. Lett. A
Molecular and Biochemical Parasitology	Mol. Biochem. Parasitol.
Molecular and Cellular Biochemistry	Mol. Cell. Biochem.
Molecular and Cellular Biology	Mol. Cell. Biol.
Molecular and Cellular Endocrinology	Mol. Cell. Endocrinol.
Molecular and Cellular Neuroscience	Mol. Cell. Neurosci.
Molecular Biology and Evolution	Mol. Biol. Evol.
Molecular Biology of the Cell	Mol. Biol. Cell
Molecular BioSystems	Mol. Biosyst.
Molecular Brain Research	Mol. Brain Res.
Molecular Cancer Therapeutics	Mol. Cancer Ther
Molecular Cell	Mol. Cell
Molecular Crystals and Liquid Crystals	Mol. Cryst. Liq. Cryst.

D4, *continued*

Molecular Endocrinology	Mol. Endocrinol.
Molecular Genetics and Genomics	Mol. Genet.
Molecular Genetics and Metabolism	Mol. Genet. Metab.
Molecular Immunology	Mol. Immunol.
Molecular Microbiology	Mol. Microbiol.
Molecular Pharmacology	Mol. Pharmacol.
Molecular Physics	Mol. Phys.
Molecular Plant-Microbe Interactions	Mol. Plant-Microbe Interact.
Molecular Reproduction and Development	Mol. Reprod. Dev.
Molecular Therapy	Mol. Ther.
Monatshefte fuer Chemie	Monatsh. Chem.
Monthly Notices of the Royal Astronomical Society	Mon. Not. R. Astron. Soc.
Mutation Research	Mutat. Res.
Nano Letters	Nano Lett.
NASA Conference Publication	NASA Conf. Publ.
Natural Product Reports	Nat. Prod. Rep
Nature (London, United Kingdom)	Nature (London, U. K.)
Nature Biotechnology	Nat. Biotechnol.
Nature Cell Biology	Nat. Cell Biol.
Nature Chemical Biology	Nat. Chem. Biol.
Nature Genetics	Nat. Genet.
Nature Immunology	Nat. Immunol.
Nature Materials	Nat. Mater.
Nature Medicine (New York, NY, United States)	Nat. Med. (N. Y., NY, U. S.)
Nature Protocols	Nat. Protoc.
Naunyn-Schmiedeberg's Archives of Pharmacology	Naunyn-Schmiedeberg's Arch. Pharmacol.
Neurochemical Research	Neurochem. Res.
Neurochemistry International	Neurochem. Int.
Neuron	Neuron
Neuropharmacology	Neuropharmacology
Neuroscience (Oxford, United Kingdom)	Neuroscience (Oxford, U. K.)
New England Journal of Medicine	N. Engl. J. Med.
New Journal of Chemistry	New J. Chem.
New Phytologist	New Phytol.
Nippon Kikai Gakkai Ronbunshu, B-hen	Nippon Kikai Gakkai Ronbunshu, B-hen
Nuclear Engineering and Design	Nucl. Eng. Des.
Nuclear Fusion	Nucl. Fusion

D4, *continued*

Nuclear Instruments & Methods in Physics Research, Section A: Accelerators, Spectrometers, Detectors, and Associated Equipment	Nucl. Instrum. Methods Phys. Res., Sect. A
Nuclear Instruments & Methods in Physics Research, Section B: Beam Interactions with Materials and Atoms	Nucl. Instrum. Methods Phys. Res., Sect. B
Nuclear Physics A	Nucl. Phys. A
Nuclear Physics B	Nucl. Phys. B
Nuclear Physics B, Proceedings Supplements	Nucl. Phys. B, Proc. Suppl.
Nucleic Acids Research	Nucl. Sci. Eng.
Nucleosides, Nucleotides & Nucleic Acids	Nucleosides, Nucleotides Nucleic Acids
Oncogene	Oncogene
Optics Communications	Opt. Commun.
Optics Letters	Opt. Lett.
Organic and Biomolecular Chemistry	Org. Biomol. Chem.
Organic Letters	Org. Lett.
Organic Process Research & Development	Org. Process Res. Dev.
Organometallics	Organometallics
Oriental Journal of Chemistry	Orient. J. Chem.
Pediatric Research	Pediatr. Res
Peptides (New York, NY, United States)	Peptides (N. Y., NY, U. S.)
Pesticide Outlook	Pestic. Outlook
Pfluegers Archiv	Pfluegers Arch.
Pharmaceutical Chemistry Journal	Pharm. Chem. J.
Pharmaceutical Research	Pharm. Res.
Pharmacological Research	Pharmacol. Res.
Pharmacology, Biochemistry and Behavior	Pharmacol., Biochem. Behav.
Pharmazie	Pharmazie
Philosophical Magazine	Philos. Mag.
Phosphorus, Sulfur and Silicon and the Related Elements	Phosphorus, Sulfur Silicon Relat. Elem.
Photochemical and Photobiological Sciences	Photochem. Photobiol. Sci.
Photochemistry and Photobiology	Photochem. Photobiol.
Physica Status Solidi A: Applied Research	Phys. Status Solidi A
Physica Status Solidi B: Basic Research	Phys. Status Solidi B
Physica Status Solidi C: Current Topics in Solid State Physics	Phys. Status Solidi C
Physical Chemistry Chemical Physics	Phys. Chem. Chem. Phys.

D4, *continued*

Physical Review A: Atomic, Molecular, and Optical Physics	Phys. Rev. A: At., Mol., Opt. Phys.
Physical Review B: Condensed Matter and Materials Physics	Phys. Rev. B: Condens. Matter Mater. Phys.
Physical Review C: Nuclear Physics	Phys. Rev. C: Nucl. Phys.
Physical Review D: Particles and Fields	Phys. Rev. D: Part. Fields
Physical Review E: Statistical, Nonlinear, and Soft Matter Physics	Phys. Rev. E: Stat., Nonlinear, Soft Matter Phys.
Physical Review Letters	Phys. Rev. Lett.
Physics Letters A	Phys. Lett. A
Physics Letters B	Phys. Lett. B
Physics of Fluids	Phys. Fluids
Physics of Plasmas	Phys. Plasmas
Physics of the Solid State	Phys. Solid State
Physiologia Plantarum	Physiol. Plant.
Physiology & Behavior	Physiol. Behav.
Phytochemistry (Elsevier)	Phytochemistry (Elsevier)
Plant and Cell Physiology	Plant Cell Physiol.
Plant and Soil	Plant Soil
Plant Cell	Plant Cell
Plant Journal	Plant J.
Plant Molecular Biology	Plant Mol. Biol.
Plant Physiology	Plant Physiol.
Plant Science (Amsterdam, Netherlands)	Plant Sci. (Amsterdam, Neth.)
Planta	Planta
Planta Medica	Planta Med.
Plasticheskie Massy	Plast. Massy
Polish Journal of Chemistry	Pol. J. Chem.
Polyhedron	Polyhedron
Polymer	Polymer
Polymer Degradation and Stability	Polym. Degrad. Stab.
Polymer Engineering and Science	Polym. Eng. Sci.
Polymer International	Polym. Int.
Polymer Journal (Tokyo, Japan)	Polym. J. (Tokyo, Jpn.)
Polymer Preprints (American Chemical Society, Division of Polymer Chemistry)	Polym. Prepr. (Am. Chem. Soc., Div. Polym. Chem.)
Polymeric Materials Science and Engineering	Polym. Mater. Sci. Eng.
Poultry Science	Poult. Sci.
Poverkhnost	Poverkhnost
Powder Technology	Powder Technol.
Pramana	Pramana

D4, *continued*

Preprints of Extended Abstracts presented at the ACS National Meeting, American Chemical Society, Division of Environmental Chemistry	Prepr. Ext. Abstr. ACS Natl. Meet., Am. Chem. Soc., Div. Environ. Chem.
Preprints of Symposia - American Chemical Society, Division of Fuel Chemistry	Prepr. Symp. - Am. Chem. Soc., Div. Fuel Chem.
Proceedings - Electrochemical Society	Proc. - Electrochem. Soc.
Proceedings of SPIE - The International Society for Optical Engineering	Proc. SPIE-Int. Soc. Opt. Eng.
Proceedings of the Chemical Society	Proc. Chem. Soc.
Proceedings of the National Academy of Sciences of the United States of America, Early Edition	Proc. Natl. Acad. Sci. U. S. A., Early Ed.
Process Biochemistry (Oxford, United Kingdom)	Process Biochem. (Oxford, U. K.)
Progress in Inorganic Chemistry	Prog. Inorg. Chem.
Progress in Organic Coatings	Prog. Org. Coat.
Progress in Solid State Chemistry	Prog. Solid State Chem.
Prostaglandins, Leukotrienes and Essential Fatty Acids	Prostaglandins, Leukotrienes Essent. Fatty Acids
Protein Expression and Purification	Protein Expression Purif.
Protein Science	Protein Sci.
Proteins: Structure, Function, and Bioinformatics	Proteins: Struct., Funct., Bioinf.
Proteomics	Proteomics
Psychopharmacology (Berlin, Germany)	Psychopharmacology (Berlin, Ger.)
Pure and Applied Chemistry	Pure Appl. Chem.
Quimica Nova	Quim. Nova
Radiation Physics and Chemistry	Radiat. Phys. Chem.
Radiation Protection Dosimetry	Radiat. Prot. Dosim.
Ranliao Huaxue Xuebao	Ranliao Huaxue Xuebao
Rapid Communications in Mass Spectrometry	Rapid Commun. Mass Spectrom.
Reaction Kinetics and Catalysis Letters	React. Kinet. Catal. Lett.
Regulatory Peptides	Regul. Pept.
Rengong Jingti Xuebao	Rengong Jingti Xuebao
Reproduction (Bristol, United Kingdom)	Reproduction (Bristol, U. K.)
Research Disclosure	Res. Discl.
Revista de Chimie (Bucharest, Romania)	Rev. Chim. (Bucharest, Rom.)
Revue Roumaine de Chimie	Rev. Roum. Chim.
RNA	RNA
Russian Chemical Bulletin	Russ. Chem. Bull.

D4, *continued*

Russian Chemical Reviews	Russ. Chem. Rev.
Russian Journal of Coordination Chemistry	Russ. J. Coord. Chem.
Russian Journal of Electrochemistry	Russ. J. Electrochem.
Russian Journal of General Chemistry	Russ. J. Gen. Chem.
Russian Journal of Genetics	Russ. J. Genet.
Russian Journal of Organic Chemistry	Russ. J. Org. Chem.
Saibo Kogaku	Saibo Kogaku
Science (Washington, DC, United States)	Science (Washington, DC, U. S.)
Science of the Total Environment	Sci. Total Environ.
Scientia Pharmaceutica	Sci. Pharm.
Scripta Materialia	Scr. Mater.
Semiconductor Science and Technology	Semicond. Sci. Technol.
Semiconductors	Semiconductors
Sensors and Actuators, A: Physical	Sens. Actuators, A
Sensors and Actuators, B: Chemical	Sens. Actuators, B
Separation and Purification Technology	Sep. Purif. Technol.
Separation Science and Technology	Sep. Sci. Technol.
Sepu	Sepu
Shipin Kexue (Beijing)	Shipin Kexue (Beijing)
Shiyou Huagong	Shiyou Huagong
Soil Biology & Biochemistry	Soil Biol. Biochem.
Soil Science Society of America Journal	Soil Sci. Soc. Am. J.
Solar Energy Materials and Solar Cells	Sol. Energy Mater. Sol. Cells
Solid State Communications	Solid State Commun.
Solid State Ionics	Solid State Ionics
Solid State Sciences	Solid State Sci.
Solid-State Electronics	Solid-State Electron.
Solvent Extraction and Ion Exchange	Solvent Extr. Ion Exch.
Special Publication - Royal Society of Chemistry	Spec. Publ. - R. Soc. Chem.
Spectrochimica Acta Part A: Molecular and Biomolecular Spectroscopy	Spectrochim. Acta, Part A
Spectrochimica Acta Part B: Atomic Spectroscopy	Spectrochim. Acta, Part B
Spectroscopy Letters	Spectrosc. Lett.
Steroids	Steroids
Structure (Cambridge, MA, United States)	Structure (Cambridge, MA, U. S.)
Studies in Surface Science and Catalysis	Stud. Surf. Sci. Catal.
Superconductor Science & Technology	Supercond. Sci. Technol.
Surface and Coatings Technology	Surf. Coat. Technol.
Surface and Interface Analysis	Surf. Interface Anal.
Surface Science	Surf. Sci.

D4, *continued*

Surface Science Letters	Surf. Sci. Lett.
Surface Science Reports	Surf. Sci. Rep.
Symposium - International Astronomical Union	Symp. - Int. Astron. Union
Synlett	Synlett
Synthesis	Synthesis
Synthetic Communications	Synth. Commun.
Synthetic Metals	Synth. Met.
Talanta	Talanta
Tanpakushitsu Kakusan Koso	Tanpakushitsu Kakusan Koso
Technical Physics	Tech. Phys.
Technical Physics Letters	Tech. Phys. Lett.
Tetrahedron	Tetrahedron
Tetrahedron Asymmetry	Tetrahedron Asymmetry
Tetrahedron Letters	Tetrahedron Lett.
Tetsu to Hagane	Tetsu to Hagane
Textile Research Journal	Text. Res. J.
The Analyst	The Analyst
THEOCHEM	THEOCHEM
Theoretical and Applied Genetics	Theor. Appl. Genet.
Theoretical Chemistry Accounts	Theor. Chem. Acc.
Theriogenology	Theriogenology
Thermochimica Acta	Thermochim. Acta
Thin Solid Films	Thin Solid Films
Thromb. Haemostasis	Top. Catal.
Thrombosis and Haemostasis	Thromb. Haemostasis
Toxicological Sciences	Toxicol. Sci.
Toxicology	Toxicology
Toxicology and Applied Pharmacology	Toxicol. Appl. Pharmacol.
Toxicology Letters	Toxicol. Lett.
Toxicon	Toxicon
Transactions of the American Foundrymen's Society	Trans. Am. Foundrymen's Soc.
Transition Metal Chemistry (Dordrecht, Netherlands)	Transition Met. Chem. (Dordrecht, Neth.)
Transplantation Proceedings	Transplant. Proc.
Trends in Pharmacological Sciences	Trends Pharmacol. Sci.
Tsvetnye Metally (Moscow)	Tsvetn. Met. (Moscow)
Ukrainskii Khimicheskii Zhurnal (Russian Edition)	Ukr. Khim. Zh. (Russ. Ed.)
Vacuum	Vacuum
Virology	Virology
Virus Research	Virus Res.
Vysokomolekulyarnye Soedineniya, Seriya A i Seriya B	Vysokomol. Soedin., Ser. A Ser. B

D4, *continued*

Water Research	Water Res.
Water Science and Technology	Water Sci. Technol.
Water, Air, and Soil Pollution	Water, Air, Soil Pollut.
Wear	Wear
Wuji Huaxue Xueba	Wuji Huaxue Xueba
Wuli Huaxue Xuebao	Wuli Huaxue Xuebao
Wuli Xuebao	Wuli Xuebao
Yaoxue Xuebao	Yaoxue Xuebao
Yingyong Huaxue	Yingyong Huaxue
Youji Huaxue	Youji Huaxue
Zairyo	Zairyo
Zeitschrift fuer Anorganische und Allgemeine Chemie	Z. Anorg. Allg. Chem.
Zeitschrift fuer Kristallographie - New Crystal Structures	Z. Kristallogr. - New Cryst. Struct.
Zeitschrift fuer Metallkunde	Z. Metallkd.
Zeitschrift fuer Naturforschung, A: Physical Sciences	Z. Naturforsch., A: Phys. Sci.
Zeitschrift fuer Naturforschung, C: Journal of Biosciences	Z. Naturforsch., C: J. Biosci.
Zhongcaoyao	Zhongcaoyao
Zhongguo Jiguang	Zhongguo Jiguang
Zhongguo Shengwu Huaxue Yu Fenzi Shengwu Xuebao	Zhongguo Shengwu Huaxue Yu Fenzi Shengwu Xuebao
Zhongguo Xitu Xuebao	Zhongguo Xitu Xuebao
Zhongguo Yiyao Gongye Zazhi	Zhongguo Yiyao Gongye Zazhi
Zhurnal Fizicheskoi Khimii	Zh. Fiz. Khim.
Zhurnal Neorganicheskoi Khimii	Zh. Neorg. Khim.

Appendix E. Tables and Conventions for Organic Chemistry

Contents

*Note: For a list of organic chemistry journals, see **List of Chemistry Journals and Their Abbreviations** in Appendix D4.*

E1 List of Organic Reactions

A

Abramovitch–Shapiro
    tryptamine synthesis
Acetoacetic ester condensation
Achmatowicz reaction
Acyloin condensation
Adams catalyst
Adkins catalyst
Adkins–Peterson reaction
Akabori amino acid reaction
Alder ene reaction
Alder–Stein rules
Aldol addition
Aldol condensation
Algar–Flynn–Oyamada reaction
Allan–Robinson reaction
Allylic rearrangement
Amadori rearrangement
Andrussov oxidation
Appel reaction
Arbuzov reaction, Arbusow
    reaction
Arens–van Dorp synthesis, Isler
    modification
Arndt–Eistert synthesis
Auwers synthesis
Azo coupling

B

Baeyer–Drewson indigo
    synthesis
Baeyer–Villiger oxidation
Baeyer–Villiger rearrangement
Bakeland process (Bakelite)
Baker–Venkataraman
    rearrangement
Baker–Venkataraman
    transformation
Bally–Scholl synthesis
Balz–Schiemann reaction

Bamberger rearrangement
Bamberger triazine synthesis
Bamford–Stevens reaction
Barbier–Wieland degradation
Bardhan–Senguph phenanthrene
    synthesis
Bartoli indole synthesis
Bartoli reaction
Barton reaction
Barton–McCombie reaction,
Barton deoxygenation
Baudisch reaction
Bayer test
Baylis–Hillman reaction
Bechamp reaction
Beckmann fragmentation
Beckmann rearrangement
Bellus–Claisen rearrangement
Belousov–Zhabotinsky reaction
Benary reaction
Benzidine rearrangement
Benzilic acid rearrangement
Benzoin condensation
Bergman cyclization
Bergmann azlactone peptide
    synthesis
Bergmann degradation
Bergmann–Zervas carbobenzoxy
    method
Bernthsen acridine synthesis
Bestmann's reagent
Betti reaction
Biginelli pyrimidine synthesis
Biginelli reaction
Birch reduction
Bischler–Möhlau indole
    synthesis
Bischler–Napieralski reaction
Blaise ketone synthesis

E1, *continued*

Blaise reaction
Blanc reaction
Blanc chloromethylation
Bodroux reaction
Bodroux–Chichibabin aldehyde
    synthesis
Bogert–Cook synthesis
Bohn–Schmidt reaction
Boord olefin synthesis
Borodin reaction
Borsche–Drechsel cyclization
Bosch–Meiser urea process
Bouveault aldehyde synthesis
Bouveault–Blanc reduction
Boyland–Sims oxidation
Boyer Reaction
Bredt's rule
Brown hydroboration
Bucherer carbazole synthesis
Bucherer reaction
Bucherer–Bergs reaction
Buchner ring enlargement
Buchner–Curtius–Schlotterbeck
    reaction
Buchwald–Hartwig amination
Bunnett reaction

C
Cadiot–Chodkiewicz coupling
Camps quinoline synthesis
Cannizzaro reaction
Carroll reaction
Catalytic reforming
CBS reduction
Chan–Lam coupling
Chapman rearrangement
Chichibabin pyridine synthesis
Chichibabin reaction
Chugaev elimination
Ciamician–Dennstedt
    rearrangement

Claisen condensation
Claisen rearrangement
Claisen–Schmidt condensation
Clemmensen reduction
Collins Reagent
Combes quinoline synthesis
Conia reaction
Conrad–Limpach synthesis
Corey–Gilman–Ganem oxidation
Cook–Heilbron thiazole
    synthesis
Cope elimination
Cope rearrangement
Corey reagent
Corey–Bakshi–Shibata reduction
Corey–Fuchs reaction
Corey–Kim oxidation
Corey–Posner, Whitesides–
    House reaction
Corey–Winter olefin synthesis
Corey–Winter reaction
Coupling reaction
Craig method
Cram's rule of asymmetric
    induction
Creighton process
Criegee reaction
Criegee rearrangement
Cross metathesis
Crum Brown–Gibson rule
Curtius degradation
Curtius rearrangement,
Curtius reaction

D
Dakin reaction
Dakin–West reaction
Danheiser Annulation
Darapsky degradation
Darzens condensation, Darzens–
    Claisen reaction, Glycidic
    ester condensation

E1, *continued*

Darzens synthesis of
    unsaturated ketones
Darzens tetralin synthesis
Delepine reaction
Demjanov rearrangement
Demjanow desamination
Dess–Martin oxidation
Diazotisation
DIBAL–H selective reduction
Dieckmann condensation
Dieckmann reaction
Diels–Alder reaction
Diels–Reese reaction
Dienol benzene rearrangement
Dienone phenol rearrangement
Dimroth rearrangement
Di-pi-methane rearrangement
Directed ortho metalation
Doebner modification
Doebner reaction
Doebner–Miller reaction, Beyer
    method for quinolines
Doering–LaFlamme carbon
    chain extension
Dötz reaction
Dowd–Beckwith ring expansion
    reaction
Duff reaction
Dutt–Wormall reaction

E

E1cB elimination reaction
Eder reaction
Edman degradation
Eglinton reaction
Ehrlich–Sachs reaction
Einhorn variant
Einhorn–Brunner reaction
Elbs persulfate oxidation
Elbs reaction
Elimination reaction
Emde degradation

Emmert reaction
Ene reaction
Epoxidation
Erlenmeyer synthesis,
    Azlactone synthesis
Erlenmeyer–Plöchl azlactone
    and amino acid synthesis
Eschenmoser fragmentation
Eschweiler–Clarke reaction
Ester pyrolysis
Étard reaction
Evans aldol

F

Favorskii reaction
Favorskii rearrangement
Favorskii–Babayan synthesis
Feist–Benary synthesis
Fenton reaction
Ferrario reaction
Ferrier rearrangement
Finkelstein reaction
Fischer indole synthesis
Fischer oxazole synthesis
Fischer peptide synthesis
Fischer phenylhydrazine and
    oxazone reaction
Fischer glycosidation
Fischer–Hepp rearrangement
Fischer–Speier esterification
Fischer Tropsch synthesis
Fleming–Tamao oxidation
Flood reaction
Forster reaction
Forster–Decker method
Franchimont reaction
Frankland synthesis
Frankland–Duppa reaction
Freund reaction
Friedel–Crafts Acylation
Friedel–Crafts Alkylation
Friedländer synthesis

E1, *continued*

Fries rearrangement
Fritsch–Buttenberg–Wiechell
   rearrangement
Fujimoto–Belleau reaction
Fukuyama coupling
Fukuyama indole synthesis

G

Gabriel ethylenimine method
Gabriel synthesis
Gabriel–Colman rearrangement,
Gabriel isoquinoline synthesis
Gallagher–Hollander
   degradation
Gassman indole synthesis
Gastaldi synthesis
Gattermann aldehyde synthesis
Gattermann–Koch reaction
Gattermann reaction
Gewald reaction
Gibbs phthalic anhydride
   process
Gilman reagent
Glaser coupling
Glycol cleavage
Gogte synthesis
Gomberg–Bachmann reaction
Gomberg–Bachmann–Hey
   reaction
Gomberg Free radical reaction
Gould–Jacobs reaction
Graebe–Ullmann synthesis
Grignard degradation
Grignard reaction
Grob fragmentation
Grubbs' catalyst
Grundmann aldehyde synthesis
Gryszkiewicz-Trochimowski
   and McCombie method
Guareschi–Thorpe condensation
Guerbet reaction
Gutknecht pyrazine synthesis

H

Haller–Bauer reaction
Haloform reaction
Hammett equation
Hammick reaction
Hammond–Leffler postulate
Hantzsch pyrrole synthesis
Hantzsch dihydropyridine
   synthesis,
Hantzsch pyridine synthesis
Hantzsch Pyridine synthesis,
   Gattermann–Skita synthesis,
   Guareschi–Thorpe conden-
   sation, Knoevenagel–Fries
   modification
Hantzsch–Collidin synthesis
Harber–Weiss reaction
Harries ozonide reaction
Haworth methylation
Haworth phenanthrene synthesis
Haworth reaction
Hay coupling
Hayashi rearrangement
Heck reaction
Helferich method
Hell–Volhard–Zelinsky
   halogenation
Hemetsberger indole synthesis
Hemetsberger–Knittel synthesis
Henkel reaction, Raecke process,
Henkel process
Henry reaction, Kamlet reaction
Herz reaction, Herz compounds
Herzig–Meyer alkimide group
   determination
Heumann indigo synthesis
Hinsberg indole synthesis
Hinsberg reaction
Hinsberg sulfone synthesis
Hoch–Campbell ethylenimine
   synthesis
Hofmann degradation,
   Exhaustive methylation

E1, *continued*

Hofmann Elimination
Hofmann Isonitrile synthesis,
    Carbylamine reaction
Hofmann produkt
Hofmann rearrangement
Hofmann–Löffler reaction,
    Löffler–Freytag reaction,
    Hofmann–Löffler–Freytag
    reaction
Hofmann–Martius
    rearrangement
Hofmann's rule
Hofmann–Sand reaction
Homo rearrangement of steroids
Hooker reaction
Horner–Wadsworth–Emmons
    reaction
Hösch reaction
Hosomi–Sakurai reaction
Houben–Fischer synthesis
Hunsdiecker reaction
Hydroboration
Hydrohalogenation

I

Ing–Manske procedure
Ipso substitution
Ivanov reagent, Ivanov reaction

J

Jacobsen rearrangement
Janovsky reaction
Japp–Klingemann reaction
Japp–Maitland condensation
Johnson–Claisen rearrangement
Jones oxidation
Jordan–Ullmann–Goldberg
    synthesis
Julia olefination
Julia–Lythgoe olefination

K

Kabachnik–Fields reaction
Kendall–Mattox reaction
Kiliani–Fischer synthesis
Kindler reaction
Kishner cyclopropane synthesis
Knoevenagel condensation
Knoop–Oesterlin amino acid
    synthesis
Knorr pyrazole synthesis
Knorr pyrrole synthesis
Knorr quinoline synthesis
Koch–Haaf reaction
Kochi reaction
Koenigs–Knorr reaction
Kolbe electrolysis
Kolbe–Schmitt reaction
Kondakov rule
Kontanecki acylation
Kornblum oxidation
Kornblum–DeLaMare
    rearrangement
Kowalski ester homologation
Krafft degradation
Krapcho decarboxylation
Kröhnke aldehyde synthesis
Kröhnke oxidation
Kröhnke pyridine synthesis
Kucherov reaction
Kuhn–Winterstein reaction
Kulinkovich reaction
Kumada coupling

L

Larock indole synthesis
Lebedev process
Lehmstedt–Tanasescu reaction
Leimgruber–Batcho indole
    synthesis
Letts nitrile synthesis
Leuckart reaction
Leuckart thiophenol reaction

E1, *continued*

Leuckart–Wallach reaction
Leuckert amide synthesis
Levinstein process
Ley Oxidation
Lieben iodoform reaction,
    Haloform reaction
Lindlar catalyst
Lobry de Bruyn–Alberda van
    Ekenstein transformation
Lossen rearrangement
Luche reduction

M

Madelung synthesis
Malaprade reaction, Periodic
    acid oxidation
Malonic ester synthesis
Mannich reaction
Markovnikov's rule,
Markownikoff rule,
Markownikow rule
Martinet dioxindole synthesis
McDougall monoprotection
McFadyen–Stevens reaction
McMurry reaction
Meerwein arylation
Meerwein–Ponndorf–Verley
    reduction
Meisenheimer rearrangement
Meissenheimer complex
Menshutkin reaction
Mentzer pyrone synthesis
Metal-ion-catalyzed σ-bond
    rearrangement
Mesylation
Merckwald asymmetric
    synthesis
Meyer and Hartmann reaction
Meyer reaction
Meyer synthesis
Meyer–Schuster rearrangement

Michael addition, Michael
    system
Michael condensation
Michaelis–Arbuzov reaction
Miescher degradation
Mignonac reaction
Milas hydroxylation of olefins
Mitsunobu reaction
Mukaiyama aldol addition
Mukaiyama reaction
Myers' asymmetric alkylation

N

Nametkin rearrangement
Nazarov cyclization reaction
Neber rearrangement
Nef reaction
Negishi coupling
Negishi Zipper reaction
Nenitzescu indole synthesis
Nenitzescu reductive acylation
Nicholas reaction
Niementowski quinazoline
    synthesis
Niementowski quinoline
synthesis
Nierenstein reaction
Nitroaldol reaction
Normant reagents
Noyori asymmetric
    hydrogenation
Nozaki–Hiyama–Kishi
Nickel/Chromium Coupling
    reaction
Nucleophilic acyl substitution

O

Ohira–Bestmann reaction
Olefin metathesis
Oppenauer oxidation
Ostromyslenskii reaction,
Ostromisslenskii reaction

E1, *continued*

Oxidative decarboxylation
Oxo synthesis
Oxy–Cope rearrangement
Oxymercuration
Ozonolysis

P–Q

Paal–Knorr pyrrole synthesis
Paal–Knorr synthesis
Paneth technique
Paolini reaction
Passerini reaction
Paterno–Büchi reaction
Pauson–Khand reaction
Pechmann condensation
Pechmann pyrazole synthesis
Pellizzari reaction
Pelouze synthesis
Perkin alicyclic synthesis
Perkin reaction
Perkin rearrangement
Perkow reaction
Petasis reaction
Petasis reagent
Peterson olefination
Peterson reaction
Petrenko–Kritschenko
    piperidone synthesis
Pfan–Plattner azulene synthesis
Pfitzinger reaction
Pfitzner–Moffatt oxidation
Pictet–Gams isoquinoline
    synthesis
Pictet–Hubert reaction
Pictet–Spengler tetrahydro-
    isoquinoline synthesis
Pictet–Spengler reaction
Piloty alloxazine synthesis
Piloty–Robinson pyrrole
    synthesis
Pinacol coupling reaction
Pinacol rearrangement

Pinner amidine synthesis
Pinner method for ortho esters
Pinner reaction
Pinner triazine synthesis
Piria reaction
Pitzer strain
Polonovski reaction
Pomeranz–Fritsch reaction
Ponzio reaction
Prelog strain
Prevost reaction
Prileschajew reaction
Prilezhaev reaction
Prins reaction
Prinzbach synthesis
Pschorr reaction
Pummerer rearrangement
Purdie methylation
    Irvine–Purdie methylation
Quelet reaction

R

Ramberg–Backlund reaction
Raney–Nickel
Rap–Stoermer condensation
Raschig phenol process
Rauhut–Currier reaction
Reed reaction
Reformatskii reaction
Reilly–Hickinbottom
    rearrangement
Reimer–Tiemann reaction
Reissert indole synthesis
Reissert reaction, Reissert
    compound
Reppe synthesis
Retropinacol rearrangement
Reverdin reaction
Riehm quinoline synthesis
Riemschneider thiocarbamate
    synthesis
Riley oxidations

E1, *continued*

Ring closing metathesis
Ring opening metathesis
Ritter reaction
Robinson annulation
Robinson–Gabriel synthesis
Robinson Schopf reaction
Rosenmund reaction
Rosenmund reduction
Rosenmund–von Braun
    synthesis
Rothemund reaction
Rowe rearrangement
Rupe reaction
Rubottom oxidation
Ruff–Fenton degradation
Ruzicka large ring synthesis
Sakurai reaction
Salol reaction
Sandheimer
Sandmeyer diphenylurea isatin
    synthesis
Sandmeyer isonitrosoacetanilide
isatin synthesis
Sandmeyer reaction
Sanger reagent
Sarett oxidation
Saytzeff rule, Saytzeff's Rule
Schiemann reaction
Schlenk equilibrium
Schlosser modification
Schlosser variant
Schlosser–Lochmann reaction
Schmidlin ketene synthesis
Schmidt degradation
Schmidt reaction
Scholl reaction
Schorigin Shorygin reaction,
Shorygin reaction, Wanklyn
    reaction
Schotten–Baumann reaction
Screttas–Yus reaction
Semidine rearrangement

Semmler–Wolff reaction
Serini reaction
Seyferth–Gilbert homologation
Shapiro reaction
Sharpless asymmetric
    dihydroxylation
Sharpless epoxidation
Sharpless oxyamination or
    aminohydroxylation
Simmons–Smith reaction
Simonini reaction
Simonis chromone cyclization
Skraup chinolin synthesis
Skraup reaction
Smiles rearrangement
SNAr nucleophilic aromatic
    substitution
SN1
SN2
SNi
Sommelet reaction
Sonn–Müller method
Sonogashira coupling
Sørensen formol titration
Staedel–Rugheimer pyrazine s
    ynthesis
Staudinger reaction
Stephen aldehyde synthesis
Stetter reaction
Stevens rearrangement
Stieglitz rearrangement
Stille coupling
Stobbe condensation
Stollé synthesis
Stork acylation
Stork enamine alkylation
Strecker amino acid synthesis
Strecker degradation
Strecker sulfite alkylation
Strecker synthesis
Stuffer disulfone hydrolysis rule
Suzuki coupling

E1, *continued*

Swain equation
Swarts reaction
Swern oxidation

T
Tamao oxidation
Tafel rearrangement
Takai olefination
Tebbe olefination
    ter Meer reaction
Thermite reactions
Thiele reaction
Thorpe reaction
Tiemann rearrangement
Tiffeneau ring enlargement
    reaction
Tiffeneau–Demjanow
    rearrangement
Tischtschenko reaction
Tishchenko reaction,
Tischischenko–Claisen reaction
Tollens reagent
Trapp mixture
Traube purine synthesis
Truce–Smiles rearrangement
Tscherniac–Einhorn reaction
Tschitschibabin reaction
Tschugajeff reaction
Twitchell process
Tyrer sulfonation process

U
Ugi reaction
Ullmann reaction
Upjohn dihydroxylation
Urech cyanohydrin method
Urech hydantoin synthesis

V
Van Slyke determination
Varrentrapp reaction
Vilsmeier reaction

Vilsmeier–Haack reaction
Voight amination
Volhard–Erdmann cyclization
von Braun amide degradation
von Braun reaction
von Richter cinnoline synthesis
von Richter reaction

W
Wacker–Tsuji oxidation
Wagner-Jauregg reaction
Wagner–Meerwein
    rearrangement
Walden inversion
Wallach rearrangement
Weerman degradation
Weinreb ketone synthesis
Wenker ring closure
Wenker synthesis
Wessely–Moser rearrangement
Westphalen–Lettré rear
    rangement
Wharton reaction
Whiting reaction
Wichterle reaction
Widman–Stoermer synthesis
Wilkinson catalyst
Willgerodt rearrangement
Willgerodt–Kindler reaction
Williamson ether synthesis
Winstein reaction
Wittig reaction
Wittig rearrangement
Wittig–Horner reaction
Wohl degradation
Wohl–Aue reaction
Wohler synthesis
Wohl–Ziegler reaction
Wolffenstein–Böters reaction
Wolff rearrangement
Wolff–Kishner reduction
Woodward cis-hydroxylation

E1, *continued*

Woodward–Hoffmann rule
Wurtz coupling, Wurtz reaction
Wurtz–Fittig reaction

Z

Zeisel determination
Zerevitinov determination,
Zerewitinoff determination

Ziegler condensation
Ziegler method
Zimmermann reaction
Zincke disulfide cleavage
Zinke nitration
Zincke reaction
Zincke–Suhl reaction

E2 Glossary of Chemistry Terms

achiral – A molecule that is superimposable on its mirror image. Achiral molecules do not rotate plane-polarized light.

acid – A proton donor or an electron pair acceptor.

acidic – Describes a solution with a high concentration of H^+ ions.

alcohol – A molecule containing a hydroxyl (OH) group. Also a functional group.

aldehyde – A molecule containing a terminal carbonyl (CHO) group. Also a functional group.

alkane – A molecule containing only carbon–hydrogen and carbon–carbon single bonds.

alkene – A molecule containing one or more carbon–carbon double bonds. Also a functional group.

alkyne – A molecule containing one or more carbon–carbon triple bonds. Also a functional group.

allylic carbon – An sp3 carbon adjacent to a double bond.

amide – A molecule containing a carbonyl group attached to a nitrogen ($-CONR_2$). Also a functional group.

amine – A molecule containing an isolated nitrogen (NR_3). Also a functional group.

anion – A molecule or atom with a negative charge.

anode – A positively charged electrode by which electrons leave an electrical device.

anti addition – A reaction in which the two groups of a reagent X–Y add on opposite faces of a carbon–carbon bond.

anti conformation – A type of staggered conformation in which the two big groups are opposite of each other in a Newman projection.

E2, *continued*

anti-aromatic – A highly unstable planar ring system with 4n pi electrons.

anticoplanar – *See* **anti-periplanar**.

anti-periplanar – The conformation in which a hydrogen and a leaving group are in the same plane and on opposite sides of a carbon–carbon single bond. The conformation required for E2 elimination.

aprotic solvents – Solvents that do not contain O–H or N–H bonds.

aromatic – A planar ring system that contains uninterrupted p orbitals around the ring and a total of 4n+2 pi electrons. Aromatic compounds are unusually stable compounds.

aryl – An aromatic group as a substituent.

atmospheres – a unit of pressure equal to 760 mmHg.

atom – the smallest component of an element that retains the chemical properties of that element

atomic number – Number of protons in an element.

Avogadro's number – Number representing the number of molecules in one mole: 6.023×10^{23}.

axial bond – A bond perpendicular to the equator of the ring (up or down), typically in a chair cyclohexane.

base – A proton acceptor or an electron pair donor.

basic – Of or denoting or of the nature of or containing a base.

benzyl group – A benzene ring plus a methylene (CH_2) unit (C_6H_5–CH_2).

benzylic position – The position of a carbon attached to a benzene ring.

benzyne – A highly reactive intermediate. A benzene ring with a triple bond.

E2, *continued*

bicyclic – A molecule with two rings that share at least two carbons.

Brønsted acid – A proton donor.

Brønsted base – A proton acceptor.

buffer solutions – A solution containing an ionic compound that resists changes in its pH.

carbanion – A negatively charged carbon atom.

carbene – A reactive intermediate, characterized by a neutral, electron-deficient carbon center with two substituents (R_2C:).

carbocation – A positively charged carbon.

carbonyl group – A carbon double-bonded to oxygen (C=O).

carboxylic acid – A molecule containing a carboxyl group (COOH). Also a functional group.

catalyst – Substance that accelerates or initiates a chemical process without changing the products of reaction

cathode – Electrode where electrons are gained (reduction) in redox reactions.

cation – An atom or molecule with a positive charge.

chair conformation – Typically, the most stable cyclohexane conformation. Resembles a chair.

charge – An excess or deficiency of electrons. Describes an object's ability to repel or attract other objects.

chemical changes – Processes or events that have altered the fundamental structure of something.

chemical equation – The written expression of a chemical reaction.

E2, *continued*

chemical shift – The location of an NMR peak relative to the standard tetramethylsilane (TMS), given in units of parts per million (ppm).

chiral center – A carbon or other atom with four nonidentical substituents.

chiral molecule – A molecule that is not superimposable on its mirror image. Chiral molecules rotate plane-polarized light.

cis – Two identical substituents on the same side of a double bond or ring.

combustion – The process by which a substance combines with oxygen to release heat and light energy.

compound – Two or more atoms joined together chemically in a definite proportion by weight.

concentration – The number of molecules of something in a specified solution or space.

configuration – The three-dimensional orientation of atoms around a chiral center; given the designation R or S.

conformation – The instantaneous spatial arrangements of atoms. Conformations can change by rotation around single bonds.

conjugate acid –Substance that is able to lose a hydrogen ion and form a base.

conjugate base – Substance is able to gain a hydrogen ion and form an acid.

conjugated double bonds – Double bonds separated by one carbon–carbon single bond. Alternating double bonds.

constitutional isomers – Molecules with the same molecular formula but with atoms attached in different ways.

coupling constant – The distance between two neighboring lines in an NMR peak (given in units of Hz).

E2, *continued*

coupling protons – Protons that interact with each other and split the NMR peak into a certain number of lines following the n+1 rule.

covalent bond – Bond in which two electrons are shared between two atoms.

d value – *See* delta value.

daughter isotope – A compound that remains after its parent isotope has undergone decay.

decay – Disintegration of an element into a different element, usually with some other particle(s) and radiation emitted.

dehydrohalogenation – Loss of a hydrohalic acid (such as HBr, HCl, and so on) to form a double bond.

delta value – The chemical shift. The location of an NMR peak relative to the standard tetramethylsilane (TMS), given in units of parts per million (ppm).

density – Mass per unit volume of a substance.

diastereomers – Stereoisomers that are not mirror images of each other.

Diels-Alder reaction – A reaction that brings together a diene and a dienophile to form bicyclic molecules and rings.

diene – A molecule that contains two alternating double bonds. A reactant in the Diels-Alder reaction.

dienophile – A reactant in the Diels-Alder reaction that contains a double bond. Dienophiles are often substituted with electron-withdrawing groups.

dipole moment – A measure of the separation of charge in a bond or molecule.

dipole-dipole forces – Intermolecular forces that are active only when the molecules are close together.

E2, *continued*

dispersion forces– Intermolecular attraction forces that exist between all molecules—the result of the movement of electrons which cause slight polar moments. Dispersion forces are generally very weak but as the molecular weight increases so does their strength.

dissociation – The breaking down of a molecular compound into its components, especially ions.

double bond – A covalent bond in which two pairs of electrons are shared between two atoms.
doublet – Describes an NMR signal split into two peaks.

E isomer – Stereoisomer in which the two highest priority groups are on opposite sides of a ring or double bond.

E1 elimination reaction – A reaction that eliminates a hydrohalic acid to form an alkene. A first order reaction that goes through a carbocation mechanism.

E2 elimination reaction – A reaction that eliminates a hydrohalic acid to form an alkene. A second order reaction that occurs in single step, in which the double bond is formed as the hydrohalic acid is eliminated.

eclipsed conformation – Conformation about a carbon–carbon single bond in which all of the bonds off of two adjacent carbons are aligned with each other.

effusion – The transmission of molecules of a gas through a small opening.

electrodes – A conductor used to make electrical contact with some part of a circuit.

electrolysis – A process by which the chemical structure of a compound is changed using electrical energy.

electromagnetic spectrum – The entire range of wavelengths that light can possess, including visible light, infrared and ultraviolet radiation, and all other types of electromagnetic radiation.

electron – An elementary particle that has a negative charge

E2, *continued*

electronegativity – A measure of the tendency of an atom to attract the electrons in a covalent bond to itself.

electrophile – A molecule that can accept a lone pair of electrons (a Lewis acid).

electrostatic forces – The forces between electrically charged objects.

element – One of the fundamental substances that consist of only one type of atom.

empirical formula – A formula that shows the simplest ratio of elements in a chemical compound.

enantiomers – Molecules that are nonsuperimposable mirror images of each other.

endothermic – Reaction that absorbs heat from its surroundings as the reaction proceeds.

energy – Ability to do work.

enthalpy – The amount of energy in a system capable of doing mechanical work

entropy – The measure of the amount of energy in a system that cannot do mechanical work.

equatorial – The bonds in a chair cyclohexane that are oriented along the equator of the ring.

equilibrium – A state that exists when a forward and reverse reaction occur at the same rate, i.e., when the reactants and products are in a constant ratio.

equilibrium constant – Value that expresses when the rate of the forward reaction equals the rate of the reverse reaction.

equilibrium expression – An expression that gives the ratio between the products and reactants.

E2, *continued*

equivalence point – The point in a reaction at which the amount of acid is the same as the amount of base.

ester – A molecule containing a carbonyl group adjacent to an oxygen (RCOOR'). Also a functional group.

ether – A molecule containing oxygen singly-bonded to two carbon atoms. Also a functional group. Often refers to diethyl ether.

exothermic – Describes a reaction that gives off heat.

fingerprint region – Region of an infrared spectrum below 1,500 cm^{-1}. The fingerprint region tends to be more distinctive for different compounds than other regions.

force – An influence that produces change in a physical quantity.

free electron – An electron that is not attached to the nucleus of an atom.

free energy – The energy of a system that is available to do work.

frequency – Number of events in a given time period, especially the number of peaks of a wave that would pass a stationary point.

functional group – A reactivity center.

gauche conformation – A type of staggered conformation in which two big groups are next to each other.

geiger counter – An instrument that measures ionizing radiation output.

Graham's law – The rate of diffusion of a gas is inversely proportional to the square root of its molecular weight.

half life – The amount of time it takes for half the atoms of an initial amount of radioactive substance to disintegrate.

halide – A molecule that contains a halogen (fluoride, chloride, bromide, iodide, or astatide). Also a functional group.

E2, *continued*

Heisenberg uncertainty principle – A principle which states that it is impossible to know the precise position and momentum of a particle at any time.

Hückel's rule – A rule that states that completely conjugated planar rings with 4n+2 pi electrons are aromatic.

hybrid orbitals – Orbitals formed from mixing together atomic orbitals, such as the spx orbitals, which result from mixing s and p orbitals.

hydrogen bonding – Strong type of intermolecular dipole-dipole attraction.

hydrolysis – A chemical reaction in which water reacts with another substance and the oxygen in water bonds with that substance.

hyperconjugation – Weak interaction (electron donation) between sigma bonds with p orbitals.

ideal gas law –Describes the relationship between pressure, volume, temperature, and moles of gas as $PV=nRT$.

inductive effects – Electron donation or withdrawal by electropositive or electronegative atoms through the sigma bond framework.

intermediate – Any species formed in a reaction on the way to making the product. Typically, intermediates are unstable.

intermolecular forces – Forces between molecules.

intramolecular forces – Forces within molecules.

ion – an electrically charged particle.

ion-dipole forces – Intermolecular force that exist between charged particles and partially charged molecules.

ionic bond – Bond in which electrons are unshared between two atoms.

ionization energy – The energy needed to remove an electron from a specific atom.

E2, *continued*

IR spectroscopy – An instrumental technique that measures infrared light absorption by molecules. Can be used to determine functional groups in an unknown molecule.

isolated double bonds – Double bonds separated by more than one carbon–carbon single bond.

isotopes – Elements with the same number of protons but different numbers of neutrons and different masses.

J value – The coupling constant between two peaks in an NMR signal. Given in units of Hz.

kelvin – The SI Unit of temperature. Equivalent to degrees Celsius plus 273.

ketone – A compound that contains a carbonyl group attached to two carbons. Also a functional group.

kinetic energy – Energy of motion.

kinetic product – The product that forms the fastest. (This product has the lowest energy of activation.)

kinetics – The study of reaction rates.

Le Chatlier's principle – States that a system at equilibrium will oppose any change in the equilibrium conditions.

Lewis acid – A substance that accepts an electron pair.

Lewis base – A substance that donates an electron pair.

Lewis structures –Representations of molecular structures based on valence electrons.

limiting reagent – In a chemical reaction, the reactant that will be consumed first.

line spectra – Spectra generated by excited substances.

E2, *continued*

Markovniknov's rule – A rule that states that electrophiles add to the less highly substituted carbon of a carbon–carbon double bond (or the carbon with the most hydrogen atoms).

mass number – The number of protons and neutrons in an atom.

mass spectrometry – An instrumental technique involving the ionization of molecules into fragments. Can be used to determine the molecular weights of unknown molecules.

meso compounds – Molecules that have chiral centers but are achiral as a result of one or more planes of symmetry in the molecule.

meta – Describes the positions of two substituents on a benzene ring that are separated by one carbon.

meta-directing substituent – Any substituent on an aromatic ring that directs incoming electrophiles to the meta position.

mixture – A substance that is composed of two or more substances, but with each retaining its original properties.

molality – The number of moles of solute per kilogram of solvent.

molarity – The number of moles of solute per liter of solution. Used to express the concentration of a solution.

mole – An expression of molecular weight in grams. Represents 6.023×10^{23} atoms or molecules.

molecular formula – An expression of the number of atoms of each element present in a molecule.

molecular ion – The fragment in a mass spectrum that corresponds to the cation radical (M+) of the molecule. The molecular ion gives the molecular mass of the molecule.

molecular orbital theory – Model for depicting the location of electrons that allows electrons to delocalize across the entire molecule.

E2, *continued*

molecular weight – The total weight of all the elements in a compound.

molecule – Two or more atoms chemically combined.

multistep synthesis – Synthesis of a compound that takes several steps to achieve.

n+1 rule – Rule for predicting the coupling for a proton in 1H NMR spectroscopy. An NMR signal will split into n+1 peaks, where n is the number of equivalent adjacent protons.

natural product – A compound produced by a living organism.

neutral – An object that has no charge.

neutron – A nuclear particle that has no electric charge.

nitrile – A compound contain a cyano group, a carbon triply-bonded to a nitrogen (CN). Also a functional group

NMR – Nuclear magnetic resonance spectroscopy. A technique that measures radiofrequency light absorption by molecules. A powerful structure-determining method.

node – A region in an orbital with zero electron density.

nucleophile – A molecule with the ability to donate a lone pair of electrons (a Lewis base).

nucleophilicity – A measure of the reactivity of a nucleophile in a nucleophilic substitution reaction.

nucleus – The central part of an atom that contains the protons and neutrons. Plural *nuclei*.

optically active – Able to rotate plane-polarized light.

orbital – The region of space in which an electron is confined.

organic compound – A carbon-containing compound.

E2, *continued*

ortho – Describes the positions of two substituents on a benzene ring that are on adjacent carbons.

ortho-para director – An aromatic substituent that directs incoming electrophiles to the ortho or para positions.

oxidation reaction – A reaction where a substance loses electrons.

oxidation-reduction-reaction – A reversible reaction in which electrons are transferred from one substance to another.

oxyacid – An acid that contains oxygen.

p bond – *See* pi bond.

para – Describes the positions of two substituents on a benzene ring that are separated by two carbons.

parent isotope – An element that undergoes nuclear decay.

partial pressures – The pressure that a specific gas exerts in a mixture.

particle – A minuscule (usually subatomic) piece of matter.

periodic table – The grouping of the known elements by their properties and in order of their atomic numbers.

pH – Measures the acidity of a solution according to the logarithm of the inverse of the concentration of the hydrogen ions.

phenyl ring – A benzene ring as a substituent. Abbreviated Ph.

photons – A quantum of electromagnetic radiation which has the properties of both a wave and a particle.

physical property – A property of a substance that can be measured without changing its chemical composition.

pi bond – A bond with electron density above and below the two atoms, but not directly between the two atoms. Found in double and triple bonds.

E2, *continued*

pK_a – The scale for defining a molecule's acidity (p$K_a = -\log K_a$).

plane of symmetry – A plane cutting through a molecule in which both halves are mirror images of each other.

plane-polarized light – Light that oscillates in a single plane.

pOH – Measures how basic a solution is based on the logarithm of the inverse of the concentration of the hydroxide ions.

polar molecule – A molecule with a partial charge.

potential energy – Stored energy; energy of composition or position.

pressure – The force per unit of area.

principal quantum number – The number related to the amount of energy an electron has and therefore describing which shell the electron is in.

product – A substance that is formed during a chemical reaction.

proportion – An equality between two relative magnitudes of quantities.

protic solvent – A solvent that contains O–H or N–H bonds.

proton – An H^+ ion. Also a positively-charged nuclear particle.

quantum – A discrete amount of something.

quantum numbers – The set of numbers used to describe the position of a specific electron.

R group – Abbreviation given to an unimportant part of a molecule. Indicates Rest of molecule.

racemic mixture – A 50:50 mixture of two enantiomers. Racemic mixtures are optically inactive (i.e., they do not rotate plane-polarized light).

radiation – Energy that is radiated or transmitted in the form of rays or waves or particles.

E2, *continued*

radical – An atom or molecule with an unpaired electron.

radioactive – Substance containing an element which decays.

ratio – The relation between things with respect to their comparative quantity.

reactants – Substances that are present at the start of a chemical reaction.

reduction reaction – A reaction in which a substance gains an electron.

resonance structures – Structures used to better depict the location of pi and nonbonding electrons on a molecule. A molecule looks like a hybrid of all resonance structures.

s bond – *See* sigma bond.

salts – Ionic compounds that are formed by replacing hydrogen in an acid by a metal.

s-cis conformation – A conformation in which the two double bonds of a conjugated diene are on the same side of the carbon–carbon single bond that connects them. The required conformation for the Diels-Alder reaction.

SI Unit – A unit of the Systeme International d'Unites, an international system that established a uniform set of measurement units.

sigma bond – A bond in which electrons are located between the nuclei of the bonding atoms. Single bonds are sigma bonds.

single bond – When an electron pair is shared by two different elements.

singlet – Describes an NMR signal consisting of only one peak.

SN1 reaction – A first order substitution reaction that goes through a carbocation intermediate.

SN2 reaction – A second order substitution reaction that takes place in one step and has no intermediates.

E2, *continued*

solute – The dissolved matter in a solution; the component of a solution that changes its state.

solution – A homogeneous mixture of two or more substances; often a liquid solution.

solvent – Liquid in which something is dissolved.

sp – A hybrid orbital made by mixing one s orbital and one p orbital.

sp2 – A hybrid orbital made by mixing one s orbital and two p orbitals.

sp3 – A hybrid orbital made by mixing one s orbital and three p orbitals.

specific heat – The amount of heat necessary for a substance to be raised by one degree Celsius.

spontaneous reaction – A reaction that will proceed without any outside energy.

staggered conformation – Conformation about a carbon–carbon single bond in which bonds of one carbon are a maximum distance apart from bonds of an adjacent carbon.

state property –A quantity that is independent of how a substance was prepared, such as altitude, pressure, volume, temperature and internal energy.

states of matter – Solid, liquid, gas, and plasma.

stereochemistry – Study of molecules in three dimensions.

stereoisomers – Molecules that have the same atom connectivity, but different orientations of those atoms in three-dimensional space.

steric hindrance – Term referring to the way that atoms can shield a site by getting in the way of the approach of a reactant.

stoichiometry – The relation between the quantities of substances that take part in a reaction or form a compound.

E2, *continued*

STP – Standard temperature and pressure, or 273.15 K and 1 atm.

s-trans conformation – The conformation in which the two double bonds of a conjugated diene are on opposite sided of the carbon–carbon single bond that connects them.

sublevel – One part of a level, each of which can hold different numbers of electrons.

substituent – An atom or group of atoms that replace a hydrogen off the main carbon chain or ring.

syn addition – A reaction in which two groups of a reagent X–Y add on the same face of a carbon–carbon double bond.

tautomers – Easily interconvertible molecules that differ in the placement of a hydrogen and double bonds. Keto and enol forms are tautomers.

term – A compound or element in a chemical equation.

thermodynamic product – The reaction product with the least energy.

thermodynamics – The study of the transfer of energy and heat in chemical reactions.

thiol – A molecule containing an SH group. Also a functional group.

titration – Reacting a solution of unknown concentration with a solution of a known concentration for the purpose of finding out more about the unknown solution.

transition state – The state of a chemical reaction that corresponds to the highest energy along the reaction coordinate.

triplet – Describes an NMR signal split into three peaks.

valence electrons – The electrons in the outermost shell of an atom that can combine with other atoms to form molecules.

E2, *continued*

van der Waals equation – An equation for non-ideal gasses that accounts for intermolecular attraction and the volumes occupied by the gas molecules.

velocity – Speed of an object; the distance travelled over time.

volume – The amount of space an object occupies.

wave – A signal which propagates through space up and down or back and forth.

wavelength – On a periodic curve, the length between two consecutive peaks or troughs.

weak acid – Substances unable to completely ionize in solution but capable of donating hydrogen ions.

weak bases – Substances unable to completely ionize in solution but capable of accepting hydrogen.

work – Measures the movement of an object against some force.

Z isomer – Isomer in which the two highest-priority substituents are on the same side of a double bond or ring.

Appendix F. Tables and Conventions for Earth Science and Environmental Science

Contents

Earth Science

Environmental Science

Earth Science

F1 Field-Specific Abbreviations

Abbreviation	Meaning
AA	Atomic Adsorption
ALK	Alkalinity
AMD	Acid Mine Drainage
ANC	Acid Neutralizing Capacity
AST	Aboveground Storage Tank
ATP	Adenosine Triphosphate
BFE	Base Flood Elevation
BMP	Best Management Practice
BOD	Biochemical (or Biological) Oxygen Demand
BOD_5	5-Day Biochemical (or Biological) Oxygen Demand
CFC	Chlorofluorocarbon
CHP	Combined Heat and Power
CMI	Crop Moisture Index
COD	Chemical Oxygen Demand
DBH	Diameter at Breast Height
DDT	1,1,1-Trichloro-2,2-Di-(4-Chlorophenyl)Ethane
DO	Dissolved Oxygen
ED_{50}	Effective Dose 50
EIS	Environmental Impact Statement
EMF	Electromagnetic Field
ESP	Electrostatic Precipitator
ET	Evapotranspiration
FIRM	Flood Insurance Rate Map
FONSI	Finding of No Significant Impact
FS	Feasibility Study
GC	Gas Chromatography
GCM	Global Circulation Model
GIS	Geographic Information System
GMW	Gram Molecular Weight
GRAS	Generally Recognized as Safe
IMP	Integrated Pest Management
LAI	Leaf Area Index
LD_{50}	Lethal Dose, 50%
LNAPLs	Lighter (Than Water) Non-Aqueous Phase Liquids

F1, *continued*

Abbreviation	Meaning
LUC	Land Use Classification
LUST	Leaking Underground Storage Tank
MBAS	Methylene Blue Active Substance
MCL	Maximum Contaminant Level
MCLG	Maximum Contaminant Level Goal
MS	Mass Spectrometry
MSY	Maximum Sustainable Yield
MTBE	Methyl Tertiary Butyl Ether
MW	Molecular Weight
NASQAN	National Stream Quality Accounting Network
NGVD	National Geodetic Vertical Datum
NOx	Nitrous Oxides
NPDES	National Pollutant Discharge Elimination System
NPL	National Priorities List
OSP	Optimum Sustainable Population
OSY	Optimum Sustainable Yield
P&I	Piping and Instrumentation
PCBs	Polychlorinated Biphenyls
PCE	Perchloroethylene
PM	Particulate Matter
POTW	Publicly Owned Treatment Works
RI	Remedial Investigation
SBR	Sequencing Batch Reactor
STP	Standard Temperature and Pressure
TDS	Total Dissolved Solids
TKN	Total Kjeldahl Nitrogen
TMDL	Total Maximum Daily Load
TOC	Total Organic Carbon
TSS	Total Suspended Solids
UIC	Underground Injection Control
UST	Underground Storage Tank
UV	Ultraviolet Radiation
VOC	Volatile Organic Compound

F2 Common Mineral Abbreviations

Mineral	Abbreviation	Mineral	Abbreviation
acmite	Acm	cassiterite	Cst
actinolite	Act	celestite	Cls
aegirine-augite	Agt	chabazite	Cbz
åkermanite	Ak	chalcocite	Cc
albite	Ab	chalcopyrite	Ccp
allanite	Aln	chlorite	Chl
almandine	Alm	chloritoid	Cld
analcime	Anl	chondrodite	Chn
anatatse	Ant	chomite	Chr
andalusite	And	chysocalla	Ccl
andradite	Adr	chrysotile	Ctl
anhydrite	Anh	clinoenstatite	Cen
ankerite	Ank	clinoferrosilite	Cfs
annite	Ann	clinohumite	Chu
anorthite	An	clinozoisite	Czo
antigorite	Atg	cordierite	Crd
anthophyllite	Ath	corundum	Crn
apatite	Ap	covelite	Cv
apophyllite	Apo	cristobalite	Crs
aragonite	Arg	cummingtonite	Cum
arfvedsonite	Arf	diaspore	Dsp
arsenopyrite	Apy	digenite	Dg
augite	Aug	diopside	Di
axinite	Ax	dolomite	Dol
barite	Brt	dravite	Drv
beryl	Brl	eckermannite	Eck
biotite	Bt	edenite	Ed
boehmite	Bhm	elbaite	Elb
bornite	Bn	enstatite (ortho)	En
brookite	Brk	epidote	Ep
brucite	Brc	fassaite	Fst
bustatite	Bst	fayalite	Fa
Ca clinoamphibole	Cam	ferroactinolite	Fac
Ca clinopyroxene	Cpx	ferroedenite	Fed
calcite	Cal	ferrosilite	Fs
cancrinite	Ccn	ferrotschermakite	Fts
carnegite	Crn	fluorite	Fl

F2, *continued*

Mineral	Abbreviation	Mineral	Abbreviation
forsterite	Fo	limonite	Lm
galena	Gn	lizardite	Lz
garnet	Grt	loellingite	Lo
gedrite	Ged	maghemite	Mgh
gehlenite	Gh	magnesiokatophorite	Mkt
gibbsite	Gbs	magnesioriebeckite	Mrb
glauconite	Glt	magnesite	Mgs
glaucophane	Gln	magnetite	Mag
goethite	Gt	margarie	Mrg
graphite	Gr	melitie	Mel
grossular	Grs	microline	Mc
grunerite	Gru	molydbenite	Mo
gypsum	Gp	monazite	Mnz
halite	Hl	monticellite	Mtc
hastingsite	Hs	montmorillonite	Mnt
haüyne	Hyn	mullite	Mul
hedenbergite	Hd	muscovite	Ms
hematite	Hem	natrolite	Ntr
hercynite	Hc	nepheline	Ne
heulandite	Hul	norbergite	Nrb
horneblende	Hbl	nosean	Nsn
humite	Hu	olivine	Ol
illite	Ill	onphacite	Omp
ilmenite	Ilm	orthoamphibole	Oam
jadeite	Jd	orthoclase	Or
johannsenite	Jh	orthoproxene	Opx
kaersutite	Krs	paragonite	Pg
kalsilite	Kls	pargasite	Prg
kaolinite	Kln	pectolite	Pct
kataphorite	Ktp	pentlandite	Pn
K feldspar	Kfs	periclase	Per
kornerupine	Krn	perovskite	Prv
kyanite	Ky	phlogopite	Phl
laumontite	Lmt	pigeonite	Pgt
lawsonite	Lws	plagioclase	Pl
lepidolite	Lpd	prehnite	Prh
leucite	Lct	protoenstatite	Pen

F2, *continued*

Mineral	Abbreviation	Mineral	Abbreviation
pumpellyite	Pmp	staurolite	St
pyrite	Py	stilbite	Stb
pyrope	Prp	stilpnomelane	Stp
pyrophyllite	Prl	strontianite	Str
pyrrhotite	Po	talc	Tlc
quartz	Qtz	thomsonite	Tms
riebeckite	Rbk	titanite	ttn
rhodochrosite	Rds	topaz	Toz
rhodonite	Rdn	tourmaline	Tur
rutile	Rt	tremolite	Tr
sandine	Sa	tridymite	Trd
sapphirine	Spr	troilite	Tro
scapolite	Scp	tschermakite	Ts
schorl	Srl	ulvöspinel	Usp
serpentine	Srp	vermiculite	Vrm
siderite	Sd	vesuviante	Ves
sillimanite	Sil	witherite	Wth
sodalite	Sdl	wollastonite	Wo
spessartine	Sps	wüstite	Wus
sphalerite	Sp	zircon	Zrn
spinel	Spl	zoisite	Zo
spodumene	Spd		

Source: Mineralogical Society of America; Kretz, Ralph, 1983, "Symbols for Rock-forming Minerals," *American Mineralogist*, v. 68, nos. 1–2, pp. 277–279.

F3 Glossary of Earth Science Terms

aa – A term originated in Hawaii describing lava flow with rough lava blocks at its surface.

abrasion – The scraping of solid particles against rock causing the rock to erode.

absolute time – An estimated age of a rock or mineral based on its level of decay.

abyssal plains – Flat or extremely gentle sloping areas of the ocean floor.

acre-foot – A unit of measurement used for measuring large-scale water resources, both man-made and naturally occurring. An acre-foot is defined as a volume of water covering one acre of land to a depth of one foot.

active volcano – A volcano which is either actively erupting or has the potential to erupt in the near future.

adiabatic rate – The change in temperature in the atmosphere caused by the rising or lowering of an air mass.

aeolian deposits – Deposits from windblown sediments.

aftershock – A tremor or series of tremors that follow after a major earthquake.

alluvium – A general term referring to material that is deposited by the flow of running water.

andesite – Volcanic rock of an intermediate growth with a fine-grained texture.

aphotic – A portion of a body of water where there is little or no light.

aquiclude – An impervious geologic structure that cannot hold or transmit fluid.

F3, *continued*

aquifer – A layer of rock where a large amount of fluid, such as water or oil, is naturally stored.

aquifer, confined – aquifers covered with a non-permeable layer that have the water table above them. Typically pressure causes the buried liquid to rise above the surface.

aquifer, perched – A small collection of water that is separated from an underlying layer of groundwater.

aquifer, secondary – An aquifer that is not the main source of water in a given area.

aquifer, unconfined – Aquifers where the upper surface part of the water table receives water from precipitation or an additional body of water such as a river or stream.

artesian well – A well in which pressure naturally forces water upward.

artificial recharge – When surface water is unnaturally added to groundwater.

ash – Fine particles of rock ejected from a volcanic explosion.

ash flow – A mixture of gas and fine rock particles ejected violently from a crater or fission.

atoll – A coral island or reef that surrounds an area of water.

avalanche – A large mass of material that falls or slides rapidly due to gravity.

basalt – Finely ground dark colored igneous rock derived from volcanic upwellings.

basic – Igneous rocks with a content low in silica and high in iron, calcium, or magnesia.

basin – An area where rocks dip towards a central point.

F3, _continued_

bed – A layer of rock laid down parallel to the surface.

bedrock – The solid layer of rock that makes up the Earth's crust.

benthic – Pertaining to the greatest depth of a body of water.

bentonite – Clay that is formed from decomposed volcanic ash.

biostratigraphy – The study of rocks based on the fossils found within them.

block – An angular chunk of solid rock ejected during a volcanic eruption.

bomb – Fragments of molten rock erupted into the air after a violent volcanic explosion.

breccia – Sedimentary rock that contains angular rock fragments naturally held together.

brine – Water with a high salt content that is extracted from the ground.

brittle-ductile transition zone – The strongest part of the Earth's crust.

butte – A small steep sided hill with a flat top.

caldera – A large volcanic crater formed by either a volcanic explosion or the collapse of a volcanic cone.

canyon – A long steep sided valley formed over a long length of time by running water.

carbonaceous – Materials that contain carbon.

channel – The bed and sides of a course of water, such as a river.

chatter mark – Scrapings on a rock caused by the movement of a glacier.

F3, *continued*

cinder cone – A cone formed from hardened lava thrown upward during a volcanic explosion.

cleavage – The property that allows a mineral to break along a smooth plane.

coal – A solid black material formed from the partial decomposition of vegetation, it is commonly burned as a source of heat and fuel.

composite volcano – A steep cone shaped volcano formed from lava flow and solidified lava.

condensation – The form water takes when it changes from a vapor to a liquid.

conglomerate – Sedimentary rock that contains smooth rock fragments naturally held together.

connate water – Groundwater that is formed from the rock itself instead of being accumulated from the surface.

continental crust – The solid outer layers of the Earth.

continental drift – The theory that the continents have drifted away from each other due to the horizontal movement of the Earth.

continental rise – The lower part of the continental shelf that leads to the abyssal plain.

continental shelf – The sloping part of the ocean floor that is closest to the surface.

continental slope – The steep section located between the continental shelf and rise.

coral reef – An aquatic ridge formed from the accumulated skeletons of coral.

crater – A steep sided depression caused by either an explosion or the impact of a landmass.

F3, *continued*

crystal – A solid form where the molecules are packed together in a regular and repeated pattern.

dacite – A light gray volcanic rock composed mostly of silica.

delta – A large collection of sediments found at the mouth of a river.

density – The measure of how tightly the atoms of a substance are packed.

density current – A flow in water maintained by gravity through a large body of water. The difference in density causes it to retain its unmixed identity.

deposition – The process in which sediment is moved and dropped onto the Earth, often moved by either wind, rain, or ice.

detachment plane – The place along a surface where a mass of land breaks away from its original portion.

detritus – Rocks that have been broken or worn down by physical means.

dew point – The level of temperature in which water vapor turns to droplets.

dormant volcano – A volcano that is presently inactive but has the potential to reactivate.

disphotic – The zone of the ocean where small quantities of light penetrate the water.

drift – Material that is deposited by glacial movement.

drumlin – A streamlined oval-shaped hill that has been shaped by flowing glacial ice.

dune – A hill formed from wind blown sand.

earthquake – A shifting of plates that causes movement to the Earth's surface.

F3, *continued*

erosion – The process in which earth and rock are worn away by natural elements such as water and wind.

erratic – Referring to large rocks and boulders being carried away by glaciers and deposited a significant distance from their source spot.

eruption – When natural material is thrown into the air by volcanic activity.

eruption cloud – The cloud of material that forms when an eruption occurs.

eruption vent – The opening of a volcano where material is able to escape.

estuary – The part of a costal river where freshwater and ocean water mix together.

euphotic – The upper levels of the ocean where the level of light penetrating the water is enough to cause photosynthesis.

evaporation – The point in which water changes from a liquid to vapor.

evolution – The theory in which living organisms change from simple to complex beings over time.

extinct volcano – A volcano that is not presently active and will most likely not become active again.

fault – A crack or fracture within the earth's crust.

firn – Ice or snow that has failed to melt from one season to the next, but has not yet reached an age to be considered a glacier.

fissures – Fractures or cracks that appear on the slopes of a volcano.

flank eruption – An eruption appearing on the sides of a volcano instead of the highest point.

flood plain – Flat or level land that is prone to flooding from the overflow of a body of water.

F3, *continued*

flood – An overflow of water in an area.

fluvial – A general term describing objects and organisms within a river.

formation – A type or grouping of rock that share similar characteristics or fossils.

fossil – A mineralized imprint of an organism preserved in rock.

fracture – A break along a mineral or rock that does not occur along a cleavage point.

fumarole – A volcanic vent in which steam, smoke, and gases escape.

geothermal energy – Energy that is obtained from the underground heat of the Earth.

glacier – A large slowly moving mass of ice that stays formed for many years.

gravel – Deposits of rock fragments that result from erosion, which are larger in size than sand.

graben – An elongated depression of crust occurring between two faults.

groundwater – Water that is stored underneath the Earth's surface.

hardness – Relating to the density and resistance of an object.

harmonic tremor – A continuous release of energy caused by the movement of underground magma.

heat transfer – The movement or transportation of heat from one source to another.

hiatus – A break or gap occurring in the geologic record of a layer of rock.

hot-spot volcanoes – Volcanoes that contain a persistent source of heat.

F3, *continued*

hydrologic cycle – The continuous circulation of water throughout the Earth.

hydrothermal reservoir – An underground area of rock containing heated water.

ice age – A period in our planet's history when the vast majority of the Earth was covered in ice and snow.

intermittent stream – A stream in which water is carried only at certain times, usually during a flood or rainy seasons.

island – A landmass that is completely surrounded by water.

isostasy – The balance between the visible portion of a mass, either made up of land or ice, and the hidden section below the surface.

kinetic energy – Energy created from the movement of an object.

lahar – A landslide formed from the buildup of debris from a volcano.

lake – A body of water surrounded by land, which can occur both naturally and artificially.

landslide – The rapid movement of land down a slope.

langley – A unit used to measure how much solar energy is distributed over a given area.

lapilli – Small stones that are ejected into the air as a result of a volcanic explosion.

lava – Magma that has reached the surface of the earth.

lava flow – An outpouring of lava onto land.

lava tube – A tube in which the outside is formed of cooled solidified lava while the inside has continuous flows of molten lava.

leeward – The side of an object or area that is facing away from the wind.

F3, *continued*

loess – Finely grained sediments that are deposited by the wind.

longshore drift – The movement of materials along a beach caused by the breaking of waves.

magma – Molten rock that is beneath the surface of the earth.

mantle – The layer of earth between the crust and the core.

magnitude – The measurement used to determine the level of energy released during an earthquake.

mass movement – The large-scale movement of material on the surface of the earth, mostly as a result of gravity.

mesa – A flat-topped landscape with steep sides.

metamorphic – Referring to rocks whose properties are changed due to extreme amounts of heat and pressure.

mid-ocean ridge – A ridge of mountain and volcanic ranges occurring on the ocean floor.

mineral – A naturally occurring inorganic solid with has a definite internal structure and chemical composition.

mountain – An elevated part of a landscape of significant mass, size and steepness.

natural gas – A mixture of hydrocarbon gases formed from the decomposition of organic material.

neritic – The shallow ocean zone that is composed from low tide to a depth of 200-meters.

nutation – The wobble of a planet as it spins on its axis.

obsidian – A black volcanic glass formed by the rapid cooling of lava.

ocean trench – A deep depression in the ocean floor.

F3, *continued*

oceanic crust – The portion of the Earth's crust that is found underneath the ocean.

ore – A mineral that is composed mostly of metallic material.

outcrop – The exposed body of a rock.

pahoehoe – A term originated in Hawaii describing lava with a smooth glassy surface.

permafrost – Soil and land that is perpetually frozen.

permeability – The measurement of a material's ability to allow liquid to pass through it.

phreatic eruption – A volcanic eruption that occurs when heated volcanic rocks interact with water.

pillow lava – Lava that has hardened into pillow-like shapes as a result of an underwater volcanic explosion.

plastic deformation – An irreversible change in the shape of material as a result of compression or expansion.

plate tectonics – The theory that geological movement such as continental movement, earthquakes, the forming of mountain ranges, and volcanic eruptions is caused from the shifting and movement of plates.

plug – Solidified lava that fills the conduit of a volcano.

pluton – A large bubble of igneous rock that is formed underground.

potential energy – Energy that is stored within a substance.

precipitation – Condensed water that falls from the atmosphere to the earth's surface.

pumice – Frothy rock that is formed from expanding gas in erupting lava.

pyroclastic – Fragmented rock that is formed from a volcanic explosion.

F3, *continued*

pyroclastic flow – Currents of hot gas, ash, and rock that rapidly moves downward after a volcanic eruption.

rhyolite – Finely grained volcanic rock that is rich in potassium and sodium.

ring of fire – A range of volcanoes that surround the Pacific Ocean.

rock flour – Finely ground rock.

seafloor spreading – The theory that new ocean crust is formed at mid-ocean ridges through volcanic activity and then gradually moves outward.

sea level – The level of the ocean where the top of the water meets the atmosphere.

sediment – Material that is formed from the erosion of rock.

seismograph – An instrument that is used to measure vibrations made by the Earth.

shield volcano – A dome shaped volcano with gently sloping sides.

sinkhole – A natural depression formed from the collapse of a surface.

soil – The topmost layer of the Earth composed of very finely grained rock and other organic material. It is instrumental in the growth of plants and other vegetation.

specific gravity – The measurement of how tightly the atoms of a substance are packed.

stratovolcano – A volcano that is composed of both lava flows and pyroclastic material.

subsidence – The large sinking of an area of the Earth's crust, occurring both naturally and artificially.

sublimation – When a solid changes into a vapor state without passing through a liquid state.

F3, *continued*

tephra – Materials of various sizes that are thrown into the air as a result of a volcanic explosion.

tide – The rising and falling of water caused by the pull of the Sun and the Moon.

till – A general term for material that is deposited by a glacier.

tsunami – A giant and destructive sea wave caused by underwater earthquakes or volcanic eruptions.

tuff – Rock formed from volcanic materials that have been cemented together.

turbidity current – An underwater current triggered by an earthquake.

upwelling – The movement of deep level cold water being raised upward by wind movement to replace warmer surface water.

vapor – Water in its gaseous state.

vein – A deposit of foreign materials following a rock fracture.

vent – An opening at the Earth's surface where volcanic material can reach the surface.

viscosity – The measurement of a liquid's resistance to flow.

volatiles – Gases that will quickly evaporate when exposed to air.

water table – The surface of underground water.

weathering – The wearing down of materials exposed to the elements.

windward – The side of an object or structure facing into the wind.

zone of ablation – The area of a glacier where the melting of snow and ice occurs faster than it can be replenished by snowfall.

zone of accumulation – The area of a glacier where material is added by snowfall occurring faster than the snow melts.

F4 Earth Science Journals and Their Abbreviations

Full Title	Abbreviation
Acta Geologica Hungarica	Acta. Geol. Hung.
Acoustics Research Letters Online	Acoust. Res Lett Online
Advances in Earthquake Engineering Series	Adv Earthquake Eng
Advances in Fluid Mechanics	Adv Fluid Mech
Advances in Space Research	Adv Space Res
Agronomie	Agronomie
American Journal of Applied Sciences	Am J Appl Sci
American Journal of Science	Am J Sci (AJS)
American Mineralogist	Am Mineral
Applied Geochemistry	Appl Geochem
Aquatic Geochemistry	Aquat Geochem
Atlantic Geology	Atl Geol
Australian Journal of Soil Research	Aust J Soil Res
Biosystems Engineering	Biosystems Eng
Bulletin of the Natural History Museum – Geology Series	Bull Nat Hist Mus Geol
Bulletin of the Natural History Museum – Zoology Series	Bull Nat Hist Mus Zool
Bulletin of Volcanology	Bull Volcanol
Canadian Journal of Earth Sciences	Can J Earth Sci
Canadian Journal of Soil Science	Can J Soil Sci
Canadian Mineralogist, The	Can Mineral
Caribbean Journal of Earth Science	Caribb J Earth Sci
Chinese Journal of Geochemistry	Chin J Geochem
Chinese Physics	Chin Phys
Clay Minerals	Clay Miner
Cold Regions Science and Technology	Cold Regions Sci Tech
Computers & Geosciences	Comput Geosci
Computational Geosciences	Comput Geosci
Culture and Agriculture	Cult Agr
Dendrochronologia	Dendrochronologia
Developments in Earth Surface Processes	Dev Earth Surf Process
Doklady Earth Sciences	Dokl Earth Sci
Earth Interaction	Earth Interact
Earth Science Digest	Earth Sci Digest
Earth Sciences History	Earth Sci Hist
Earth-Science Reviews	Earth Sci Rev
Earth Surface Processes	Earth Surf Process
eEarth	eEarth

F4, *continued*

Economic Geology and the Bulletin of the Society of Economic Geologists	Econ Geol Bull Soc Econ Geol
Ecos, Transactions, American Geophysical Union	Eos Trans Am Geophys Union
Energy & Fuels	Energ Fuel
Environmental and Engineering Geoscience	Environ Eng Geosci
European Mineralogical Union Notes in Mineralogy	Eur Mineral Union Notes Mineral
Geographical and Environmental Modelling	Geogr Environ Model
GeoJournal	GeoJournal
Geológica Acta	Geol Acta
Geology for Economic Development	Geol Econ Dev
Geophysical Research Letters	Geophys Res Lett
Geotechnical & Geological Engineering	Geotech Geol Eng
Ground Water	Ground Water
Historical Biology	Hist Biol
Hydrology and Earth System Sciences	Hydrol Earth Syst Sci
IEEE Geoscience and Remote Sensing Letters	IEEE Geosci Rem Sens Lett
IEEE Transactions on Antennas and Propagation	IEEE Trans Antenn Propag
IEEE Transactions on Geoscience and Remote Sensing	IEEE Trans Geosci Rem Sens
International Journal of Coal Geology	Int J Coal Geol
International Journal for Numerical and Analytical Methods in Geomechanics	Int J Numer Anal Meth Geomech
International Journal of Primatology	Int J Primatol
Irish Journal of Earth Sciences	Ir J Earth Sci
Journal of African Earth Sciences	J Afr Earth Sci
Journal of Applied Sciences Research	J Appl Sci Res
Journal of Geodynamics	J Geodyn
Journal of Geophysical Research	J Geophys Res (JGR)
Journal of Land Use Science	J Land Use Sci
Journal of Low Frequency Noise, Vibration and Active Control	J Low Freq Noise Vib Active Contr
Journal of Metamorphic Geology	J Metamorph Geol
Journal of Quaternary Science	J Quaternary Sci
Journal of the Science of Food and Agriculture	J Sci Food Agr
Journal of South American Earth Sciences	J S Am Earth Sci

F4, *continued*

Journal of Southeast Asian Earth Sciences	J Southeast Asian Earth Sci
Journal of Volcanology and Geothermal Research	J Volcanol Geoth Res
Lecture Notes in Earth Sciences	Lect Notes Earth Sci
Mapping Sciences and Remote Sensing	Mapp Sci Rem Sens
Marine Geology	Mar Geol
Marine and Petroleum Geology	Mar Petrol Geol
Mineralogical Record, The	Mineral Rec
Moscow University Geology Bulletin	Moscow Univ Geol Bull
Nature and Science	Nat Sci
Natural Hazards	Nat Hazards
New Technology Magazine	New Tech Mag
New Zealand Journal of Geology and Geophysics	New Zeal J Geol Geophys
Nuclear Geophysics	Nucl Geophys
Optics and Lasers in Engineering	Optic Laser Eng
Ore Geology Reviews	Ore Geol Rev
Palaios	Palaios
Physics and Chemistry of the Earth - Part A – Solid Earth and Geodesy	Phys Chem Earth A Solid Earth Geodes
Precambrian Research	Precambrian Res
Quaternary Research	Quaternary Res
Radio Science	Radio Sci
Reviews in Mineralogy	Rev Mineral
Reviews in Mineralogy and Geochemistry	Rev Mineral Geochem
Romanian Journal of Mineralogy	Rom J Mineral
Russian Journal of Earth Science	Russ J Earth Sci
Soil Biology and Biochemistry	Soil Biol Biochem
Soil Dynamics and Earthquake Engineering	Soil Dynam Earthquake Eng
Soil Technology	Soil Tech
Soil Use & Management	Soil Use Manag
Solar System Research	Sol Syst Res
Strength, Fracture and Complexity	Strength Fract Complex
Surveys in Geophysics	Surv Geophys
Western Pacific Earth Sciences	West Pac Earth Sci
Zambia Journal of Applied Earth Sciences	Zambia J Appl Earth Sci
Zeitschrift für Geologische Wissenschaften (Journal for the Geological Sciences)	Z geol Wiss

Environmental Science

F5 Common Units Used in Environmental Science

Unit	Definition	Common Usage
°F	degrees Fahrenheit	temperature
°R	degrees Rankine	temperature
acre-ft	volume of water equivalent to one acre of land covered by one foot of water	water volume
acre-ft/day	acre-foot per day	water flow
atm	atmosphere	pressure
BTU	British thermal unit	energy or power
cal	calorie	heat
cfm	cubic feet per minute	water flow
cfs	cubic feet per second	water flow
cfsm	cubic feet per second per square mile	water flow per area
cp	centipoise	viscosity
dyn	dyne	force
erg	erg	work or energy
ft-lb or ft-lb$_f$	foot-pound	force
gpd	gallons per day	water flow
gpm	gallons per minute	water flow
in Hg	inches of mercury	pressure
JTU	Jackson Turbidity Unit	turbidity of water
mg g^{-1}	milligrams of analyte per gram of sample	substances in soil or in the human body
mg kg^{-1}	milligrams of analyte per kilogram of sample	substances in soil or in the human body
mg L^{-1}	milligrams of analyte per liter of sample	substances in water
mgd	million gallons per day	water flow
mm Hg	millimeters of mercury	pressure
NTU	Nephelometric Turbidity Unit	turbidity of water
ppb	parts of analyte per billion parts of sample	substances in water, soil, or air

F5, *continued*

Unit	Definition	Common Usage
ppb_w	parts of analyte per billion parts of sample on a weight basis	solid or liquid substances in soil
ppm	parts of analyte per million parts of sample	substances in water, soil, or air
ppm_v	parts of analyte per million parts of sample on a volume basis	gases in air
ppm_w	parts of analyte per million parts of sample on a weight basis	solid or liquid substances in soil
ppt	parts of analyte per thousand parts of sample	substances in water, soil, or air
psi	pounds per square inch	pressure
psia	pounds per square inch absolute	absolute pressure
$\mu g \; g^{-1}$	micrograms of analyte per gram of sample	substances in soil or in the human body
$\mu g \; kg^{-1}$	micrograms of analyte per kilogram of sample	
$\mu g \; L^{-1}$	micrograms of analyte per liter of sample	substances in water

• In general, use of the solidus (or slash, as in "mg/L") is limited to the simplest unit expressions.

• Negative index expressions (as in $mg \; L^{-1}$) are generally preferred in publishing and are typically required for complex units (e.g., $mg \; L^{-1} s^{-1}$).

• A space or (less commonly) a middle dot ($\cdot$) is used between multiple unit symbols.

F6 Common Alphabetic Symbols for Variables

Symbol	Variable
A	area or acceleration
F	force
H	enthalpy
L	length
M	molarity
m	mass
N	normality
n	number of moles
P	pressure
Q	heat or flow rate
S	entropy
T	temperature
t	time
V	volume
v	velocity
x or y	mole fraction
μ	absolute viscosity
ρ	density
υ	kinematic viscosity

As many symbols may have multiple meanings (e.g., Q represents heat in thermodynamics, but flow rate in hydrology), the context of the subject matter must be carefully considered when using symbols in earth science and environmental science publications.

F7 Units in Environmental Science Diagrams

Two of the most common types of diagrams encountered in earth and environmental sciences are biogeochemical cycles and material balances. Biogeochemical cycle diagrams depict the transfer of elements (such as carbon, water, or other substances) between different components of the environment (e.g., the earth and the atmosphere). Quantitative cycle diagrams include numerical values that should have consistent units. Material balances illustrate the flow of materials (matter) in and out of nonradioactive processes (e.g., an air pollution control system).

It is essential to use consistent units within a diagram and between the diagram and the text.

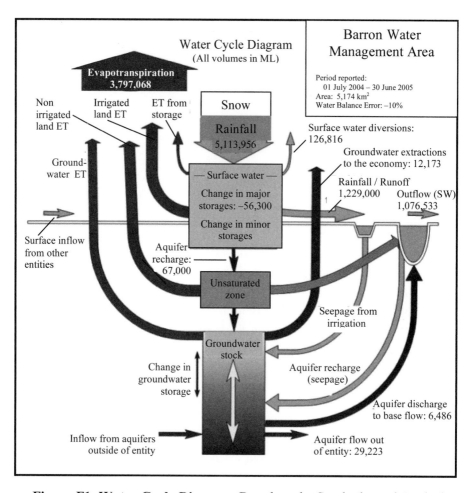

Figure F1 Water Cycle Diagram. Based on the Stocktake and Analysis of Australia's Water Accounting Practice program (Bureau of Rural Sciences *Water 2010*), the diagram describes the link between groundwater, the unsaturated zone, and surface water. Notice that, although the diagram does not state units of measurements with its figures within the chart, "ML" is identified in the headline as the main unit of measurement. In any following text, "ML" would be used consistently with any data mentioned. Text for a diagram such as this, should also describe the time period reported in the diagram, the water balance format used, the components of the report, classifying data quality, and water balance reports for other important areas.

   When sending in a manuscript, journals often ask that any diagrams, charts, and tables be sent in separate files from the actual article. It is thus important that any illustrative materials present the data consistently and clearly enough to allow the diagram to stand on its own.

F8 Abbreviations, Signs, and Symbols for Scientific and Engineering Terms

absolute	abs
absolute value	| |
absorbance	A
absorptivity	a
absorptivity, molar	ε
acceleration, angular	α
acceleration, linear	a
acre-foot (feet)	acre-ft
activity, chemical (absolute)	λ
activity, chemical (relative)	a
activity coefficient	γ
affinis	aff.
alternating current	ac / $\leftrightharpoons$ / $\rightleftharpoons$
alternating-current	a-c
altitude	alt
ampere	A
analytical variability	ξ_a
angle	$\angle$
angle between	$\wedge$
angle between a_o and b_o in the unit cell	γ
angle between a_o and c_o in the unit cell	β
angle between b_o and c_o in the unit cell	α
angle between the two optic axes of a biaxial mineral	2V
angstrom	Å
angular frequency	ω
angular velocity	ω
anhydrous	anhyd
antilogarithm	antilog
approaches	$\rightarrow$
approximate	approx
approximately	$\approx$
aqueous	aq
area (land area)	a
astronomical unit	au

asymptotically equal to	$\simeq$
atmosphere	atm
atomic mass	m_a or m
atomic mass of species X	m (X) of m_x
atomic number	at no or Z
atomic number of species X	Z (X) or Z_x
atomic weight	at wt or M
atomic weight of species X	M (X) or M_x
automatic data processing	ADP
average	avg
average	$^-$ or $\langle\,\rangle$
Avogadro's number	N or N_A
avoirdupois	avdp
azimuth	az or α
bar	bar
barn (area)	b
barometer	bar.
barrel	bbl
barrel per day	bbl/d
base of natural logarithms	e
baud	Bd
baumé	°Bé
becquerel	Bq
before present (dates before 1950, in thousands of years)	B.P.
bench mark (in illustrations)	BM
bench mark (in text)	B.M.
Bernoulli number	B
Bessel Function (first kind, zero order)	$J_o(x)$
Bessel Function, hyperbolic (first kind, zero order)	$I_o(x)$
bias	δ
billon gallons per day	Ggal/d
billion years	b.y.
binary coded decimal	BCD

F8, *continued*

boiling point	bp	circular (shape)	cir
Boltzmann constant	*k*	citrate-extractable	
Boltzmann function	H	heavy metal	cxHM
biochemical oxygen		coefficient	coef
demand	BOD	cold-extractable copper	cxCu
bit, byte	b	collection (s)	colln(s).
Bohrmagneton	μB	cologarithm	colog
bottom-withdrawal		compressibility	κ
tube	BW-tube	concentrate	conc
braces	{ }	concentrated	concd
brackets	[ ]	concentration	concn *or* c
Bragg angle, glancing angle		conductance	G
(2θ is twice the glancing		conductivity	γ
angle in X-ray diffraction)	θ	confer (to be compared to)	cf.
breadth (width)	*b or B*	confidence limit, lower,	
Brinell hardness number	Bhn	for the population mean	μL
British thermal unit	Btu	confidence limit, upper,	
bushel	bu	for the population mean	μU
byte	B	constant	const
calculated	calc	constant as defined	K
calorie	cal	continued	con.
candela	cd	Coordinated Universal	
candela-hour	c·h	Time	UTC
capacitance	C	corner	cor.
carat	kt	correlation coefficient	ρ *or r*
Cartesian coordinates	*x,y,z*	cosecant	csc
cathode ray	CR	cosecant, hyperbolic	csch
cathode-ray tube	CRT	cosine	cos
Celsius (used with degree		cosine, hyperbolic	cosh
symbol)	°C	cotangent	cot
centimeter	cm	cotangent, hyperbolic	coth
centimeter-gram-second		coulomb	C
(system)	CGS	counts per minute	c/min
centimeter-gram-second		critical	crit
(unit)	cgs	Cross, Iddings, Pirsson,	
central processing unit	CPU	and Washington	CIPW
chemical oxygen demand	COD	cross section of atoms	
chemical potential	μ	and nuclei	σ
chi-square statistic	χ^2	crystallographic axes	*a, b, c*
circa (about)	ca.	cubic centimeter	cm^3
circle	○	cumulative frequency	c.f.

F8, *continued*

curie	Ci
cutting point in a hypothesis test	Ω
cycle (radio)	c
cycles per minute	c/min
cycles per second	c/s
cylinder	cyl
darcy, darcies	D
day	d
debye unit	D
decay constant	λ
decay constant based on alpha emission	λ_a
decay constant based on negative beta emission	$\lambda_{\beta-}$
decay constant based on orbital electron capture	$\lambda_\in$ *or* λ_{EC}
decay constant based on positron emission	$\lambda_{\beta+}$
decay constant based on spontaneous fission	λ_{SF}
decible	dB
degree	°
degree Celsius	°C
degree Fahrenheit	°F
degree Rankine	°R
degree réamur	°R
degrees of freedom	d.f.
delta (finite change, incremental variations, difference)	Δ *or* δ
density (mass)	ρ
density (relative)	d
depth	h
deuterium	D *or* ^2H
deutron	d
diameter	diam, D, *or* d
dielectric constant (permittivity)	$\in$
dielectric flux	ψ
differential, partial	∂

differential thermoanalysis	dta
differential, total	d *or* d
dilute	dil
direct current	dc *or* $\rightarrow$
direction of extraordinary ray	E
direction of flow	$\rightarrow$
direction ordinary ray	O
discharge; total water discharge; rate of discharge; recharge	Q
disintegrations per minute	d/min
disintegrations per second	d/s
dissociation constant	K
dissociation constant, negative logarithm of; -log K	pK
dissolved oxygen	DO
dissolved solids	DS
distilled	dist
ditto (the same)	do.
divided by	÷
dozen	doz
dram	dr
dropping mercury electrode	dme
dry basis	DB
dyne	dyn
efficiency	eff
electric current	I
electric-current density	J, j
electric-field strength	E
electric potential	V
electromagnetic unit	emu
electromotive force	emf *or* E'
electron	e *or* e
electron mass	m_e
electron-spin resonance	esr
electronvolt	eV
electrostatic flux	ψ
electrostatic unit	esu
elementary charge	e
elevation	elev

F8, *continued*

emendatio (emended)	emend.
end point	EP
energy	E
energy (kinetic)	E_k
energy (potential)	E_v
enthalpy	H
entropy	S
entropy (standard state of)	$S°$
ephemeris time	ET
equal to	$=$
nearly equal to	$\approx$
not equal to	$\neq$
equation(s)	eq (s)
equilibrium constant	K
equivalent	equiv.
equivalent conductivity	$\wedge$
equivalent uranium	eU
equivalent weight	equiv. wt.
error function	erf
error function	
(complement to)	erfc
Euler number	E
ex grupo	ex gr.
exchange	↓↑
exchangeable-potassium-	
percentage	EPP
exchangeable-sodium-	
percentage	ESP
excited hydrogen atom	H*
exponential of	exp. E
factorial product	!
Fahrenheit (used with	
degree symbol)	°F
farad	F
Faraday's constant (faraday)	F
foot, feet	ft.
footcandle	fc
footlambert	fL
foot (feet) per	
second cubed	ft/s^3
foot-pound	ft·lb

foot pound-second	
(system)	FPS
force	F
force (moment)	M
formality	f
freezing point	fp
frequency	f or v
frequency (spectroscopy)	v
friction, coefficient of	μ or f
Froude number	F
F-statistic for equality of	
variances	F
fugacity	f
function of x	$f(x)$
fusion point	fnp
gallon	gal
gallons per day	gal/day or gpd
gallons per	
minute	gal/min or gpm
gamma function	Γ
gas, as in H_2O (g)	(g)
gas constant	R
gas liquid partition	
chromatography	glpc
gauss	G
Geiger-Müller (unit	
modifier)	G-M
Gibbs free energy, Gibbs	
function	G
Gibbs free energy, (standard	
state)	$G°$
gradient	∇
grain	gr
gram	g
gravitational acceleration,	
acceleration of free fall,	
local acceleration due to	
gravity	g
gravitational constant	G
gray (unit of measure for	
absorbed dose)	Gy

F8, *continued*

greater than	$>$
much greater than	$\gg$
not greater than	$\ngtr$
greater than or equal to	
	$\geq \ or \ \geqq$
Greenwich mean astronomical time	G.m.a.t.
Greenwich mean time	G.m.t.
gross	gr
gross weight	gr. wt.
half-life	$T_{1/2}$
half-life reduced	$fT_{1/2}$
haversine	hav
head, total	H
heat capacity	C
heat capacity at constant pressure	C_P
heat capacity at constant volume	C_V
hectare	ha
height	h
Helmholtz free energy	A
henry, henries	H
hertz	HZ
high-pressure (unit modifier)	h-p
high-pressure metal vapor	HPMV
horsepower	hp
hour	h
hydrogen-ion concentration, negative $\log_{10}$ of	pH
hyperbolic functions, inverse	ar
hypothesis (alternative)	H_1
hypothesis (null)	H_0
identical/not identical	$\equiv / \not\equiv$
imaginary square root of -1	i *or* j
inch (period may be used if abbreviation might be confused with preposition "in")	in
inch-pound	in-lb

indeterminate	indet.
index of refraction	n
indices of reflection for biaxial crystals	$n_x, n_y,$ and n_z *or* $\alpha, \beta,$ and γ
indices of reflection for uniaxial crystals	n_O and n_E *or* ω and $\in$
inductance (mutual)	M
inductance (self)	L
infinity	∞
infrared	ir
inside diameter	id
integral	$\int$
integral, closed (circuital or contour)	ϕ
intensity of X-rays reflected from crystallographic planes	I
intermediate-pressure (unit modifier)	i-p
intersection or logical product	$\cap$
ionization constant	K or K_i
irrigation-water classification: C denotes conductivity (electrical); S denotes sodium-adsorption ratio (SAR); numbers denote respective numerical quality classes	$C2$-$S3$
Jackson turbidity unit	Jtu
joule	J
joule per kelvin	J/K
Joule-Thomson coefficient	μ
kelvin (no degree symbol)	K
kilobyte	K
kilogram	kg
kilohm	$k\Omega$
kilowatthour	kWh
K-meson	K

F8, *continued*

knot	kn
lambert	L
langley	ly
Laplacian operator	∇^2 *or* Δ
latitude (abbreviation used only with figures)	lat
length	l
less then	$<$
much less then	$\ll$
not less then	$\nless$
less than or equal to	$\leqq$ *or* $\leq$
limit of $f(x)$	$\lim f(x)$
linear alkylsulfonate	LAS
liquefied petroleum gas	LPG
liquid	liq
liquid oxygen	lox
liter	L
locality, localities (abbreviation only with numbers)	loc (s).
logarithm (common)	log
logarithm (natural)	$\log_e$ *or* ln
logical product or intersection	$\cap$
logical sum or union	$\cup$
longitude (abbreviation used only with figures; omit period when "long" is used with "lat"; use period if abbreviation may be confused with the adjective "long")	long
longitudinal velocity; P-wave velocity	v_P
low frequency	LF
low-pressure (unit modifier)	l-p
lumen	lm
luminous flux	Φ
lux	lx
magnetic-field strength or intensity	H
magnetic flux	Φ

mass number	A
magnetic induction	B
Manning's roughness (resistance) coefficient	n
mass	m
mass number of species X	$A\,(X)$ *or* A_x
matrix; for example	$\| a_{ij} \|$ *or* (a_{ij}) *or* A $\|\ \|$ *or* $(\ )$ *or* A
matrix, cofactor of element	$a_{ij}\,A_{ij}$
matrix, conjugate	A
matrix, determinate of; for example $|a_{ij}|$	$|\ \ |$
matrix, identity	I
matrix, inverse	A^{-1}
matrix, transpose	A^{T}
maximum	max
maxwell	Mx
mean (statistical)	μ *or* m
mean life	τ
mean of a linear combination q	μ_q
mean of the lognormal distribution	α
mean of the negative binomial distribution	θ
mean of sample means	$\mu_{\bar{w}}$
mean of the variance of sample means	μs
mean sea level	m.s.l.
mean square error	M.S.E.
megabyte	Mb
megohm	MΩ
melting point	mp
member of (used with a set and its elements)	$\in$
meta (in organic compounds)	m
meter	m
metric ton	t

F8, *continued*

microGal	µG
micron	µ
mile	mi
miles per hour	mi/h
or mph	
Miller indices	*hkl*
millimeter of mercury	mmHg
million	M
million gallons per day	Mgal/d
million years	m.y.
minimum	min
minus	−
minus or plus	∓
minute	min
minute; prime; foot	′
mixture melting point	mmp
Modified Mercalli	MM
molality, molal (concentration)	*m*
molar concentration of substance B	c_B
molar mass of substance B	M_B
molarity, molar (concentration)	*M*
mole	mol
molecular concentration	*C*
molecular weight	mol wt
month	mo
motorship	MS
multiplied by	× *or* •
multiplying factor for the geometric mean of lognormality distributed observations	ψ_n
multiplying factor for the variance of lognormally distributed observations	ϕ_n
multispectral scanner	MSS
muon	µ

nabla; del; differential vector operator	▽
natural variabiltity	ξ_n
nautical mile	nmi
neutrino	ν
neutron	n
new genus	n. gen.
new series	new ser.
new species	n. sp.
new variety	n. var.
newton	N
newton meter	N·m
Newtonian gravitational constant	*G*
no data	n.d.
no record, not reported	n.r.
nomen, nudum	nom. nud.
normality, normal (concentration)	*N*
not available, not applicable	NA
not determined	n.d.
nucleon number	*A*
number of observations in a population	*N*
number of observations (sample size)	*n*
number of samples	*k*
observations	*w*
observed frequency of observations	*O*
oersted	O_e
ohm	Ω
ohm centimeter	Ω·cm
ohm meter	Ω·m
optical directions in a crystal; also rays of light in these directions and pleochroic colors in these directions	X, Y, Z
ortho (in organic compounds)	*o*
ounce	oz

F8, *continued*

outside diameter	od	population standard	
oven-dry basis	ODB	deviation	σ
oxidation-reduction		population variance	σ^2
potential	Eh	posterior distribution of a	
para (in organic		parameter θ	$D_1(\theta)$
compounds)	p	potassium-adsorption	
parsec	pc	ratio	PAR
part(s)	pt(s).	potential difference	V *or* U
part(s) per billion	ppb	pound (mass)	lb
part(s) per million	ppm	pound avoirdupois	lb avdp
part(s) per thousand	ppt *or* ‰	pound-force	lbf
partial pressure of oxygen		pound-force per	
or carbon dioxide:		square inch	lbf/in^2

$$P_{O_2}, P_{O_2}, \text{ or } P(O_2), P(CO_2)$$

particle-size diameter	ϕ	precipitate	ppt *or* $\downarrow$
partition function	Z	preparation variability	ξ_P
pascal	Pa	pressure	P *or* p
pascal second	Pa s	primary wave	P-wave
peck	pk	prior distribution of	
percentage risk of		a parameter θ	$D_O(\theta)$
type I error	α	probability of the	
percentage risk of		event A	$P(A)$
type II error	β	product of a series	II
period	T	proportion	: :
phase	ph	proportion of a success in a	
phenyl	Ph	binomial population	θ
phot	ph	protium	^1H
photon	γ	proton	p
pint	pt	quantity of electric charge	
pi (mathematical constant)	π	or electricity; quantity of	
pion	π	heat; quantity of light	Q
planck constant	h	quart	qt
plus	$+$	rad	rd
plus or minus	$\pm$	radian	rad
poise	P	radiance	B
Poisson ratio	v *or* μ	radiant emissivity	J
pooled sample variance		radiant energy	Q
population coefficient of		radiant energy density	u
variation and skewness	γ	radiant exposure	H
population mean	μ	radiant flux	Φ
		radiant intensity	I

F8, *continued*

radiation absorbed dose	rad
radical	√
radio detection and ranging	radar
radio frequency	RF
radius	r *or* R
random fluctuation of "experimental error"	e
random access memory	RAM
Range(s) (legal land divison)	R(s).
rankine (used with degree symbol)	°R
ratio; is to (when solidus is used, the word "ratio" should follow, for example, Cu/Ni ratio)	: *or* /
reactance	X
read-only memory	ROM
réamur (used with degree symbol)	°R
refractive index at 20°C, sodium (*D*) line	n_D^{20}
relative cumulative frequency	r.c.f.
repeating decimal; bar covers part repeated	$1.\overline{3}$
resistance	R
resistivity	ρ
return beam vidicon	RBV
reversible reaction	⇌
revolutions per minute	r/min *or* rpm
revolutions per second	r/s *or* rps
Reynolds number	R
roentgen (used with degree symbol)	°R
roentgen equivalent, man or mammal	rem
roentgen equivalent, physical	rep

root	√
root mean square	rms
rubidium acid phthalate	RAP
Rydberg constant	R *or* Ry
Rydberg constant for infinite mass	R_∞
salinity (parts per thousand)	‰
sample coefficient of variation	$\dfrac{C}{w}$
sample mean	
sample standard deviation	s
sample variance	s^2
sampling variability	ξ_s
saturated calomel electrode	sce
secant	sec
secant, hyperbolic	sech
second (time)	s
second; double prime; inch	"
second-foot	s·ft
secondary wave	S–wave
section(s) (subdivision of Township and Range)	sec(s)
sensu lato	s.l.
sensu stricto	s.s.
shear velocity; S-wave velocity	υ_S
siemens	S
sievert	Sv
sine	sin
sine, hyperbolic	sinh
sine of the amplitude (an elliptic function)	sn
skewness of frequency distribution	sk
sodium line in spectrum	D
sodium-adsorption ratio	SAR
solid, as in AgCl(s)	(s)
solid angle	ω
solidus (also called virgule, slash)	/
soluble	sol

F8, *continued*

		sum	Σ
solution	soln	sum of squares	SS
sound navigation and		sum of squares of the	
ranging	sonar	replication totals	T_r^2
spacing of Bragg planes		sums of squares of the	
in a crystal	d	treatment totals	Tt^2
species (singular)	sp.	sum totals of observations	
(plural)	spp.	in a sample	T
specific gravity	sp gr	surface tension	γ *or* σ
specific heat	sp ht	tangent	tan
specific heat capacity	c	tangent, hyperbolic	tanh
specific volume	sp vol	temperature	temp
square	sq	temperature, in degrees	
square centimeter	cm^2	Celsius	t
square root	$\sqrt{}$	temperature, in kelvins;	
standard	std	absolute temperature;	
standard deviation	σ	thermodynamic	
standard error of		temperature	T
laboratory means	s_x	tesla	T
standard mean ocean		theoretical frequency of	
water	SMOW	observations	T
standard state	$^\circ$	thermogravimetric analysis	tga
standard state Gibbs		thickness	t *or* d
free energy	G°	thin-layer chromatography	tlc
standard temperature		thousand	k
and pressure	STP	thus	sic
standardized normal		time	t
distribution	s.n.d.	ton, metric (tonne)	t
station(s) (abbreviation used only		total (grand) of observations	
with numbers)	sta(s).	squared	G^2
steradian (solid angle)	sr	Townships(s) (legal	
stokes	St	land division)	T., Tps.
strain, normal, or linear	$\in$	trace	tr.
strain, shear	γ	trace of a matrix (math)	tr
stress, normal	σ	transformed observation	u
stress, shear	τ	transmittance	T
subgenus subgen.		triangle	Δ
subset of; is contained in	$\subset$	trigonometric functions,	
subspecies	subsp.	inverse	arc
"Suggestions to		tritium	T *or* ^{3}H
Authors"	STA	tritium unit	TU

F8, *continued*

triton	t
true mean	t.m.
ultraviolet	uv
undetermined	undet.
unified atomic mass unit	u
union or logical sum	$\cup$
unit-cell edges	$a_0, b_0, and\ c_0$
United States (abbreviation used as adjective only)	U.S. *or* US
United States Geological Survey	USGS
U.S. Government Printing Office	GPO
United States National Museum	USNM
universal time	u.t.
Universal Time, Coordinated	UTC
Universal Transverse Mercator	UTM
vacuum	vac
vapor pressure	vp
variance, statistic to estimate the variance of lognormally distributed observations	V^2
variance of quantity q	σ_q^2
variance of lognormal distribution	β^2
variance of negative binomial distribution	k
variation operator	δ
variations; varies as	$\propto$
variety	var.
velocity	v *or* u
velocity of light (in vacuo)	c
velocity, P-wave	v_P
velocity, S-wave	v_S
versed sine	vers
versus (standard usage)	vs.

vertical angle elevation bench mark	VABM
vinculum (above letter; for example $\overline{w}$)	$\overline{}$
viscosity, dynamic	η
viscosity, kinematic	ν
volt	V
voltampere	VA
volume, chemical and physical use	V
volume strain, bulk strain	θ
watt	W
watthour	Wh
wavelength	λ
wavenumber	σ *or* ν
weber	Wb
weight	wt
weight per volume	w/v
weight per weight	w/w
yard	yd
year	yr
yields	$\rightarrow$
Young's modulus of elasticity	E

F9 Glossary of Environmental Science Terms

abiotic – Non-living factors within the environment.

absorption – The ability to take in and hold liquid.

acid rain – Precipitation in which the current PH level is lower than its naturally occurring level, usually brought on by the release of pollution into the atmosphere.

adulterants – Any chemical substance that is artificially added to another substance.

adaptation – A feature in an organism that assists in its survival of its environment.

adsorption – When a gas or a liquid accumulates on the surface of another surface.

aerobic organism – An organism that breathes oxygen to survive.

aerosols – Tiny solid and liquid particles that are suspended in the atmosphere.

afforestation – Planting new forests on lands that have not been recently forested.

agent – An entity, which can be biological, chemical, or physical in nature, which is harmful to an organism.

air – The gaseous part of the atmosphere that is breathed by organisms.

air pollution – The modification of the natural characteristics of the atmosphere by a chemical, particulate matter, or biological agent.

algal bloom – A sudden increase in the algae population of a aquatic area.

anaerobic – Not requiring air or oxygen; used in reference to decomposition processes that occur in the absence of oxygen.

aquaculture – The growing and or raising of aquatic organisms in a controlled environment.

F9, *continued*

aquifer – A bed or layer yielding water for wells and springs etc.; an underground geological formation capable of receiving, storing and transmitting large quantities of water. Aquifer types include: confined (sealed and possibly containing "fossil" water); unconfined (capable of receiving inflow); and Artesian (an aquifer in which the hydraulic pressure will cause the water to rise above the upper confining layer).

atmosphere – The layer of gases that surround a planetary mass.

bacteria – Microscopic organisms found in every environment on earth.

barrier island – A long narrow island running parallel to a mainland, often protecting it from storms and large waves.

biodegradable – The ability of a component to be broken down into the environment.

biological oxygen demand (BOD) – The measurement of pollution levels in water.

biomass – The living material in a given environmental area.

biome – A community of plants and animals that are commonly found in similar environmental regions.

biosphere – The physical area of the earth that is able to sustain life.

blackwater – Water contaminated with a high concentration of pollutions.

bog – A soggy wetland that is composed mainly of moss and peat.

brackish – Water that is a mixture of fresh and salt water.

by-product – Waste material produced by an organism, such as feces.

carbon dioxide – A gas found in nature that is exhaled by organisms after oxygen is taken in.

carbon monoxide – An odorless poisonous gas formed by incomplete carbon combustion.

F9, *continued*

carcinogen – An agent that can cause cancer in organisms.

carnivore – An organism that primarily feeds on the flesh of other organisms to survive.

carrying capacity – The maximum level of inhabitants that a given area can support.

chlorinated hydrocarbons – A poisonous compound often used as an insecticide that can build up in the tissues of organisms. These may linger within its enviornment and travel up through the food chain.

climate – The long-term weather pattern of a geological area.

climate change – A change in weather over time or in a region; usually relating to changes in temperature, wind patterns, and rainfall; although it may be natural or anthropogenic, it generally refers to anthropogenic global warming.

cloud forest – A highly elevated tropical forest covered with a layer of fog throughout the year. Also known as a fog forest.

community – A group of organisms living together in the same ecological region.

competition – The health and survival of an organism being threatened by the presence of another organism.

confined aquifer – Aquifers that have the water table above their upper boundary and are typically found below unconfined aquifers.

contaminant – A substance that causes harm to the environment and the inhabiting life forms.

cyclone – Intense low pressure weather systems; mid-latitude cyclones are atmospheric circulations that rotate clockwise in the Southern Hemisphere and anti-clockwise in the Northern Hemisphere and are generally associated with stronger winds, unsettled conditions, cloudiness and rainfall. Tropical cyclones (which are called hurricanes in the Northern Hemisphere) cause storm surges in coastal areas.

F9, *continued*

dead zone – An area in the ocean that is unable to sustain life due to depleted oxygen levels.

dichloro-diphenyl-trichloroethane (DDT) – A synthetic chemical used as a pesticide that can cause great harm to organisms, both to those that the substance is directly used on and those that eat the poisoned organisms.

decomposition – The breakdown of organic material.

ecological succession – A change in the ecosystem of an environment brought on by the gradual replacement of one community of organisms to another.

ecosystem – A community that consists of living organisms and non-living materials that interact together.

ecotone – A transition zone between two communities of organisms.

El Niño – The unusual warming of the water in the Pacific Ocean, occurring every four to twelve years, brought on by cold water not rising to the surface. This occurrence forms unusual weather patterns throughout the world.

environment – The external factors that surround an organism.

endangered species – Species that are threatened with becoming extinct.

estuary – A semi-enclosed coastal body of water with one or more rivers or streams flowing into it, and with a free connection to the open sea.

eutrophication – The deoxidization of a body of water brought on by excessive plant growth and decay in the water.

extinct – When an organism dies out and no longer exists.

fire ecology – The study of the ecological effects of fire on the ecosystem.

F9, *continued*

food chain – A sequence in which lesser organisms are eaten by more advanced ones. For example, plankton are eaten by fish, and fish are eaten by bears.

food web – A series of interconnecting food chains within an eco-system.

forest management – The man-made changing of a forest, often through plant removal, the introduction of new plant life, controlled fires, and roadway construction.

fossil fuels – The fossilized remains of plants and animals found within the Earth's crust that can be used as a form of energy. Examples are coal, petroleum, and natural gas.

Gaia hypothesis – The theory of viewing all living and unliving parts of the Earth as a single organism.

Gause's principle – A theory in which two species that compete for the same resources cannot coexist together.

global warming – The observable increase in global temperatures considered to be caused by the human induced enhanced greenhouse effect trapping the sun's heat in the Earth's atmosphere.

greenhouse effect – The insulating effect of atmospheric greenhouse gases (e.g., water vapor, carbon dioxide, methane, etc.) that keeps the Earth's temperature about 60 °F (16 °C) warmer than it would be otherwise.

habitat – A place in the environment that has ideal living conditions for specific organisms.

habitat degradation – The process of a habitat becoming unable to support the life forms, which currently inhabit it. Most times, this is a result of human interaction.

hazardous waste – A byproduct of manufacturing that causes harm to the environment.

F9, *continued*

herbivore – An organism that primarily feeds on plant life to survive.

hurricane – A powerful storm that originates in the ocean and often moves over landmasses.

humidity – The amount of moisture in the atmosphere.

hydrosphere – All the earth's water, including water found in seas, streams, lakes and other bodies, the soil, groundwater, and in the air.

insectivore – An organism whose main diet consists of eating insects.

integrated pest management (IPM) – A strategy of pest-control that strives to use natural remedies, such as introducing crop-friendly insects to kill crop-killing insects, instead of pesticides.

interspecific competition – Competition for natural resources in an ecosystem between different species.

intraspecific competition – Competition for natural resources in an ecosystem between members of the species.

jet streams – Narrow fast moving currents of air.

keystone species – A species that shapes and impacts its environment based on its function rather than its abundance.

land use – The manipulation of natural land for human use.

life span – The longest age that a species is expected to live.

limiting factors – Physical, chemical, and environmental factors that hinder the growth of an individual organism or the size of a population.

lightning – A discharge of electricity within the atmosphere.

marsh – Low-lying, waterlogged land that is often a meeting point between dry land and water.

mesosphere – The layer of the Earth's atmosphere between 50 to 90 kilometers above the Earth's surface.

F9, *continued*

methane – A naturally occurring gas formed from the decomposition of organic material.

mitigation – The repairing of a damaged ecosystem, formed from natural means or through human contamination.

monsoon – A seasonal reversal of wind direction, bringing heavy rain and increased wind activity.

mutation – An abnormal change in the structure of a creature, brought on by environmental contamination or alterations in its DNA.

mutualism – A symbiotic relationship between two organisms in which both benefit from the other.

natural resource – A substance in nature that is useful to humans.

natural selection – The theory that organisms with dominant traits will thrive and weaker ones will die out.

nitrogen-fixing bacteria – Bacteria that changes nitrogen in the air into forms that are beneficiary to plant life.

niche – The function of an organism or group of organisms in an ecological community.

non-renewable resources – Materials that will not renew themselves within a time frame needed for further usage, such as coal and oil.

nutrient cycle – The intake, use, and outtake of nutrients in plants and the environment.

omnivore – An organism whose diet consists of eating both plants and animal flesh.

oligotrophic – An environmental area that is naturally unfit to sustain life.

overshoot – When the population of a community of organisms exceeds the carrying capacity of its environment.

F9, *continued*

ozone – A bluish, poisonous, oxygenic gas formed from the reaction of ultraviolet light on oxygen.

ozone layer – A layer of ozone surrounding the earth that protects its inhabitants from overexposure to ultraviolet light.

parasite – An organism that lives on, or in another organism, and obtains nutrition from feeding off the host organism.

pesticide – Any substance, both naturally occurring and artificially created, that is used in the extermination of insects.

photosynthesis – The process in which light is converted into energy in plant life.

pioneer species – Any species that first inhabits a previously uninhabited area. Often they are grass, moss, algae, and microbes.

poachers – Individuals who illegally hunt plants and animals, often ones that are endangered or protected.

pollutant – Referring to anything that can pollute the environment.

pollution – Occurrences natural (a volcano blast) and artificial (smog from a smoke stack) that cause undesirable change in the environment.

precipitation – Moisture in the atmosphere that falls to the surface of the Earth. This can occur as fog, frost, hail, rain, sleet, or snow.

prescribed burning – The usage of fire to clean out dead and decaying plant life in a forest, also known as controlled burning.

rain – Precipitation that falls to the earth in the form of water.

rainforest – A dense forest with high levels of humidity and daily occurrence of rainfall.

red tide – An outbreak of dinoflagellates that produces a red coloring to the ocean surface of tropical waters.

F9, *continued*

renewable resource – Any resource that will naturally replenish itself after it is used by other organisms.

salinity – The amount of salt content in water and soils.

secondary succession – A population of plant or animal life inhabiting and dominating an area after the previous dominating life forms were driven out, often as a result of a natural disaster.

silviculture – The study and practice in using trees for human benefit.

smog – Fog that had been polluted by smoke.

species – The classification of a group of organisms capable of interbreeding with each other.

species diversity – The measurement of the number of different species in a given area.

temperate – Moderate temperatures, weather, or climate; free from extremes; mean annual temperature between $0 - 20°C$.

temperature inversion – The increase of temperature in the air coinciding with the rise in atmosphere.

thermosphere – The utmost level of the atmosphere before it breaks into outerspace, located above the mesosphere.

thunder – A shock wave in the air, characterized by a crashing sound, brought on by the heating of air after the discharge of lightening.

tornado – A violent, funnel-shaped column of air brought on by severe thunderstorms.

trace metals – Metals of a microscopic level found in the cells of plants and animals.

weather – Natural, daily-changing conditions of the atmosphere, brought on by temperature, precipitation, humidity, air pressure, cloud cover, etc.

wetlands – Land that is naturally saturated with water.

F10 Environmental Science Journals and Their Abbreviations

Name of Journal	Abbreviation
Acta Biotechnologica	*Acta Biotechnol*
Accident Analysis & Prevention	*Accid Anal Prev*
Advanced Synthesis & Catalysis	*Adv Synth Catal*
Advances in Air Pollution Series	*Adv Air Pollut*
Advances in Atmospheric Sciences	*Adv Atmos Sci*
Advances in Environmental Research	*Adv Environ Res*
Advances in Water Resources	*Adv Water Resour*
Aerobiologia	*Aerobiologia*
African Journal of Aquatic Science	*Afr J Aquat Sci*
African Journal of Marine Science	*Afr J Mar Sci*
Agriculture and Environment	*Agr Environ*
Agricultural and Forest Meteorology	*Agric For Meteorol*
American Journal of Primatology	*Am J Primatol*
Antonie van Leeuwenhoek	*Antonie Leeuwenhoek*
Applied and Environmental Microbiology	*Appl Environ Microbiol*
Applied Herpetology	*Appl Herpetol*
Applied Microbiology and Biotechnology	*Appl Microbiol Biotechnol*
Aquatic Ecology	*Aquat Ecol*
Aquatic Ecosystem Health and Management	*Aquat Ecosys Health Manag*
Aquatic Sciences	*Aquat Sci*
Aquatic Toxicology	*Aquat Toxicol*
Asian Journal of Water Environment and Pollution	*Asian J Water Environ Pollut*
Atmospheric Environment	*Atmos Environ*
Atmospheric Chemistry and Physics	*Atmos Chem Phys (ACP)*
Atmospheric Research	*Atmos Res*
Atmospheric Science Letters	*Atmos Sci Lett*
Behavioral Ecology	*Behav Ecol*
Biochemical Systematics and Ecology	*Biochem Systemat Ecol*
Biodiversity & Conservation	*Biodiversity & Conservation*
Biological Conservation	*Biol Conservat*
Biological Control	*Biol Contr*

F10, *continued*

Biosystems Engineering	*Biosystems Eng*
Biotechnology and Bioengineering	*Biotechnol Bioeng*
Building and Environment	*Build Environ*
Bulletin of the American Meteorological Society, The	*B Am Meteorol Soc*
Bulletin de l'Institut Pasteur	*Bull Institut Pasteur*
Climatic Change	*Climatic Change*
Chemistry and Ecology	*Chem Ecol*
Chinese Journal of Atmospheric Sciences	*Chin J Atmos Sci*
City – Analysis of Urban Trends, Culture, Theory, Policy, Action	*City Anal Urban Trends Cult Theor Pol Action*
Climate Dynamics	*CD*
Conservation Genetics	*Conservat Genet*
Critical Reviews in Biotechnology	*Crit Rev Biotechnol*
EcoHealth	*EcoHealth*
Ecological Applications	*Ecol Appl*
Ecological Economics	*Ecol Econ*
Ecological Modelling	*Ecol Model*
Environment, Development and Sustainability	*Environ Dev Sustain*
Environmental Biology of Fishes	*Environ Biol Fish*
Environmental Conservation	*Environ Conservat*
Environmental Engineering and Policy	*Environ Eng Pol*
Environmental Engineering Science	*Environ Eng Sci*
Environmental History	*Environ Hist*
Environmental Modelling & Software	*Environ Model Software*
Environmental Policy and Law	*Environ Pol Law*
Environmental Quality Management	*Environ Qual Manag*
Environmental Science & Policy	*Environ Sci Pol*
Environmental Toxicology	*Environ Toxicol*
Environmental Toxicology and Water Quality	*Environ Toxicol Water Qual*
Environmental Fluid Mechanics	*Environ Fluid Mech*
Environmetrics	*Environmetrics*
Ethics, Place & Environment	*Ethics Place Environ*
European Environmental Law Review	*Eur Environ Law Rev*
European Journal of Lipid Science and Technology	*Eur J Lipid Sci Tech*

F10, *continued*

Experimental & Applied Acarology	*Exp Appl Acarol*
Fisheries Oceanography	*Fish Oceanogr*
Fisheries Research	*Fish Res*
Food and Chemical Toxicology	*Food Chem Toxicol*
Food Science and Technology Research	*Food Sci Tech Res*
Forest Pathology	*Forest Pathol*
Fresenius' Journal of Analytical Chemistry	*Fresen J Anal Chem*
Global Change & Human Health	*Global Change Hum Health*
Global Change, Peace & Security	*Global Change Peace Secur*
Green Chemistry	*Green Chem*
Human Ecology	*Hum Ecol*
Indian Journal of Biotechnology	*Indian J Biotechnol*
Innovative Food Science & Emerging Technologies	*Innovat Food Sci Emerg Tech*
Instrumentation Science & Technology	*Instrum Sci Tech*
International Journal of Environment and Pollution	*Int J Environ Pollut*
International Journal of Environmental Science and Technology	*Int J Environ Sci Tech*
International Journal of Environmental Studies	*Int J Environ Stud*
International Journal of Green Energy	*Int J Green Energ*
International Journal of Heat and Fluid Flow	*Int J Heat Fluid Flow*
International Journal of Marine and Coastal Law, The	*Int J Mar Coast Law*
International Journal of Mining, Reclamation and Environment	*Int J Min Reclamat Environ*
International Journal of Oceans and Oceanography (IJOO)	*Int J Oceans Oceanogr*
International Journal of Thermal Sciences	*Int J Therm Sci*
International Journal of Tropical Insect Science	*Int J Trop Insect Sci*
Irrigation and Drainage	*Irrigat Drain*
Journal for Nature Conservation	*J Nat Conservat*
Journal of Agricultural and Environmental Ethics	*J Agr Environ Ethics*
Journal of Applied Sciences & Environmental Management	*J Appl Sci Environ Manag*
Journal of Applied Meteorology and Climatology	*J Appl Meteorol (JAM)*
Journal of the Atmospheric Sciences	*J Atmos Sci (JAS)*
Journal of Biochemical and Molecular Toxicology	*J Biochem Mol Toxicol*

F10, *continued*

Journal of Chemical Technology & Biotechnology	*J Chem Tech Biotechnol*
Journal of Climate, The	*J Climate (JCLI)*
Journal of Environmental Planning and Management	*J Environ Plann Manag*
Journal of Environmental Policy and Planning	*J Environ Pol Plann*
Journal of Environmental Quality	*J Environ Qual*
Journal of Environmental Radioactivity	*J Environ Radioact*
Journal of Environmental Sciences	*J Environ Sci*
Journal of Hydrology	*J Hydrol*
Journal of Industrial Ecology	*J Ind Ecol*
Journal of Industrial Microbiology and Biotechnology	*J Ind Microbiol Biotechnol*
Journal of International Wildlife Law and Policy	*J Int Wildl Law Pol*
Journal of Fish Diseases	*J Fish Dis*
Journal of Marine Science and Technology	*J Mar Sci Tech*
Journal of Materials Processing Technology	*J Mater Process Tech*
Journal of Microbiological Methods	*J Microbiol Meth*
Journal of Natural History	*J Nat Hist*
Journal of Oceanography	*J Oceanogr*
Journal of Physical Oceanography	*J Phys Oceanogr (JPO)*
Journal of Plankton Research	*J Plankton Res*
Journal of Polymers and the Environment	*J Polymer Environ*
Journal of Water Resources Planning and Management	*J Water Resour Plann Manag*
Land Use Policy	*Land Use Pol*
Land Degradation & Development	*Land Degrad Dev*
Landscape and Ecological Engineering	*Landsc Ecol Eng*
Landscape Research	*Landsc Res*
Leisure Sciences	*Leisure Sci*
Limnology & Oceanography	*Limnol Oceanogr*
Low Temperature Science	*Low Temp Sci*
Marine Ecology Progress Series	*Mar Ecol Progr*
Marine Policy	*Mar Pol*
Meteorologische Zeitschrift	*Meteorol Z*
Microbial Ecology in Health and Disease	*Microb Ecol Health Dis*
Microbiological Research	*Microbiol Res*

F10, *continued*

Molecular Breeding	*Mol Breed*
Molecular Ecology	*Mol Ecol*
Natural Hazards Review	*Nat Hazards Rev*
Nature Biotechnology	*Nat Biotechnol*
New Review of Academic Librarianship	*New Rev Acad Librarian*
Nigerian Journal of Health and Biomedical Sciences	*Niger J Health Biomed Sci*
Ocean & Coastal Management	*Ocean Coast Manag*
Oceanographic Literature Review	*Oceanogr Lit Rev*
Oil and Chemical Pollution	*Oil Chem Pollut*
Opolis	*Opolis*
Optimal Control Applications and Methods	*Optim Contr Appl Meth*
Pacifica Review	*Pacifica Review*
Paleoceanography	*Paleoceanography*
Papers in Meteorology and Geophysics	*Paper Meteorol Geophys*
Physics and Chemistry of the Earth - Part B – Hydrology, Oceans and Atmosphere	*Phys Chem Earth B Hydrol Oceans Atmos*
Physics of Wave Phenomena	*Phys Wave Phenom*
Planning Theory & Practice	*Plann Theor Pract*
Polar Biology	*Polar Biol*
Polymers for Advanced Technologies	*Polymer Adv Tech*
Population and Environment	*Popul Environ*
Process Control and Quality	*Process Contr Qual*
Progress in Energy and Combustion Science	*Progr Energ Combust Sci*
Progress in Industrial Ecology	*Progr Ind Ecol*
Quarterly Journal of the Royal Meteorological Society, The	*Q J Roy Meteorol Soc*
Regional Environmental Change	*Reg Environ Change*
Reviews in Environmental Science and Biotechnology	*Rev Environ Sci Biotechnol*
Russian Journal of Bioorganic Chemistry	*Russ J Bioorg Chem*
Russian Journal of Ecology	*Russ J Ecol*
Science of the Total Environment, The	*Sci Total Environ*
Separation and Purification Technology	*Separ Purif Tech*
Separations Technology	*Sep Tech*
Simulation Practice and Theory	*Simulat Pract Theor*

F10, *continued*

Stochastic Environmental Research and Risk Assessment	*Stoch Environ Res Risk Assess*
Tellus - Series A – Dynamic Meteorology and Oceanography	*Tellus Dyn Meteorol Oceanogr*
Theoretical and Applied Climatology	*Theor Appl Climatol*
Trends in Ecology & Evolution	*Trends Ecol Evol*
Urban Ecosystems	*Urban Ecosyst*
Water, Air, and Soil Pollution	*Water Air Soil Pollut*
Water Resources Research (WRR)	*Water Resour Res*
Weather and Forecasting	*Weather Forecast*
Wind Energy	*Wind Energ*
World Development	*World Dev*
Water Quality and Ecosystem Modeling	*Water Qual Ecosys Model*
World Journal of Microbiology and Biotechnology	*World J Microbiol Biotechnol*
World Water and Environmental Engineering	*World Water Environ Eng*

Appendix G. Table and Conventions in Life Science

Contents

G1 Chromosome Abbreviations

Abbreviation	Meaning
AI	First meiotic anaphase
AII	Second meiotic anaphase
Ace	Acentric fragment
Add	Additional material of unkown origin
b	Break
V	Constitutional anomaly
cen	Centromere
chi	Chimera
chr	Chromosome
cht	Chromatid
cp	Composite karyotype
cx	Complex chromatid interchanges
del	Deletion
de novo	Chromosome abnormality not inheritd
der	Derivative
dia	diakinesis
dic	Dicentric
dip	Diplotene
dir	Direct
dis	Distal
dit	Dictyotene
dmin	Double minute
dup	Duplication
e	Exhange
end	Endoreplication
fem	Female
fis	Fission
fra	Fragile site
g	Gap
h	Heterochromatin
hsr	Homogeneously staining region
i	Isochromosome
idem	Stemline karyotype in subclones
ider	Isoderivaive
idic	Isodicentric
inc	Incomplete karyotype
ins	Insertion

G1, *continued*

Abbreviation	Meaning
inv	Inversion or inverted
mar	Marker chromosome
mat	Maternal origin
med	Medial
min	Minute acentric fragment
ml	Mainline
n	Modal number
mos	Mosaic
oom	Oogonial metaphase
P	Short arm
PI	Firs meiotic prophase
c	Pachyne
pcc	Paternal orgin
pcd	Premature chromosome condensation
prx	Premature centromere division
psu	Proximal
pvz	Pseudo-
q	Pulverization
qdp	Ong arm
qr	Quadruplicaiton
r	Quadriradial
rcp	Rign form
rea	Reciprocal
rec	Rearrangement
rob	Recombinant chromosome
roman numerals	Robertsonian translocation
I	Univalent
II	Bivalent
III	Trivalent

G1, *continued*

Abbreviation	Meaning
IV	Quadrivalent
s	Satellite
sce	Sister chromatid exchange
sct	Secondary constricion
sdl	Sideline
sl	stemline
spm	Spermatogonial metaphase
stk	Satellite stalk
t	Translocation
tan	Tandem
tas	Telomeric associaton
tel	Telomere
ter	Terminal
tr	Triradial
trc	Tricentric
trp	triplication
upd	Uniparental disomy
v	variant or variable region
sma	Chiasmat(ta)
zyg	Zygotene
:	Break
::	Break and reunion
;	Separates chromosomes and chromosome bands in structural rearrangements involving 2 or more chromosomes
→	Fro m-to

G1, *continued*

Abbreviation	Meaning
+	Gain
-	Loss
~	Intervals in a chromosome segment where breakpoint is uncertain
<>	Brackets for ploidy
[ ]	Brackets for number of cells
=	Number of chiasmata
X (multiplication sign)	Multiple copies
?	Questionable identification
/	Separates clones

From AMA Manual of Style, 9th edition, pages 397-399

G2 Ranks In Plant Taxonomy

Rank	Abbreviation	Meaning
biovar	bv.	biological variety
combination nova	comb. nov.	new combination
corrigendum	corrig.	to be corrected
cultivar	cv.	cultivated variety
emendavit	emend.	he or she corrected
familia nova	fam. nov.	new family
forma [c]	f. [c]	form
forma specialis	f.sp.	special form
genus novum	gen.nov.	new genus
genus approbatum	nom. approb.	approved name
nomen conservadum	nom. nov.	name to be conserved
nomen novum	nom. rej.	new name
noven nudum	nom. nud.	bare name
nomen rejiciendum	nom. rej.	name to be rejected
nomen revictum	nom rev.	revived name

G2, *continued*

Abbreviation	Rank	Meaning
pv.	pathovar	pathogenic variety
sp.	specis [singular]	species
sp. nov.	species novum	new species
spp.	species [plural]	species
subsp.	subspecies	subspecies
var.[c]	varietas [c]	variety
var. nov. [c]	Varietas novum [c]	new variety

G3 Common Viral Abbreviations

Abbreviation	Organism
AMV	avian myeloblastosis virus
CMV	cucumber mosaic virus, cytomegalovirus
CPV	*cytoplasmic polyhedrosis virus*
FMDV	*foot-and-mouth disease virus*
GV	*granulosis virus*
HBV	*human hepatitis B virus*
HDV	*human hepatitis D virus*
HIV-1	*human immunodeficiency virus 1*
HIV-2	*human immunodeficiency virus 2*
HTLV-1 (or I)	*human T-cell leukemia virus*
HTLV-2 (or II)	*human T-cell leukemia virus*
NPV	*nuclear polydedrosis virus*
SIV	*simian immunodeficiency virus*

G4 Taxonomic Name Endings

Rank	Plants	Algae	Fungi	Animals	Bacteria
Division/ Phylum	-phyta		-mycota		
Subdivision/ Subphylum	-phytina		-mycotina		
Class	-opsida	-phyceae	-mycetes		-ia
Subclass	-idae	-phycidae	-mycetidae		-idae
Superorder	-anae				
Order	-ales				-ales
Suborder	-ineae				-ineae
Infraorder	-aria				
Superfamily	-acea			-oidea	
Epifamily				-oidae	
Family	-aceae			-idae	-aceae
Subfamily	-oideae			-inae	-oideae
Infrafamily				-odd	
Tribe	-eae			-ini	-eae
Subtribe	-inae			-ina	-inae
Infratribe				-ad	

G5 Life Science Glossary

abdomen – In vertebrates, the area of the body between the diaphragm and the pelvis where the stomach, intestines, kidneys and liver are located. In invertebrates such as insects, it is the entire rear portion of the animal.

abiogenesis – The theory that life can spontaneously form from non-living material.

Adenosine Triphosphate (ATP) – A chemical component that transmits energy from one cell to another.

adrenaline – A hormone secreted into the body during times of extreme stress which causes the organism's body to avoid the effects of fatigue and increase chances of survival.

aerobic organism – An organism that depends on oxygen to survive.

algology – The study of algae, also known as Phycology.

amino acids – Chemical compounds which produce proteins.

anaerobic organism – An organism that survives due to the absence of oxygen.

anther – The part of the stamen section of a plant that contains pollen.

asexual – An organism that has no determined sex.

asexual reproduction – Reproduction that occurs with the involvement of only one parent, often as a result of budding or fission.

atavism – A trait showing up in a present organism that was once common in an earlier ancestor.

bark – The rough covering of the trunk, branches, roots and stems of certain plants, most prominently trees.

bigeneric – A hybrid plant formed from the crossbreeding of plants from different genera.

binary fission – The splitting of a cell into two equal cells, seen as a form of asexual reproduction.

biogenesis – The theory that life can only form from preexisting life.

biosphere – The sum total of all life on the planet.

biosynthesis – The process in which chemical compounds are formed by a series of reactions in enzymes.

biota – The total collection of organisms that make up a particular geological place or time period.

blade – The broad flat section of a leaf or grass.

bog plant – A plant that lives in continually wet soil and is not completely submerged in water.

bolting – The premature growing of flowers and fruit on a plant.

botany – The branch of science devoted to the studying of plant life.

bud – A underdeveloped portion of a plant, often protected by outside scales, that will mature and blossom into a flower or a leaf.

bulb – A flesh like plant bud, often formed underground, in which a new plant will blossom from.

calyx – The collected sepal of a flower.

cellulose – A chemical compound that forms the skeletal structure of a plant.

chlorophyll – A green pigment found in plant cells that converts the energy from light to turn carbon dioxide to carbohydrates.

chloroplasts – The organelles in plant cells that contain Chlorophyll.

chromosome – The threadlike structures in cells that carry hereditary information of an organism.

cilia – Hairlike structures that stick out of a cell and help move fluid over the cell.

G5, *continued*

clone – A cell or organism that is created asexually and is identical to its host.

commensalism – A symbiotic relationship in which one organism gains a benefit and the other is neither helped nor harmed.

compost – Decomposed organic material that is beneficial to soil.

cross-pollination – The transferring of pollen from one plant to another.

cub – Relating to the young of select carnivorous animals, such as big cats and wolves.

cuticle – A tough flexible covering that provides protection to parts of an organism. In humans it is the layer of skin appearing where the nail grows from the finger.

cytology – A branch of biology dealing with the study of cells.

cytoplasm – Part of a cell outside the nucleus but within the membrane.

deciduous – Referring to the losing of a part once its use is no longer needed. Examples of this are leaves falling from a tree in autumn and children losing their first set of teeth.

deoxyribonucleic Acid (DNA) – Double strained nucleic acid that carries the genetic information of an individual. Found in all organisms with the exception of viruses.

diffusion – The movement of material from one cell to the next.

dormancy – A plant ceasing growth and activity.

dwarf – An organism that is smaller then it normally would be.

embryo – An organism that is in the early stages of development.

endogenous – Relating to influencing factors of an organism that originate and occur within the organism.

endorphins – Pleasure causing hormones that are released into the blood-stream at times when an organism's body is in trauma as a way to block out pain.

endoskeleton – A skeleton system that is internally within the body, found in vertebrate organisms.

enzymes – Protein catalysts that speed up specific chemical reactions within an organism.

epidermis – The outer layer of a plant or animal that offers it protection.

epiphytic – A plant that attaches itself to another plant for structural support and not for nutritional support.

epizoic – An organism that attaches itself to the surface of another organism.

eukaryotic – A cell or organism that has a distinct nucleus.

exogenous – Relating to influencing factors of an organism that originate and occur outside of the organism.

exoskeleton – A hard protective covering that supports the body of many forms of invertebrates, such as clams, arachnids and insects.

exotic – A plant or animal that is not native to the area it is found in.

extinction – A species or group fading out of existence. Extinction is both a natural occurrence and one that is brought on by artificial factors, such as over hunting and deforestation.

extracellular – Relating to activity and material that is outside of a cell.

fauna – A group of animals found in a specific time and region.

fermentation – The process in which sugars and carbohydrates are broken down and used as energy without the involvement of oxygen.

fertile – Describing a plant or animal that had the ability to produce life.

G5, *continued*

filament – The stalk of a plant that supports the anther.

flagella – Whiplike structures, longer and less numerous then Cilia, that stick out of a cell and help move the cell through fluid.

fight-or-flight Response – A response to potentially harmful situations that is hardwired into a organism, resulting in it either confronting an adversary or fleeing from it.

flora – The group of plants found in a specific time and region.

fruit – The ripened ovaries of a seed bearing plant.

gene – A chromosomal unit that carry information of specific hereditary traits. A person with red hair would carry genes that would pass this trait onto their offspring.

germination – The beginning stage in a plant seed's development.

glucose – A monosaccharides which acts as a major source of fuel within the body of many organisms, including humans.

glycolysis – The first stage of cellular respiration in which glucose is broken down into pyruvate acid.

gymnosperm – A group of plants whose seeds are exposed to the environment as opposed to being enclosed within fruit.

hand pollination – the pollination of plants through unnatural means.

homeostasis – The process of an organism maintaining a constant and stable internal environment by regulating itself to any changes in its environment.

humus – A dark shapeless matter formed from the decomposition of plants and animals.

hybrid – An organism that is a cross between two different parents.

hydrolysis – The decomposition of a chemical compound through the interaction of water.

G5, *continued*

inbreeding – The breeding of organisms that are directly related to each other, such as siblings having children.

infant – Offspring in the very early stages of life. Most often used to describe human children in the first year of life.

inflorescence – A cluster of flowers that are often characteristic to a specific plant.

instars – The state of an arthropod between molts.

instinct – A behavior or mannerism that is genetically bred into an organism and does not have to be taught.

invertebrates – An organism that does not have a spinal column or backbone.

kernel – The seed of a cereal or grain plant.

lactate (1) – The salt of a lactic acid.

lactate (2) – The ability for a organism to produce milk, which is used to feed their young.

larva – The early stage of a newly hatched animal that has yet to go through metamorphosis.

leaf – A flattened structure growing on the step of a plant in which photosynthesis takes place in.

life cycle – The functional and morphological stages through which an organism passes through.

limb – An appendage of a organism, such as an arm, leg, or wing.

meiosis – The division of cells in which the number of chromosomes is reduced to half of the original number.

metabolism – The measurement in which fuel (food) is broken down and converted into energy in a body.

G5, *continued*

metamorphosis – The complete change of an organism changing from one form to another, such as caterpillar becoming a butterfly.

microbe – A microorganism that typically causes disease.

microorganism – An organism that is microscopic or submicroscopic.

mitosis – The division of cells in which the number of chromosomes remains the same in both and a nucleus is present in both cells. Also known as Nuclear Division.

monosaccharides – Simple carbohydrates that act as the building blocks of complex carbohydrates and cannot be broken down into simpler sugars through hydrolysis.

molt – When the outer covering of an organism is shed and a new covering grows into its place.

monocarpic – A plant that produces flowers or fruit only once and then dies.

mulch – Organic covering that protects plants from weeds and losing water.

mutation – An abnormal change in the structure of a creature, brought on by environmental contamination or alterations in the DNA of a creature.

mutualism – A symbiotic relationship between two organisms in which both benefit from the other.

nucleoplasm – The fluid within a cell that surround the nucleus.

nucleus – The structure within a cell that acts as its "brain."

nymph – An insect that undergoes incomplete metamorphosis.

organelles – A part of a cell that causes a certain function to take place.

organism – An individual form of life.

G5, *continued*

osmosis – The movement of fluid inside and through a semipermeable membrane. The fluid will move from a more concentrated side to a less concentrated side until equilibrium is established.

ovary – The reproductive organ that produces eggs in organisms.

ovules – The cells in a ovary of a plant that develop into seeds after fertilization.

parasitism – A relationship between two organisms where one is harmed through the benefit of the other, such a tick on a deer.

peat – Partially decomposed vegetative material found in low temperate waterlogged areas.

photosynthesis – The process in plants in which sunlight is transformed into energy.

phycology – see Algology.

phyletic Evolution – The evolution of a particular group or species.

pistil – The reproductive organ of a flower.

plankton – Microscopic single celled organisms, both plant (phytoplankton) and animal (zooplankton) that float in large groups on the surfaces of oceans and freshwater.

pollen – Miniature grains, which are the male fertilization factor of a flowering plant.

pollen grain – The individual grains of pollen.

pollination – The transfer of pollen from one plant to another.

polycarpic – A plant that produces flowers and fruit over many seasons.

polyp – An organism that becomes stationary by attaching itself to a object.

G5, *continued*

polysaccharide – A class of complex carbohydrates characterized by a large number of monosaccharides.

prokaryotes – An organism or cell that lacks a distinctive nucleus.

protoplasm – An older term for the material that is found within a cell.

queen – A fertile member of a social insect colony who's job is to continually lay eggs.

reagent – A chemical substance that measures, detects, and produces other substances.

regeneration – The ability of a organism to regrow a bodily part it has lost, such as a spider regrowing a extra limb.

rhizoid – A protrusion in certain plants that act similarly to roots, found in fungi and mosses.

Root – The organ that anchors a plant to the ground and absorbs water and nutrients.

runner – A stem of a plant that lies along the ground and is capable of forming a new plant.

seed – An ovule that is fertilized and contains a plant embryo.

seedling – An infant plant that had formed from a seed.

self-Pollination – When a plant is fertilized from its own pollen as opposed from the pollen of other plants.

sepal – The outer leaves of a flower that protects it from external dangers.

sessile (1) – Referring to leaves and or flowers of a plant that grow directly from the stem instead of from the stalk.

sessile (2) – Animals that are not able to move by themselves and become attached to objects.

G5, *continued*

shoot – The bud or sprout of a plant.

shrub – A short plant that doesn't have a definitive trunk.

spores – Dust sized materials that are the reproductive structure of lesser plants.

sprout – Referring to the growth of a young plant.

stamen – The reproductive organ of a plant that produces pollen.

stamina – The ability of a organism to withstand fatigue and disease.

stem – The structural part of a plant in which leaves and flowers grow from.

stigma – The female reproductive part of a plant in which pollen is deposited into.

stoma – Pores in the epidermis of a plant in which gases are exchanged through.

style – The section of a flowering pistil where the stigma and the ovary are housed.

succulent – A plant whose leaves or stems store water, such as a cactus.

symbiosis – When a group of organisms work together for the benefit of each party.

taproot – The primarily root of a plant that is usually longer then the surrounding roots.

tissue – A grouping of cells that share similar structures and functions. Muscle tissue is the same throughout the body.

transpiration – The process in which water escapes a plant through its pores.

tuber – An underground plant organ that is a source for both food storage and reproduction.

G5, *continued*

vertebrates – An organism that has a spinal column or backbone present.

worker – A member of a social insect colony that performs a select task, such as building the nest and collecting food.

yolk – The portion of a egg that provides nutrition for the developing young of egg laying organisms.

zoophyte – An invertebrate that outwardly resembles a plant, such as a sea sponge and coral.

zygote – An egg cell of a plant or animal that has been fertilized.

G6 Life Science Journals and Their Abbreviations

Abbreviations for journals should follow the US National Library of Medicine's *Index Medicus.* Journal names should be italicized.

Full Title	**Abbreviation**
Academic Medicine	Acad. Med.
Advances in Experimental Medicine & Biology	Adv. Exp. Med. Biol.
American Journal of Human Genetics, The	Am. J. Hum. Genet.
American Journal of the Medical Sciences, The	Am. J. Med. Sci.
American Journal Of Pathology	Am. J. Pathol.
American Journal of Science	Am. J. Sci.
American Naturalist, The	Am. Nat.
Annals of the New York Academy of Sciences	Ann. N.Y. Acad. Sci.
Annual Reports in Medicinal Chemistry	Annu. Rep. Med. Chem.
Annual Review of Cell Biology	Annu. Rev. Cell. Biol.
Annual Review of Materials Science	Annu. Rev. Mater. Sci.
Annual Review of Microbiology	Annu. Rev. Microbiol.
Annual Review of Neuroscience	Annu. Rev. Neurosci.
Antiviral Research	Antivir. Res.
Applied Microbiology	Appl. Microbiol.
Applied Microbiology and Biotechnology	Appl. Microbiol. Biotechnol.
Archives of Microbiology	Arch. Microbiol.
Archives of Virology	Arch. Virol.
Basic Life Sciences	Basic Life Sci.
Biochemistry and Cell Biology	Biochem. Cell Biol.
Biochemistry & Molecular Biology International	Biochem. Mol. Biol. Int.
Biochemical Pharmacology	Biochem. Pharmacol.
Biological Reviews	Biol.Rev.
Biology of Reproduction	Biol. Reprod.
Biomedical Letters	Biomed. Lett.
Biomedical Research on Trace Elements	Biomed. Res. Trace Elem.
Bioorganic & Medicinal Chemistry	Bioorg. Med. Chem.
Bioorganic & Medicinal Chemistry Letters	Bioorg. Med. Chem. Lett.
Bioscience Biotechnology & Biochemistry	Biosci. Biotechnol. Biochem.
Biotechniques	Biotechniques
Biotechnology	Biotechnology
Biotechnology Program	Biotechnol. Prog.

G6, *continued*

Biotechnology Techniques	Biotechnol. Tech.
Canadian Journal of Biochemistry	Can. J. Biochem.
Canadian Journal of Chemistry	Can. J. Chem.
Canadian Journal of Microbiology	Can. J. Microbiol.
Cell	Cell
Cell Calcium	Cell Calcium
Cell Growth & Differentiation	Cell Growth Differ.
Cellular Immunology	Cell Immunol.
Cellular & Molecular Biology (Oxford)	Cell Mol. Biol.
Cell Communication and Signaling	Cell Signalling
Cell and Tissue Research	Cell Tissue Res.
Cellulose Chemistry and Technology	Cellul. Chem. Technol.
Central European Journal of Biology	Cent. Eur. J. Biol.
Chemotherapy	Chemotherapy
Clinical Science	Clin. Sci.
Cold Spring Harbor Symposia on Quantitative Biology	Cold Spring Harbor Symp. Quant. Biol.
Current Biology	Curr. Biol.
Current Genetics	Curr. Genet.
Current Microbiology	Curr. Microbiol.
Current Opinion in Cell Biology	Curr. Opin. Cell. Biol.
Current Opinion in Structural Biology	Curr. Opin. Struct. Biol.
Developmental Biology	Dev. Biol.
Developmental Brain Research	Dev. Brain Res.
Digestive Diseases and Sciences	Dig. Dis. Sci.
DNA and Cell Biology	DNA Cell Biol.
Drug Development and Industrial Pharmacy	Drug Dev. Ind. Pharm.
Drug Metabolism Reviews	Drug Metab. Rev.
Drugs	Drugs
Drugs under Experimental and Clinical Research	Drugs Exp. Clin. Res.
Drugs of the Future	Drugs Future
Enzyme and Microbial Technology	Enzyme Microb. Technol.
European Journal of Biochemistry	Eur. J. Biochem.
European Journal of Cell Biology	Eur. J. Cell Biol.
European Journal of Clinical Chemistry and Clinical Biochemistry	Eur. J. Clin. Chem. Clin. Biochem.
European Journal of Medicinal Chemistry	Eur. J. Med. Chem.
European Journal of Pharmacology	Eur. J. Pharmacol.
Europhysics Letters	Europhys. Lett.
Experimental Cell Research	Exp. Cell Res.
Experimental Neurology	Exp. Neurol.
FEBS Letters	FEBS Lett.

G6, *continued*

FEMS Immunology and Medical Microbiology	FEMS Immunol. Med. Microbiol.
FEMS Microbiology Letters	FEMS Microbiol. Lett.
Genes	Genes
Genes & Development	Gene Dev.
Genetics	Genetics
Genome	Genome
Genomics	Genomics
General Pharmacology	Gen. Pharmacol.
Geriatrics	Geriatrics
Human Genetics	Hum. Genet.
Human Molecular Genetics	Hum. Mol. Genet.
Human Mutation	Hum. Mutat.
Immunogenetics	Immunogenetics
International Journal of Biological Sciences	Int. J. Biol. Sci.
International Journal of Experimental Pathology	Int. J. Exp. Pathol.
International Journal of Radiation Biology	Int. J. Radiat. Biol.
International Journal of Radiation Oncology, Biology, Physics	Int. J. Radiat. Oncol. Biol. Phys.
Journal of Antibiotics (English ed.)	J. Antibiot.
Journal of Antimicrobial Chemotherapy	J. Antimicrob. Chemother.
Journal of Applied Bacteriology	J. Appl. Bacteriol.
Journal of Bacteriology	J. Bacteriol.
Journal of Biological Chemistry	J. Biol. Chem.
Journal of Biotechnology	J. Biotechnol.
Journal of Cellular Biochemistry	J. Cell. Biochem.
The Journal of Cell Biology	J. Cell. Biol.
Journal of Cellular Physiology	J. Cell. Physiol.
Journal of Cell Science	J. Cell. Sci.
Journal of Clinical Endocrinology and Metabolism	J. Clin. Endocrinol. Metab.
Journal of Clinical Investigation	J. Clin. Invest.
Journal of Clinical Microbiology	J. Clin. Microbiol.
Journal of Clinical Pathology	J. Clin. Pathol.
The Journal of Clinical Pharmacology	J. Clin. Pharmacol.
Journal of Experimental Biology	J. Exp. Biol.
Journal of Experimental Medicine	J. Exp. Med.
Journal of General Microbiology	J. Gen. Microbiol.
Journal of General Physiology	J. Gen. Physiol.
Journal of General Virology	J. Gen. Virol.
Journal of Lipid Research	J. Lipid Res.
Journal of Membrane Biology	J. Membr. Biol.

G6, *continued*

Journal of Molecular Biology	J. Mol. Biol.
Journal of Molecular and Cellular Cardiology	J. Mol. Cell. Cardiol.
Journal of Molecular Structure	J. Mol. Struct.
Journal of the National Cancer Institute	J. Natl. Cancer Inst.
Journal of Neuroimmunology	J. Neuroimmunol.
Journal of Neurophysiology	J. Neurophysiol.
Journal of Neuroscience	J. Neurosci.
Journal of Neuroscience Research	J. Neurosci. Res.
Journal of Nuclear Medicine	J. Nucl. Med.
Journal of Pharmacokinetics and Biopharmaceutics	J. Pharmacokinet. Biopharm.
Journal of Pharmacology and Experimental Therapeutics	J. Pharmacol. Exp. Ther.
Journal of Pharmaceutical and Biomedical Analysis	J. Pharm. Biomed. Anal.
Journal of Pharmacy and Pharmacology	J. Pharm. Pharmacol.
Journal of Pharmaceutical Sciences	J. Pharm. Sci.
Journal of Protein Chemistry	J. Protein Chem.
Journal of Structural Biology	J. Struct. Biol.
Journal of Theoretical Biology	J. Theor. Biol.
Journal of Virology	J. Virol.
Journal of Virological Methods	J. Virol. Methods
Neuroscience and Behavioral Physiology	Neurosci. Behav. Physiol.
Neuroscience Letters	Neuroci. Lett.
Nucleic Acids Research	Nucleic Acids Res.
Nuclear Medicine and Biology	Nucl. Med. Biol.
Oncogene	Oncogene
Pediatric Research	Pediatr. Res .
Peptides	Peptides
PLoS Biology	PLoS Biol.
Proceedings of the American Association for Cancer Research	Proc. Am. Assoc. Cancer Res.
Proceedings of the National Academy of Sciences of the United States of America	Proc. Natl. Acad. Sci. U.S.A.
Proceedings of the Society for Experimental Biology and Medicine	Proc. Soc. Exp. Biol. Med.
Progress in Biotechnology	Prog. Biotechnol.
Progress in Lipid Research	Prog. in Lipid Res.
Progress in Nucleic Acid Research and Molecular Biology	Prog. Nucleic Acid Res. Mol. Biol.
Protein Engineering	Protein Eng.
Protein Science	Protein Sci.
Proteins: Structure, Function, and Bioinformatics	Proteins: Struct., Funct., Bioinf.

G6, *continued*

Proteins: Structure, Function, and Genetics	Proteins: Struct., Funct., Genet.
Research Communications in Chemical Pathology and Pharmacology	Res. Commun. Chem. Pathol. Pharmacol.
Research Communications in Molecular Pathology and Pharmacology	Res. Commun. Mol. Pathol. Pharmacol.
Scientific American	Sci. Am.
Science	Science
Society for Neuroscience Abstract	Soc. Neurosci. Abstr.
Steroids	Steroids
Structural Biology	Struct. Biol.
Synapse	Synapse
Theoretical and Applied Genetics	Theor. Appl. Genet.
Trends in Biochemical Sciences	Trends Biotechnol.
Trends in Biotechnology	Trends Biotechnol.
Trends in Genetics	Trends Genet.
Trends in Neuroscience	Trends Neurosci.
Trends in Pharmacological Sciences	Trends Pharmacol. Sci.
Virology	Virology
Western Journal of Medicine	West. J. Med.
Xenobiotica	Xenobiotica

Appendix H. Tables and Conventions in Medical Science

Contents

H1 Abbreviations of Academic Degrees, Certifications, and Honors

Abbreviation	Meaning
ART	accredited record technician
BPharm	bachelor of pharmacy
BS, BCh, BC, CB, or ChB	bachelor of surgery
BSN	bachelor of science in nursing
CHES	certified health education specialists
CIH	certified industrial hygienist
CNM	certified nurse midwife
CNMT	certified nuclear medicine technologist
CO	certified orthoptist
COMT	certified ophthalmic medical technologist
CPFT	certified pulmonary function technologist
CRNA	certified registered nurse anesthetist
CRTT	certified respiratory therapy technician
CTR	certified tumor registrar
DC	doctor of chiropractic
DCh or ChD	doctor of surgery
DDS	doctor of dental surgery
DHL	doctor of humane letters
DMD	doctor of dental medicine
DME	doctor of medical education
DMSc	doctor of medical science
DNE	doctor of nursing education
DNS or DNSc	doctor of nursing science
DO or OD	doctor of optometry
DO	doctor of osteopathy
DPH or DrPH	doctor of public health, doctor of public hygiene
DPharm	doctor of pharmacy
DPM	doctor of podiatric medicine
DSW	doctor of social work
DTM&H	diploma in tropical medicine and hygiene
DTPH	diploma in tropical pediatric hygiene
DVM, DMV, VMD	doctor of veterinary medicine

H1, *continued*

DVMS	doctor of veterinary medicine and surgery
DVS or DVSc	doctor of veterinary science
EdD	doctor of education
ELS	editor in the life sciences
EMT	emergency medical technician
EMT-P	emergency medical technician-paramedic
FCGP	fellow of the College of General Practitioners
FCPS	fellow of the College of Physicians and Surgeons
FFA	fellow of the faculty of Anaesthetists of the Royal College of Surgeons
FNP	family nurse practitioner
FRACP	fellow of the Royal Australian College of Physicians
FRCA	fellow of the Royal College of Anesthesia
FRCGP	fellow of the Royal College of General Practitioners
FRCOG	fellow of the royal College of Obstetricians and Gynaecologists
FRCP	fellow of the Royal College of Physicians
FRCPath	fellow of the Royal College of Pathologists
FRCPC	fellow of the Royal College of Physicians of Canada
FRCP(Glasg)	fellow of the Royal College of Physicians and Surgeons of Glasgow qua Physician
FRCP or FRCP(Edin)	fellow of the Royal College of Physicians of Edinburgh
FRCPI or FRCO(Ire)	fellow of the Royal College of Physicians of Ireland
FRCR	fellow of the Royal College of Radiologists
FRCS	fellow of the Royal College of Surgeons
FRCSC	fellow of the Royal College of Surgeons of Canada

H1, *continued*

FRCSE or FRCS(Edin)	fellow of the Royal College of Surgeons of Edinburgh
FRCS(Glasg)	fellow of the Royal College of Physicians and Surgeons of Glasgow qua Surgeon
FRCSI or FRCS(Ire)	fellow of the Royal College of Surgeons of Ireland
FRCVS	fellow of the Royal College of Veterinary Surgeons
FRS	fellow of the Royal Society
GNP	gerontologic or geriatric nurse practitioner
JD	doctor of jurisprudence
LLB	bachelor of laws
LLM	master of laws
LPN	licensed practical nurse
LVN	licensed visiting nurse; licensed vocational nurse
M(ASCP)	registered technologist in microbiology (American Society of Clinical Pathologists)
MA or AM	master of arts
MB or BM	bachelor of medicine
MBA	master of business administration
MBBS or MB, BS	bachelor of medicine, bachelor of surgery
MD or DM	doctor of medicine
Med	master of education
MFA	master of fine arts
MHA	master of hospital administration
MLS	master of library science
MMM	master of medical management
MN	master of nursing
MPA	master of public administration
MPH	master of public health
MPharm	master of pharmacy
MPhil	master of philosophy
MPPA	master of public policy administration
MRCP	member of the Royal College of Physicians

H1, *continued*

MRCS	member of the Royal College of Surgeons
MS, MSc, or SM	master of science
MS, SM, MCh, or MSurg	master of surgery
MSN	master of science in nursing
MSPH	master of science in public health
MStat	master of statistics
MSW	master of social welfare; master of social work
MT	medical technologist
MTA	medical technical assistant
MT(ASCP)	registered medical technologist (American Society of Clinical Pathologists)
MUS	master in urban studies
ND	naturopathic doctor
OT	occupational therapist
OTR	occupational therapist, registered
PA	physician assistant
PA-C	Physician assistant, certified
PharmD, DP, or PD	doctor of pharmacy
PhD or DPhil	doctor of philosophy
PhG	graduate in pharmacy
PNP	pediatric nurse practitioner
PsyD	doctor of psychology
PT	physical therapist
RD	registered dietitian
RN	registered nurse
RNA	registered nurse anesthetist
RNC or RN,C	registered nurse, certified
RPFT	registered pulmonary function technologist
RPh	registered pharmacist
RPT	registered physical therapist
RRL	registered record librarian
RT	radiologic technologist, respiratory therapist
RTR	recreational therapist, registered
SCd, DSc, or DS	doctor of science
STD	doctor of systematic theology
ThD or DTh	doctor of theology

H2 Style Guide to Clinical, Technical, and Other Common Terms Used in Medical Writing

Abbreviation	Expanded Form
AAA	abdominal aortic aneurysm
ABC	avidin-biotin complex
AC	alternating current
ACE	angiotensin-converting enzyme
ACS	acute coronary syndromes
ACTH	*corticotropin* (previously adrenocorticotropic hormone)
AD	Alzheimer disease
ADH	antidiuretic hormone
ADHD	attention-deficit disorder
ADL	activities of daily living
aDNA	ancient DNA
ADP	adenosine diphosphate
ADPase	adenosine diphosphatase
AED	automated external defibrillator
AF	atrial fibrillation
AFP	α-fetoprotein
AIDS	acquired immunodeficiency syndrome
ALL	acute lymphoblastic leukemia; acute lymphocytic leukemia
allo-SCT	allogeneic stem cell transplantation
ALS	amyotrophic lateral sclerosis
ALT	alanine aminotransferase (previously SGPT)
AML	acute monocytic leukemia; acute myeloblastic leukemia; acute myelocytic leukemia
AMP	adenosine monophosphate
ANA	antinuclear anti-body
ANCOVA	analysis of covariance
ANLL	acute nonlymphocytic leukemia
ANOVA	analysis of variance
AOR	adjusted odds ratio
APACHE	Acute Physiology and Chronic Health Evaluation
APB	atrial premature beat
APC	atrial premature contraction
ARC	use *symptomatic HIV infection* (previously AIDS-related complex)
ARDS	acute respiratory distress syndrome

H2, *continued*

ARMD	age-related macular degeneration
ARR	absolute risk reduction
ART	antiretroviral therapy
ASC	adult stem cell
ASC-US	atypical squamuos cells of uncertain significance
ASD	atrial septal defect; autistic spectrum disorder
AST	aspartate aminotransferase (previously SGOT)
ATP	adenosine triphosphate
ATPase	adenosine triphosphatase
AUC	area under the curve
AUROC	area under the receiver operating characteristic curve
BAC	blood alcohol concentration
BADL	basic activities of daily living
BAER	brainstem auditory evoked response
BCG	bacilli Calmette-Guérin (do not expand as a drug)
BDI	Beck Depression Inventory
bid	twice a day (do not abbreviate)
BMD	bone mineral density
BMI	body mass index
BMT	bone marrow transplant
BP	blood pressure
BPD	bronchopulmonary dysplasia
BPH	benign prostatic hyperplasia
BPRS	Brief Psychiatric Rating Scale
BSA	body surface area
BSE	bovine spongiform encephalopathy; breast self-examination
BUN	blood urea nitrogen (use *sera urea nitrogen*)
C…	complement (used preceding a number, e.g., C3, C4)
CABG	coronary artery bypass graft
CAD	coronary artery disease
CAGE	cut down, annoyed, guilty, eye opener
CAM	complementary and alternative medicine
cAMP	cyclic adenosine monophosphate
CARS	compensatory anti-inflammatory response syndrome
CART	combination antiretroviral therapy
CBC	complete blood (add *cell*) count
CCU	cardiac care unit; critical care unit

H2, *continued*

CD*	clusters of differentiation (not to be confused with the common abbreviation for *compact disc*)
cDNA	complementary DNA
CD-ROM	compact disc read-only memory
CEA	carcinoembryonic antigen; cost-effective analysis
CEU	continuing education unit
cf	Compare (generally used in footnotes)
CF	cystic fibrosis
CFS	chronic fatigue syndrome
CFT	complement fixation test
CFU	colony-forming unit
cGMP	cyclic guanosine monophosphate
CHD	coronary heart disease
CHF	congestive heart failure
CI	confidence interval
CIN	cervical intraepithelial neoplasia
CIS	carcinoma in situ
CJD	Greutzfeldt-Jakob disease
CK	creatine kinase
CK-BB	creatine kinase BB (BB is the isozyme)
CK-MB	creatine kinase MB
CK-MM	creatine kinase MM
CL	confidence limit
CLIA	Clinical Laboratory Improvement Amendments
CME	continuing medical education
CMI	cell-mediated immunity
CML	chronic myelocytic leukemia
CMV	cytomegalovirus
CNS	central nervous system
CONSORT	Consolidated Standards of Reporting Trials
COPD	chronic obstructive pulmonary disease
COX-2	cyclooxygenase 2
CPAP	continuous positive airway pressure
CPD	continuing professional development
CPK	use *creatine kinase*
CPR	cardiopulmonary resuscitation
CPT	*Current Procedural Terminology*
CQI	continuous quality improvement
CRF	corticotropin-releasing factor

H2, *continued*

cRNA	complementary RNA
CRP	C-reactive protein
CSF	cerebrospinal fluid, colony-stimulating factor
CT	cost-utility anaysis
CVS	chorionic villus sampling
DALY	disability-adjusted life-year
damp	deoxyadenosine monophosphate (deoxyadenylate)
D&C	dilation and curettage
DC	direct current
DCIS	ductal carcinoma in situ
DDD	defined daily dose
DDT	dichorodiphenyltichloroethane (chlorophenothane)
DE	dose equivalent
DEV	duck embryo vaccine
DFA	direct fluorescence assay
dGMP	deoxyguanosine monophosphate (deoxyguanylate)
DIC	disseminated intravascular coagulation
DIF	direct immunofluorescence
DNA	deoxyribonucleic acid
DNAR	do not attempt resuscitation
DNase	deoxyribonuclease
DNH	do not hospitalize
DNR	do not resuscitate
DOS	disk operating system
DOT	directly observed therapy
DOTS	directly observed therapy, short course
dpi	dots per inch (generally a printing resolution term)
DRE	digital rectal examination
DRG	diagnosis related group
DS	duplex-sonography
DSM-II	*Diagnostic and Statistical Manual of Mental Disorders* (Third Edition)
DSM-III-R	*Diagnostic and Statistical Manual of Mental Disorders* (Third Edition Revised)
DSM-IV	*Diagnostic and Statistical Manual of Mental Disorders* (Fourth Edition)
DSM-IV-TR	*Diagnostic and Statistical Manual of Mental Disorders* (Fourth Edition, Text Revision)
DSMB	data and safety monitoring board

H2, *continued*

DT	delirium tremens
DTaP	diptheria and tetanus toxoids and pertussis [vaccine]
DXA	dual-energy x-ray absorptiometry
EBM	evidence-based medicine
EBV	Epstein-Barr virus
EC	ejection click
ECA	epidemiologic catchment area
ECG	electrocardiogram; electrocardiographic
ECMO	extracorporeal membrane oxygenation
ECT	electroconvulsive therapy
ED	effective dose; emergency department
ED_{50}	median effective dose
EDTA	ethylenediaminetetraacetic acid
EEE	eastern equine encephalomyelitis
EEG	electroencephalogram; electroencephalographic
EGD	esophagogastroduodenoscopy
EIA	enzyme immunoassay
ELISA	enzyme-linked immunosorbent assay
EM	electron microscope; electron microscopic; electron microscopy
EMG	electromyogram; electromyographic
EMIT	enzyme-multiplied immunoassay technique
EMS	electrical muscle stimulation; emergency medical services; eosinophilia-myalgia syndrome
EMT	emergency medical technician
ENG	electronystagmogram; electronystagmographic
EOG	electro-oculogram; electro-oculographic
ERCP	endoscopic retrograde cholangiopancreatography
ERG	electroretinogram; electroretinographic
ESBC	extended-spectrum β-lactamases
ESC	embryonic stem cell
ESR	erythrocyte sedimentation rate
ESRD	end-stage renal disease
ESWL	extracorporeal shock wave lithotripsy
EVR	evoked visual response
F	French (add *catheter*; is always preceded by a number)
$FEF_{25\%-75\%}$	forced expiratory flow; midexpiratory phase
FEV	forced expiratory volume

H2, *continued*

FEV_1	forced expiratory volume in the first second of expiration
FIO_2	fraction of inspired oxygen
FISH	fluorescence in situ hybridization
FLAIR	fluid-attenuated inversion recovery
FSH	follicle-stimulating hormone
FTA	fluorescent treponemal antibody
FTA-ABS	fluorescent treponemal antibody absorption (add *test*)
FUO	fever of unknown origin
FVC	forced vital capacity
GABA	γ-aminobutyric acid
GAD	generalized anxiety disorder
GAF	Global Assessment of Functioning [Scale]
GB	Gigabyte (a computer term)
GCS	Glasgow Coma Scale
G-CSF	granulocyte colony-stimulating factor
GDP	guanosine diphosphate
GDS	Geriatric Depression Scale
GED	General Education Development
GERD	gastroesophageal reflux disease
GFR	glomerular filtration rate
GH	growth hormone
GI	gastrointestinal
GIFT	gamete intrafallopian transfer
GLC	gas-liquid chromatography
GM-CSF	granulocyte-macrophage colony stimulating factor
GMP	guanosine monophosphate
GMRI	gated magnetic resonance imaging
GMT	geometric mean titer
GnRH	gonadotropin-releasing hormone
GSC	germline stem cell
GU	genitourinary
GUI	graphical user interface
GVHD	graft-vs-host disease
HAART	highly active antiretroviral therapy
HALE	health-adjusted life expectancy
HAV	hepatitis A virus
HbA_{1c}	hemoglobin A_{1c}
Hbco	carboxyhemoglobin; oxygenated hemoglobin

H2, *continued*

Hbo$_2$	oxyhemoglobin; oxygenated hemoglobin
HbS	sickle cell hemoglobin
HBsAg	hepatitis B surface antigen
HBSS	Hanks balanced salt solution
HBV	hepatitis B virus
hCG	human chorionic gonadotropin
HCV	hepatitis C virus
HDL	high-density lipoprotein
HDL-C	high-density lipoprotein cholesterol
HDRS	Hamilton Depression Rating Scale
hGH	human growth hormone
HHV	human herpesvirus
Hib	*Haemophilus influenzae* type b
HIPAA	Health Insurance Portability and Accountability Act
HIV	human immunodeficiency virus
HL	hearing level
HLA	human leukocyte antigen
HMO	health maintenance organization
HPF	high-power field
HPLC	high-performance liquid chromatography
HPV	human papillomavirus (use hyphen and numeral to indicate specific type)
HR	hazard ratio
HRQOL	health-related quality of life
HSC	hematopoietic stem cell
HSIL	high-grade squamous intraepithelial lesion
HSV	herpes simplex virus
HT	hormone therapy
5-HT	Use *serotonin* or 5-hydroxytryptamine
HTLV	human T-lymphotropic virus (used with hyphen and arabic numeral to indicate specific type)
HTML	hypertext markup language
http	hypertext transfer protocol (used as: http://...)
HUS	hemolytic uremic syndrome
IADL	instrumental activities of daily living
ICD	implantable cardioverter-defibrillator
ICD-9	*International Classification of Diseases, Ninth Revision*
ICD-9-CM	*International Classification of Diseases, Ninth Revision, Clinical Modification*

H2, *continued*

ICD-10	*International Classification of Diseases, Tenth Revision*
ICD-10-CM	*International Classification of Diseases, Tench Revision, Clinical Modification*
ICU	injecting drug user; injection drug user
ID	infective dose
IFN	interferon
Ig	immunoglobulin
IGF-1	insulinlike growth factor
IL	interleukin (abbreviate only when indicating a specific protein factor)
IM	intramuscular; intramuscularly
IND	investigational new drug
INR	international normalized radio
IOP	intraocular pressure
IPA	intimate partner abuse
IPV	intimate partner violence
IQ	intelligence quotient
IRB	institutional review board
IRMA	immunoradiometric assay
ISBN	International Standard Book Number
ISG	immune serum globulin
ISSN	International Standard Serial Number
ITI	intratubal insemination
ITP	idiopathic thrombocytopenic purpura
ITT	intention to treat
IUD	intrauterine device
IUGR	intrauterine growth retardation
IUI	intrauterine insemination
IV	intravenous; intravenously
IVF	in vitro fertilization
IVIG	intravenous immunoglobulin
IVP	intravenous pyelogram
JPEG	Joint Photographic Experts Group
kB	kilobyte
KUB	kidneys, ureter, bladder
LA	left atrium
LAD	left anterior descending coronary artery
LAO	left anterior oblique coronary artery

H2, *continued*

LASEK	laser epithelial keratomileusis
LASIK	laser in situ keratomileusis
LAV	lymphadenopathy-associated virus
LBW	low birth weight (low-birth-rate infant)
LCA	left coronary artery
LCR	locus control region
LCX, CX	left circumflex coronary artery
LD	lethal dose
LD$_{50}$	median lethal dose
LDH	lactate dehydrogenase
LDL	low-density lipoprotein
LDL-C	low-density lipoprotein cholesterol
LGA	large for gestational age
LH	luteinizing hormone
LHRH	luteinizing hormone-releasing hormone (*gonadorelin* as diagnostic agent)
LMW	low molecular weight
LOCF	last observation carried forward
LOD	logarithm of odds
logMAR	logarithm of the minimum angle of resolution
LOS	length of stay
LR	likelihood ratio
LSD	lysergic acid diethylamide
LSIL	low-grade squamous intraepithelial lesion
LV	left ventricle; left ventricular
LVEDV	left ventricular end-diastolic volume
LVEF	left ventricular ejection fraction
LVOT	left ventricular outflow tract
m–	meta- (used in chemical formulas and names)
MAOI	monoamine oxidase inhibitor
MAPC	multipotent adult progenitor cell
MB	megabyte
MBC	minimum bactericidal concentration
MCH	mean corpuscular hemoglobin
MCHC	mean corpuscular hemoglobin concentration
MCO	managed care organization
MCV	mean corpuscular volume
MD	muscular dystrophy
MDR	multidrug-resistant

H2, *continued*

MEC	mean effective concentration
MEM	minimal essential medium
MEN	multiple endocrine neoplasia
MeSH	Medical Subject Headings [of the US National Library of Medicine]
MET	metabolic equivalent task
MGUS	monoclonal gammopathy of uncertain significance
MHC	major histocompatibility complex
MI	mitralinsufficiency; myocardial infarction
MIC	minimum inhibitory concentration
MICU	medical intensive care unit
MMPI	Minnesota Multiphasic Personality Inventory
MMR	measles-mumps-rubella [vaccine]
MMSE	Mini-Mental State Examination
MODS	multiple-organ dysfunction syndrome
MOOSE	Meta-analysis of Observational Studies in Epidemiology
MPS	Mortality Probability Score
MRA	magnetic resonance angiography
MRI	magnetic resonance imaging
mRNA	messenger RNA
MRSA	methicillin-resistant *Staphylococcus aureus*
MS	mitral stenosis; multiple sclerosis
MSA	metropolitan statistical area
MSC	mesenchymal stem cell
MSET	multistage exercise test
MVC	motor vehicle crash
NAD	nicotinamide adenine dinucleotide
NADP	nicotinamide adenine dinucleotide phosphate
nb	*nota bene* (note well—used to place emphasis)
NDA	new drug application
Nd:YAG	neodymium:yttrium-aluminum-garnet [laser]
NEC	necrotizing enterocolitis
NF	*National Formulary*
NICU	neonatal intensive care unit
NK	natural killer (add *cells*)
NMN	nicotinamide mononucleotide
NNH	number needed to harm
NNS	number needed to screen

H2, *continued*

NNT	number needed to treat
NOS	not otherwise specified
npo	nothing by mouth (do not abbreviate in clinical setting)
NPV	negative predictive value
NS	not significant
NSAID	nonsteroidal anti-inflammatory drug
NSC	neural stem cell
NSTE	non-ST-segment elevation
o-	ortho- (use only in chemical formulas)
OC	oral contraceptive
OCD	obsessive-compulsive disorder
OD	oculus dexter (use only with a number)
OGTT	oral glucose tolerance test
OR	odds ration
OS	oculus sinister (use only with a number)
OS	opening snap
OSA	obstructive sleep apnea
OU	oculus unitas (use only with a number)
p-	para- (use only in chemical formulas or names)
PA	posteroanterior; pulmonary artery
PAC	premature atrial contraction; pulmonary artery catheter
$PaCO_2$	partial pressure of carbon dioxide, arterial
PaO_2	partial pressure of oxygen, arerial
PAO_2	partial pressure of oxygen in the alveoli
PAD	peripheral artery disease
PAS	periodic acid-Schiff
PAT	paroxysmal atrial tachycardia
PBS	phosphate-buffered saline
PBSC	peripheral blood stem cell
PCI	percutaneous coronary intervention
PCO_2	partial pressure of carbon dioxide
PCP	*Pneumocystis jiroveci* pneumonia (previously *Pneumocystis carinii* pneumonia)
PCR	polymerase chain reaction
PCT	practical clinical trial; pragmatic clinical trial
PCW	pulmonary capillary wedge
PDA	personal digital assistant
PDF	portable document format
PDR	*Physician's Desk Reference*

H2, *continued*

PE	pulmonary embolism
PEEP	positive end-expiratory pressure
PEG	percutaneous endoscopic gastrostomy; pneumoencephalographic; pneumoencephalography
PEP	postexposure prophylaxis
PET	positron emission tomographic; positron emission tomography
PFGE	pulsed-field gel electrophoresis
PGF	placental growth factor
pH	negative logarithm of hydrogen ion concentration
PICC	peripherally inserted central catheter
PICU	pediatric intensive care unit
PID	pelvic inflammatory disease
PKU	phenylketonuria
PMS	premenstrual syndrome
po	orally (do not abbreviate)
PO_2	partial pressure of oxygen
POAG	primary open-angle glaucoma
PPD	purified protein derivative (tuberculin)
PPO	preferred provider organization
PPROM	preterm premature rupture of membranes
PPV	positive predictive value
prn	as needed (do not abbreviate)
PRO	peer review organization; professional review organization
PROM	premature rupture of membranes
PSA	prostate-specific antigen
$PsqO_2$	subcutaneous tissue oxygen tension
PSRO	professional standards review organization
PST	Pacific standard time
PSVT	paroxysmal supraventricular tachycardia
PT	physical therapy; prothrombin time
PTCA	percutaneous transluminal coronary angioplasty
PTSD	posttraumatic stress disorder
PTT	partial thromboplastin time
PUFA	polyunsaturated fatty acid
PUVA	psoralen–UV-A
PVC	premature ventricular contraction

H2, *continued*

PVR	peripheral vascular resistance; pulmonary vascular resistance
PVS	permanent vegetative state; persistent vegetative state
QA	quality assurance
QALY	quality-adjusted life-year
QC	quality control
qd	every day (do not abbreviate)
QI	quality improvement
qid	4 times daily (do not abbreviate)
qod	every other day (do not abbreviate)
QOL	quality of life
QUOROM	Quality of Reporting of Meta-analyses
RA	rheumatoid arthritis
RAM	random access memory
RAST	radioallergosorbent test
RBC	red blood cell
RBRVS	resource-based relative value scale
RCA	right coronary artery
RCT	randomized clinical trial; randomized controlled trial
RDA	recommended daily (or dietary) allowance;
RDC	Research Diagnostic Criteria
rDNA	ribosomal DNA
RDS	respiratory distress syndrome
REM	rapid eye movement
RFLP	restriction fragment length polymorphism
RFP	radiofrequency pulse
rh	recombinant human
Rh	rhesus
rhNGF	recombinant human nerve growth factor
RIA	radioimmunoassay
RIND	reversible ischemic neurological deficit
RNA	ribonucleic acid
RNAi	RNA interference
ROC	receiver operating characteristic [curve]
ROM	read-only memory
ROP	retinopathy of prematurity
RPR	rapid plasma reagin
RR	relative risk; risk ratio
RSV	respiratory syncytial virus

H2, *continued*

RT-PCR	reverse transcription–polymerase chain reaction; reverse transcriptase–polymerase chain reaction
RV	right ventricle; right ventricular
RVEF	right ventricular ejection fraction
RVOT	right ventricular outflow tract
SAD	seasonal affective disorder
SADS	Schedule for Affective Disorders and Schizophrenia
SAH	subarachnoid hemorrhage
SAPS	Simplified Acute Physiology Score
SARS	sever acute respiratory syndrome
SAS	Statistical Analysis System
SCID	severe combined immunodeficiency; Structured Clinical Interview for *DSM* (use with *DSM* edition number)
SD	standard deviation (abbreviate only when used with a number or in Mean [SD] construction in table stubs and headings)
SE	standard error (abbreviate only when used with a number)
SEM	standard error of the mean (abbreviate only when used with a number)
SEM	scanning electron microscope; systolic ejection murmer
SF-36	36-Item Short Form Health Survey
SGA	small for gestational age
SGML	standardized general markup language
SGOT	Use *aspartate aminotransferase* (for serum glutamic-oxaloacetic transaminase)
SGPT	Use *alanine aminotransferase* (for serum glutamic-pyruvic transaminase)
SIADH	syndrome of inappropriate secretion of antidiuretic hormone
SICU	surgical intensive care unit
SIDS	sudden infant death syndrome
SIL	squamous intraepithelial lesion
SIP	Sickness Impact Profile
siRNA	small interfering RNA
SIRS	systemic inflammatory response syndrome
SLE	St Louis encephalitis; systemic lupus erythematosus
SNP	single-nucleotide polymorphism

H2, *continued*

SPECT	single-photon emission computed tomography
SPF	sun protection factor
SPSS	Statistical Product and Service Solutions (previously Statistical Package for the Social Sciences)
SSC	somantic stem cell
SSC	standard saline citrate
SSNRI	selective serotonin-norepinephrine reuptake inhibitor
SSPE	sodium chloride, sodium phosphate, EDTA [buffer]
SSPE	subacute sclerosing panencephalitis
SSRI	selective serotonin reuptake inhibitor
STARD	Standards for Reporting Diagnostic Accuracy
STD	sexually transmitted disease
STEMI	ST-segment elevation myocardial infarction
STI	sexually transmitted infection; structured treatment interruption
SUN	serum urea nitrogen
SVR	systemic vascular resistance
$T_{1/2}$	half-life
T_3	triiodothyronine
T_4	thyroxine
TAHBSO	total abdominal hysterectomy with bilateral salpingo-oophorectomy
TAT	Thematic Apperception Test
TB	terabyte
TB	tuberculosis
TBI	traumatic brain injury
TBSA	total body surface area
TCA	tricyclic antidepressant
TCD_{50}	median tissue culture dose
TE	echo time
THA	total-hip arthroplasty
TI	inversion time
TIA	transient ischemic attack
TIBC	total iron-binding capactity
tid	3 times a day (do not abbreviate)
TIFF	Tag(ged) Image File Format
TLC	thin-layer chromatography; total lung capacity
TNF	tumor necrosis factor
TNM	tumor, node, metastasis

H2, *continued*

tPA	tissue plasminogen activator
TPN	total parenteral nutrition
TQM	total quality management
TR	repetition time
TRH	thyrotropin-releasing hormone (*protirelin* as diagnostic agent)
tRNA	transfer RNA
TRP	tyrosine-related protein
TRUS	transrectal ultrasonography
TSH	Use *thyrotropin* (previously thyroid-stimulating hormone)
TSS	toxic shock syndrome; toxic "strep" [streptococcal] syndrome
TTP	thrombotic thrombocytopenic purpura
UHF	ultrahigh frequency
ul	uniformly labled (used within parentheses)
URI	uniform resource identifier
URL	uniform resource locator
URN	uniform resource name
URTI	upper respiratory tract ifection
US	ultrasonography; ultrasound
USAN	United States Adopted Names [Council]
USP	United States Pharmacopeia
USSC	unrestricted somatic stem cell
UV	ultraviolet
UV-A	ultraviolet A
UV-B	ultraviolet B
UV-C	ultraviolet C
VAIN	vaginal intraepithelial neoplasia
vCJD	variant Creutzfeldt-Jakob disease
VDRL	Venereal Disease Research Laboratory (add *test*)
VEGF	vascular endothelial growth factor
VEP	visual evoked potential
VER	visual evoked response
VHDL	very high-density lipoprotein
VHF	very high frequency; viral hemorrhagic fever
VLBW	very low birth weight (very low-birth-rate infant)
VLDL	very low density lipoprotein
$\dot{V}O_2$	oxygen consumption per unit time [with dot over V]

H2, *continued*

$\dot{V}O_{2max}$	maximum oxygen consumption [with dot over V]
VPB	ventricular premature beat
$\dot{V}/Q$	ventilation-perfusion [ratio or scan]—[with dot over V]
vs	versus (use *v* for legal references)
VSD	ventricular septal defect
VT	ventricular tachycardia; tidal volume
VZV	varicella zoster virus
WAIS	Wechsler Adult Intelligence Scale
WBC	white blood cell
WEE	western equine encephalomyelitis
WISC-R	Wechsler Intelligence Scale for Children
XML	extensible markup language
YLD	years living with disability
YPLL	years of potential life loss

H3 Medical Laboratory Tests and Conversion Factors

Analyte	Specimen	Reference Range, Conventional Unit	Conven-tional Unit	Conversion Factor (Multiply by)	Reference Range, SI Unit	SI Unit
Acetaminophen	Serum, plasma	10–30	µg/mL	6.614	66–200	µmol/L
Acetoacetate	Serum, plasma	<1	mg/dL	97.95	<100	µmol/L
Acetone	Serum, plasma	<1.0	mg/dL	0.172	<0.17	mmol/L
Acid phosphatase	Serum	<5.5	U/L	16.667	<90	nkat/L
Activated partial thromboplastin time (APTT)	Whole blood	25–40	s	1.0	25–40	s
Adenosine deaminase	Serum	11.5–25.0	U/L	16.667	190–420	nkat/L
Adrenocorticotropic hormone (ACTH)	Plasma	<120	pg/mL	0.22	<26	pmol/L
Alanine	Plasma	1.87–5.89	mg/dL	112.2	210–661	µmol/L
Alanine aminotransferase (ALT)	Serum	10–40	U/L	0.0167	0.17–0.68	µkat/L
Albumin	Serum	3.5–5.0	g/dL	10	35–50	g/L
Alcohol dehydrogenase	Serum	<2.8	U/L	16.667	<47	nkat/L
Aldolase	Serum	1.0–7.5	U/L	0.0167	0.02–0.13	µkat/L
Aldosterone	Serum, plasma	2–9	ng/dL	27.74	55–250	pmol/L
Alkaline phosphatase	Serum	30–120	U/L	0.0167	0.5–2.0	µkat/L

Alprazolam	Serum, plasma	10-50	ng/mL	3.24	32–162	nmol/L
Amikacin	Serum, plasma	20-30	µg/mL	1.708	34–52	µmol/L
α-Aminobutyric acid	Plasma	0.08-0.36	mg/dL	96.97	8–35	µmol/L
δ-Aminolevulinic acid	Serum	15-23	µg/dL	0.0763	1.1–8.0	µmol/L
Amiodarone	Serum, plasma	0.5-2.5	µg/mL	1.55	0.8–3.9	µmol/L
Amitriptyline	Plasma	120-250	ng/mL	3.605	433–903	nmol/L
Ammonia (as nitrogen)	Serum, plasma	15-45	µg/dL	0.714	11–32	µmol/L
Amobarbital	Serum	1-5	µg/mL	4.42	4–22	µmol/L
Amphetamine	Serum, plasma	20-30	ng/mL	7.4	148–222	nmol/L
Amylase	Serum	27-131	U/L	0.01667	0.46–2.23	µkat/L
Androstenedione	Serum	75-205	ng/dL	0.0349	2.6–7.2	nmol/L
Angiotensin I	Plasma	<25	pg/mL	0.772	<15	pmol/L
Angiotensin II	Plasma	10-60	pg/mL	0.957	0.96–58	pmol/L
Angiotensin-converting enzyme	Serum	<40	U/L	16.667	<670	nkat/L
Anion gap $Na^+-(Cl^-+HCO_3^-)$	Serum, plasma	8-16	mEq/L	1.0	8–16	mmol/L
Antidiuretic hormone (ADH)	Plasma	1-5	pg/mL	0.923	0.9–4.6	pmol/L
Antithrombin III	Plasma	21-30	mg/dL	10	210–300	mg/L
$α_1$-Antitrypsin	Serum	78-200	mg/dL	0.184	14.5–36.5	µmol/L

H3 Medical Laboratory Tests and Conversion Factors, *continued*

Analyte	Specimen	Reference Range, Conventional Unit	Conven-tional Unit	Conversion Factor (Multiply by)	Reference Range, SI Unit	SI Unit
Apolipoprotein A-I	Serum	80-151	mg/dL	0.01	0.8-1.5	g/L
Apolipoprotein B	Serum, plasma	50-123	mg/dL	0.01	0.5-1.2	g/L
Arginine	Serum	0.37-2.40	mg/dL	57.05	21-138	µmol/L
Arsenic	Whole blood	<2-23	µg/L	0.0133	0.03-0.31	µmol/L
Ascorbic acid (see Vitamin C)						
Asparagine	Plasma	0.40-0.91	mg/dL	75.689	30-69	µmol/L
Aspartate aminotransferase (AST)	Serum	10-30	U/L	0.0167	0.17-0.51	µkat/L
Apartic acid	Plasma	<0.3	mg/dL	75.13	<25	µmol/L
Atrial natriuretic hormone	Plasma	20-77	pg/mL	0.325	6.5-2.5	pmol/L
Bands (see White blood cell count)						
Basophils (see White blood cell count)						
Base excess	Whole blood	−2 to 3	mEq/L	1.0	−2 to 3	mmol/L
Bicarbonate	Serum	21-28	mEq/L	1.0	21-28	mmol/L
Bile acids (total)	Serum	0.3-2.3	µg/mL	2.448	0.73-5.63	µmol/L
Bilirubin, total	Serum	0.3-1.2	mg/dL	17104	5.0-21.0	µmol/L
Bilirubin. direct (conjugated)	Serum	0.1-0.3	mg/dL	17.104	1.7-5.1	µmol/L
Biotin	Serum	200-500	pg/mL	0.00409	0.82-2.05	nmol/L
Bismuth	Whole blood	1-12	µg/L	4.785	4.8-57.4	nmol/L

		Specimen	Conventional range	Conventional unit	Factor	SI range	SI unit
Blood gases	Cabon dioxide, PCO$_2$	Arterial blood	35–45	mm Hg	0.133	4.7–5.9	kPa
	pH	Arterial blood	7.35–7.45		1.0	7.35–7.45	
	Oxygen, PO$_2$	Arterial blood	80–100	mm Hg	0.133	11–13	kPa
Brain–type natriuretic peptide (BNP)		Plasma	<167	pg/mL	1.0	<167	ng/L
Bromide (toxic)		Serum	>1250	μg/mL	0.0125	>15.6	mmol/L
C1 esterase inhibitor		Serum	12–30	mg/dL	10	120–300	mg/L
C3 complement		Serum	1200–1500	μg/mL	0.001	1.2–1.5	g/L
C4 complement		Serum	350–600	μg/mL	0.001	0.35–0.60	g/L
Cadmium		Whole blood	0.3–1.2	μg/L	8.896	2.7–10.7	nmol/L
Caffeine		Serum, plasma	3–15	μg/L	0.515	2.5–7.5	μmol/L
Calcitonin		Plamsa	3–26	pg/mL	0.292	0.8–7.6	pmol/L
Calcium, ionized		Serum	4.60–5.08	mg/dL	0.25	1.15–1.27	mmol/L
Calcium, total		Serum	8.2–10.2	mg/dL	0.25	2.05–2.55	mmol/L
Cancer antigen (CA) 125		Serum	<35	U/mL	1.0	<35	kU/L
Carbamazepine		Serum, plasma	8–12	μg/mL	4.233	34–51	μmol/L
Carbon dioxide (total)		Serum, plasma	22–28	mEq/L	1.0	22–28	mmol/L
Carboxyhemoglobin (toxic)		Whole blood	>20	%	0.01	>0.2	Proportion of 1.0
Carcinoembryonic antigen (CEA)		Serum	<3.0	ng/mL	1.0	<3.0	μg/L

H3 Medical Laboratory Tests and Conversion Factors, *continued*

Analyte		Specimen	Reference Range, Conventional Unit	Conventional Unit	Conversion Factor (Multiply by)	Reference Range, SI Unit	SI Unit
β–Carotene		Serum	10–85	μg/dL	0.01863	0.2–1.6	μmol/L
Carotenoids		Serum	50–300	μg/dL	0.01863	0.9–5.6	μmol/L
Ceruloplasmin		Serum	20–40	mg/dL	10	200–400	mg/L
Chloramphenicol		Serum	10–25	μg/mL	3.095	31–77	μmol/L
Chlordiazepoxide		Serum, plasma	0.4–3.0	μg/mL	3.336	1.3–10.0	μmol/L
Chloride		Serum, plasma	96–106	mEq/L	1.0	96–106	mmol/L
Chlorpromazine		Plasma	50–300	ng/mL	3.126	157–942	nmol/L
Chlorpropamide		Plasma	75–250	mg/L	3.61	270–900	μmol/L
Cholecalciferol (see Vitamin D)							
Cholesterol (total)	Desirable	Serum, plasma	<200	mg/dL	0.0259	<5.18	mmol/L
	Borderline high	Serum, plasma	200–239	mg/dL	0.0259	5.18–6.18	mmol/L
	High	Serum, plasma	≥240	mg/dL	0.0259	≥6.21	mmol/L
Cholesterol, high-density (HDL) (low level)		Serum, plasma	<40	mg/dL	0.0259	<1.03	mmol/L
Cholesterol, low-density(LDL) (high level)		Serum, plasma	>160	mg/dL	0.0259	4.144	mmol/L
Cholinesterase		Serum	5–12	mg/L	2.793	14–39	nmol/L
Chorionic gonadotropin (β–hCG) (nonpregnant)		Serum	5.0	mIU/mL	1.0	5.0	IU/L

Chromium	Whole blood	0.7–28.0	µg/L	19.232	13.4–538.6	nmol/L
Citrate	Serum	1.2–3.0	mg/dL	52.05	60–160	µmol/L
Citrulline	Plasma	0.2–1.0	mg/dL	57.081	12–55	µmol/L
Clonazepam	Serum	10–50	ng/mL	0.317	0.4–15.8	nmol/L
Clonidine	Serum, plasma	1.0–2.0	ng/mL	4.35	4.4–8.7	nmol/L
Clozapine	Serum	200–350	ng/mL	0.003	0.6–1.0	µmol/L
Coagulation factor I (Fibrinogen)	Plasma	0.15–0.35	g/dL	29.41	4.4–10.3	µmol/L
	Plasma	150–350	mg/dL	0.01	1.5–3.5	g/L
Coagulation factor II (prothrombin)	Plasma	70–130	%	0.01	0.70–1.30	Proportion of 1.0
Coagulation factor V	Plasma	70–130	%	0.01	0.70–1.30	Proportion of 1.0
Coagulation factor VII	Plasma	60–140	%	0.01	0.60–1.40	Proportion of 1.0
Coagulation factor VIII	Plasma	50–200	%	0.01	0.50–2.00	Proportion of 1.0
Coagulation factor IX	Plasma	70–130	%	0.01	0.70–1.30	Proportion of 1.0
Coagulation factor X	Plasma	70–130	%	0.01	0.70–1.30	Proportion of 1.0
Coagulation factor XI	Plasma	70–130	%	0.01	0.70–1.30	Proportion of 1.0
Coagulation factor XII	Plasma	70–130	%	0.01	0.70–1.30	Proportion of 1.0
Cobalt	Serum	4.0–10.0	µg/L	16.968	67.9–169.7	nmol/L
Cocaine (toxic)	Serum	>1000	ng/mL	3.297	>3300	nmol/L
Codeine	Serum	10–100	ng/mL	3.34	33–334	nmol/L
Coenzyme Q10 (ubiquinone)	Plasma	0.5–1.5	µg/mL	1.0	0.5–1.5	mg/L

H3 Medical Laboratory Tests and Conversion Factors, *continued*

Analyte	Specimen	Reference Range, Conventional Unit	Conventional Unit	Conversion Factor (Multiply by)	Reference Range, SI Unit	SI Unit
Copper	Serum	70–140	µg/dL	0.157	11–22	µmol/L
Coproporphyrin	Urine	<200	µg/24 h	1.527	<300	µmol/d
Corticotropin	Plasma	<120	pg/mL	0.22	<26	pmol/L
Cortisol	Serum, plasma	5–25	µg/dL	27.588	140–690	nmol/L
Cotinine	Plasma	0–8	µg/L	5.675	0–45	nmol/L
C–peptide	Serum	0.5–2.5	ng/mL	0.331	0.17–0.83	nmol/L
C–reactive protein	Serum	0.08–3.1	mg/L	9.524	0.76–28.5	nmol/L
Creatine	Serum	0.1–0.4	mg/dL	76.25	8–31	µmol/L
Creatine kinase (CK)	Serum	40–150	U/L	0.0167	0.67–2.5	µkat/L
Creatine kinase–MB fraction	Serum	0–7	ng/mL	1.0	0–7	µg/L
Creatinine	Serum, plasma	0.6–1.2	mg/dL	88.4	53–106	µmol/L
Creatinine clearance	Serum, plasma	75–125	mL/min/1.73 m^2	0.0167	1.24–2.08	mL/s/m^2
Cyanide (toxic)	Whole blood	>1.0	µg/mL	23.24	>23	µmol/L
Cyclic adenosine monophosphate (cAMP)	Plasma	4.6–8.6	ng/mL	3.04	14–26	nmol/L
Cyclosporine	Serum	100–400	ng/mL	0.832	83–333	nmol/L
Cystine	Plasma	0.40–1.40	mg/dL	41.615	16–60	µmol/L
D–dimer	Plasma	<0.5	µg/mL	5.476	<3.0	nmol/L

Dehydroepiandroster-one (DHEA)	Serum	1.8–12.5	ng/mL	3.47	6.2–43.3	nmol/L
Dehydroepiandroster-one sulfate (DHEA–S)	Serum	50–450	µg/dL	0.027	1.6–12.2	µmol/L
Deoxycorticosterone	Serum	2–19	ng/dL	30.5	61–576	nmol/L
Desipramine	Serum, plasma	50–200	ng/mL	3.754	170–700	nmol/L
Diazepam	Serum, plasma	100–1000	ng/mL	0.0035	0.35–3.51	µmol/L
Digoxin	Plasma	0.5–2.0	ng/mL	1.281	0.6–2.6	nmol/L
Diltiazem	Serum	<200	mg/L	2.412	<480	µmol/L
Disopyramide	Serum, plasma	2.8–7.0	µg/mL	2.946	8.3–22.0	µmol/L
Dopamine	Plasma	<87	pg/mL	6.528	<475	pmol/L
Doxepin	Serum, plasma	30–150	ng/mL	3.579	108–538	nmol/L

Electrophoresis (protein)

Proportion of total protein

Albumin	Serum	52–65	%	0.01	0.52–0.65	Proportion of 1.0
α_1–Globulin	Serum	2.5–5.0	%	0.01	0.025–0.05	Proportion of 1.0
α_2–Globulin	Serum	7.0–13.0	%	0.01	0.07–0.13	Proportion of 1.0
β–Globulin	Serum	8.0–14.0	%	0.01	0.08–0.14	Proportion of 1.0
γ–Globulin	Serum	12.0–22.0	%	0.01	0.12–0.22	Proportion of 1.0

H3 Medical Laboratory Tests and Conversion Factors, *continued*

Analyte	Specimen	Reference Range, Conventional Unit	Conventional Unit	Conversion Factor (Multiply by)	Reference Range, SI Unit	SI Unit
Electrophoresis (protein), *continued*						
Albumin	Serum	3.2–5.6	g/dL	10.0	32–56	g/L
α_1–Globulin	Serum	0.1–0.4	g/dL	10.0	1–10	g/L
α_2–Globulin	Serum	0.4–1.2	g/dL	10.0	4–12	g/L
β–Globulin	Serum	0.5–1.1	g/dL	10.0	5–11	g/L
γ–Globulin	Serum	0.5–1.6	g/dL	10.0	5–16	g/L
Eosinophils (*see* White blood cell count)						
Ephedrine (toxic)	Serum	>2	μg/mL	6.052	>12.1	μmol/L
Epinephrine	Plasma	<60	pg/mL	5.459	<330	pmol/L
Erythrocyte count (*see* Red blood cell count)						
Erythrocyte sedimentation rate	Whole blood	0–20	mm/h	1.0	0–20	mm/h
Erythropoietin	Serum	5–36	IU/L	1.0	5–36	IU/L
Estradiol (E$_2$)	Serum	30–400	pg/mL	3.671	110–1470	pmol/L
Estriol (E$_3$)	Serum	5–40	ng/mL	3.467	17.4–138.8	nmol/L
Estrogens (total)	Serum	60–400	pg/mL	1.0	60–400	ng/L
Estrone (E$_1$)	Serum, plasma	1.5–25.0	pg/mL	3.698	5.5–92.5	pmol/L
Ethanol (ethyl alcohol)	Serum, whole blood	<20	mg/dL	0.2171	<4.3	mmol/L
Ethchlorvynol (toxic)	Serum, plasma	>20	μg/mL	6.915	>138	μmol/L

Concentration

Analyte	Specimen					
Ethosuximide	Serum	40–100	mg/L	7.084	280–700	µmol/L
Ethylene glycol (toxic)	Serum, plasma	>30	mg/dL	0.1611	>5	mmol/L
Fatty acids (nonesterified)	Serum, plasma	8–25	mg/dL	0.0355	0.28–0.89	mmol/L
Fecal fats (as stearic acids)	Stool	2.0–6.0	g/d	1.0	2–6	g/24 h
Fenfluramine	Serum	0.04–0.30	µg/mL	4.324	0.18–1.30	µmol/L
Fentanyl	Serum	0.01–0.10	µg/mL	2.972	0.02–0.30	µmol/L
Ferritin	Serum	15–200	ng/mL	2.247	33–450	pmol/L
α_1-Fetoprotein	Serum	<10	ng/mL	1.0	<10	µg/L
Fibrin degradation products	Plasma	<10	µg/mL	1.0	<10	mg/L
Fibrinogen	Plasma	200–400	mg/dL	0.0294	5.8–11.8	µmol/L
Flecanide	Serum, plasma	0.2–1.0	µg/mL	2.413	0.5–2.4	µmol/L
Fluoride	Whole blood	<0.05	mg/dL	0.5263	<0.027	mmol/L
Fluoxetine	Serum	200–1100	ng/mL	0.00323	0.65–3.56	µmol/L
Flurazepam (toxic)	Serum, plasma	>0.2	µg/mL	2.5	>0.5	µmol/L
Folate (folic acid)	Serum	3–16	ng/mL	2.266	7–36	nmol/L
Follicle–stimulating hormone (FSH)	Serum, plasma	1–100	mIU/mL	1.0	1–100	IU/L
Fructosamine	Serum	36–50	mg/L	5.581	200–280	mmol/L
Fructose	Serum	1–6	mg/dL	55.506	55–335	µmol/L
Galactose	Serum, plasma	<20	mg/dL	0.0555	<1.10	mmol/L
Gastrin	Serum	25–90	pg/mL	0.481	12–45	pmol/L
Gentamicin	Serum	6–10	µg/mL	2.090	12–21	µmol/L

H3 Medical Laboratory Tests and Conversion Factors, *continued*

Analyte	Specimen	Reference Range, Conventional Unit	Conventional Unit	Conversion Factor (Multiply by)	Reference Range, SI Unit	SI Unit
Glucagon	Plasma	20–100	pg/mL	1.0	20–100	ng/L
Glucose	Serum	70–110	mg/dL	0.0555	3.9–6.1	mmol/L
Glucose-6-phosphate dehydrogenase	Whole blood	10–14	U/g hemoglobin	0.0167	0.17–0.24	nkat/g hemoglobin
Glutamic acid	Plasma	0.2–2.8	mg/dL	67.967	15–190	μmol/L
Glutamine	Plasma	6.1–10.2	mg/dL	68.423	420–700	μmol/L
γ–Glutamyltransferase (GGT)	Serum	2–30	U/L	0.01667	0.03–0.51	μkat/L
Glutethimide	Serum	2–6	μg/mL	4.603	9–28	μmol/L
Glycerol (free)	Serum	0.3–1.72	mg/dL	0.1086	0.32–0.187	mmol/L
Glycine	Plasma	0.9–4.2	mg/dL	133.2	120–560	μmol/L
Gold	Serum	<10	μg/dL	50.770	<500	nmol/L
Growth hormone (GH)	Serum	0–18	ng/mL	1.0	0–18	μg/L
Haloperidol	Serum, plasma	6–24	ng/mL	2.66	16–65	nmol/L
Haptoglobin	Serum	26–185	mg/dL	10	260–1850	mg/L
Hematocrit	Whole blood	41–50	%	0.01	0.41–0.50	Proportion of 1.0
Hemoglobin	Whole blood	14.0–17.5	g/dL	10.0	140–175	g/L

Mean corpuscular hemoglobin (MCH)	Whole blood	26–34	pg/cell	1.0	26–34	pg/cell
Mean corpuscular hemoglobin concentration (MCHC)	Whole blood	33–37	g/dL	10	330–370	g/L
Mean corpuscular volume (MCV)	Whole blood	80–100	μm^3	1.0	80–100	fL
Hemoglobin A_{1c} (glycated hemoglobin)	Whole blood	4–7	% total of hemoglobin	0.01	0.04–0.07	Proportion of total hemo—globin
Hemoglobin A_2	Whole blood	2.0–3.0	%	0.01	0.02–0.03	Proportion of 1.0
Histamine	Plasma	0.5–1.0	μg/L	8.997	4.5–9.0	nmol/L
Histidine	Plasma	0.5–1.7	mg/dL	64.45	32–110	μmol/L
Homocysteine	Plasma	0.68–2.02	mg/L	7.397	5–15	μmol/L
Homovanillic acid	Urine	1.4–8.8	mg/24 h	5.489	8–48	μmol/d
Hydrocodone	Serum	<0.02	μg/mL	3.34	<0.06	μmol/L
Hydromorphone	Serum	0.008–0.032	μg/mL	3504	28–112	nmol/L
β-Hydroxybutyric acid	Plasma	<3.0	mg/dL	96.06	<300	μmol/L

H3 Medical Laboratory Tests and Conversion Factors, *continued*

Analyte	Specimen	Reference Range, Conventional Unit	Conven-tional Unit	Conversion Factor (Multiply by)	Reference Range, SI Unit	SI Unit
5-Hydroxy-indoleacetic acid (5-HIAA)	Urine	2–6	mg/24 h	5.23	10.4–31.2	μmol/d
Hydroxyproline	Plasma	<0.55	mg/dL	76.266	<42	μmol/L
Ibuprofen	Serum	10–50	μg/mL	4.848	50–243	μmol/L
Imipramine	Plasma	150–250	ng/mL	3.566	536–893	nmol/L
Immunoglobulin A (IgA)	Serum	40–350	mg/dL	10	400–3500	mg/L
Immunoglobulin D (IgD)	Serum	0–8	mg/dL	10	0–80	mg/L
Immunoglobulin E (IgE)	Serum	0–1500	μg/L	0.001	0–1.5	mg/L
Immunoglobulin G (IgG)	Serum	650–1600	mg/dL	0.01	6.5–16.0	g/L
Immunoglobulin M (IgM)	Serum	54–300	mg/dL	10	550–3000	mg/L
Insulin	Serum	2.0–20	μIU/mL	6.945	14–140	pmol/L
Insulinlike growth factor	Serum	130–450	ng/mL	0.131	18–60	nmol/L
Iodine	Serum	58–77	μg/L	7.880	450–580	nmol/L
Iron	Serum	60–150	μg/dL	0.179	10.7–26.9	μmol/L

Iron–binding capacity	Serum	250–450	µg/dL	0.179	44.8–80.6	µmol/L
Isoleucine	Plasma	0.5–1.3	mg/dL	76.236	40–100	µmol/L
Isoniazid	Plama	1–7	µg/mL	7.291	7–51	µmol/L
Isopropanol (toxic)	Serum, plasma	>400	mg/L	0.0166	>6.64	mmol/L
Kanamycin	Serum, plasma	25–35	µg/mL	2.08	52–72	µmol/L
Ketamine	Serum	0.2–6.3	µg/mL	4.206	0.8–26	µmol/L
17-Ketosteroids	Urine	3–12	mg/24 h	3.33	10–42	µmol/d
Lactate	Plasma	5.0–15	mg/dL	0.111	0.6–1.7	mmol/L
Lactate dehydrogenase (LDH)	Serum	100–200	U/L	0.0167	1.7–3.4	µkat/L
LDH isoenzymes LD_1	Serum	17–27	%	0.01	0.17–0.27	Proportion of 1.0
LD_2	Serum	27–37	%	0.01	0.27–0.37	Proportion of 1.0
LD_3	Serum	18–25	%	0.01	0.18–0.25	Proportion of 1.0
LD_4	Serum	3–8	%	0.01	0.03–0.08	Proportion of 1.0
LD_5	Serum	0–5	%	0.01	0–0.05	Proportion of 1.0
Lead	Serum	<10–20	µg/dL	0.0483	<0.5–1.0	µmol/L

H3 Medical Laboratory Tests and Conversion Factors, *continued*

Analyte	Specimen	Reference Range, Conventional Unit	Conven-tional Unit	Conversion Factor (Multiply by)	Reference Range, SI Unit	SI Unit
Leucine	Plasma	1.0–2.3	mg/dL	76.237	75–175	μmol/L
Leukocytes (*see* White blood cell count)						
Lidocaine	Serum, plasma	1.5–6.0	μg/mL	4.267	6.4–25.6	μmol/L
Lipase	Serum	31–186	U/L	0.01667	0.5–3.2	μkat/L
Lipoprotein(a) [Lp(a)]	Serum	10–30	mg/dL	0.0357	0.35–1.0	μmol/L
Lithium	Serum	0.6–1.2	mEq/L	1.0	0.6–1.2	mmol/L
Lorazepam	Serum	50–240	ng/mL	3.114	156–746	nmol/L
Luteinizing hormone (LH)	Serum, plasma	1–104	mIU/mL	1.0	1–104	IU/L
Lycopene	Serum	0.15–0.25	mg/L	1.863	0.28–0.46	μmol/L
Lymphocytes (*see* White blood cell count)						
Lysergic acid diethylamide	Serum	<0.004	μg/mL	3726	<15	nmol/L
Lysine	Plasma	1.2–3.5	mg/dL	68.404	80–240	μmol/L
Lysozyme	Serum, plasma	0.4–1.3	mg/dL	10	4–13	mg/L
Magnesium	Serum	1.3–2.1	mEq/L	0.50	0.65–1.05	mmol/L
Manganese	Whole blood	10–12	μg/L	18.202	182–218	nmol/L
Maprotiline	Plasma	200–600	ng/mL	1.0	200–600	μg/L
Melatonin	Serum	10–15	ng/L	4.305	45–66	pmol/L
Meperidine	Serum, plasma	400–700	ng/mL	4.043	1620–2830	nmol/L
Mercury	Serum	<5	μg/L	4.985	<25	nmol/L
Metanephrine (total)	Urine	<1.0	mg/24 h	5.07	<5	μmol/d

Metformin	Serum	1–4	µg/mL	7.742	8–30	µmol/L
Methadone	Serum, plasma	100–400	ng/mL	0.00323	0.32–1.29	µmol/L
Methamphetamine	Serum	0.01–0.05	µg/mL	6.7	0.07–0.34	µmol/L
Methanol	Plasma	<200	µg/mL	0.0312	<6.2	mmol/L
Methaqualone	Serum, plasma	2–3	µg/mL	4.0	8–12	µmol/L
Methemoglobin	Whole blood	<0.24	g/dL	155	<37.2	µmol/L
Methemoglobin	Whole blood	<1.0	% of total hemoglobin	0.01	<0.01	proportion of total hemoglobin
Methicillin	Serum	8–25	mg/L	2.636	22–66	µmol/L
Methionine	Plasma	0.1–0.6	mg/dL	67.02	6–40	µmol/L
Methotrexate	Serum, plasma	0.04–0.36	mg/L	2200	90–790	nmol/L
Methyldopa	Plasma	1–5	µg/mL	4.735	5.0–25	µmol/L
Metoprolol	Serum, plasma	75–200	ng/mL	3.74	281–748	nmol/L
β_2-Microglobulin	Serum	1.2–2.8	mg/L	1.0	1.2–2.8	mg/L
Morphine	Serum, plasma	10–80	ng/mL	3.504	35–280	nmol/L
Myoglobin	Serum	19–92	µg/L	0.0571	1.0–5.3	nmol/L
Naproxen	Serum	26–70	µg/mL	4.343	115–300	µmol/L
Niacin (nicotinic acid)	Urine	2.4–6.4	mg/24 h	7.30	17.5–46.7	µmol/d
Nickel	Whole blood	1.0–28.0	µg/L	17.033	17–476	nmol/L
Nicotine	Plasma	0.01–0.05	mg/L	6.164	0.062–0.308	µmol/L

H3 Medical Laboratory Tests and Conversion Factors, *continued*

Analyte	Specimen	Reference Range, Conventional Unit	Conventional Unit	Conversion Factor (Multiply by)	Reference Range, SI Unit	SI Unit
Nitrogen (nonprotein)	Serum	20–35	mg/dL	0.714	14.3–25.0	mmol/L
Nitroprusside (as thiocyanate)		6–29	μg/mL	17.2	103–500	μmol/L
Norepinephrine	Plasma	110–410	pg/mL	5.911	650–2423	pmol/L
Nortriptyline	Serum, plasma	50–150	ng/mL	3.797	190–570	nmol/L
Ornithine	Plasma	0.4–1.4	mg/dL	75.666	30–106	μmol/L
Osmolality	Serum	275–295	mOsm/kg	1.0	275–295	mmol/kg
Osteocalcin	Serum	3.0–13.0	ng/mL	1.0	3.0–13.0	μg/L
Oxalate	Serum	1.0–2.4	mg/mL	11.107	11–27	μmol/L
Oxazepam	Serum, plasma	0.2–1.4	μg/mL	3.487	0.7–4.9	μmol/L
Oxycodone	Serum	10–100	ng/mL	3.171	32–317	nmol/L
Oxygen, partial pressure (Po_2)	Arterial blood	80–100	mm Hg	0.133	11–13	kPa
Paraquat	Whole blood	0.1–1.6	μg/mL	5.369	0.5–8.5	μmol/L
Parathyroid hormone	Serum	10–65	pg/mL	0.1053	10–65	ng/L
Pentobarbital	Serum, plasma	1–5	μg/mL	4.439	4.0–22	μmol/L
Pepsinogen	Serum	28–100	ng/mL	1.0	28–100	μg/L
pH (*see* Blood gases)						
Phencyclidine (toxic)	Serum, plasma	90–800	ng/mL	4.109	370–3288	nmol/L

Analyte	Specimen	Conventional Range	Conventional Units	Factor	SI Range	SI Units
Phenobarbital	Serum, plasma	15–40	µg/mL	4.31	65–172	µmol/L
Phenylalanine	Plasma	0.6–1.5	mg/dL	60.544	35–90	µmol/L
Phenyl-propanolamine	Serum	0.05–0.10	µg/mL	6613	330–660	nmol/L
Phenytoin	Serum, plasma	10–20	mg/L	3.968	40–79	µmol/L
Phosphorus (inorganic)	Serum	2.3–4.7	mg/dL	0.323	0.74–1.52	mmol/L
Placental lactogen	Serum	0.5–1.1	µg/mL	46.296	23–509	nmol/L
Plasminogen (antigenic)	Plasma	10–20	mg/dL	0.113	1.1–2.2	µmol/L
Plasminogen activator inhibitor	Plasma	4–40	ng/mL	19.231	75–750	pmol/L
Platelet count (thrombocytes)	Whole blood	150–350	$\times 10^3/\mu L$	1.0	150–350	$\times 10^9/L$
Porphyrins (total)	Urine	20–120	µg/L	1.203	25–144	nmol/L
Potassium	Serum	3.5–5.0	mEq/L	1.0	3.5–5.0	mmol/L
Prealbumin	Serum	19.5–35.8	mg/dL	10	195–358	mg/L
Pregnanediol	Urine	<2.6	mg/24 h	3.12	<8	µmol/d
Pregnanetriol	Urine	<2.5	mg/24 h	2.972	<7.5	µmol/d
Primidone	Serum, plasma	5–12	µg/mL	4.582	23–55	µmol/L
Procainamide	Serum, plasma	4–10	µg/mL	4.25	17–42	µmol/L
Progesterone	Serum	0.15–25	ng/mL	3.18	0.5–79.5	nmol/L
Prolactin	Serum	3.8–23.2	µg/L	43.478	90–140	pmol/L
Proline	Plasma	1.2–3.9	mg/dL	86.858	104–340	µmol/L
Propoxyphene	Plasma	0.1–0.4	µg/mL	2.946	0.3–1.2	µmol/L

H3 Medical Laboratory Tests and Conversion Factors, *continued*

Analyte	Specimen	Reference Range, Conventional Unit	Conven-tional Unit	Conversion Factor (Multiply by)	Reference Range, SI Unit	SI Unit
Propranolol	Serum	50–100	ng/mL	3.856	193–386	nmol/L
Prostate-specific antigen	Serum	<4.0	ng/mL	1.0	<4.0	µg/L
Protein (total)	Serum	6.0–8.0	g/dL	10.0	60–80	g/L
Prothrombin time (PT)	Plasma	10–13	s	1.0	10–13	s
Protoporphyrin	Red blood cells	15–50	µg/dL	0.0178	0.27–0.89	µmol/L
Protriptyline	Serum, plasma	70–250	µg/dL	3.787	266–950	nmol/L
Pyridoxine (*see* Vitamin B$_6$)						
Pyruvate	Plasma	0.5–1.5	mg/dL	113.56	60–170	µmol/L
Quinidine	Serum	2.0–5.0	µg/mL	3.082	6.2–15.4	µmol/L
Red blood cell count	Whole blood	3.9–5.5	×10^6/µL	1.0	3.9–5.5	×10^{12}/L
Renin	Plasma	30–40	pg/mL	0.0237	0.7–1.0	pmol/L
Reticulocyte count	Whole blood	25–75	×10^3/µL	1.0	25–75	×10^9/L
Reticulocyte count	Whole blood	0.5–1.5	% of red blood cells	0.01	0.005–0.015	Proportion of red blood cells
Retinol (*see* Vitamin A)						
Riboflavin (*see* Vitamin B$_2$)						
Rifampin	Serum	4–40	mg/L	1.215	5–49	µmol/L
Salicylates	Serum, plasma	150–300	µg/mL	0.0724	1086–2172	µmol/L
Selenium	Serum, plasma	58–234	µg/L	0.0127	0.74–2.97	µmol/L

Serine	Plasma	0.7–2.0	mg/dL	95.156	65.193	µmol/L
Serotonin (5-hydroxytryptamine)	Whole blood	50–200	ng/mL	0.00568	0.28–1.14	µmol/L
Sex hormone–binding globulin	Serum	1.5–2.0	µg/mL	8.896	13–17	nmol/L
Sodium	Serum	136–142	mEq/L	1.0	136–142	mmol/L
Somatomedin C (Insulinlike growth factor)	Serum	130–450	ng/mL	0.131	18–60	nmol/L
Somatostatin	Plasma	<25	pg/mL	0.6110	<15	pmol/L
Streptomycin	Serum	7–50	mg/L	1.719	12–86	µmol/L
Strychnine	Whole blood	<0.5	mg/L	2.99	<1.5	µmol/L
Substance P	Plasma	<240	pg/mL	0.742	<180	pmol/L
Sulfate	Serum	10–32	mg/L	31.188	310–990	µmol/L
Sulfmethemoglobin	Whole blood	<1.0	% of total hemoglobin	0.01	<0.010	Proportion of total hemoglobin
Taurine	Plasma	0.3–2.1	mg/dL	79.91	24.168	µmol/L
Testosterone	Serum	300–1200	ng/dL	0.0347	10.4–41.6	nmol/L
Tetrahydrocannabinol	Serum	<0.20	µg/mL	3.180	<0.60	µmol/L
Theophylline	Serum, plasma	10–20	µg/mL	5.55	56–111	µmol/L
Thiamine (see Vitamin B$_2$)						
Thiopental	Serum, plasma	1–5	µg/mL	4.144	4.1–20.7	µmol/L

H3 Medical Laboratory Tests and Conversion Factors, *continued*

Analyte	Specimen	Reference Range, Conventional Unit	Conven-tional Unit	Conversion Factor (Multiply by)	Reference Range, SI Unit	SI Unit
Thioridazine	Serum, plasma	1.0–1.5	μg/mL	2.699	2.7–4.1	μmol/L
Threonine	Plasma	0.9–2.5	mg/dL	84	75–210	μmol/L
Thrombin time	Plasma	16–24	s	1.0	16–24	s
Thrombocytes (*see* Platelet count)						
Thyroglobulin	Serum	3–42	ng/mL	1.0	3–42	μg/L
Thyroid-stimulating hormone (TSH)	Serum	0.4–4.2	mIU/L	1.0	0.4–4.2	mIU/L
Thyroxine, free (FT$_4$)	Serum	0.9–2.3	ng/dL	12.871	12.30	pmol/L
Thyroxine, total (T$_4$)	Serum	5.5–12.5	μg/dL	12.871	71–160	nmol/L
Thyroxine-binding globulin	Serum	16.0–24.0	μg/mL	17.094	206–309	nmol/L
Tissue plasminogen activator	Plasma	<0.04	IU/mL	1000	<40	IU/L
Tobramycin	Serum, plasma	5.10	μ/mL	2.139	10–21	μmol/L
Tocainide	Serum	4–10	μg/mL	5.201	21–52	μmol/L
α-Tocopherol (*see* Vitamin E)						
Tolbutamide	Serum	80–240	μg/mL	3.70	296–888	μmol/L
Transferrin	Serum	200–400	mg/dL	0.0123	2.5–5.0	μmol/L
Triglycerides	Serum	<160	mg/dL	0.0113	1.8	mmol/L
Triiodothyronine, free (FT$_3$)	Serum	130–450	pg/dL	0.0154	2.0–7.0	pmol/L

Analyte	Specimen	Reference range	Units	Factor	SI reference range	SI units
Triiodothyronine, total (T$_3$)	Serum	60–180	ng/dL	0.0154	0.92–2.76	nmol/L
Troponin I	Serum	0–0.4	ng/mL	1.0	0–0.4	µg/L
Troponin T	Serum	0–0.1	ng/mL	1.0	0–0.1	µg/L
Tryptophan	Plasma	0.5–1.5	mg/dL	48.967	25–73	µmol/L
Tyrosine	Plasma	0.4–1.6	mg/dL	55.19	20–90	µmol/L
Urea nitrogen	Serum	8–23	mg/dL	0.357	2.9–8.2	mmol/L
Uric acid	Serum	4.0–8.0	mg/dL	59.485	240–480	µmol/L
Urobilinogen	Urine	1–3.5	mg/24 h	1.7	1.7–5.9	µmol/d
Valine	Plasma	1.7–3.7	mg/dL	85.361	145–315	µmol/L
Valproic acid	Serum, plasma	50–100	µg/mL	6.934	346–693	µmol/L
Vancomycin	Serum, plasma	20–40	µg/mL	0.690	14–28	µmol/L
Vanillylmandelic acid (VMA)	Urine	2.1–7.6	mg/24 h	5.046	11–38	µmol/d
Vasoactive intestinal polypeptide	Plasma	<50	pg/mL	0.2960	<15	pmol/L
Vasopressin	Plasma	1.5–2.0	pg/mL	0.923	1.0–2.0	pmol/L
Verapamil	Serum, plasma	100–500	ng/mL	2.20	220–1100	nmol/L
Vitamin A (retinol)	Serum	30–80	µg/dL	0.0349	1.05–2.80	µmol/L
Vitamin B$_1$ (thiamine)	Serum	0–2	µg/dL	29.6	0–75	nmol/L
Vitamin B$_2$ (riboflavin)	Serum	4–24	µg/dL	26.6	106–638	nmol/L
Vitamin B$_3$	Whole blood	0.2–1.8	µg/mL	4.56	0.9–8.2	µmol/L
Vitamin B$_6$ (pyridoxine)	Plasma	5–30	ng/mL	4.046	20–121	nmol/L

H3 Medical Laboratory Tests and Conversion Factors, *continued*

Analyte	Specimen	Reference Range, Conventional Unit	Conven-tional Unit	Conversion Factor (Multiply by)	Reference Range, SI Unit	SI Unit
Vitamin B_{12}	Serum	160–950	pg/mL	0.7378	118–701	pmol/L
Vitamin C (ascorbic acid)	Serum	0.4–1.5	mg/dL	56.78	23–85	µmol/L
Vitamin D (1,25 dihydroxyvitamin D)	Serum	25–45	pg/mL	2.6	60–108	pmol/L
Vitamin D (25-hydroxyvitamin D)	Plasma	14–60	ng/mL	2.496	35–150	nmol/L
Vitamin E (α-tocopherol)	Serum	5–18	µg/mL	23.22	12–42	µmol/L
Vitamin K	Serum	0.13–1.19	ng/mL	2.22	0.29–2.64	nmol/L
Warfarin	Serum, plasma	1.0–10	µg/mL	3.247	3.2–32.4	µmol/L
White blood cell count	Whole blood	4500–11 000	/µL	0.001	4.5–11.0	$\times 10^9$/L
Differential count — Neutrophils-segmented	Whole blood	1800–7800	/µL	0.001	1.8–7.8	$\times 10^9$/L
Differential count — Nuetrophils-bands	Whole blood	0–700	/µL	0.001	0–0.70	$\times 10^9$/L
Differential count — Lymphocytes	Whole blood	1000–4800	/µL	0.001	1.0–4.8	$\times 10^9$/L
Differential count — Monocytes	Whole blood	0–800	/µL	0.001	0–0.80	$\times 10^9$/L
Differential count — Eosinophils	Whole blood	0–450	/µL	0.001	0–0.45	$\times 10^9$/L
Differential count — Basophils	Whole blood	0–200	/µL	0.001	0–0.20	$\times 10^9$/L

White blood cell count, *continued*						
Neutrophils-segmented	Whole blood	56	%	0.01	0.56	Proportion of 1.0
Nuetrophils-bands	Whole blood	3	%	0.01	0.03	Proportion of 1.0
Lymphocytes	Whole blood	34	%	0.01	0.34	Proportion of 1.0
Monocytes	Whole blood	4	%	0.01	0.04	Proportion of 1.0
Eosinophils	Whole blood	2.7	%	0.01	0.027	Proportion of 1.0
Basophils	Whole blood	0.3	%	0.01	0.003	Proportion of 1.0
Zidovudine	Serum, plasma	0.15–0.27	μg/mL	3.7	0.56–1.01	μmol/L
Zinc	Serum	75–120	μg/dL	0.153	11.5–18.5	μmol/L

(Differential count (number fraction))

H4 Glossary of Medical Terms

Acquired Immune Deficiency Syndrome (AIDS). A disease that causes the body's immune system to break down, making it unable to fight off other diseases.

adrenaline. A hormone that is excreted into the body during times of extreme stress, including an increase in heart rate, increased oxygen flow throughout the body and dilated pupils. Also known as epinephrine.

allergen. A substance that causes an allergic reaction within the body, usually foods, dust, pollen or drugs.

allergic reaction. A body having an adverse reaction to a foreign substance that normally will not cause harm to the body. The effects on the body can range to relatively harmless (excess sneezing) to life threatening (anaphylactic shock).

Alzheimer's disease. A progressive disease in which a person's memory is severely impaired due to the dying off of brain cells.

amino acids. Organic chemical compounds made from a joining of amino groups and acidic carboxyl groups, which act as the building blocks of proteins.

analgesics. A medicine and or drug that helps to relieve pain.

anaphylactic shock. A severe and immediate allergic reaction resulting in a sudden drop in blood pressure, irregular heart beats and difficulty in breathing.

anesthetics. A drug that makes a person's body loses all feeling and sensation, and at times consciousness. Is used to induce a controlled sleep so patients can be operated on during surgery.

Angiotensin–Converting Enzyme Inhibitors (ACE Inhibitors). A group of drugs that are used to lower blood pressure and ease stress on the heart after a heart attack.

H4, *continued*

antacids. Medicine that regulates excessive stomach acid activity.

antibiotics. Medicine that is used to treat and prevent bacterial infections within a person's body.

antibodies. T–shaped naturally occurring proteins that bond to and deactivate antigens.

anticoagulant. A substance that prevents or slows down blood clotting.

antidepressant. A drug that is used to treat and manage depression.

antigen. A foreign substance that causes the body harm.

antihistamines. Drugs that are used to treat symptoms brought on by an overabundance of histamine within the body.

antipsychotic drugs. Drugs that are used to treat and manage psychosis.

antiseptic. A drug that destroys microorganisms. Often used to treat skin abrasions and to sterilize an area before a needle is inserted there.

anorexia. The eating disorder in which a person goes not eat or eats less then is needed to remain healthy and active.

arterioles. A small blood vessel that acts as a bridge between an artery and capillaries.

artery. A large blood vessel that carries blood away from the heart and circulates it throughout the body.

asthma. An inflammatory disease in which a person's airways narrow and breathing becomes difficult. Can be brought on by the breathing in of allergens, extreme physical activity, or anxiety.

autoimmune disease. A disease that causes the body's immune system to attack and kill cells within its own body.

barbiturate. A drug that depresses the nervous system and acts as a sedative. Used to help manage seizures and to calm people during times of heightened anxiety.

H4, *continued*

benign. A condition, disease, or abnormality that is nonthreatening to the human body.

benzodiazepines. A class of drugs that slow down the central nervous system. Used as a tranquilizer and or muscle relaxer.

beta–blockers. A class of drugs that decrease the number of contractions within the heart and the force of the contraction itself.

biopharmaceuticals. A drug that is produced through the use of biotechnology and are not created from a natural biological source.

biopsy. The removal of cells or tissue from a source sight with the intent to examine and test them.

biotechnology. Technology that develops new microorganisms and gives new traits to preexisting organic material.

blood pressure. Referring to the pressure the heart excerpts as blood is pushed through it, measured by the number of times the heart beats per minute.

bone marrow. Tissue that fills cavities within bones and produces blood cells, specifically red blood cells.

bulimia. An eating disorder in which a person will gorge on high calorie food and then forcefully induce vomiting to rid it from their system.

calcium channel blockers. A drug that stops calcium from entering the cells of already developed muscles.

cancer. A disease in which cells grow abnormally at one point in the body and later spread out and break down material throughout the body.

carcinogen. Any substance that can cause cancer to develop.

carcinoma. A tumor that lines the walls of an organ and is a precursor to cancer.

catheter. A device that is inserted in a person's body to allow for a pathway for various fluids to flow through.

H4, *continued*

central nervous system. Referring to the section of the nervous system that consists of the brain and spinal cord.

cerebral edema. Referring to an excess amount of fluid accumulating in and around the brain.

chemotherapy. A cancer treatment in which anticancer chemicals are used in an attempt to destroy the cancer.

cholesterol. A waxy substance that strengthens cell membranes and helps form vitamin D, steroid hormones, and bile. An overabundance of cholesterol can blocks blood flow and lead to a infarction.

cleft palate. A disorder in which a fissure forms at the roof of one's mouth during the development of an embryo.

Computerized Axial Tomography Scan (CAT Scan). A device that takes multiple x–rays of a subject and turns it into a computerized three–dimensional image.

constipation. A condition in which a person has difficulty voiding their bowels.

contraceptives. Devises that are used to prevent pregnancy during intercourse.

contractions. Referring to the movement of muscles. More commonly is used to refer to the painful widening of the uterus during childbirth.

coronary arteries. Two arteries that supply blood to the heart.

cytoplasm. Referring to the parts of a cell that are outside of the nucleus but within the cell membrane.

defecation. The removal of solid waste from the body.

delirium. A mental condition in which a person's ability to think and reason becomes impaired.

depression. A medical condition in which feelings of great sadness are intense and long lasting.

H4, *continued*

diabetes. A progressive disease in which the body is unable to produce insulin or maintain insulin levels within the body.

diuretic. A drug that is used to increase the formation of urine in an effort to reduce excessive water within a person's body.

dyspnea. Having difficulty breathing or breathing being painful and labored. Also known as shortness of breath. Can be a temporary or permanent condition.

endemic area. An area on the body in which a certain disease has formed.

endorphins. Hormones that are released into the body at times of extreme stress or trauma, such as when the body has suffered an injury or when a person's body has endured extreme physical activity, which is commonly known as a "runner's high."

enzyme. A protein that speeds up the reactions between substances.

epinephrine. Another phrase for adrenaline. Additionally a generic name for a class of drugs that synthetically perform the effects of adrenaline.

epidemic. Referring to a disease that can infect many members of a population at the same time.

epilepsy. A medical condition in which a person is plagued by frequent seizures.

fiber (1). Cells that form the physical shape and structure of internal organs. **(2).** Material found in plants that helps food move throughout the digestion track and helps in defecation.

fight–or–flight response. In prehistoric times a response to potentially harmful situations, resulting in either confronting an adversary or fleeing from it. In modern times this results more in a person's need to deal with a stressful situation, either in fight (aggressive behavior, arguing with others) or flight (social withdrawal, social support, substance abuse).

frostbite. The freezing of living tissue brought on from exposure to extreme cold.

H4, *continued*

gamma globulin. An antibiotic rich protein that helps to fight foreign substances within the body.

gangrene. The death and decay of tissue and or organs, brought on by a lack of blood flow and oxygen.

heart attack. An infarction within the heart that causes blood flow to become blocked. Can severely damage the tissue within the heart and often leads to death.

histamine. An organic compound that causes the constriction of muscles and the inflammation of airways when it is released into the body.

homeostasis. The regulation of internal systems within the body so they remain at a constant despite changing external factors.

hormones. Chemicals that regulate activity within organs and aid in growth and development.

hypothermia. A condition referring to the abnormal lowering of the body's core temperature, most times brought on by prolonged exposure to freezing temperatures.

immune system. A system of functions within the body that protect us from infection and disease.

immunization. The process of making the body immune to select diseases and or ailments.

in vitro. Referring to a process taking place in an artificial controlled environment.

in vitro fertilization. The fertilization of an egg that takes place outside of the body.

in vivo. Referring to activity or a process that takes place within an organism.

infarction. The damaging and dying of muscle within the body due to the loss of blood flow to the tissue, usually caused by a blood clot.

H4, *continued*

inflammation. A localized protective measure of the body when a specific area is injured or infected. This results in the area becoming swollen, reddened, and painful.

inoculation. Introducing a foreign agent to an organism in order for it to build up an immunity to it.

interferon. A naturally occurring and synthetic protein that aids in the prevention of a virus multiplying within the body.

joints. Referring to the meeting point between bones within the body.

laxative. A drug that helps regulates and increases bowel movements. Often are used to treat constipation.

lipids. Referring to material that is insoluble in water and helps form the structure of cellular material. Examples of which are fats and oils.

lumen. The hollow space within a tubular structure, such as the space within intestines.

Magnetic Resonance Imaging (MRI). A machine that uses magnetic and radio pulses to create vivid images of the human body.

malocclusion. The misalignment of the top and bottom set of teeth when they might down.

menopause. Referring to the point on a women's life when her body stops menstruating.

metabolism. The process in which material within the body is broken down.

metastasis. Referring to the growth of cancer cells that have transferred from their original growing site.

narcotics. A drug that dulls the reactions of the central nervous system. Also used to describe drugs that a person would illegally take.

nausea. An unpleasant feeling within a person's stomach that make a person want to throw up.

H4, *continued*

nerve. A bundle of fibers that allow nerve impulses to move throughout the body.

nerve impulse. Electrical signals that transmit information throughout the nervous system.

nervous system. The internal system that allows an organism to interact with and adapt to its environment.

neuron. A cell that produces a nerve impulse.

neurotransmitter. A chemical that allows information to move between the gaps between neurons.

neurotoxins. Toxins that specifically attack nerve cells.

nitroglycerin. A drug that acts as a vasodilator to blood vessels within the heart.

Nonsteroidal Anti–Inflammatory Drugs (NSAIDs). Drugs that treat inflammation and pain that do not contain steroids within them.

obesity. A medical condition in which a person body weight is 20% or above there ideal weight.

Obsessive Compulsive Disorder (OCD). A disorder in which a person feels compelled to immediately do certain activities in a very careful and select manner.

pain. An unpleasant sensation the body feels when it is harmed. Is an indicator that something is wrong with the body.

pacemaker. An electrical device surgically implanted into a person's chest that regulates the heartbeats of a damaged heart.

pandemic. An epidemic that has spread over a great area such as a country or continent.

panic attack. An intense feeling of fear and anxiety resulting in hyperventilation, dizziness, sweating, trembling, dyspnea, chest pain and paresthesia.

H4, *continued*

pathogen. An agent that has the potential to carry disease.

Penicillin. A drug formed from mold that is used as an antibiotic.

peptide. A short grouping of amino acids.

peristalsis. Referring to the contractions of digestive muscles that allow food to pass through.

plaque (1). A film of food particles and bacteria that form on the surfaces of teeth.

plaque (2). Deposits of fat that form in the walls of blood vessels, which can slow and or block blood flow.

polypeptides. A peptide that contains a large number of amino acids.

polyps. A tumor that develops on the inside of a hollow mucus lined organ or opening, such as the intestines, rectum, or the inside of the nose.

presbyopia. A condition in which a person's eye is less able to view close objects sharply.

prosthesis. An artificial device used to replace a missing body part, such as a missing limb or tooth.

proteins. A large grouping of amino acids that make up the majority of mass within internal structures.

psychosis. A severe mental disorder in which a patient loses touch with reality, becoming unable to differentiate between what is actually happening around them and what is fake or imaginary.

puberty. Referring to the physical changes a person's body goes through as they change from children to adult.

saturated fat. Fats that raises cholesterol levels within the blood, often coming from animal products.

scalpel. A small knife with a razor sharp edge that is used for making incisions during surgery.

H4, *continued*

seizures. A loss of motor function and feeling, followed by uncontrol–lable shaking and spasms, caused by misfiring nerve cells in the brain.

Sexually Transmitted Diseases (STDs). Diseases that are transferred from one person to the other through sexual activity.

shock. A state a person enters into when blood flow is not sufficient enough to meet the body's basic needs.

somnolence. A state in which the body is usually tired or drowsy.

spasm. A sudden and involuntary contraction of a muscle.

steroid (1) Referring to a group of substances that share similar chemical structures. **(2)** A drug (often illegally used) to increase muscle mass.

stroke. The sudden dying of brain cells as a result of blood flow to the brain being blocked.

synapse. A junction where information is passed from one nerve cell to another or from a nerve cell to an organ.

tissue Plasminogen Activator (tPA). An enzyme that helps to dissolve clots.

tissue. A group of cells that perform the same specific function.

tourniquet. A device that is used to stop blood flow to a limb when an extreme injury occurs. Should only be used in when applying pressure to the wound will not stop blood flow.

trauma. Referring to physical and chemical injury to the body.

tranquilizer. A drug that calms the central nervous system, which can result in unconsciousness in high doses.

uric acid. The end product from the breakdown of protein that is eliminated from the body through urine.

ulcer. A depressed erosion of skin, most commonly found to occur within the walls of the intestine.

H4, *continued*

ultrasound. A devise that used ultrasonic sound waves to form an image of an internal structure. Most commonly used during pregnancy to form an image of the developing fetus.

vaccinate. Medicine that is injected into a person, usually at a young age, to prevent them from obtaining certain diseases.

vascular system. Referring to the total system of arteries, veins and capillaries within the body.

vasodilators. A drug that is used to dilute blood vessels.

vertigo. A type of dizziness in which a person feels as if their surroundings are swaying even though the environment if stationary.

vomit. Referring to partially digested food that has been ejected from the body through the mouth.

wart. A contained growth on he skin caused by a virus.

yawn. A deep intake of oxygen meant to aid the body when it becomes tired or overworked.

H5 Abbreviations of Health Agencies and Organizations

Abbreviation	Organization
AAAAI	American Academy of Allergy, Asthma, and Immunology
AAAS	American Association for the Advancement of Science
AABB	American Association of Blood Banks
AACAP	American Academy of Child and Adolescent Psychiatry
AACC	American Association of Clinical Chemists
AACIA	American Association for Clinical Immunology
AACN	American Association of Colleges of Nursing; American Association of Critical–Care Nurses
AAD	American Academy of Dermatology
AAFP	American Academy of Family Physicians
AAFPRS	American Academy of Facial Plastic and Reconstructive Surgery
AAHSLD	Association of Academic Health Science Library Directors
AAI	American Association of Immunologists
AAMC	Association of American Medical Colleges
AAMCH	American Association of Maternal and Child Health
AAN	American Academy of Neurology; American Academy of Neuropathologists; American Academy of Nursing
AANA	American Association of Nurse Anesthetists
AANP	American Academy of Nurse Practitioners
AANS	American Association of Neurological Surgeons
AAO	American Academy of Ophthalmology
AAOHNS	American Academy of Otolaryngology–Head and Neck Surgery
AAOS	American Academy of Orthopaedic Surgeons
AAP	American Academy of Pediatrics; American Association of Pathologists
AAPA	American Academy of Physician Assistants; American Association of Pathologist's Assistants
AAPHP	American Association of Public Health Physicians
AAPM	American Academy of Pain Medicine; American Association of Physicists in Medicine
AAPMR	American Academy of Physical Medicine and Rehabilitation
AAPS	American Association of Plastic Surgeons
AARP	American Association of Retired Persons
AATM	American Academy of Tropical Medicine
AATS	American Association of University Professors

H5, *continued*

AAWR	American Association for Women Radiologists
ABA	American Bar Association
ABMS	American Board of Medical Specialties
ACA	American College of Allergists; American College of Anesthetists
ACAAI	American College of Allergy, Asthma, and Immunology
ACC	American College of Cardiology
ACCME	Accreditation council for Continuing Medical Education
ACCP	American College of Chest Physicians
ACEP	American College of Emergency Physicians
ACG	American College of Gastroenterology
ACGME	Accreditation Council for Graduate Medical Education
ACHA	American College Health Association
ACIP	Advisory Committee on Immunization Practices
ACLM	American College of Legal Medicine
ACMQ	American College of Medical Quality
ACNM	American College of Nuclear Medicine; American College of Nurse–Midwives
ACNP	American College of Nuclear Physicians
ACOEM	American College of Occupational and Environmental Medicine
ACOG	American College of Obstetricians and Gynecologists
ACP	American College of Physicians
ACPE	American College of Physician Executives
ACPM	American College of Preventive Medicine
ACR	American College of Radiology; American College of Rheumatology
ACS	American Cancer Society; American Chemical Society; American College of Surgeons
ACSM	American College of Sports Medicine
ADA	American Dental Association; American Dermatological Association; American Diabetes Association; American Dietetic Association
ADRDA	Alzheimer's Disease and Related Disorders Association
AERS	Adverse Event Reporting System; (US Food and Drug Administration)
AES	American Epilepsy Society
AFAR	American Federation for Aging Research
AFCR	American Federation for Clinical Research

H5, *continued*

AFIP	Armed Forces Institute of Pathology
AFS	American Fertility Society
AGA	American Geriatrics Society
AHA	American Heart Association; American Hospital Association
AHrA	American Healthcare Radiology Administrators
AHRQ	Agency for Healthcare Research and Quality
AJCC	American Joint Committee on Cancer
ALA	American Library Association; American Lung Association
ALROS	American Laryngological, Rhinological and Otological Society
AMA	Aerospace Medical Association; American Management Association; American Marketing Association; American Medical Association; Australian Medical Association
AMDA	American Medical Directors Association
AMPA	American Medical Publishers' Association
AMSA	American Medical Student Association
AMSUS	Association of Military Surgeons of the United States
AMWA	American Medical Women's Association; American Medical Writers Association
ANA	American Neurological Association; American Nurses Association
ANSI	American National Standards Institute
AOA	Alpha Omega Alpha; American Orthopaedic Association; American Osteopathic Association
AOMA	American Occupational Medicine Association
AONE	American Organization of Nurse Executives
AORN	Association of Operating Room Nurses
AOS	American Otological Society
AOWHN	American Organization of Women's Health Nurses
APA	Ambulatory Pediatrics Association; American Pharmaceutical Association; American Psychiatric Association; American Psychological Association
APHA	American Public Health Association
APM	Academy of Physical Medicine
APS	American Physical Society; American Physiological Society; American Psychological Society
ARA	American Rheumatism Association
ARC	American Red Cross
ARENA	Applied Research Ethics National Association

H5, *continued*

ARRS	American Roentgen Ray Society
ARVO	Association for Research in Vision and Ophthalmology
ASA	American Society of Anesthesiologists
ASAM	American Society of Addiction Medicine
ASCN	American Society of Clinical Nutrition
ASCO	American Society of Clinical Oncology; American Society of Clinical Ophthalmology
ASCP	American Society of Clinical Pathologists; American Society of Consultant Pharmacists
ASCPT	American Society of Clinical Pharmacology and Therapeutics
ASCRS	American Society of Cataract and Refractive Surgery; American Society of Colon and Rectal Surgeons
ASDR	American Society of Diagnostic Radiology
ASDS	American Society for Dermatologic Surgery
ASG	American Society for Genetics
ASHG	American Society of Human Genetics
ASLME	American Society of Law, Medicine & Ethics
ASM	American Society for Microbiology
ASMT	American Society of Medical Technologists
ASPRS	American Society of Plastic and Reconstructive Surgeons
ASTHO	Association of State and Territorial Health Officers
ASTMH	American Society of Tropical Medicine and Hygiene
ASTRO	American Society for Therapeutic Radiology and Oncology
ASTS	American Society of Transplant Surgeons
ATA	American Thyroid Association
ATS	American Thoracic Society
AUA	American Urological Association
BMA	British Medical Association
CAP	College of American Pathologists
CDC	Centers for Disease Control and Prevention
CMA	Canadian Medical Association
CMS	Centers for Medicare &Medicaid Services
CNS	Child Neurology Society
CSE	Council of Science Editors
DHHS	Department of Health and Human Services
EASE	European Association of Science Editors
ECDC	European Centre for Disease Prevention and Control
ECFMG	Educational Commission for Foreign Medical Graduates
EEOC	Equal Employment Opportunity Commission

H5, *continued*

EIS	Epidemic Intelligence Service; (US centers for disease control and Prevention)
EPA	Environmental Protection Agency
EU	European Union
FASEB	Federation of American Societies for Experimental Biology
FCC	Federal Communications Comission
FASEB	Federation of American Societies for Experimental Biology
FCC	Federal Communications Commission
FDA	Food and Drug Administration
FTC	Federal Trade Commission
GLMA	Gay and Lesbian Medical Association
GSA	Gerontological Society of America
IARC	International Agency for Research on Cancer
ICAAC	Interscience Conference on Antimicrobial Agents and Chemotherapy
ICMJE	International Committee of Medical Journal Editors
ICN	International Council of Nurses
ICRC	International Committee of the Red Cross
ICS	International College of Surgeons
IDSA	Infectious Diseases Society of America
IEEE	Institute of Electrical and Electronics Engineers
IOM	Institute of Medicine
IPPNW	International Physicians for the Prevention of Nuclear War
ISBT	International Society of Blood Transfusion
ISO	International Organization for Standardization
JCAHO	Joint Commission on Accreditation of Healthcare Organizations
MGMA	Medical Group Management Association
MLA	Medical Library Association
MRC	Medical Research Council
MSF	Médecins Sans Frontières
NAME	National Association of Medical Examiners
NAMS	North American Menopause Society
NAS	National Academy of Sciences
NASA	National Aeronautics and Space Administration
NBME	National Board of Medical Examiners
NCBI	National Center for Biotechnology Information
NCCAM	National Center for Complementary and Alternative Medicine
NCHS	National Center for Health Statistics
NCI	National Cancer Institute

H5, *continued*

NCOA	National Committee on Quality Assurance
NCRR	National Center for Research Resources
NEI	National Eye Institute
NHGRI	National Human Genome Research Institute
NHLBI	National Heart, Lunch, and Blood Institute
NHO	National Hospice Organization
NIA	National Institute on Aging
NIAAA	National Institute on Alcohol Abuse and Alcoholism
NIAID	National Institute of Allergy and Infectious Diseases
NIAMS	National Institute of Arthritis and Musculoskeletal and Skin Diseases
NIBIB	National Institute of Biomedical Imaging and Bioengineering
NICHD	National Institute of Child Health and Human Development
NIDA	National Institute on Drug Abuse
NIDCD	National Institute on Deafness and Other Communication Disorders
NIDCR	National Institute of Dental and Craniofacial Research
NIDDK	National Institute of Diabetes and Digestive and Kidney Diseases
NIEHS	National Institute of Environmental Health Sciences
NIGMS	National Institute of General Medical Sciences
NIH	National Institutes of Health
NIMH	National Institute of Mental Health
NINDS	National Institute of Neurological Disorders and Stroke
NINR	National Institute of Nursing Research
NIOSH	National Institute for Occupational Safety and Health
NISO	National Information Standards and Organization
NLM	National Library of Medicine
NLN	National League for Nursing
NMA	National Medical Association
NMHA	National Mental Health Association
NRC	National Research Council; Nuclear Regulatory Commission
NRMP	National Resident Matching Program
NSF	National Science Foundation
NSPB	National Society for the Prevention of Blindness
OMAR	Office of Mental Applications of Research
ONS	Oncology Nursing Society
OPRR	Office for Protection from Research Risks
ORI	Office of Research Integrity

H5, *continued*

ORWH	Office of Research on Women's Health
OSHA	Occupational Safety and Health Administration
PAHO	Pan American Health Organization
PHR	Physicians for Human Rights
PHS	Public Health Service
PSR	Physicians for Social Responsibility
PSRO	Professional Standards Review Organization
RDCRN	Rare Diseases Clinical Research Network
RPB	Research to Prevent Blindness
RSNA	Radiological Society of North America
SAMBA	Society for Ambulatory Anesthesia
SAMHSA	Substance Abuse and Mental Health Services Administration
SCCM	Society of Critical Care Medicine
SEC	Securities and Exchange Commission
SID	Society for Investigative Dermatology
SMCAF	Society of Medical Consultants to the Armed Forces
SNM	Society of Nuclear Medicine
SSA	Social Security Administration
SSO	Society of Surgical Oncology
SSP	Society for Scholarly Publishing
STC	Society for Technical Communication
STS	Society of Thoracic Surgeons
UICC	International Union Against Cancer (Union Internationale Contre le Cancer)
UN	United Nations
UNHCR	United Nations High Commissioner for Refugees
UNICEFF	United Nations Children's Fund
UNOS	United Network for Organ Sharing
USAN	United States Adopted Names [Council]
VA	Department of Veterans Affairs
WAME	World Association of Medical Editors
WFP	World Food Program
WHO	World Health Organization
WIC	Special Supplemental Nutrition for Women, Infants, and Children
WMA	World Medical Association

H6 The National Library of Medicine (NLM) List of Accepted Abbreviations

Word	Abbreviation	Word	Abbreviation
Abnormal	Abnorm	Anaesthesia	Anaesth
Abuse	Abuse	Anaesthetist	Anaesthetist
Academia	Acad	Anaesthetists	Anaesth
Academy	Acad	Analgesia	Analg
Acoustical	Acuost	Anatomical	Anat
Actions	Actions	Anatomy	Anat
Acupuncture	Acupunct	Andrology	Androl
Acute	Acute	Anesthesia	Anesth
Addiction	Addict	Anesthesiology	Anesthesiol
Addictions	Addict	Angiology	Angiol
Additives	Addit	Angle	Angle
Administration	Adm	Animal	Anim
Adolescence	Adolescence	Ankle	Ankle
Adolescent	Adolesc	Annals	Ann
Advanced	Adv	Annual	Annu
Advancement	Adv	Anthropology	Anthropol
Advances	Adv	Antibiotics	Antibiot
Adverse	Adverse	Anticancer	Anticancer
Aesthetic	Aesthetic	Antigens	Antigens
Affairs	Aff	Antimicrobial	Antimicrob
Affective	Affective	Antiviral	Antiviral
African	Afr	Apheresis	Apheresis
Age	Age	Appetite	Appetite
Ageing	Ageing	Applied	Appl
Agents	Agents	Archives	Arch
Aging	Aging	Argentina	Argent
Air	Air	Arizona	Ariz
Alabama	Ala	Arkansas	Ark
Alaska	Alaska	Army	Army
Alcohol	Alcohol	Arteriosclerosis	Arterioscl
Alcoholism	Alcohol	Artery	Artery
Allergy	Allergy	Arthritis	Arthritis
Allied	Allied	Artificial	Artif
America	Am	Asian	Asian
American	Am	Assessment	Assess

H6, *continued*

Association	Assoc	Breast	Breast
Asthma	Asthma	British	Br
Audiology	Audiol	Bulletin	Bull
Audiovisual	Audiov	Burns	Burns
Auditory	Aud	Calcified	Calcif
Australia	Aust	Calcium	Calcium
Australian	Aust	Canadian	Can
Autism	Autism	Cancer	Cancer
Autonomic	Auton	Carbohydrate	Carbohydr
Avian	Avian	Carcinogenesis	Carcinog
Aviation	Aviat	Carcinogenic	Carcinog
Bacteriology	Bacteriol	Cardiography	Cardiogr
Bangladesh	Bangladesh	Cardiology	Cardiol
Basic	Basic	Cardiovascular	Cardiovasc
Behavior	Behav	Care	Care
Behavioral	Behav	Caries	Caries
Behaviors	Behav	Catheterization	Cathet
Biochemical	Biochem	Cell	Cell
Biochemistry	Biochem	Cells	Cells
Biocommunications	Biocomm	Cellular	Cell
Biofeedback	Biofeedback	Central	Cent
Biological	Biol	Cephalalgia	Cephalalgia
Biology	Biol	Cerebral	Cereb
Biomaterials	Biomater	Ceylon	Ceylon
Biomechanical	Biomech	Chemical	Chem
Biomedical	Biomed	Chemicals	Chem
Biometrics	Biometrics	Chemistry	Chem
Biophysical	Biophys	Chemists	Chem
Biophysics	Biophys	Chemotherapy	Chemother
Bioscience	Biosci	Chest	Chest
Biosocial	Biosoc	Child	Child
Biosystems	Biosystems	Childhood	Child
Biotechnological	Biotechnol	Children	Child
Biotechnology	Biotechnol	Childs	Childs
Birth	Birth	Chinese	Chin
Blood	Blood	Chromatographic	Chromatogr
Bone	Bone	Chromatography	Chromatogr
Brain	Brain	Chronic	Chronic
Brazillian	Braz	Chronicle	Chron

H6, *continued*

Circulation	Circ	Crybiology	Crybiol
Circulatory	Circ	Culture	Cult
Cleft	Cleft	Current	Curr
Cleveland	Cleve	Currents	Curr
Clinic	Clin	Cutaneous	Cutan
Clinical	Clin	Cutis	Cutis
Clinics	Clin	Cybernetics	Cybern
Cognition	Cogn	Cyclic	Cyclic
Collagen	Coll	Cytogenetics	Cytogenet
College	Coll	Cytology	Cytol
Colon	Colon	Cytometry	Cytometry
Colorado	Colo	Dairy	Dairy
Communicable	Commun	Danish	Dan
Communication	Commun	Deaf	Deaf
Communications	Commun	Decision	Decis
Community	Community	Defects	Defects
Comparative	Comp	Deficiency	Defic
Complement	Complement	Delivery	Deliv
Comprehensive	Compr	Demography	Demogr
Computerized	Comput	Dental	Dent
Computers	Comput	Dentistry	Dent
Connecticut	Conn	Dependencies	Dependencies
Connective	Connect	Dermatitis	Dermatitis
Consulting	Consult	Dermatological	Dermatol
Contact	Contact	Dermatology	Dermatol
Contaminants	Contam	Dermatopathology	Dermatopathol
Contamination	Contam	Detection	Detect
Contemporary	Contemp	Development	Dev
Contributions	Contrib	Devices	Devices
Control	Control	Diabetes	Diabetes
Controlled	Control	Diagnosis	Diagn
Copenhagen	Copenh	Diagnostic	Diagn
Cornea	Cornea	Dialysis	Dial
Cornell	Cornell	Diarrhoeal	Diarrhoeal
Corps	Corps	Dietetic	Diet
Cortex	Cortex	Differentiation	Differ
Council	Counc	Digestion	Digestion
Craniofacial	Craniofac	Digestive	Dig
Critical	Crit	Dimensions	Dimens

H6, *continued*

Directions	Dir	Epidemiology	Epidemiol
Directors	Dir	Ergology	Ergol
Discussions	Discuss	Ergonomics	Ergonomics
Disease	Dis Enteral	Essays	Essays
Diseases	Dis	Ethics	Ethics
Disorders	Disord	Eugenics	Eugen
Disposition	Dispos	European	Eur
DNA	DNA	Evaluation	Eval
Drug	Drug	Exceptional	Except
Drugs	Drugs	Exercise	Exerc
Ear	Ear	Experimental	Exp
Early	Early	Eye	Eye
East African	East Afr	Factors	Factors
Economic	Econ	Family	Fam
Ecotoxicology	Ecotoxicol	Federation	Fed
Educational	Educ	Fertility	Fertil
Egyptian	Egypt	Finnish	Finn
Electrocardiology	Electrocardiol	Fitness	Fitness
Electroencephalo-graphy	Electroence-phalogr	Florida	Fla
Electromyography	Electromyogr	Food	Food
Electron	Electron	Foot	Foot
Electrotherapeutics	Electrother	Forensic	Forensic
Embryo	Embryo	Foundation	Found
Embryology	Embryol	Function	Funct
Emergency	Emerg	Fundamental	Fundam
Endocrine	Endocr	Gastroenterology	Gastroenterol
Endocrinological	Endocrinol	Gastrointestinal	Gastrointest
Endocrinology	Endocrinol	Gene	Gene
Endoscopy	Endosc	General	Gen
Engineering	Eng	Genetic	Genet
Enteral	Enteral	Genetics	Genetics
Entomology	Entomol	Genitourinary	Genitourin
Environmental	Environ	Geographical	Geogr
Enzyme	Enzyme	Georgia	Ga
Enzymology	Enzymol	Geriatric	Geriatr
Epidemiologic	Epidemiol	Geriatrics	Geriatr

H6, *continued*

Gerontologist	Gerontologist	Hybridoma	Hybridoma
Gerontology	Gerontol	Hygiene	Hyg
Group	Group	Hypertension	Hypertens
Groups	Groups	Hypnosis	Hypn
Growth	Growth	Hypotheses	Hypotheses
Gut	Gut	Imaging	Imaging
Gynaecological	Gynaecol	Immunity	Immun
Gynaecology	Gynaecol	Immunoassay	Immunoassay
Gynecologic	Gynecol	Immunobiology	Immunobiol
Gynecology	Gynecol	Immunogenetics	Immunogenet
Haematology	Haematol	Immunological	Immunol
Haemostasis	Haemost	Immunology	Immunol
Hastings Center	Hastings Cent	Immunopharma-cology	Immunopharma-col
Hawaii	Hawaii	Immunotherapy	Immunother
Head	Head	Implant	Implant
Headache	Headache	Including	Incl
Health	Health	India	India
Hearing	Hear	Indian	Indian
Heart	Heart	Indiana	Indiana
Hematological	Hematol	Industrial	Ind
Hematology	Hematol	Infection	Infect
Hemoglobin	Hemoglobin	Infectious	Infect
Hemostasis	Hemost	Inflammation	Inflamm
Hepatology	Hepatol	Informatics	Inform
Heredity	Hered	Information	Inf
Hip	Hip	Inherited	Inherited
Histochemical	Histochem	Injury	Inj
Histochemistry	Histochem	Inorganic	Inorg
Histology	Histol	Inquiry	Inquiry
Histopathology	Histopathol	Institutes	Inst
History	Hist	Instrumentation	Instrum
Homosexuality	Homosex	Insurance	Insur
Horizons	Horiz	Intellectual	Intellect
Hormone	Horm	Intelligence	Intell
Hormones	Horm	Intensive	Intensive
Hospital	Hosp	Interactions	Interact
Hospitals	Hospitals	Interferon	Interferon
Human	Hum	Internal	Intern
Humans	Hum	International	Int

H6, *continued*

Internist	Internist	Library	Libr
Interventional	Intervent	Life	Life
Intervirology	Intervirol	Life–threatening	Life Threat
Intraocular	Intraocul	Lipid	Lipid
Invasion	Invasion	Lipids	Lipids
Invertebrate	Invertebr	Literature	Lit
Investigation	Invest	Louisiana	La
Investigational	Investig	Lung	Lung
Investigations	Invest	Lymphokine	Lymphokine
Investigative	Invest	Lymphology	Lymphol
In Vitro	In Vitro	Madagascar	Madagascar
In Vivo	In Vivo	Magnesium	Magnesium
Iowa	Iowa	Magnetic	Magn
Irish	Ir	Main	Main
Isotopes	Isot	Making	Making
Isozymes	Isozymes	Malaysia	Malaysia
Israel	Isr	Management	Manage
Issues	Issues	Manipulative	Manipulative
Istanbul	Istanbul	Marital	Marital
Japanese	Jpn	Maritime	Marit
Joint	Joint	Maryland	Md
Journal	J	Mass	Mass
Kansas	Kans	Mathematical	Math
Kentucky	Ky	Maxillofacial	Maxillofac
Kidney	Kidney	Measurement	Meas
Kinetics	Kinet	Mechanics	Mech
Laboratory	Lab	Media	Media
Language	Lang	Medical	Med
Laparoendoscopic	Laparoendosc	Medicinal	Med
Laryngology	Laryngol	Medicine	Med
Larynx	Larynx	Membrane	Membr
Lasers	Lasers	Mental	Ment
Law	Law	Metabolic	Metab
Lectures	Lect	Metabolism	Metab
Legal	Leg	Metastasis	Metastasis
Leprosy	Lepr	Methods	Methods
Letters	Lett	Mexico	Mex
Leukocyte	Leukoc	Michigan	Mich
Leukotriene	Leukotriene	Microbial	Microb
Leukotrienes	Leukotrienes	Microbiological	Microbiol

H6, *continued*

Microbiology	Microbiol	Neuroendocrinology	Neuroendocrinol
Microcirculation	Microcirc	Neurogenetics	Neurogenet
Microscopy	Mictrosc	Neuroimmunology	Neuroimmunol
Microvascular	Microvasc	Neurologic	Neurol
Microwave	Microw	Neurological	Neurol
Military	Milit	Neurology	Neurol
Mineral	Miner	Neuropathology	Neuropathol
Minnesota	Minn	Neuropediatrics	Neuropediatr
Mississippi	Miss	Neuropeptides	Neuropeptides
Missouri	Mo	Neuropharmacology	Neuropharmacol
Modification	Modif	Neurophysiology	Neurophysiol
Molecular	Mol	Neuropsychobiology	Neuropsycobiol
Monographs	Monogr	Neuropsychology	Neuropsychol
Morphology	Morphol	Neuropsychopharma-cology	Neuropsycho-pharmacol
Motility	Motil	Neuroradiology	Neuroradiol
Muscle	Muscle	Neuroscience	Neurosci
Mutagenesis	Mutagen	Neurosurgery	Neurosurg
Mutation	Mutat	Neurosurgical	Neurosurg
Mycobacterial	Mycobact	Neurotoxicology	Neurotoxicol
Narcotics	Narc	Neurotrauma	Neurotrauma
National	Natl	New	N
Natural	Nat	New England	N Engl
Nature	Nat	New Jersey	N J
Naval	Nav	New Orleans	New Orleans
Nebraska	Nebr	New York	N Y
Neck	Neck	New Zealand	N Z
Neglect	Negl	North America	North Am
Neonate	Neonate	North Carolina	N C
Nephrology	Nephrol	Nose	Nose
Nephron	Nephron	Nuclear	Nucl
Nervosa	Nerv	Nucleotide	Nucleotide
Nervous	Nerv	Nurse	Nurse
Netherlands	Neth	Nursing	Nurs
Neural	Neural	Nutrition	Nutr
Neurobehavioral	Neurobehav	Nutritional	Nutr
Neurobiology	Neurobiol	Obesity	Obes
Neurochemistry	Neurochem	Obstetric	Obstet
Neurocytology	Neurocytol	Obstetrics	Obstet

H6, *continued*

Occupational	Occup	Pediatrician	Pediatrician
Ocular	Ocul	Pediatrics	Pediatr
Official	Off	Pennsylvania	Pa
Ohio	Ohio	Peptide	Pept
Oklahoma	Okla	Peptides	Pept
Oncology	Oncol	Perception	Perception
Ophthalmic	Opthalmic	Perceptual	Percept
Ophthalmological	Ophthalmol	Perinatal	Perinat
Ophthalmology	Ophthalmol	Perinatology	Perinatol
Optical	Opt	Periodontal	Periodont
Optics	Opt	Periodontology	Periodontol
Optometric	Optom	Personality	Pers
Optometry	Optom	Perspectives	Perspect
Oral	Oral	Pharmaceutical	Pharm
Organization	Organ	Pharmacokinetics	Pharmacokinet
Organs	Organs	Pharmacology	Pharmacol
Orthodontics	Orthod	Pharmaco–psychiatry	Pharmaco–psychiatry
Orthodontist	Orthod	Pharmacotherapy	Pharmacother
Orthopaedic	Orthop	Pharmacy	Pharm
Orthopsychiatry	Orthopsychiatry	Philosophical	Philos
Orthotics	Orthotics	Phosphorylation	Phosphorylation
Osaka	Osaka	Photobiology	Photobiol
Oslo	Oslo	Photochemistry	Photochem
Osteopathic	Osteopath	Photodermatology	Photodermatol
Otolaryngology	Otolaryngol	Photography	Photogr
Otology	Otol	Physical	Phys
Otorhinolaryn-gology	Otorhinolaryn-gol	Physician	Physician
Pace	Pace	Physicians	Physicians
Paediatric	Paediatr	Physics	Phys
Paediatrics	Paediatr	Physiological	Physiol
Palate	Palate	Physiology	Physiol
Panama	Panama	Placenta	Placenta
Pan American	Pan Am	Planning	Plann
Paper	Pap	Plastic	Plast
Papua New Guinea	Papua New Guinea	Podiatric	Podiatr
Parasite	Parasite	Podiatry	Podiatry
Parasitology	Parasitol	Poisoning	Poisoning
Parenteral	Parenter	Policy	Policy
Pathology	Pathol	Politics	Polit

H6, *continued*

Pollution	Pollut	Psychosocial	Psychosoc
Population	Popul	Psychosomatic	Psychosom
Postgraduate	Postgrad	Psychosomatics	Psychosom
Poultry	Poult	Psychotherapy	Psychother
Practice	Pract	Public	Public
Practitioners	Pract	Puerto Rico	P R
Pregnancy	Pregnancy	Quantitative	Quant
Prenatal	Prenat	Quarterly	Q
Preparative	Prep	Radiation	Radiat
Prevention	Prev	Radiography	Radiogr
Primary	Primary	Radioisotopes	Radioisotopes
Primatology	Primatol	Radiologists	Radiol
Proceedings	Proc	Radiology	Radiol
Process	Process	Rational	Ration
Processes	Processes	Reactions	React
Products	Prod	Recombinant	Recomb
Programs	Programs	Reconstructive	Reconstr
Progress	Prog	Record	Rec
Prostaglandin	Prostaglandin	Rectum	Rectum
Prostaglandins	Prostaglandins	Regional	Reg
Prostate	Prostate	Regulation	Regul
Prosthetic	Prosthet	Regulatory	Regul
Prosthetics	Prosthet	Rehabilitation	Rehabil
Protein	Protein	Renal	Renal
Protozoology	Protozool	Report	Rep
Psyche	Psyche	Reports	Rep
Psychiatric	Psyciatr	Reproduction	Reprod
Psychiatry	Psychiatry	Reproductive	Reprod
Psychoactive	Psychoactive	Research	Res
Psychoanalysis	Psychoanal	Residue	Reisdue
Psychoanalytic	Psychoanal	Resonance	Reson
Psycholinguistic	Psycholinguist	Respiration	Respir
Psychologist	Psychol	Respiratory	Respir
Psychology	Psychol	Response	Response
Psychoneuroendocrinology	Psychoneruoendocrinol	Resuscitation	Resuscitation
Psychopathology	Psychopathol	Retardation	Retard
Psychopharmacology	Psychopharmacol	Retina	Retina
Psychophysiology	Psychophysiol	Review, Reviews	Rev

H6, *continued*

Rheumatism, Rheumatic	Rheum	Sports	Sports
Rheumatology	Rheumatol	Stain	Stain
Rhinology	Rhinol	Standardization	Stand
Rhode Island	R I	Standards	Stand
Safety	Safety	Statistical	Stat
Scandinavian	Scand	Steroid	Steroid
Scanning	Scan	Steroids	Steroids
Schizophrenia	Schizophr	Stockholm	Stockh
School	Scho	Strabismus	Strabismus
Science	Sci	Stress	Stress
Sciences	Sci	Stroke	Stroke
Scientific	Sci	Structure	Struct
Scottish	Scott	Studies	Stud
Security	Secur	Subcellular	Subcell
Seminars	Semin	Submicroscopic	Submicrosc
Series	Ser	Substance	Subst
Service	Serv	Suicide	Suicide
Sex	Sex	Superior	Super
Sexual	Sex	Support	Support
Sexually	Sex	Surgeon	Surg
Shock	Shock	Surgeons	Surg
Singapore	Singapore	Surgery	Surg
Skeletal	Skeletal	Surgical	Surg
Sleep	Sleep	Swedish	Swed
Social	Soc	Symposia	Symp
Societies	Soc	Symposium	Symp
Society	Soc	System	Syst
Sociological	Sociol	Systems	Syst
Sociology	Sociol	Technical	Tech
Somatic	Somatic	Technology	Technol
Somatosensory	Somatosens	Tennessee	Tenn
South African	S Afr	Teratogenesis	Teratogenesis
South Carolina	S C	Teratology	Teratol
South Dakota	S D	Thailand	Thai
Southeast	Southeast	Theoretical	Theor
Southern	South	Therapeutics	Ther
Space	Space	Therapies	Ther
Spectrometry	Spectrom	Therapy	Ther
Speech	Speech	Thermal	Therm

H6, *continued*

Spine	Spine	Thoracic	Thorac
Thorax	Thorax	Ulster	Ulster
Throat	Throat	Ultramicroscopy	Ultramicrosc
Thrombosis	Thromb	Ultrasonic	Ultrason
Thromboxane	Thromboxane	Ultrasonics	Ultrasonics
Thymus	Thymus	Ultrasound	Ultrasound
Tissue	Tissue	Ultrastructural	Ultrasctruct
Today	Today	Ultrastructure	Ultrastruct
Tokyo	Tokyo	Undeasea	Undersea
Tomography	Tomogr	Union	Union
Topics	Top	Uremia	Uremia
Total	Total	Vision	Vis
Toxicologic	Toxicol	Visual	Vis
Toxicological	Toxicol	Vital	Vital
Toxicology	Toxicol	Vitamin	Vitam
Traditional	Tradit	Vitaminology	Vitaminol
Transactions	Trans	Vitamins	Vitam
Transfer	Transfer	Vitro	Vitro
Transfusion	Transfusion	Vivo	Vivo
Transmission	Transm	Welfare	Welfare
Transmitted	Transm	Western	West
Transplant	Transplant	West Indian	West Indian
Transplantation	Transplantation	West Virginia	W Va
Traumatic	Trauma	Wildlife	Wildl
Tropical	Trop	Wisconsin	Wis
Tuberculosis	Tuberc	Women	Women
Tumor	Tumor	Women's	Womens
Tunis	Tunis	Zoology	Zool
Turkish	Turk	Zoonoses	Zoonoses

H7 Medical Journals and Their Abbreviations

Academic Medicine	Acad. Med.
Advances in Cancer Research	Adv. Cancer Res.
AJR: American Journal of Roentgenology	AJR Am. J. Roentgenol.
American Family Physician	Am. Fam. Physician
American Heart Journal	Am. Heart J.
American Journal of Cardiology, The	Am. J. Cardiol
American Journal of Clinical Nutrition	Am. J. Clin. Nutr.
American Journal of Clinical Pathology	Am. J. Clin. Pathol.
American Journal of Human Genetics, The	Am. J. Hum. Genet.
American Journal of the Medical Sciences, The	Am. J. Med. Sci.
American Journal of Medicine, The	Am. J. Med.
American Journal of Nursing	Am. J. Nurs.
American Journal of Obstetrics and Gynecology	Am. J. Obstet. Gynecol.
American Journal of Ophthalmology	Am. J. Ophthalmol.
American Journal Of Pathology	Am. J. Pathol.
American Journal of Psychiatry, The	Am. J. Psychiatry
American Journal of Public Health	Am. J. Public Health
American Journal of Respiratory and Critical Care Medicine	Am. J. Respir. Crit. Care Med.
American Journal of Surgery, The	Am. J. Surg.
Anesthesiology	Anesthesiology
Annals of Emergency Medicine	Ann. Emerg. Med.
Annals of Internal Medicine	Ann. Intern. Med.
Annals of Neurology	Ann. Neurol.
Annals of Physics	Ann. Phys.
Annals of Plastic Surgery	Ann. Plast. Surg.
Annals of Surgery	Ann. Surg.
Annals of Thoracic Surgery, The	Ann. Thorac. Surg.
Annual Reports in Medicinal Chemistry	Annu. Rep. Med. Chem.
Annual Review of Immunology	Annu. Rev. Immunol.
Annual Review of Physiology	Annu. Rev. Physiol.
Anticancer Research	Anticancer Res.
Applied Microbiology	Appl. Microbiol.
Archives of Dermatology	Arch. Dermatol.
Archives of Disease in Childhood	Arch. Dis. Child.
Archives of Family Medicine	Arch. Fam. Med.
Archives of Internal Medicine	Arch. Intern. Med.
Archives of Neurology	Arch. Neurol.

H7, *continued*

Archives of Ophthalmology	Arch. Ophthalmol.
Archives of Pathology & Laboratory Medicine	Arch. Pathol. Lab. Med.
Archives of Pharmacology	Arch. Pharmacol.
Archives of Surgery	Arch. Surg.
Archives of Virology	Arch. Virol.
Arthritis & Rheumatism	Arthritis. Rheum.
Biochemical Pharmacology	Biochem. Pharmacol.
Biology of Reproduction	Biol. Reprod.
Biomedical Letters	Biomed. Lett.
Biomedical Research on Trace Elements	Biomed. Res. Trace. Elem.
Bioorganic & Medicinal Chemistry	Bioorg. Med. Chem.
Bioorganic & Medicinal Chemistry Letters	Bioorg. Med. Chem. Lett.
Blood	Blood
Brain	Brain
Brain Research	Brain Res.
British Journal of Cancer	Br. J. Cancer
British Journal of Clinical Pharmacology	Br. J. Clini. Pharmacol.
British Journal of Haematology	Br. J. Haematol.
British Journal of Obstetrics and Gynaecology	Br. J. Obstet. Gynaecol.
British Journal of Pharmacology	Br. J. Pharmacol.
British Journal of Radiology	Br. J. Radiol.
British Journal of Rheumatology	Br. J. Rheumatol.
British Journal of Surgery	Br. J. Surg.
British Medical Journal	Br. Med. J.
CA: A Cancer Journal for Clinicians	CA Cancer J. Clin.
Canadian Medical Association Journal	Can. Med. Assoc. J.
Cancer	Cancer
Cancer Chemotherapy and Pharmacology	Canc. Chemother. Pharmacol.
Cancer Chemotherapy Reports	Canc. Chemother. Rep.
Cancer Letters	Canc. Lett.
Cancer Research	Canc. Res.
Canadian Journal of Physiology and Pharmacology	Can. J. Physiol. Pharmacol.
Carcinogenesis	Carcinogenesis
Cardiovascular Research	Cardiovasc. Res.
Cellular Immunology	Cell Immunol.
Challenges in Modern Medicine	Challenges Mod. Med.

H7, *continued*

Chemotherapy	Chemotherapy
Circulation Journal of the American Heart Association	Circ. J. Am. Heart Assoc.
Circulation Research	Circ. Research
Circulation	Circulation
Clinical and Experimental Immunology	Clin. Exp. Immunol.
Clinical Immunology and Immunopathology	Clin. Immunol. Immunopathol.
Clinical Investigation	Clin. Invest.
Clinical Orthopaedics	Clin. Orthop.
Clinical Pharmacology Therapy	Clin. Pharmacol. Ther.
Clinical Physiology	Clin. Physiol.
Clinical Pediatrics	Clin. Pediatr.
Clinical Science	Clin. Sci.
Critical Care Medicine	Crit. Care Med.
Current Genetics	Curr. Genet.
Current Problems in Surgery	Curr Probl Surg
Developmental Biology	Dev Biol
Developmental Brain Research	Dev Brain Res
Development	Development
Diabetes	Diabetes
Digestive Diseases and Sciences	Dig. Dis. Sci.
Drug Development and Industrial Pharmacy	Drug Dev. Ind. Pharm.
Drugs	Drugs
Drugs Under Experimental and Clinical Research	Drugs Exp. Clin. Res.
Drugs of the Future	Drugs Future
Endocrine	Endocrine
Endocrinology	Endocrinology
European Journal of Cancer	Eur. J. Cancer
European Journal of Clinical Chemistry and Clinical Biochemistry	Eur. J. Clin. Chem. Clin. Biochem.
European Journal of Clinical Pharmacology	Eur. J. Clin. Pharmacol.
European Journal of Endocrinology	Eur. J. Endocrinol.
European Journal of Immunology	Eur. J. Immunol.
European Journal of Medicinal Chemistry	Eur. J. Med. Chem.
Experimental and Clinical Psychopharmacology	Exp. Clin. Psychopharmacol.

H7, *continued*

Experimental Nephrology	Exp. Neurol.
Family Medicine	Fam. Med.
Family Planning Perspectives	Fam. Plann. Perspect.
Family Practice Research Journal	Fam. Pract. Res. J.
Gastroenterology	Gastroenterology
Genes	Genes
Genes and Development	Genes Dev.
Genetics	Genetics
Genome	Genome
Genomics	Genomics
General Pharmacology	Gen. Pharmacol.
Geriatrics	Geriatrics
Gut	Gut
Health Physics	Health Phys.
Heart	Heart
Heart and Lung	Heart Lung
Histochemistry	Histochemistry
Hospital Practice	Hosp. Pract.
Human Genetics	Hum. Genet.
Human Immunology	Hum. Immunol.
Human Molecular Genetics	Hum. Mol. Genet.
Human Mutation	Hum. Mutat.
Human Reproduction	Hum. Reprod.
Immunity	Immunity
Immunogenetics	Immunogenetics
Immunology Letters	Immunol. Lett.
Immunology	Immunology
Immunology Today	Immunol. Today
International Immunology	Int. Immunol.
International Immunology	Int. Immunol.
International Journal of Cancer	Int. J. Cancer
International Journal of Experimental Pathology	Int. J. Exp. Pathol.
International Journal of Health Services	Int. J. Health Serv.
International Journal of Phamaceutics	Int. J. Pharm.
International Journal of Radiation Biology	Int. J. Radiat. Biol.
International Journal of Radiation Oncology, Biology, Physics	Int. J. Radiat. Oncol. Biol. Phys.
International Journal of STD and AIDS	Int. J. STD AIDS
Journal of Abnormal Child Psychology	J. Abnorm. Child Psychol.

H7, *continued*

Journal of Abnormal Psychology	J. Abnorm. Phychol.
Journal of Allergy and Clinical Immunology, The	J. Allergy Clin. Immunol.
Journal of the American College of Cardiology	J. Am. Coll. Cardiol.
Journal of the American College of Surgeons	J. Am. Coll. Surg.
Journal of Antibiotics (English ed.)	J. Antibiot.
Journal of Antimicrobial Chemotherapy	J. Antimicrob. Chemother.
Journal of Bone and Joint Surgery. American Volume	J. Bone Joint Surg Am
Journal of Bone and Joint Surgery. British Volume	J Bone Joint Surg. Br.
Journal of Burn Care and Rehabilitation	J. Burn Care Rehab.
Journal of Burn Care and Research	J. Burn Care Res.
Journal of Cancer Education	J. Cancer Educ.
Journal of Cancer Research and Clinical Oncology	J. Cancer Res. Clin. Oncol.
Journal of Cardiovascular Pharmacology	J. Cardiovasc. Pharmacol.
Journal of Clinical Endocrinology and Metabolism	J. Clin. Endocrinol. Metab.
Journal of Clinical Investigation	J. Clin. Invest.
Journal of Clinical Pharmacology, The	J. Clin. Pharmacol.
Journal of Endocrinology	J. Endocrinol.
Journal of Experimental Medicine	J. Exp. Med.
Journal of Family Practice	J. Fam. Pract.
Journal of General Microbiology	J. Gen. Microbiol.
Journal of General Physiology, The	J. Gen. Physiol.
Journal of Immunology, The	J. Immunol.
Journal of Infectious Diseases	J. Infect. Dis.
Journal of Laboratory and Clinical Medicine, The	J. Lab. Clin. Med.
Journal of Molecular and Cellular Cardiology	J. Mol. Cell. Cardiol.
Journal of the National Cancer Institute	J. Natl. Cancer. Inst.
Journal of Nervous and Mental Disease, The	J. Nerv. Ment. Dis.
Journal of Neurophysiology	J. Neurophysiol.
Journal of Neurosurgery	J. Neurosurg.
Journal of Nuclear Medicine	J. Nucl. Med.
Journal of Pediatrics, The	J. Pediatr.
Journal of Pharmacology and Experimental Therapeutics	J. Pharmacol. Exp. Ther.
Journal of Surgical Research	J. Surg. Res.

H7, *continued*

Journal of Thoracic and Cardiovascular Surgery	J. Thorac. Cardiovasc. Surg.
Journal of Trauma, The	J. Trauma
Journal of Urology, The	J. Urol.
Kidney International	Kidney Int.
Mayo Clinic Proceedings	Mayo Clin. Proc.
Medical Letter on Drugs and Therapeutics	Med. Lett. Drugs Ther.
Medicine	Medicine
Medicinal Research Reviews	Med. Res. Rev.
Medical Science Research	Med. Sci. Res.
Molecular and Cellular Endocrinology	Mol. Cell. Endocrinol.
Molecular and General Genetics	Mol. Gen. Genet.
Molecular Reproduction and Development	Mol. Reprod. Dev.
Nature Genetics	Nat. Genet.
New England Journal Of Medicine	N. Engl. J. Med.
Neurology	Neurology
Nursing Research	Nurs. Res.
Obstetrics & Gynecology	Obstet. Gynecol.
Oncogene	Oncogene
Pediatric Clinics of North America	Pediatr. Clin. North Am.
Pediatrics	Pediatrics
Pediatric Research	Pediatr. Res.
Physiology & Behavior	Physiol. Behav.
Physiological Reviews	Physiol. Rev.
Physical Therapy	Phys. Ther.
Progress in Clinical and Biological Research	Prog. Cardiovasc. Dis.
Public Health Reports	Public Health Rep.
Radiologic Clinics of North America	Radiol. Clin. North Am.
Radiology	Radiology
Reproduction, Nutrition, Development	Reprod. Nutr. Dev.
Surgery	Surgery
Surgical Clinics of North America	Surg. Clin. North Am.
Surgery, Gynecology & Obstetrics	Surg. Gynecol. Obstet.
Transplantation	Transplantation
Transplantation Proceedings	Transplant. Proc.
Trends in Genetics	Trends Genet.
Ultrasound in Medicine and Biology	Ultrasound Med. Biol.
Urologic Clinics of North America	Urol. Clin. North Am.
Vaccine	Vaccine
Western Journal of Medicine	West J. Med.

Appendix I. Further Reading and Resources

This appendix presents an annotated bibliography of the works used to assemble and inform the chapters and sections of this work. Following the basic bibliographic listing, arranged alphabetically by author, a brief comment appears (in italics) concerning the strengths of the work and the uses to which it may be put in various types of science writing. Many of the works listed contain bibliographies of their own and readers are encouraged to consult those works for further titles and information. Readers are also referred to the final sections of Chapters 4 and 5 in this work for more detailed bibliographic material on citation styles and the legalities of copyrights and permissions.

AIP Style Manual (4th edition). American Institute of Physics: New York. 1990. *A downloadable manual that is updated frequently, particularly with regard to the PACS—the Physics and Astronomy Classification Scheme (a necessary component of any paper submitted to a serious journal). Both the manual and PACS are at: http://www.aip.org/pubservs/style/4thed/toc.html.*
The AIP Manual does not cover all areas of modern physics and should thus be used with: The Physical Review Style and Notation Guide, edited by Waldron, Judd, and Miller, revised 2005, *and sponsored by the American Physical society, available online at:* http://authors.aps.org/STYLE/
and by The Review of Modern Physics Style Guide, *a particularly useful author guideline with an excellent essay by Karie Friedman, the Guide's editor, on "Writing a Better Scientific Article," Available online from APS at: rmp.aps.org/files/rmpguide.pdf*

Alley, Michael. *The Craft of Scientific Writing* (3rd edition). Springer: New York, 1996. *A popular guidebook oriented heavily to physics, with useful direction on effective use of illustration. Contains many memorable quotes and aphorisms.*

AMA Manual of Style: A Guide for Author and Editors (10th edition). JAMA and Oxford University Press: New York. 2007. *A monumental work of over 1000 pages, covering medical science thoroughly, but in need of greater focus on pure life science areas. The work (and sponsor) seem preoccupied with legal and ethical issues.*

AMS Brief Guide for Authors. The American Meteorological Society: Boston. 2000. *One of the better author guidelines, available online at: www.ametsoc.org/PUBS/Authorsguide/pdf_vs/authguide.pdf.*

ASA Publications Handbook and Style Manual. Produced jointly by American Society of Agronomy (ASA), Crop Science Society of America (CSSA), and Soil Science Society of America (SSSA), and available online (where it is continually updated), this is perhaps the best of the society-sponsored author guides available. It may be down loaded at: https://www.agronomy.org/publications/style/

Barrass, Robert. *Scientists Must Write: A Guide to Better Writing for Scientists, Engineers, and Students* (second edition). Routledge-Falmer: New York. 2002. *Pays special attention to what a scientist should do during research in anticipation of the day the writing will begin. A work students have found very useful and inspirational.*

Bates, Robert L., Marla D. Adkins-Aeljeson and Rex C. Bichanan, *Geowriting: A Guide to Writing, Editing, and Printing in earth Science* (fifth edition). American Geological Institute: Alexandria, VA. 1995. *A concise and well-organized summary of the basic elements of science writing, focusing on geoscience, but applicable to science in general. A most useful little book to have.*

Blum, Deborah, Mary Knudson and Robin Marantz Henig. *A Field Guide for Science Writers: The Official Guide of the National Association of Science Writers* (second edition). Oxford University Press: New York. 2005. *A collection of original essays by many of the world's leading science writers about virtually every aspect of the field, from practical mechanics to professional matters of dealing with publishers and institutions. A must-have for any freelancer.*

Briscoe, Mary Helen. *Preparing Scientific Illustrations: A Guide to Better Posters, Presentations, and Publications* (second edition). Springer-Verlag: New York. 1996. *Deals with the artwork and how to integrate it with the text. Extremely useful for students and anyone explaining or presenting scientific material to a live audience.*

Bishop, C.T. *How to Edit a Scientific Journal.* ISI Press: Philadelphia. 1984. *Deals concisely with the mechanics and economics of journal management. Vital information someone on the journal staff must have if the journal is to be economically viable.*

Carosso, Rebecca B., *Technical Communication*. Wadsworth: New York. 1986. *A careful presentation of the techniques of manual writing and the preparation of reports, surveys, and slide presentations.*

Chicago Manual of Style (15th edition). University of Chicago Press: Chicago. 2003. *The most widely used style guide for academic and popular publishing, CMS is still (even after so many editions) in need of better coverage of science and technology.*

Cleveland, William S. *Visualizing Data*. Hobart Press: Summit, NJ. 1993. *An advanced guide to the graphic presentation of statistical data. A difficult area for which even practiced researchers are forced to seek outside expertise.*

Coghill, Anne M., and Lorrin R. Garson. *The ACS Style Guide* (3rd edition). American Chemical Society and Oxfrod University Press: Washington, DC. 2006. *A remarkably useful and comprehensive work (though under 500 pages), with contributions by experts in science communication, illustration, and electronic media.*

Cone, John D., and Sharon L. Foster. *Dissertations and Theses from Start to Finish*. American Psychological Association: Washington, DC. 1993. *Widely used by graduate students for many years, which accounts for the mnay times it has been reprinted. Could stand updating.*

Council of Science Editors. *Scientific Style and Format: The CSE Manual for Authors, Editors, and Publishers* (7th edition). CSE and Rockefeller University Press: Reston, VA. 2006. *An important guide that remains burdened by its life-science orientation (CSE was formerly the Council of Biology Editors), an idiosyncratic organization, and a British bent. Some readers prefer the 6th edition, which was published by Cambridge University Press in 1994.*

Davis, Martha. *Scientific Papers and Presentations*. Academic Press: New York. 1997. *A thorough and highly readable survey with excellent guidance on integrating illustrations of all kinds to enhance the effectiveness of the writing.*

Day, Robert A. *Scientific English* (2nd edition). Oryx: Phoenix. 1995. *An excellent companion to his 1988 work,* How to Write and Publish a Scientific Paper *(also published by Oryx); and his 2007 work,* The Craft of Scientific Presentations *(Springer: New York).*

Friedman, Sharon M., *et al. Scientists and Journalists: Reporting Science and News.* AAAS: Washington, DC. 1988. *A collection of still-relevant essays on a vital subject, with an excellent bibliography on public communication of science.*

Galison, Peter, *How Experiments End.* University of Chicago Press: Chicago. 1987. *A detailed analysis of life in a physics laboratory, with special attention paid to the role of writing—grant proposals; technical memoranda; progress reports; journal submissions, from preliminary letters to full-blown papers—in the research process.*

Gibaldi, Joseph. *The MLA Handbook for Writers of Research Papers* (6th edition). Modern Language Association: New York. 2003. *By the author and publisher of the perennial university favorite,* The MLA Style Manual. *Very popular with college students and geared to university-level writing.*

Goldbort, Robert. *Writing for Science.* Yale University Press: New Haven & London. 2006. *A comprehensive guide that covers many different types of science writing, from popular essays and laboratory notes to research-level journal articles. A thorough presentation of the "science as writing" approach that is filled with valuable guidance and many instructive examples.*

Gopen, George D., and Judith A Swan. "The Science of Scientific Writing," *American Scientist,* Vol. 78, no 6: 550-558. *A classic study that revolutionized science writing.*

Hale, Constance, and Jesse Scanlon. *Wired Style: Principles of English Usage in the Digital Age* (second edition). Broadway Books: New York. 1999. *A vast improvement over a previous edition and, hopefully, the first of many more editions as this field becomes more fully realized and organized. Particularly important for authors of electronic content.*

Hammerich, Irene, and Claire Harrison. *Developing Online Content: The Principles of Writing and Editing for the Web.* John Wiley & Sons: New York. 2002. *A comprehensive and practical guide to developing content for electronic publications of all kinds. For authors and editors in all areas, but especially in any scientific field.*

Hansen, Wallace R., *Suggestions to Authors of the Reports of the United States Geological Survey* (seventh edition). US Government Printing Office: Washington, DC. 1991. *Covers a remarkably wide area in all of science, especially earth, environmental, meteorological, and agronomic. Quaintly valuable precisely because it assumes it will be the only style guide an author will have at his or her disposal.*

Hoffmann, Roald. *The Same and Not the Same.* Columbia University Press: New York. 1995. *An early advocate of the "science as writing" school, with interesting insights into the development of the modern science (chemistry) journal.*

IEEE Information for Authors (second edition). IEEE: Piscataway, NJ. 2000. *Important for any writing in electronic or information science. IEEE Computer Society also posts an extensive and useful style guide for mathematics (both general and computer-related) at:*
http://www.computer.org/portal/site/ieeecs/menuitem.c5efb9b8ade9096b 8a9ca0108bcd45f3/index.jsp?&pName=ieeecs_level1&path=ieeecs/publ ications/author/style&file=abc.xml&xsl=generic.xsl&;jsessionid=JyRJ2 qWxDy53J6hskdpkv9Y6Spj71Y44h2G9CXRYmY1K5b10MJws!41717591

Katz, Michael J. *The Elements of the Scientific Paper.* Yale University Press: New Haven. 1985. *Clear analysis of the parts of an effective paper and the aims and character of each section.*

Laszlo, Pierre. *Communicating Science: A Practical Guide.* Springer: New York. 2006. *A scientist's engaging plea for humanizing science through humanizing the communication of science. Laszlo puts his approach into practice in popular works,* Salt: Grain of Life *(Harper, 2002) and* Citrus: A History *(Univ. of Chicago Press: 2008).*

Lightman, Alan, *The Discoveries: Great Breakthroughs in Twentieth-Century Science.* Pantheon: New York. 2005. *Available as a Vintage paperback, this book is on this list because it contains the original papers behind the most important scientific discoveries of the past century in virtually every field of science.*

Lide, David R., *CRC Handbook of Chemistry and Physics.* CRC Press/Taylor & Francis: Boca Raton, Florida. Annual—2010 marks its 90th edition. *For nearly a century, no research laboratory could be considered adequately equipped without a copy of "the CRC" on hand. Covers mathematics, physics, chemistry, organic chemistry, health science, astronomy, and laboratory methods.*

Locke, David. *Science as Writing.* Yale University Press: New Haven. 1992. *A landmark work that mapped out with great clarity and conviction the importance of good writing in creating and advancing science.*

Maggio, Rosalie. *The Bias-Free Word Finder: A Dictionary of Nondiscriminatory Language.* Beacon Press: Boston. 1992. *A useful and practical guide through a field filled with pitfalls.*

Microsoft Manual of Style for Technical Publications (third edition). Microsoft Press: Redmond, WA. 2004. *A manual geared to programmers and IT professionals, but useful to science writers in other fields. Vastly improved over previous editions.*

Three works by the former CIBA Foundation editor of life-science symposia, and for ten years a leading force in the European Association of Science Editors, are highly regarded by many scientists:
O'Connor, Maeve. *The Scientist as Editor: Guidelines for Editors of Books and Journals.* John Wiley & Sons: New York. 1979.
— *Writing Successfully in Science.* Taylor & Francis: London. 1991.
— *How to Copyedit Scientific Books and Journals.* ISI Press: Philadelphia. 1986

O'Neil, Maryadele J. *The Merck Index: An Encyclopedia of Chemicals, Drugs, and Biologicals* (14th edition). Merck and John Wiley & Sons: West Point, PA, and New York. 2006. *Not to be confused with* The Merck Manual, *which is a physician's reference tool, this is the pharmacological companion work to the CRC.*

Paradis, James G., and Muriel Zimmerman. *MIT Guide to Science and Engineering Writing* (second edition). MIT Press: Cambridge, MA. 2002. *A surprisingly engaging survey of techniques for preparing electronic manuscripts and journal articles.*

Publication Manual of the American Psychological Association (fifth edition). APA: Washington, DC. 2001. *A widely used and well-regarded style manual, especially useful in areas that cross over from the physical sciences into the social sciences.*

Rubens, Philip, ed. *Science and Technical Writing: A Manual of Style* (2nd edition). Routledge: New York. 2001. *A useful work that covers some areas exceedingly well and others superficially. The emphasis on audience analysis seems unnecessary.*

Schoenfeld, R. *The Chemist's English* (3rd edition). VCH: Weinheim, Germany. 1990. *Instructive and entertaining essays on specific areas of writing in the chemical sciences.*

Selzer, Jack, ed. *Understanding Scientific Prose.* University of Wisconsin Press: Madison, WI. 1993. *Strong advocate of the need to write as a means of becoming an astute and discerning reader.*

Skillin, Marjorie E., and Robert M. Gay. *Words into Type* (third edition). Prentice-Hall: Englewood Cliffs, NJ. 1974. *Although completely superceded by the Chicago Manual, worn volumes of this classic can still be found on the shelves of many editors, particular in the sciences. Considered very strong (and clear) in explaining grammar.*

Swanson, Ellen, *Mathematics into Type: Copy Editing and Proofreading of Mathematics for Editorial Assistants and Authors.* (second edition). American Mathematical Society: Providence, RI. 1991. *Swanson's "little yellow booklet" was a cherished editorial tool for many years. Now downloadable at www.ams.org .*

Taylor, B.N. *Guide for the Use of the International System of Units (SI).* National Institute of Standards and Technology: Gaithersberg, MD. 1995. *A continuously updated guide to SI— downloadable at:* http://www.iso.org/iso/publications_and_e-products/si-guide.htm *Remarkably, the jury is still out on SI, but this guidebook makes the best of the system and explains it clearly and succinctly.*

Tichy, H.J., and Sylvia Fourdrinier. *Effective Writing for Engineers, Managers, and Scietists* (second edition). John Wiley & Sons: New York. 1988. *A practical guide that speaks to people in the field who must write as part of their work (as opposed to academics, for whom writing is a central professional element). Contains particularly useful advice on how to begin.*

Tufte, Edward R. *The Visual Display of Quantitative Information* (2nd edition). Graphics Press: New Haven. 2001. *Along with Tufte's two companion works,* Envisioning Information *(1990) and* Visual Explanations (1997), *Tufte has inspired many new developments in scientific graphics, much to the betterment of scientific discourse. His influence is apparent in works that try to present his principles systematically, such as Wolff and Yeager's* Visualization of Natural Phenomena *(see below).*

Wilkins, George A. *The IAU Style Manual: The Preparation of Astro-nomical Papers and Reports.* International Astronomical Union: Paris. 1989. *A rare instance in which a style manual was not updated frequently only because of the discipline's incredibly fast pace. IAU is currently preparing a revised manual with NASA/Smithsonian.*

Williams, Joseph M., *Style: Lessons in Clarity and Grace* (ninth edition). Longman. 2006. *A careful analysis of what makes a sentence work and what prevents it from communicating. A classic must-read for anyone who writes (and has already absorbed Strunk and White).*

Wolff, Robert S., and Larry Yaeger. *Visualization of Natural Pheno-mena.* Springer: New York. 1993. *A well-illustrated (with an elec-tronic support component) that applies many of Tufte's principles (see above) systematically and in a usable way.*

Zeiger, Mimi, *Essentials of Writing Biomedical Research Papers* (second edition). McGraw-Hill: New York. 2000. *Directed at the laboratory researcher with much helpful advice on navigating the journal submission process. A "from-the-ground-up" approach that science researchers in all fields will find useful.*

Zinseer, William. *On Writing Well: The Classic Guide to Writing Non-fiction* (6th edition). Harper-Perennial: New York. 1998. *One of the very few general writing guides that devotes some attention to the need for and means of writing science well.*

A Final Note:

The works listed in this appendix deal only with editorial and publishing issues as they relate to the sciences. Any writer, researcher, or editor who wishes to participate in the science publishing enterprise should be familiar with the material contained in these books, to be sure. More important, however, is a familiarity and an appreciation of the excellent writing that has been done by a long list of authors that stretches from Isaac Asimov to Anthony Zee. That a compendium of great books in science—of all disciplines, for all levels of reader, and to different levels of sophistication and complexity (including textbooks)—is still lacking, in spite of the many people engaged in science profes-sionally or as an avocation, indicates how sorely is needed an appre-ciation of good science writing by the public.

INDEX

935